NETWORKS AND ALGORITHMS

NETWORKS AND ALGORITHMS

AN INTRODUCTORY APPROACH

ALAN DOLAN and JOAN ALDOUS
The Open University, Milton Keynes, UK

Based on The Open University course
Graphs, Networks and Design

JOHN WILEY & SONS
Chichester · New York · Brisbane · Toronto · Singapore

This book is based upon The Open University course TM 361: *Graphs, Networks and Design*, first published in 1981. Material is reproduced by permission of The Open University

Other Wiley Editorial Offices

John Wiley & Sons, Inc., 605 Third Avenue,
New York, NY 10158-0012, USA

Jacaranda Wiley Ltd, G.P.O. Box 859, Brisbane,
Queensland 4001, Australia

John Wiley & Sons (Canada) Ltd, 22 Worcester Road,
Rexdale, Ontario M9W 1L1, Canada

John Wiley & Sons (SEA) Pte Ltd, 37 Jalan Pemimpin #05-04,
Block B, Union Industrial Building, Singapore 2057

Library of Congress Cataloging-in-Publication Data

Dolan, Alan.
Networks and algorithms : an introductory approach / Alan Dolan
and Joan Aldous.
p. cm.
Includes bibliographical references and index.
ISBN 0 471 93992 7 (cloth) : ISBN 0 471 93993 5 (paper)
1. Systems analysis. 2. Algorithms. 3. Mathematical models.
I. Aldous, Joan. II. Title.
T56.6.D65 1993
511′.5—dc20 93–491
CIP

British Library Cataloguing in Publication Data

A catalogue record for this book is available from the British Library

ISBN 0 471 93992 7 (cloth)
ISBN 0 471 93993 5 (paper)

Typeset in Times 10/12 pt by Pure Tech Corporation, Pondicherry, India
Printed and bound in Great Britain by Biddles Ltd, Guildford, Surrey

CONTENTS

PREFACE

Network theory has in recent years become a valuable aid to planning and analysis in an increasingly wide range of applications in industry, commerce and government. Aspects of the topic have been taught under many guises—as operational research in management studies; as linear programming, optimization or combinatorics in mathematics; and as network theory or algorithms in computer science.

The aim of this book is to present a reasonably comprehensive account of such topics at an approachable level and style suitable for undergraduate students. As such we believe that it will satisfy the need for an accessible teaching text in this area.

This book was developed from part of a British Open University distance-learning course on Graphs, Networks and Design which has been successfully completed by several thousand students since it first appeared in 1981. Open University courses are written for students who work on their own with only a small amount of tutorial support. Consequently, when the course team were producing Graphs, Networks and Design, they paid particular attention to the presentation, clarity and arrangement of the material in order that students would be able to learn from it by independent study. An important part of the philosophy (retained in this book) was to provide problems (with full worked solutions given at the back of the book) for students to attempt at appropriate points in the text so that they could assess their understanding and progress before moving on to the next topic. We feel that this type of structured teaching material in book form is also admirably suited for use with conventional teaching. In addition to the problems in the body of the text, a selection of exercises is given at the ends of most of the chapters. Whereas the problems are for the students' own use, and so are provided with full worked solutions, the exercises are intended to be set by the instructor as assignments and are not provided with solutions.

This volume is the second book based on Graphs, Networks and Design. The first volume, on Graphs, by R. J. Wilson and J. J. Watkins appeared in 1990. Each book is self-contained and is suitable for a university course on discrete mathematics or computer science. The approach, terminology and notation are the same in both books, so an instructor wishing to teach both graphs and networks will find that they can be used concurrently.

The main contributors of material to the Open University course Graphs, Networks and Design which has been used in this book were: Joan Aldous, Keith Cavanagh, Chris Dillon, Alan Dolan, Chris Earl, Yin-Seong Ho, Roy Nelson, Richard Scott, and Robin Wilson.

Study paths for the book are shown in the following diagram.

Each of the blocks of chapters in the middle of the diagram can be studied independently of the others, except that it is preferable to study Chapter 3 before Chapter 7, and to study Chapter 8 before Chapter 10. The first two chapters should be studied before any of the others.

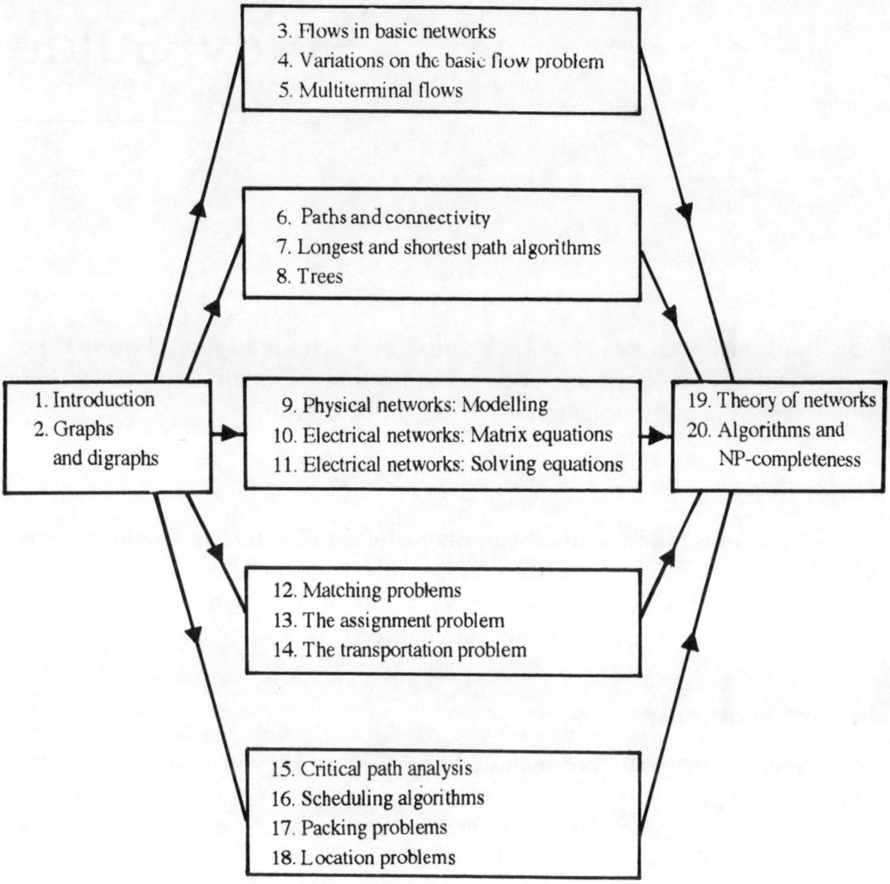
1. Introduction
2. Graphs and digraphs
3. Flows in basic networks
4. Variations on the basic flow problem
5. Multiterminal flows
6. Paths and connectivity
7. Longest and shortest path algorithms
8. Trees
9. Physical networks: Modelling
10. Electrical networks: Matrix equations
11. Electrical networks: Solving equations
12. Matching problems
13. The assignment problem
14. The transportation problem
15. Critical path analysis
16. Scheduling algorithms
17. Packing problems
18. Location problems
19. Theory of networks
20. Algorithms and NP-completeness

Study Paths

Study Guide

This book has been designed to be used as a self-study text. Two features in particular have been included to aid study by an individual working alone. These are the objectives and the in-text problems.

Objectives

These are to be found at the beginning of each chapter, and indicate what you should be able to do after reading the chapter. They may be used as a revision aid to check that you have learnt all the items listed, and also as an extended contents list for the chapter.

In-text Problems

A number of problems are included in the text. These are designed to test the previous part of the text, and we consider that it is very important for you to do the problems as you come to them in your reading of the book. They will enable you to check that you have properly understood the text so far, and will also help you to consolidate your learning.

We also strongly recommend that after attempting the problems you look at our solutions (given at the end of the book) even if you are able to do the problems unaided. This is because the solutions sometimes contain informative comments which relate directly to the following text.

CHAPTER 1

INTRODUCTION

After reading this chapter, you should be able to:

- understand the terms *graph*, *vertex*, *edge*, *digraph*, *arc*, *network*, *weight* and *algorithm*;
- classify network problems as *existence*, *construction*, *enumeration* or *optimization* problems;
- understand how networks can be used as an aid to the analysis of practical problems, and explain some of the limitations of this process;
- appreciate the need for efficient algorithms for network problems.

1.1 BASIC CONCEPTS

In this chapter we introduce the two main themes of this book—*networks* and *algorithms*. To convey something of the flavour and scope of the subject, we give a number of examples of practical network problems, all of which are discussed more fully in later chapters.

Networks

The term *network* is used in its everyday meaning in expressions such as *telephone network* or *railway network*. We shall also use the term *network* in a more specialized sense to mean a type of diagram called a *graph* to which are added the numerical values of some quantity.

For example, we can represent part of the North Kent railway network by the following graph, in which the stations are represented by points, called *vertices*, and the connections between them by lines, called *edges*.

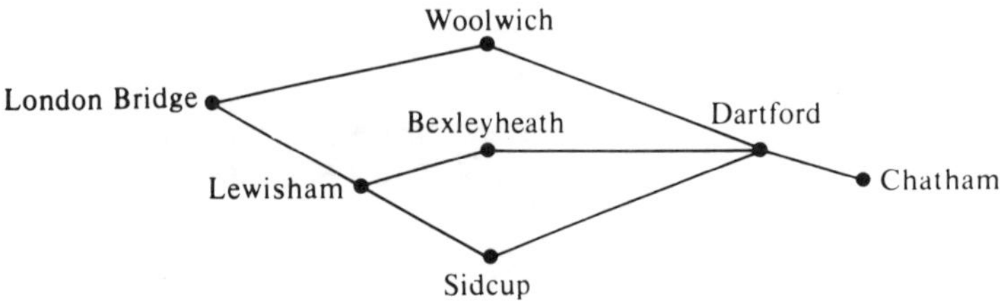

If we indicate on the graph the distances between the stations by inserting their numerical values beside the corresponding edges, the diagram becomes a network in the mathematical sense of the word. The numbers associated with each edge (or sometimes with each vertex) of a network are called *weights*.

Notice that a graph represents only the structure (or topology) of a system, and gives no further information about its elements. A network, on the other hand, conveys some additional, quantitative, information—in this case, the distances between the railway stations.

In a graph, the edges do not imply any preference for one direction over another. If direction is important, for example in representing a one-way traffic road system, we can indicate this by inserting a direction arrow on each edge, as shown on the following diagram.

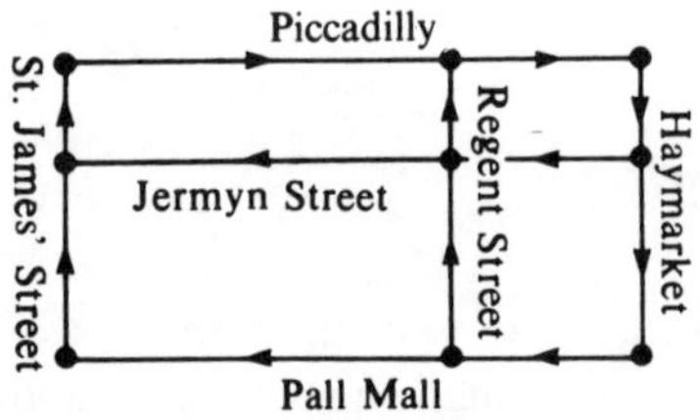

Such a diagram is called a *directed graph*, usually abbreviated to *digraph*, and the directed edges are called *arcs*. Direction is often important also in networks, so the underlying structure of a network is often a digraph rather than a graph.

In a network, each edge or arc can be considered as representing a channel for some kind of flow—such as traffic along a road, water through a pipe, or electric current through a resistor. The numerical values or weights give some information relating to the flow—for example, the amount of flow, the *capacity* of the channel (that is, the maximum flow possible), the length of the channel, or some other quantity. Weights may also be assigned to vertices—for example, to indicate the maximum flow possible through each vertex. In the example of a network given below, the weight of each arc is indicated by the number next to the arc, and the weight of each vertex is indicated by the circled number next to the vertex.

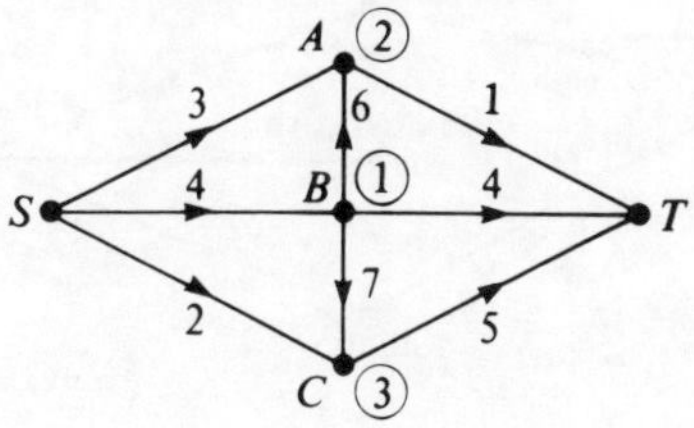

Algorithms

Network problems involving only a small number of vertices can usually be solved fairly easily by trial-and-error methods. However, such methods are not satisfactory for large-scale problems because they take far too long, and in many cases the time involved would be so great as to make such methods totally impractical. Also, for problems involving optimization, even if a solution is obtained, we usually cannot be sure that it is the best one.

What we need is a systematic step-by-step procedure which can be applied to large networks. Such a procedure for solving a particular type of problem is called an *algorithm*, and consists of a list of precise instructions or rules, together with a list of what input data is required. When we apply the algorithm to an instance of the problem, we take the appropriate data for that instance and apply the instructions step by step.

A particularly important feature of algorithms is that they are in a form which lends itself to conversion to computer programs. Many network problems which occur in industry or commerce are far too large or complex to be dealt with by hand, so the algorithms must be implemented by a computer. This means that the various instructions must be precisely and unambiguously stated, and the algorithm must terminate after a finite number of steps.

1.2 EXAMPLES OF NETWORKS

The following examples introduce a number of different network problems; they have been chosen to illustrate some of the main types of network discussed in this book.

Railway Networks

A large number of commuters travel regularly to London on the North Kent lines of British Rail. The main line runs from Chatham to Dartford, from where there are three routes to London Bridge. We can represent this part of the railway network by a digraph:

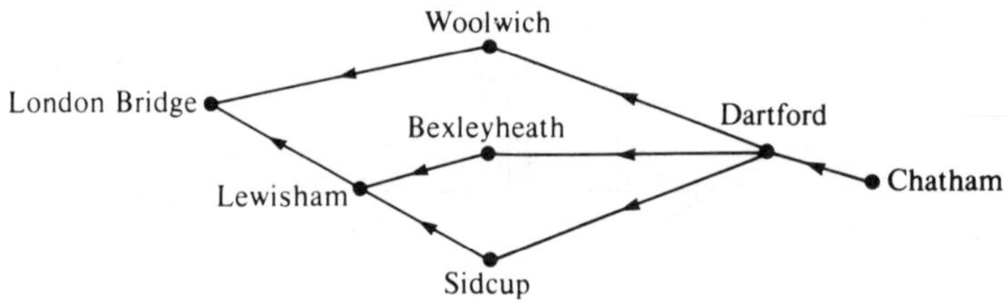

Now each section of the line has a *capacity*, in that there is a limit to the number of trains per hour which can travel along it in safety. Suppose that each section of the line can take up to 10 trains per hour; then this number is the capacity of each arc of the digraph, resulting in the following network representation:

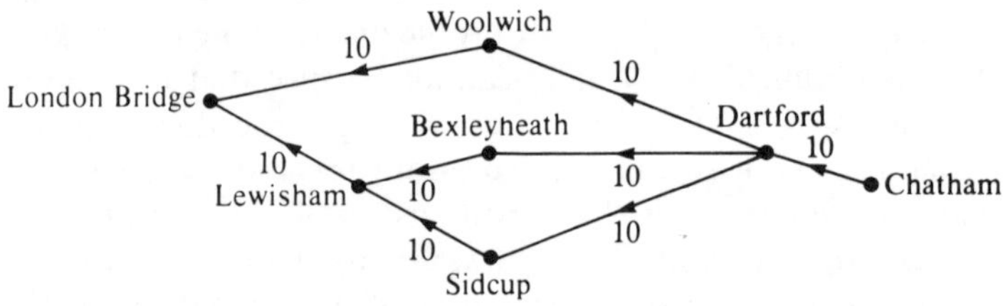

Problem 1.1

(a) If each line in the above network can take up to 10 trains per hour, how many trains per hour can run from Dartford to London Bridge?
(b) How can this number be increased by modifying the capacity of just one arc?

Note that if the number of trains through a particular station (Lewisham, for example) is also limited, then we can represent this by assigning a capacity to the corresponding vertex.

In order to work with a simple network diagram, we considered a much simplified version of the actual railway system in North Kent. In reality there are many additional factors which need to be included in a full-scale analysis, such as additional lines feeding trains into the system, the fact that in practice the capacities (weights) are different for different sections of track, and so on. This is an example of a network flow problem of the type discussed in Chapter 3.

Telecommunication Networks

In Chapter 6 we look at telecommunication networks, as represented by graphs whose vertices correspond to subscribers or exchanges and whose edges correspond to interconnections between these exchanges. It would be impractical to link every subscriber directly to every other one, and so a complicated system of exchanges and switching centres is employed. The disadvantage of this is that the total number of calls which can be carried by any part of the network is limited. Thus the problem of finding the maximum number of simultaneous calls between any two parts of the system can be thought of as a network problem rather than a graph problem. The capacity of any exchange (vertex) or interconnection (edge) is the maximum number of calls it can carry at any given time.

Road Networks

We can represent a system of trunk roads and the major towns which they connect by representing each town by a vertex of a graph and each road by an edge. Now suppose that we wish to find the shortest route, using trunk roads, between two particular towns. In this problem we are not concerned with the amount of traffic that each road can take; instead we are concerned only with the length of each section of the road, so we assign the length as the weight on the corresponding edge of our graph.

The problem of finding the shortest route between any two towns on our map is thus reduced to that of finding the shortest route between the two corresponding vertices on our graph—that is, the route with the lowest total weight.

Problem 1.2 Find the shortest route from S to T in the following network:

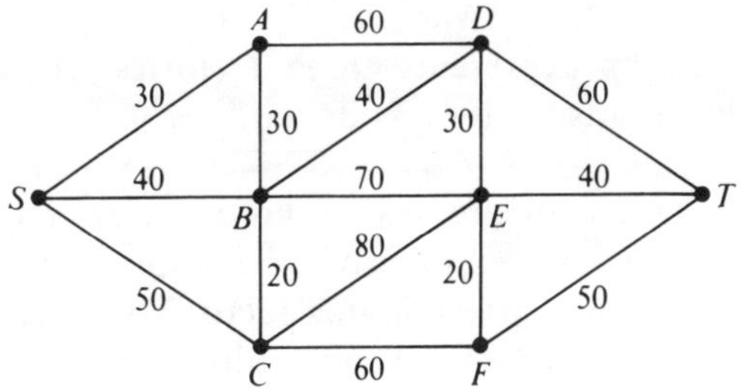

(*Hint*: First find the shortest routes from A to T, B to T, C to T.)

A simple shortest path problem like the one above may be solved by inspection, but for larger scale problems we need an algorithm. We describe such an algorithm in Chapter 7.

Physical Networks

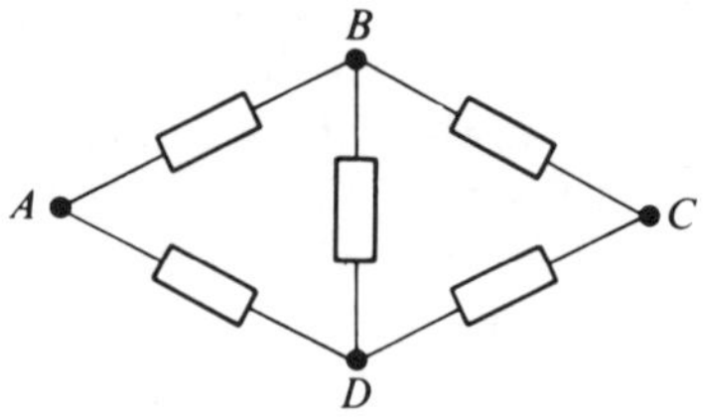

The above diagram shows a network of electrical resistors which is used in some measuring instruments. The resistor connecting B and D represents the resistance of a meter, and a voltage is applied across terminals A and C. Normally the circuit is used to determine the value of one of the resistances in the arms when the values of the others are known. A graphical representation can be used to analyse this circuit and to determine the values of the currents flowing in the resistors.

Similarly, a system of fluid distribution pipes may be represented by a network diagram which can be used to calculate pressure at various points in the system and the consequent flow in each pipe: the higher the pressure difference between any two points, the greater the flow between them. In physical systems like this, the pressures at the various points, which drive the flow through the system, are often called *potentials*.

Problems such as these are discussed in Chapters 9, 10 and 11.

Transportation Networks

The following diagram represents three factories, A, B and C, which are connected by roads, railways or airlines to two warehouses, M and N.

The vertices of the network represent the factories and warehouses. The circled number next to a vertex indicates the amount of supply or demand for the commodity at the corresponding factory or warehouse. The edges represent the routes along which the commodity can be sent, and the weight on each edge (indicated by a circled number on the edge) represents the cost of sending one unit of the commodity along that edge.

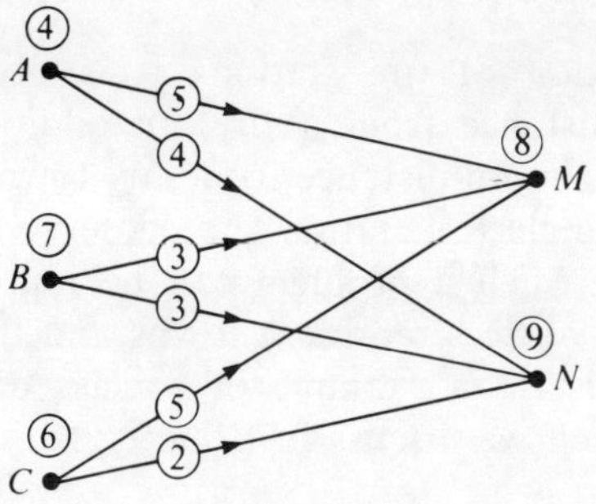

The problem is to determine which factories should supply which warehouses so that the total transportation cost is a minimum. Such problems, called *transportation problems*, are analysed in Chapter 14.

Activity Networks for Scheduling Industrial Projects

Activity networks are a type of network diagram used in industry to aid the scheduling of projects which may involve hundreds of separate activities. In one type of activity network, each activity is represented by a vertex, and an arc joining a vertex X to a vertex Y indicates that the activity X must be completed before activity Y can be started. In the following example, the number next to each arc represents the time taken to complete the activity represented by the vertex from which the arc originates.

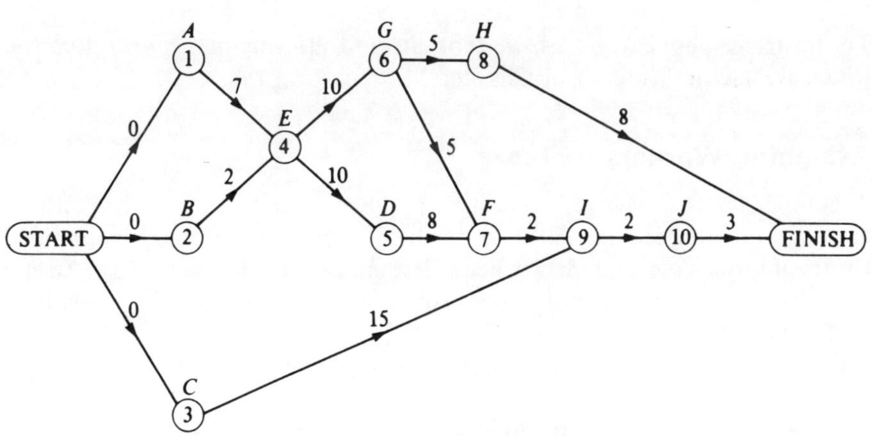

activity network for building construction

Scheduling problems are discussed in Chapters 15 and 16.

Location Problems

In deciding the location of fire stations it is obviously important to minimize the likely distance from a fire. In addition, a planner would probably specify a maximum distance from any town to a fire station, and also require that no single fire station should serve more than a specified maximum population. Such a problem can be represented as a network problem in which the vertices represent towns and the edges represent the connecting roads. This is an example of a class of problems known as *location problems*, which we discuss in Chapter 18.

1.3 CLASSIFICATION OF NETWORK PROBLEMS

Network theory can be considered to be a branch of combinatorics, which is concerned with the arrangement, classification and enumeration of objects within a finite or discrete system. The subject of combinatorics is a wide-ranging one, and includes such topics as graph theory, permutations and combinations, and the structure of geometrical and other systems. Most combinatorial problems can be classified under one or more of the following headings:

existence problems	does there exist . . .? is it possible to . . .?
construction problems	if . . . exists, how can you construct it?
enumeration problems	how many . . . are there? (counting problem) can you list them all? (listing problem)
optimization problems	if there are several possibilities, which one is best?

To illustrate these categories, we look at the following *assignment problem*, which we tackle in detail in Chapter 12.

Assigning Workers to Tasks

A company has a new project which involves four tasks, A, B, C and D, and has four workers available to do them. However, worker 1 is qualified to do only tasks A and B; worker 2 is qualified to do tasks A, B, C and D, worker 3 is qualified to do only tasks A and B, and worker 4 is qualified to do only task B.

In considering this situation, the following questions arise:

Existence problem: Is it possible to assign all four workers to tasks for which they are qualified? (In fact, such an assignment is not possible, since only worker 2 can do tasks C and D. However, an assignment of three workers, one of which must be worker 2, is possible.)

Construction problem: How can we construct an assignment of three workers to tasks for which they are qualified?

Enumeration problem: How many different assignments are there of three workers to tasks for which they are qualified? Can we list them?

Optimization problem: Which assignment of three workers is the best one? (To answer this, we need some additional information—for example, that worker X would prefer to do task A rather than task B.)

Problem 1.3 Classify each of the following problems as an existence, construction, enumeration or optimization problem.

(a) A salesman wishes to travel along every street in a small town. Is it possible for him to find a route which passes along each street once and only once?
(b) A hiker going on a journey can carry only a certain total weight in his knapsack. He has a number of items he would like to take with him, each of which has a particular value for the journey. Which items should he pack so that the total value is a maximum, subject to the weight restriction?
(This is an example of a type of problem known as the *knapsack problem*.)
(c) Four applicants all apply for the same four jobs. How many different ways are there of allocating the applicants to the jobs?
(d) A modern telephone network consists of many tens of thousands of interconnections. Which arrangement of interconnections ensures the maximum number of simultaneous telephone calls at minimum cost?
(e) A salesman wishes to visit a number of towns. How can a route be constructed (assuming one exists) which passes through each town once and only once?

Let us look at each type of problem in turn.

Existence Problems

Faced with a problem, it is natural to ask: does a solution exist? This is not always an easy question to answer. If the answer to an existence problem is YES, we may be able to substantiate it by simply producing an appropriate solution. On the other hand, if the answer is NO, then actually proving that there are no solutions may be very difficult indeed. The following is an example of an existence problem.

Networks with Lower and Upper Capacities

Suppose that a company is designing a pipeline network to transport oil. Each pipe in the network has a minimum flow of oil (the lower capacity) for it to be economically justifiable. Also, each pipe has a maximum possible flow (the upper capacity) depending on the size and length of the pipe. The company may wish to calculate the maximum value of flow in the network, but the first question which arises is an existence problem—does there exist a flow which satisfies the constraints of the lower and upper capacities on the pipelines? This problem is analysed in Chapter 4, where a method of solution is given in the form of an algorithm.

Construction Problems

Once we know that a solution exists, we should like to find a way of actually constructing it. For example, it is not particularly helpful to know that it is possible to get out of a maze if you are stuck in the middle of it—what you really need is a method for getting out! As we saw above, we can often show that an existence problem can be solved by actually constructing a solution, but this is not always the case. In many problems we know that solutions exist because there are theoretical reasons for this, but these reasons may not give us any clue as to how a solution can be constructed.

Examples of construction problems we shall meet in this book include finding a flow in a network with lower and upper capacities (provided that a solution exists) and finding the currents and voltages in an electrical network, where we know that there must always be a solution, but finding it in a complex network may not be a simple matter.

Enumeration Problems

Once we know that a particular problem has a solution, and we can actually construct such solutions, the next questions are: *how many solutions are there*? and *what are they*? We distinguish between two types of problem corresponding to these two questions. A *counting* problem is one in which we want to know *how many* objects of a certain kind there are; a *listing* problem is one in which we want to produce a *list* of all these objects.

For example, suppose that three people, A, B and C, apply for the same three jobs, X, Y and Z. Consider the following problems.

(a) How many different ways are there of allocating the three applicants to the three jobs?
(b) What are all the possible allocations?

Problem (a) is a *counting* problem. We can allocate applicant A to any of the 3 jobs. For each such allocation, we can then give applicant B either of the remaining 2 jobs. In each case, this leaves one job for applicant C. So there are a total of $3 \times 2 \times 1 = 6$ possible allocations.

Problem (b) is a *listing* problem. The possible allocations of A, B and C to jobs X, Y and Z are:

XYZ	XYZ	XYZ	XYZ	XYZ	XYZ
ABC	ACB	BAC	CAB	BCA	CBA

The problems of counting and listing may be very closely related: in some problems the easiest way of counting something may be to construct a list of all possibilities and then to count how many there are. For example, one possible way of obtaining the answer 6 in problem (a) is to list the various possibilities (as we have done in problem (b)) and then to count them. Indeed, for some counting problems this may be the *only* known method of solution—for example, when counting the number of times the digit 3 appears in the first hundred significant figures of π.

In many instances the listing problem is very much harder than the corresponding counting problem. In fact, there are many problems for which the answer to the counting problem is known but no one has been able or willing to list all the possibilities. As an example, consider a typical gas pipe-line network off the coast of Louisiana:

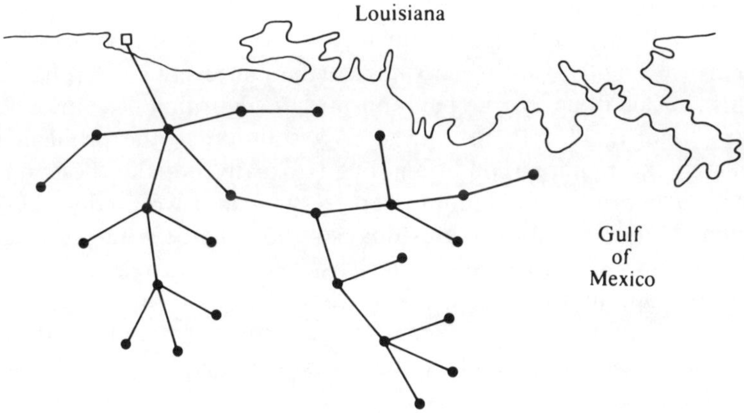

The gas is fed along pipe-lines from the offshore gas fields to an onshore terminal. No unnecessary pipe-lines are laid, so the pipe-line network has the branching form illustrated above; this is called a *tree structure*. There are very many possible such tree structures—for the 25 gas fields shown there are in fact 26^{24} or about 9.1×10^{33} different tree structures connecting the gas fields to the terminal. However, although we know how many tree structures there are, actually listing them is not practicable.

The solution to this counting problem for n vertices (including the onshore terminal) is known to be n^{n-2} but, as we have seen, it is in practice impossible to list them all except for very small values of n. (This enumeration problem and the design of efficient gas pipe-line networks are discussed further in Chapter 8.)

One approach to optimization problems is to simply list all the possibilities and then choose the best one. However, before following this method, it is important to first count the number of possibilities—in many cases there are far too many possibilities for it to be practicable to list them all. When investigating the efficiency of an algorithm for an optimization problem, an upper bound on the time taken to solve a problem can sometimes be obtained by counting all the possibilities which may need to be examined by the algorithm. (See, for example, the discussion in Chapter 20 of the efficiency of an algorithm for the knapsack problem.)

Optimization Problems

In many combinatorial problems it is not enough to know that a given problem has a solution. It may not even be enough to be able to construct a solution using an efficient algorithm, or to list all the possible solutions. In some cases we need to find the 'best' solution—in a commercial context, this is usually the cheapest one.

Problems of this kind occur frequently in the study of networks. In such problems we usually want to maximize or minimize some given parameter, and this is often done by using an appropriate algorithm. The method used in many algorithms is to first guess at a solution of the problem which satisfies all the required conditions but is usually not the 'best' one. We then try to improve this solution step by step until we finally obtain the optimum solution. As long as this final solution can be obtained in a finite (and, preferably, small) number of steps, and we can tell when we have reached the optimum solution, this method is a good one.

For many optimization problems, it is a simple matter to find suitable algorithms which can be applied quickly and efficiently, but unfortunately there are a large number of problems for which this is not the case. An important example of such a problem is the *travelling salesman problem.*

THE TRAVELLING SALESMAN PROBLEM

A salesman wishes to visit a number of cities, selling his goods as he goes. He wants to select the route which involves the least possible total distance. Which one does he choose?

In principle, we can solve this problem by looking at all possible routes and choosing the one which involves the least total distance. For example, consider the following four cities.

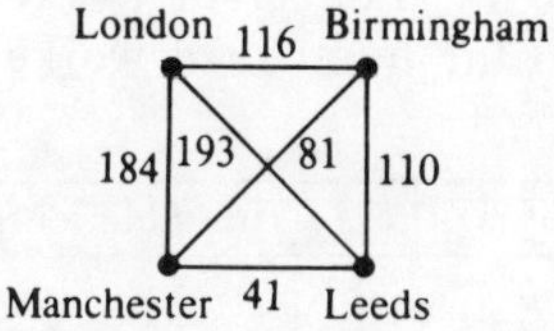

In this example it is easy to show by trial and error that a solution of the travelling salesman problem is to follow the route London–Birmingham–Manchester–Leeds–London (in either direction) with total distance

$$116 + 81 + 41 + 193 = 431 \text{ miles.}$$

But as soon as we increase the number of cities, we run into difficulties, since there is no known algorithm which provides a simple and efficient solution. Although there are several *ad hoc* procedures which can be used to give approximate solutions, a full solution effectively involves looking at all possible routes and choosing the shortest. This is feasible if there are ten cities, since the number of possible routes is then 362 880, and a computer sorting through these at a rate of one million per second will find the best route in about 0.36 seconds. On the other hand, if there are twenty cities, then the number of possible routes is about 1.22×10^{17}, and a computer sorting through them at the same rate would take almost four thousand years! The rapid increase in the number of possible routes as the number of cities increases is an example of what is known as the *combinatorial explosion*. We shall meet this phenomenon many times in this book.

1.4 EFFICIENCY OF ALGORITHMS

It is important for an algorithm to be efficient. To understand what we mean by efficiency, and why it is so important, consider the following example.

Suppose that five workers are available to carry out five different tasks in a project. The workers are not all equally good at each task; this can be expressed by giving a cost to each possible assignment of a worker to a task. The problem is to find the best assignment of all the workers to the tasks—that is, the assignment with the lowest value of total cost.

One way of tackling this problem is simply to examine all possible assignments and then to choose the one with lowest total cost. For five workers, the total number of possibilities is

$$5 \times 4 \times 3 \times 2 \times 1 = 5! = 120,$$

so to do this by hand would not take very long. But what if we have a larger number of workers? The following table shows how the number of possibilities increases with the number of workers, assuming the number of tasks is always equal to the number of workers.

number of workers	possible assignments
5	120
6	720
7	5 040
8	40 320
10	3 628 800
15	1.3×10^{12}
20	2.4×10^{18}

Even with only 7 workers, the number of possibilities is over five thousand; with 10 workers, it is quite out of the question to examine all possibilities by hand. This rapid increase in the number of possible assignments is another example of the combinatorial explosion.

Even with a computer, it soon becomes impractical to examine all possibilities. If a computer examined one million possibilities in a second, then a problem with 10 workers would take less than four seconds, but a problem with 20 workers would take over seventy thousand million years!

What is needed is a more efficient algorithm—that is, an algorithm for which the time taken to reach a solution does not increase so rapidly with the size of the problem. In Chapter 13 we describe such an algorithm (called the *Hungarian algorithm*) which takes a time proportional to n^2, where n is the number of workers. The exact value of the constant of proportionality depends on the details of the algorithm and the computer. If we assume that for a problem with 5 workers it takes 1000 times as long as our previous algorithm which examined all possibilities, we can draw up the following table comparing the times in seconds, taken by the two algorithms.

number of workers	time taken by algorithm examining all possibilities	time taken by Hungarian algorithm
5	1.2×10^{-4}	0.12
6	7.2×10^{-4}	0.17
7	5×10^{-3}	0.24
8	4×10^{-2}	0.31
10	3.6	0.48
15	1.3×10^{4} years	1.1
20	7.6×10^{10} years	1.9

The differences between these times for the larger values of n demonstrate the importance of having an efficient algorithm. (Efficiency in this sense will be discussed further in Chapter 20.) In this book we describe algorithms for a variety of network problems. It is important always to bear in mind that although they may seem unnecessarily complicated and perhaps sometimes tedious to work through by hand, they are indispensable for solving problems involving larger networks on a computer.

1.5 THE PROCESS OF NETWORK ANALYSIS

Many practical problems arising in industry, science or commerce can be formulated as network problems. The network problem can be tackled by mathematical methods such as those described in this book, but network analysis must be used with care: it is important to keep in mind the relation between the original practical problem and the mathematical network problem.

In the physical and engineering sciences, the process of expressing a problem in mathematical language is commonly referred to as *mathematical modelling*. The equations which describe the behaviour of a mechanical system or of the flow of electric current in a circuit are described as a *mathematical model* of the corresponding physical situation.

However, this language is not so appropriate for industrial and commercial networks where we are often dealing with complex non-deterministic systems rather than with simple deterministic mechanical ones, and where it is often difficult to describe the problems precisely.

To see what is involved in network analysis, it is helpful to divide the process into a number of stages.

1. *The first stage is the recognition that a practical problem exists.*

2. *The practical problem must be identified and carefully formulated in ordinary language.*

At this stage, interpretation, judgement and opinion are involved, so this description may not be a true reflection of the original problem. Also, the description may well be incomplete.

3. *The above formulation of the practical problem is represented as a network problem.*

This stage often involves simplifications which may or may not be justified. Sometimes, only one aspect of the practical problem can be represented as a network problem.

4. *The network problem is analysed mathematically.*

However, once the problem reaches the mathematician, it may prove too difficult for him or her to solve, and it may be necessary to simplify it even further before it will yield to mathematical techniques. Alternatively, the mathematician may try to obtain an approximate solution to the problem rather than an exact one. An approximate solution may well be good enough in practice, in which case the network analysis will have proved its value. On the other hand, there may have been so many simplifications that the eventual result is of no practical use.

5. *The mathematical solution is interpreted in terms of the original problem.*

If the conclusions are not satisfactory, it may be necessary to work through one or more of the above stages several times before an acceptable result is achieved. This process often involves a dialogue between technologist and mathematician with each making suggestions to be adopted by the other. The mathematician will often have to learn enough about the practical problem to be able to appreciate the difficulties involved in formulating it as a network problem, and the technologist may have to learn enough mathematics to be able to appreciate the difficulties involved in solving the resulting mathematical problem and in interpreting the results.

We illustrate the above comments by the following example. Consider the problem of siting fire-stations in order to give the best possible service.

In ordinary language, the problem is: *where should we site the fire-stations so that they give the most efficient service to the people in the area they serve?*

The problem is an optimization problem which may be formulated more precisely as follows: *where should we site the fire-stations so that a fire-engine may reach every house within the area covered in at most ten minutes?*

This formulation is not particularly helpful, because it involves the prediction of traffic movement, and this will depend on the time of day, roadworks, school holidays, special local events, accidents, and so on.

A mathematical formulation of the problem may well be: *where should we site the fire-stations so that every house in the area is within six miles of a fire-station?*

We now have a location problem which we can solve, but the mathematical solution may not be satisfactory. For example, we may find that the optimum location for a fire-station lies in the middle of a large lake, in which case some compromise solution will have to be reached.

In view of what has been said, it is clearly important to be cautious when interpreting the mathematical solution in practical terms. Network analysis is a powerful tool, and can give insight into many complex problems, but the conclusions drawn from it may need to be modified by additional practical considerations.

CHAPTER 2

GRAPHS AND DIGRAPHS

After reading this chapter, you should be able to:

For graphs:

- explain the terms *graph*, *vertex*, *vertex-set*, *edge*, *edge-list*, *multiple edges*, *loop*, *simple graph*, *subgraph*, *adjacent*, *incident*, *degree*, *degree-sequence*, *regular graph* and *isolated vertex*;
- explain what it means for two graphs to be *the same*, and what it means for them to be *isomorphic*;
- state and use the handshaking lemma;
- explain the terms *walk*, *trail*, *path*, *cycle*, *connected graph*, *Eulerian trail* and *Hamiltonian cycle*;
- explain the terms *complete graph*, *cycle graph*, *bipartite graph* and *tree*;
- find the incidence matrix and the adjacency matrix of a given graph.

For digraphs:

- explain the terms *digraph*, *vertex*, *vertex-set*, *arc*, *arc-list*, *multiple arcs*, *loop*, *simple digraph*, *subdigraph*, *underlying graph*, *adjacent*, *incident to*, *incident from*, *in-degree*, *out-degree*;
- explain what it means for two digraphs to be *the same*, and what it means for them to be *isomorphic*;
- explain the terms *walk*, *trail*, *path*, *cycle*, *connected digraph* and *strongly connected digraph*;
- find the incidence matrix and the adjacency matrix of a given digraph.

Introduction

In the previous chapter we introduced the concepts of *graph* and *digraph*, and explained some basic terminology. In this chapter we investigate some of the properties of graphs and digraphs, and define a number of terms which will be needed for our study of networks. This chapter may be regarded as a source of reference for the theory of graphs and digraphs which is required for this book.

2.1 GRAPHS AND ISOMORPHISMS

In Chapter 1 we described how graphs can be used to represent relationships between certain objects. We obtained a pictorial representation in which the objects are represented by points (*vertices*) and the relationships are represented by lines (*edges*) joining them. Note that we can use many different diagrams to represent the same graph. For example, suppose that three factories A, B and C supply three stores X, Y and Z. The supply routes can be represented by either of the following graphs:

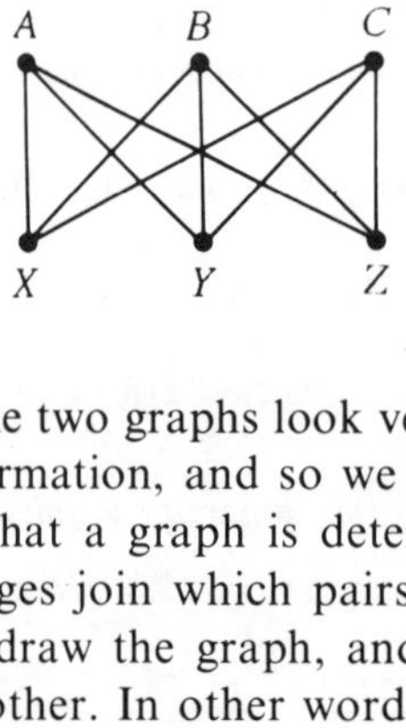

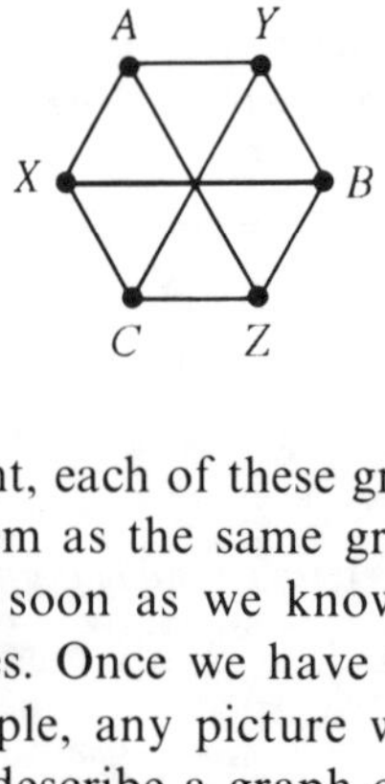

Although the two graphs look very different, each of these graphs conveys the same information, and so we regard them as the same graph.

It is clear that a graph is determined as soon as we know its vertices, and which edges join which pairs of vertices. Once we have this information, we can draw the graph, and in principle, any picture we draw is as good as any other. In other words, we can describe a graph completely by listing its vertices and edges in any order, and the actual way in which the vertices and edges are drawn is not important.

For example, each of the above graphs may be described completely by the following lists:

Vertices: A, B, C, X, Y, Z
Edges: AX, AY, AZ, BX, BY, BZ, CX, CY, CZ

We now formally define a graph in terms of these lists.

Definitions *A **graph** G consists of a set of elements, called **vertices**, and a list of unordered pairs of these elements, called **edges**. The set of vertices is called the **vertex-set** of G, and the list of edges is called the **edge-list** of G. A **subgraph** of G is a graph all of whose vertices belong to the vertex-set of G and all of whose edges belong to the edge-list of G.*

*If v and w are vertices of G, then an edge of the form vw is said to **join** v and w. Two or more edges joining the same pair of vertices are called **multiple edges**, and an edge joining a vertex to itself is called a **loop**. A graph with no loops or multiple edges is called a **simple graph**.*

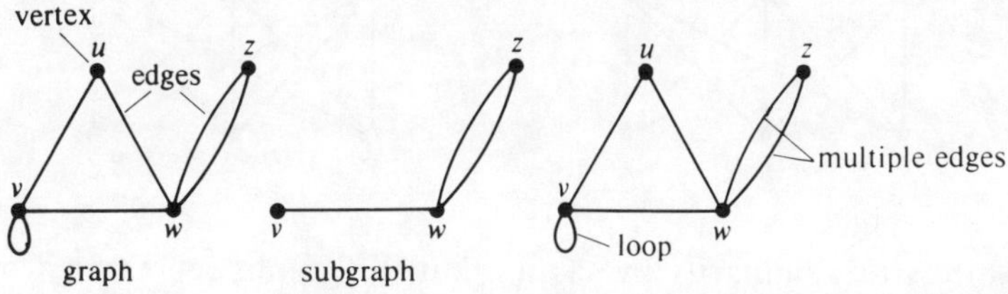

(Note that in graph theory the terminology is not completely standard; for example, some authors use the term *node* or *point* for what we have called a *vertex*, and *arc* or *line* for what we have called an *edge*. Any of these choices of terminology is acceptable as long as it is used consistently.)

Problem 2.1 Write down the vertex-set and edge-list of each of the following graphs:

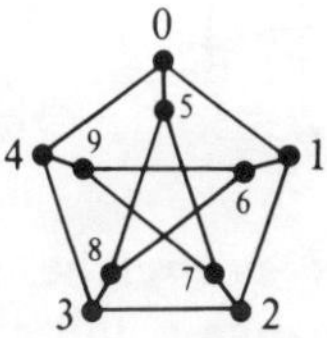

(a)

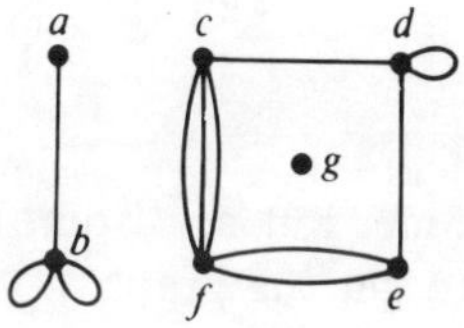

(b)

Problem 2.2 Which of the following graphs are subgraphs of the graph (a) in Problem 2.1?

(a)

(b)

(c)

(d)

Isomorphic Graphs

We have seen that it is possible for two graph diagrams to look very different, but to represent the same graph. On the other hand, it is possible for two graph diagrams to look very similar, but to represent different graphs. For example, the diagrams below look very similar, but are clearly not the same graph (since in the second diagram A and Y are joined, whereas in the first graph they are not).

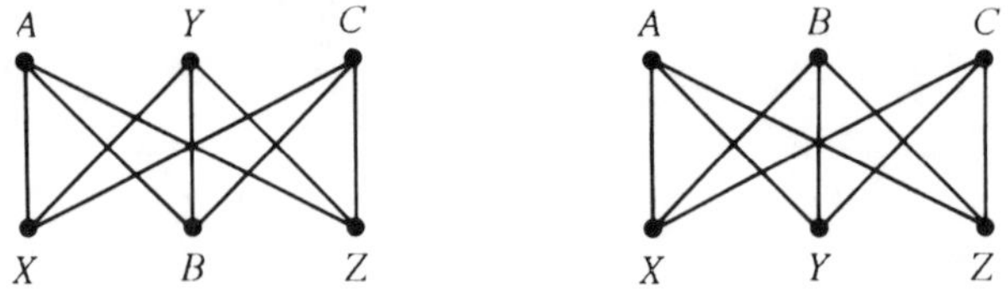

We express this similarity by saying that the graphs represented by these two diagrams are *isomorphic*. This means that *the two graphs have essentially the same structure*: we can relabel the vertices in the first one to get the second one: simply replace B by Y, and conversely. Similarly, the graphs represented by the diagrams

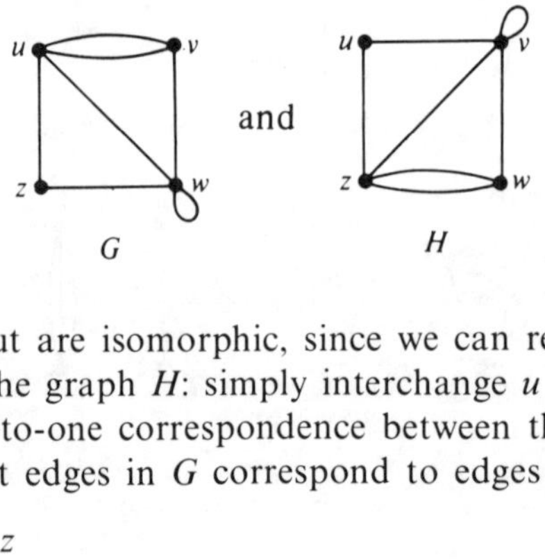

are not the same, but are isomorphic, since we can relabel the vertices in the graph G to get the graph H: simply interchange u and z, and v and w. This gives us a one-to-one correspondence between the vertices of G and those of H such that edges in G correspond to edges in H—namely

$$\begin{array}{lcccc} G: & u & v & w & z \\ & \updownarrow & \updownarrow & \updownarrow & \updownarrow \\ H: & z & w & v & u \end{array}$$

Note that

the two edges uv in G correspond to the two edges zw in H;
the edge uw in G corresponds to the edge zv in H;
the loop at w in G corresponds to the loop at v in H;
and so on.

This leads to the following definition.

Definition *Two graphs G and H are* ***isomorphic*** *if H can be obtained from G by relabelling the vertices—that is, if there is a one-to-one correspondence between the vertices of G and those of H, such that the number of edges joining any pair of vertices in G is equal to the number of edges joining the corresponding pair of vertices in H. This one-to-one correspondence is called an* ***isomorphism****.*

Note that in checking whether or not two graphs are *the same*, we must check carefully that all the labels on the vertices correspond. However, in checking whether or not two graphs are *isomorphic*, we may ignore the symbols used to label the vertices, since the vertices can be relabelled as necessary. In view of this, we often drop the labels altogether when they are not relevant to the problem in hand, and say, for example, that the 'unlabelled graph'

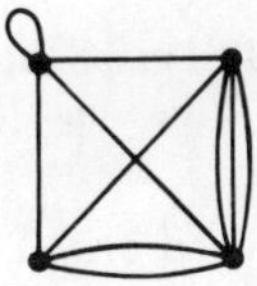

refers to any of the following isomorphic graphs:

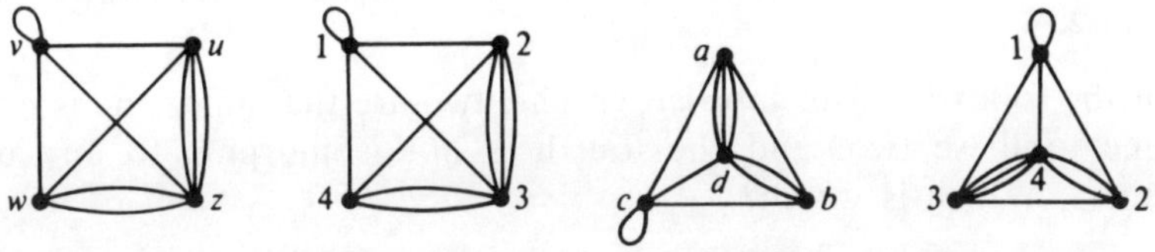

It follows that two *unlabelled* graphs, such as

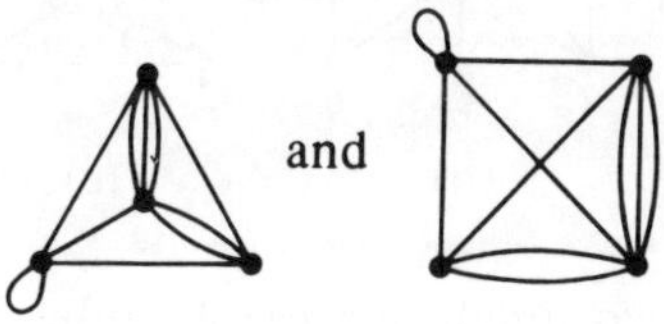

are isomorphic if labels can be attached to their vertices so that they become the same graph. (When we say that two *unlabelled* graphs are the same, we mean that they are isomorphic.)

We can summarize the above as follows.

Labelled graphs

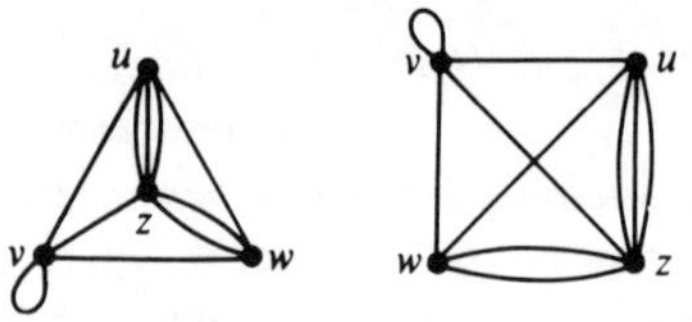

these labelled graphs are the same
(and so are isomorphic)

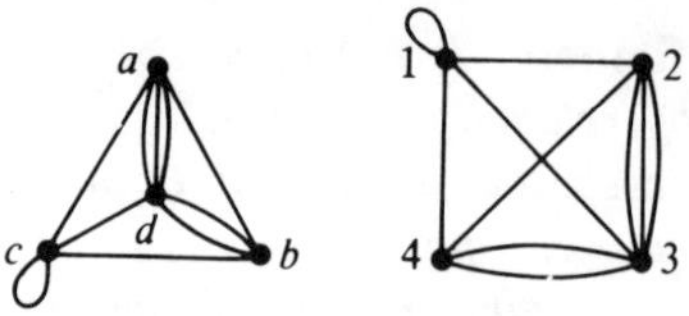

these labelled graphs are not the same,
but are isomorphic

Unlabelled graphs

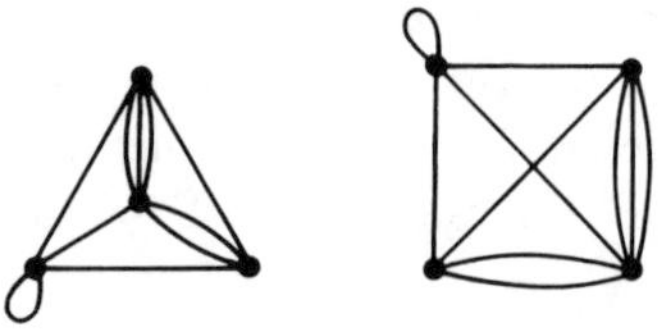

these unlabelled graphs are isomorphic
(the same)

In future, whenever we use the word *graph*, it will be clear from the context whether we are refering to labelled or unlabelled graphs. It there is any possibility of confusion, we shall insert the word *labelled* or *unlabelled*, as appropriate.

Problem 2.3

(i) Of the following four labelled graphs, two are the same, one is isomorphic to these two, and the fourth is not isomorphic to any of the others. Which is which?

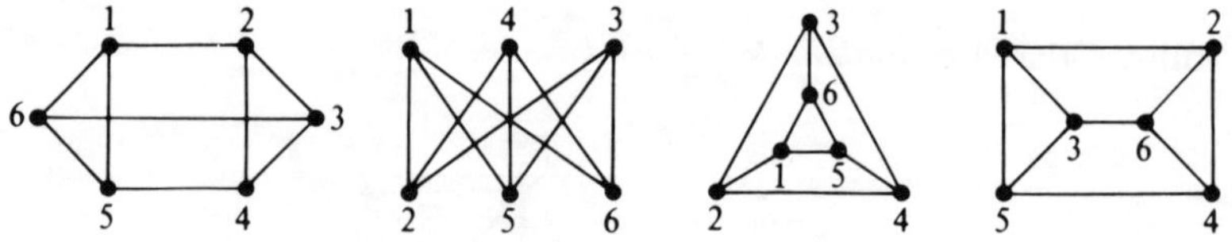

(a) (b) (c) (d)

(ii) Are the following two unlabelled graphs isomorphic?

Counting Unlabelled Graphs

When counting unlabelled graphs, we distinguish between graphs which are not isomorphic. For example, there are just four non-isomorphic simple graphs with three vertices:

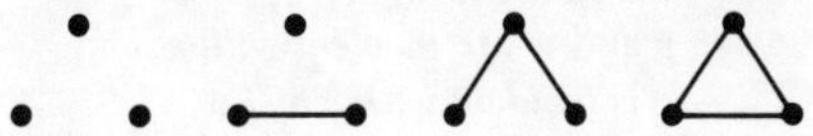

Counting Labelled Graphs

When counting labelled graphs, we distinguish between labelled graphs which are not the same. For example, there are just eight different simple labelled graphs with three vertices:

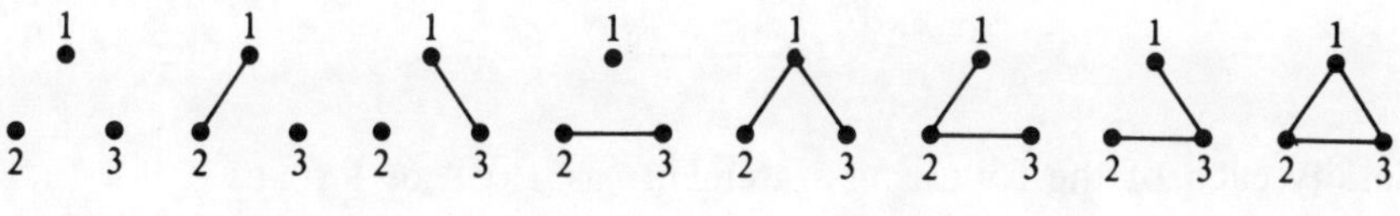

The following table lists the numbers of unlabelled and labelled simple graphs with up to eight vertices.

	number of vertices							
	1	2	3	4	5	6	7	8
unlabelled graphs	1	2	4	11	34	156	1 044	12 346
labelled graphs	1	2	8	64	1 024	32 768	2 097 152	268 435 456

In general, counting problems for labelled graphs are easier to solve than their analogues for unlabelled graphs, and in fact for certain types of graph the former problem has been solved while the latter problem remains unsolved.

2.2 ADJACENCY AND INCIDENCE FOR GRAPHS

Since graph theory is primarily concerned with interconnections between objects of various kinds, we shall find it convenient to introduce some terminology which tells us when certain vertices and edges occur next to each other in a graph. This terminology applies to all graphs, and can be used equally well for wires connecting terminals in an electrical network, or roads connecting towns on a road map.

Definitions *Let v and w be vertices of a graph. If v and w are joined by an edge e, then v and w are said to be* **adjacent**. *Moreover, v and w are said to be* ***incident with e***, *and e is said to be* ***incident with v and w***.

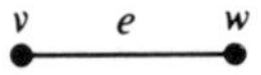

v and w are *adjacent*;
v and w are *incident with e*;
e is *incident with v and w*.

Problem 2.4 In the following graph, u, v, w and z are vertices, and a, b, c, d, e and f are edges:

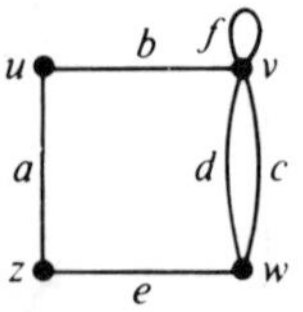

Classify each of the following statements as TRUE or FALSE:

(a) u and z are adjacent; (b) v and z are adjacent;
(c) d and z are adjacent; (d) v is incident with d;
(e) u is incident with c; (f) b is incident with d;
(g) e is incident with w; (h) v is incident with f.

Using the concept of incidence, we now give formal definitions of the terms *degree* and *degree-sequence*.

Definitions *Let G be a graph without loops, and let v be a vertex of G. The* ***degree*** *of v is the number of edges incident with v, and is denoted by* **deg v**. *The* **degree-sequence** *of G is the sequence obtained by listing these vertex-degrees in non-decreasing order. If every vertex of G has the same degree (r, say), then G is said to be* ***regular of degree r***.

Although these terms are defined for graphs without loops, they can easily be extended to graphs in which loops are present. We do this by requiring that *each loop contributes 2 to the degree of a vertex*.

For example, the graph on the right has the following vertex-degrees:

deg $u = 1$, deg $v = 4$, deg $w = 3$,

deg $x = 0$, deg $y = 8$, deg $z = 4$,

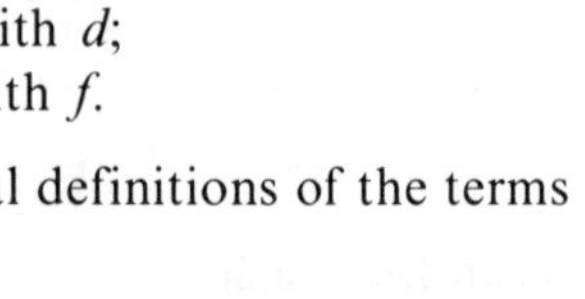

and the degree sequence is (0, 1, 3, 4, 4, 8).

A vertex of degree 0 (such as x) is called an **isolated vertex**.

Problem 2.5

(i) Write down the degree-sequence of each of the following graphs:

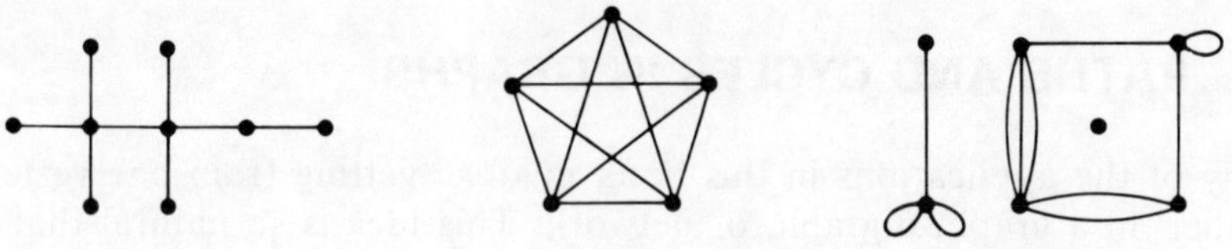

(a) (b) (c)

(ii) For each of the graphs in part (i), write down

(1) the number of edges;
(2) the sum of the degrees of all the vertices.

Do you notice any connection between your answers to parts (1) and (2)? If so, can you explain why this connection arises?

In the solution to the above problem, we saw that the sum of the vertex-degrees of each graph in part (i) is exactly twice the number of edges. A corresponding result holds for all graphs, and is usually called the *handshaking lemma*.

The handshaking lemma *In any graph, the sum of all the vertex-degrees is equal to twice the number of edges.*

Proof Since each edge has two ends, it must contribute exactly 2 to the sum of the degrees. The result follows immediately. □

The name *handshaking lemma* arises from the fact that a graph can be used to represent a group of people shaking hands at a party. In such a graph, the people are represented by vertices, and an edge is included whenever the corresponding people have shaken hands. With this interpretation, the number of edges represents the total number of handshakes, the degree of a vertex is the number of hands shaken by the corresponding person, and the sum of the degrees is the total number of hands shaken. So the handshaking lemma states simply that the total number of hands shaken is equal to twice the number of handshakes—the reason being, of course, that exactly two hands are involved in each handshake.

There are some important consequences of the handshaking lemma.

Consequences of the handshaking lemma

1. *In any graph, the sum of all the vertex-degrees is an even number.*
2. *In any graph, the number of vertices of odd degree is even.*
3. *If G is a graph which has n vertices and is regular of degree r, then G has exactly* $\frac{1}{2}nr$ *edges.*

2.3 PATHS AND CYCLES IN GRAPHS

Many of the applications in this book involve 'getting from one vertex to another' in a graph, digraph, or network. This idea is so natural that you have already met it several times, possibly without realizing it. For example, in the previous chapter we mentioned the problem of finding the shortest route between one town and another. Other examples include the routeing of a telephone call between one subscriber and another, the flow of current between two terminals of an electrical network, and the supply of goods from the manufacturer to the retailer. Our aim in this section is to make this idea precise by means of a few definitions. We start by defining a *walk* in a graph.

Definition *A* ***walk of length k*** *in a graph G is a succession of k edges of G of the form*

$$uv,\ vw,\ wx,\ \ldots,\ yz.$$

We denote this walk by uvwx . . . yz, and refer to it as a ***walk between u and z****.*

Note that the 'second vertex' of each edge is the same as the 'first vertex' of the next.

Intuitively we can think of this as a walk from u to v, then to w, then to x, and so on, until we end up eventually at the vertex z.

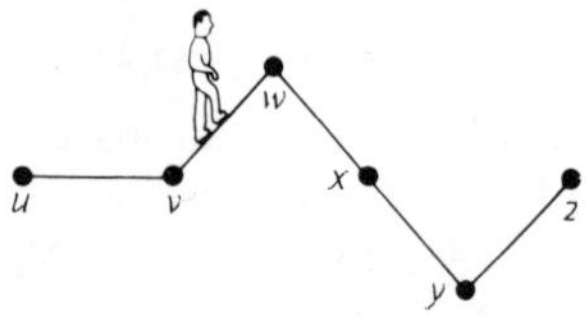

Alternatively, since the edges are undirected, we can think of it as a walk from z to y and on, eventually, to x, w, v and u. So we can also denote this walk by $zy \ldots xwvu$, and refer to it as a ***walk between z and u***.

Note that we do not require all the edges or vertices in a walk to be different. For example, in the following graph, $uvwxywvzzy$ is a walk of length 9 between u and y, which includes the edge vw twice, and the vertices v, w, y and z twice.

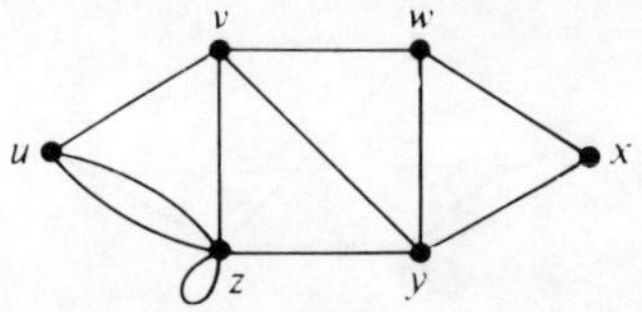

This leads us to make the following definitions.

Definitions *If all the edges (but not necessarily all the vertices) of a walk are different, then the walk is called a* ***trail****. If, in addition, all the vertices are different, then the trail is called a* ***path****.*

In the diagram above, the walk $vzzywxy$ is a trail which is not a path (since the vertices y and z both occur twice), whereas the walk $vwxyz$ has no repeated vertices, and is therefore a path. A walk such as $uzvuz$ is also a trail as along as the two occurrences of uz refer to the two different edges joining u and z.

(Note that the terminology in this subject is not standard—a *walk* appears in the literature as an *edge-sequence*, *route* or *path*; a *trail* appears as a *path* or *chain*; and a *path* appears as a *chain*, *arc*, *simple path* or *simple chain*!)

We can use the concept of a path to explain exactly what is meant by a *connected graph*. A graph is said to be connected if it is 'in one piece'. For example, the following graph is not connected, but can be split into four connected pieces.

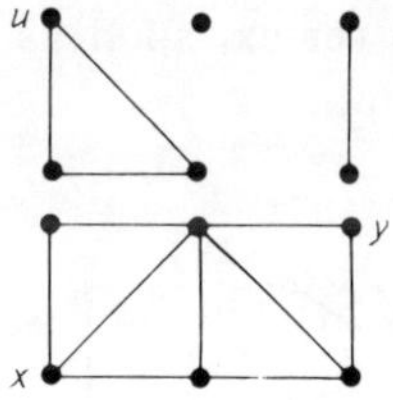

The observation that there is a path between x and y (which lie in the same piece), but not between u and y (which lie in different pieces) leads us to make the following definitions.

Definitions *A graph G is* ***connected*** *if there is a path in G between any given pair of vertices, and* ***disconnected*** *otherwise. Every disconnected graph can be split up into a number of connected subgraphs, called* ***components****.*

The above graph is a disconnected graph with four components.

Problem 2.6 Complete the following statements concerning the graph

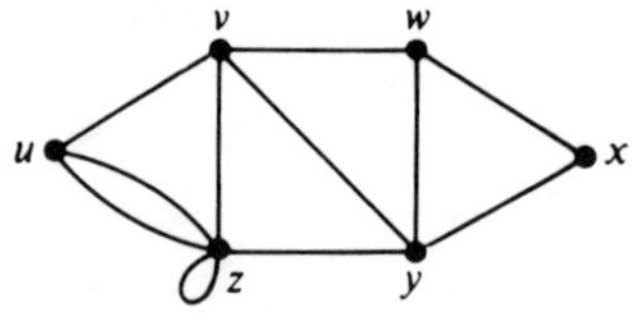

(a) *xyzzvy* is a □ of length □ between □ and □;

(b) *vuvzv* is a □ of length □ between □ and □;

(c) *vw* is a □ of length □ between □ and □;

(d) *uvwxyzu* is a □ of length □ between □ and □.

Problem 2.7 Find all the paths between *s* and *z* in the following graph:

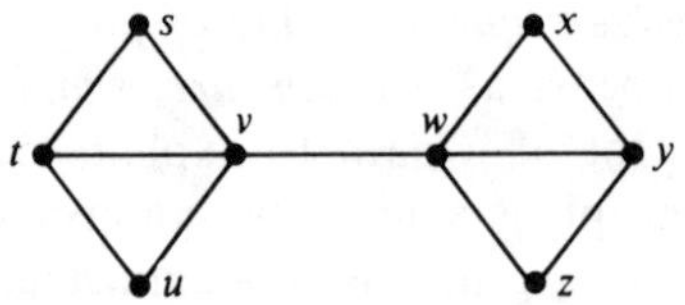

It is also useful to have a special term for those walks or trails which start and finish at the same vertex, such as the walk *vuvzv* and the trail *uvwxyzu* in the following graph:

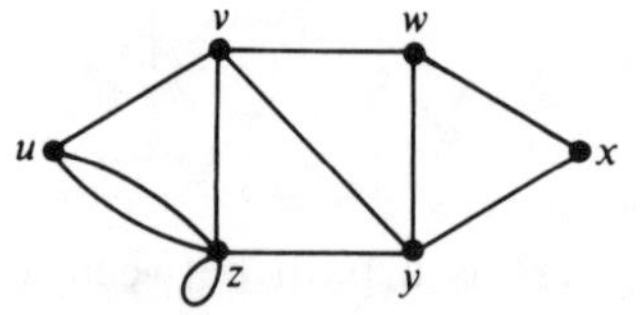

This leads us to make the following definitions.

Definitions *A* ***closed walk*** *in a graph G is a succession of edges of G of the form*

$$uv,\ vw,\ wx,\ \ldots,\ yz,\ zu.$$

If all of these edges are different, then the walk is called a **closed trail**. *If, in addition, the vertices u, v, w, x, . . ., y, z are all different, then the trail is called a* **cycle**.

(In the literature a *cycle* is often called a *circuit*).

In the above graph, the closed walk *uvwyyvzu* is a closed trail which is not a cycle (since the vertex *v* occurs twice), whereas the closed trails *zz*, *vwxyv* and *vwxyzv* are all cycles. A cycle of length three, such as *vwyv* or *wxyw*, is called a **triangle** (for obvious reasons). In describing closed walks we can allow any vertex to be the starting vertex. For example, the triangle *vwyv* can equally well be described by the letters *wyvw* or *yvwy* or (since the direction is immaterial) by *vywv*, *wvyw* or *ywvy*.

Problem 2.8 In the graph on the right, find:

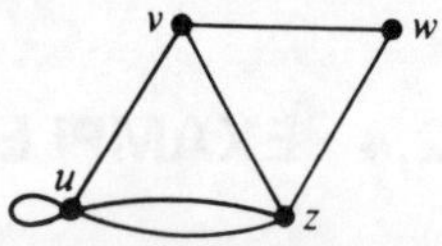

(a) a closed walk which is not a closed trail;
(b) a closed trail which is not a cycle;
(c) all the cycles of lengths 1, 2, 3 and 4.

Eulerian Trails and Hamiltonian Cycles

Two historically important problems concerning closed trails and cycles are the following.

THE EXPLORER'S PROBLEM
An explorer wishes to explore all the roads between a number of cities. Can he arrange his tour so as to traverse each road only once?

THE TRAVELLER'S PROBLEM
A traveller wishes to visit a number of cities. Can he arrange his tour so as to visit each city only once?

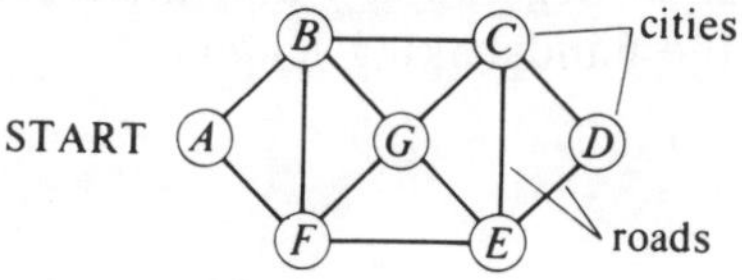

To appreciate the difference between these two problems, consider the above road map. The explorer wishes to find a tour which starts at *A*, takes him along each road exactly once, and ends back at *A*. Examples of such a tour are

ABCDEFBGCEGFA and *AFGCDEGBCEFBA*.

The traveller wishes to find a tour which starts at A, takes him to each city exactly once, and ends back at A. Examples of such a tour are

$$ABCDEGFA \quad \text{and} \quad AFEDCGBA$$

Note that the explorer travels along each road just once but may visit a particular city several times, whereas the traveller visits each city just once but may omit several of the roads on the way.

Let us regard the road map as a graph whose vertices correspond to the cities and whose edges correspond to the roads.

The explorer's problem is the problem of *finding a closed trail which includes every edge of the graph*; such a trail is called an **Eulerian trail**.

The traveller's problem is that of *finding a cycle which includes every vertex of the graph*; such a cycle is called a **Hamiltonian cycle**.

2.4 EXAMPLES OF GRAPHS

In this section we introduce some important types of graph which will feature in later applications.

Complete Graphs

A **complete graph** is a graph in which each pair of distinct vertices are joined by exactly one edge. The complete graph with n vertices is denoted by K_n.

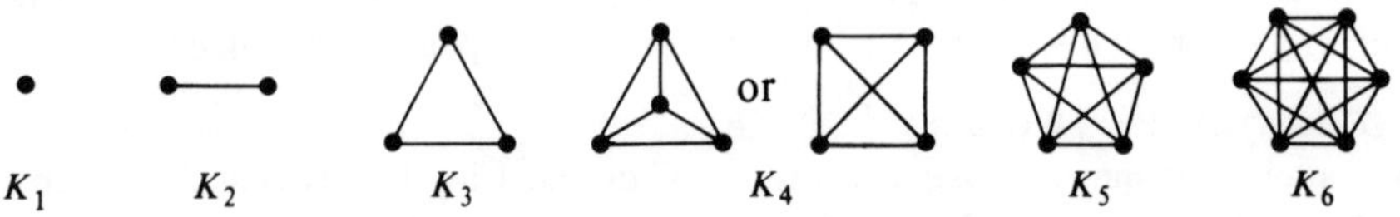

Note that K_n is regular of degree $n - 1$ and therefore has $\frac{1}{2}n(n - 1)$ edges, by consequence 3 of the handshaking lemma.

Cycle Graphs

A **cycle graph** is a graph consisting of a single cycle. The cycle graph with n vertices is denoted by C_n.

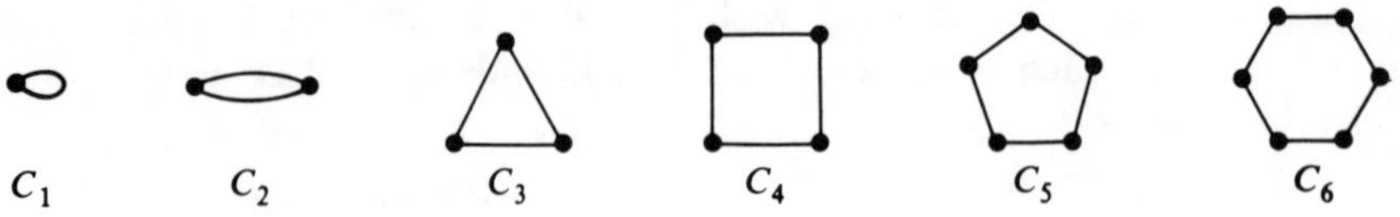

Note that C_n is regular of degree 2, and has n edges.

Bipartite Graphs

Of particular importance in applications to network theory are the bipartite graphs. A **bipartite graph** is a graph whose vertex-set can be split into two sets A and B in such a way that each edge of the graph joins a vertex in A to a vertex in B. We can distinguish the vertices in A from those in B by drawing the former in black and the latter in white, so that each edge is incident with a black vertex and a white vertex. Some examples of bipartite graphs are:

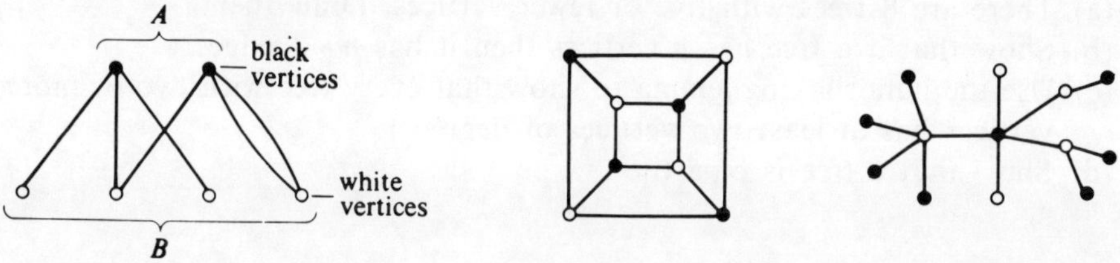

A **complete bipartite graph** is a bipartite graph in which each black vertex is joined to each white vertex by just one edge. The complete bipartite graph with r black vertices and s white vertices is denoted by $K_{r,s}$. Some examples of complete bipartite graphs are:

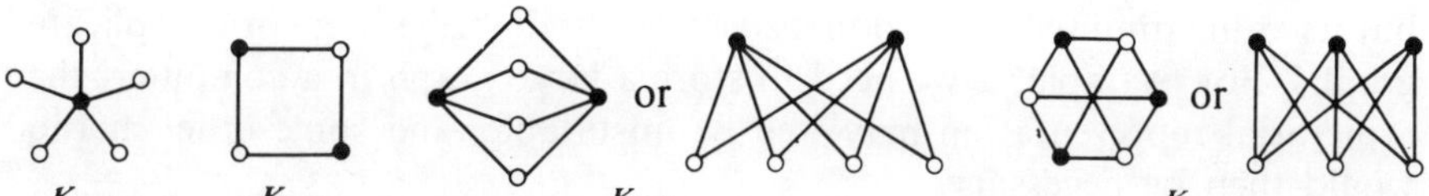

$K_{1,5}$ $K_{2,2}$ $K_{2,4}$ $K_{3,3}$

Note that $K_{r,s}$ has $r + s$ vertices (r vertices of degree s, and s vertices of degree r), and rs edges. Also note that $K_{r,s} = K_{s,r}$. It is usual, but not necessary, to put the smaller of r and s first.

Problem 2.9

(a) Draw the graphs C_8, K_8 and $K_{4,4}$.
(b) Show that in any bipartite graph each cycle has even length.

Trees

A **tree** is a connected graph which contains no cycles. Some examples of trees are shown overleaf.

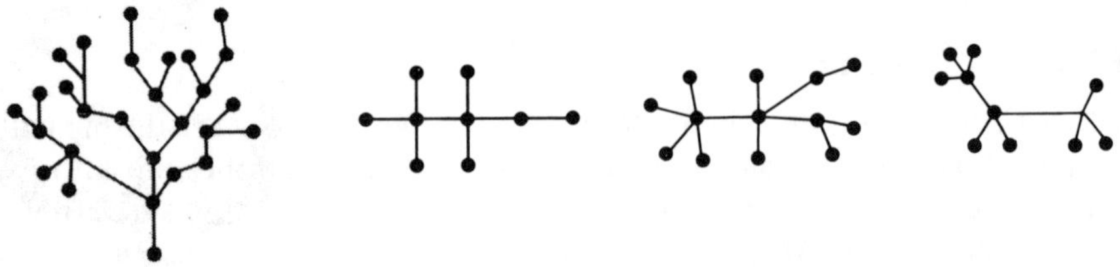

Problem 2.10

(a) There are 8 trees with five or fewer vertices. Draw them.
(b) Show that if a tree has n vertices then it has $n - 1$ edges.
(c) Use the handshaking lemma to show that every tree with two or more vertices has at least two vertices of degree 1.
(d) Show that a tree is bipartite.

2.5 REPRESENTING A GRAPH

We have already seen two ways of representing a graph—as a diagram consisting of points joined by lines, and as a set of vertices and a list of edges. The pictorial representation is very useful in many situations, especially when we want to examine the structure of the graph as a whole, but its value diminishes as soon as we wish to describe large or complicated graphs. For example, if we need to store a large graph in a computer, then a pictorial representation may well be unsuitable, and some other method would then be necessary.

One possibility is to store the set of vertices and the list of edges, and this method is often used in practice, especially when the graph is 'sparse'—that is, it has a lot of vertices but not very many edges. Another method is to take each vertex in turn and list those vertices which are adjacent to it. By joining each vertex to its 'neighbours', we can easily reconstruct the graph. Yet another method is to draw up a table indicating which pairs of vertices are adjacent or a table indicating which vertices are incident with which edges.

Each of these methods has its advantages, but the last one is particularly useful in a number of practical applications. Using this method, we represent a graph by a rectangular array of numbers, called a *matrix*. Such matrices lend themselves to mechanical manipulation, and in several applications of graph theory they are the most natural way of formulating a given problem. There are various types of matrix that can be used to specify a given graph. Here we describe the simplest ones—the *incidence matrix* and the *adjacency matrix*. For convenience, we restrict our attention to graphs without loops.

The Incidence Matrix

The incidence matrix involves the incidence of vertices and edges. To see what is involved, consider the following example. In this graph we have circled the labels of the vertices to distinguish them from the labels of the edges.

Example 1

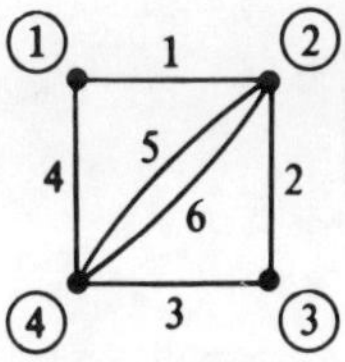

$$\begin{array}{l} \\ \text{row ①} \rightarrow \\ \text{row ②} \rightarrow \\ \text{row ③} \rightarrow \\ \text{row ④} \rightarrow \end{array} \begin{array}{cccccc} \text{col. 1} & \text{col. 2} & \text{col. 3} & \text{col. 4} & \text{col. 5} & \text{col. 6} \\ \downarrow & \downarrow & \downarrow & \downarrow & \downarrow & \downarrow \end{array}$$

$$\begin{bmatrix} 1 & 0 & 0 & 1 & 0 & 0 \\ 1 & 1 & 0 & 0 & 1 & 1 \\ 0 & 1 & 1 & 0 & 0 & 0 \\ 0 & 0 & 1 & 1 & 1 & 1 \end{bmatrix}$$

On the left we have a graph with four vertices and six edges, and on the right we have a matrix with four rows and six columns. Each of the numbers appearing in the matrix is either 1 or 0, depending on whether or not the corresponding vertex and edge are incident with each other. For example,

vertex ① is incident with edge 4, so 1 appears in row 1 column 4;
vertex ② is not incident with edge 4, so 0 appears in row 2 column 4.

We can generalize this idea, as follows.

Definition *Let G be a graph without loops, with n vertices labelled ①, ②, ③, . . . , ⓝ and m edges labelled 1, 2, 3, . . ., m.* The ***incidence matrix A(G)*** *is the n × m matrix in which the entry in row i and column j is 1 if the vertex ⓘ is incident with the edge j, and 0 otherwise.*

Note that the incidence matrix depends on the particular way in which the vertices and edges are labelled, and we obtain one incidence matrix from another by interchanging rows (corresponding to relabelling the vertices) and columns (corresponding to relabelling the edges). We use the letter **A** to denote the incidence matrix rather than the adjacency matrix, because this is standard practice in many of the applications we shall be considering.

Problem 2.11 Write down the incidence matrix of the following graph:

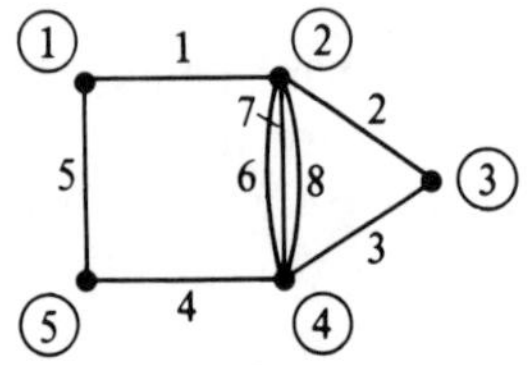

Problem 2.12 Draw the graph whose incidence matrix is

$$\begin{bmatrix} 1 & 1 & 1 & 1 & 0 & 0 & 0 & 0 \\ 1 & 1 & 0 & 0 & 1 & 1 & 0 & 0 \\ 0 & 0 & 0 & 0 & 0 & 0 & 0 & 0 \\ 0 & 0 & 0 & 1 & 0 & 1 & 1 & 1 \\ 0 & 0 & 1 & 0 & 1 & 0 & 1 & 1 \end{bmatrix}$$

The Adjacency Matrix

Consider the following example.

Example 2

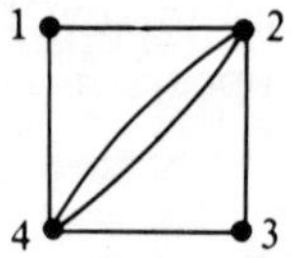

	col. 1	col. 2	col. 3	col. 4
row 1 →	0	1	0	1
row 2 →	1	0	1	2
row 3 →	0	1	0	1
row 4 →	1	2	1	0

On the left we have a graph with four vertices, and on the right we have a matrix with four rows and four columns. The numbers appearing in the matrix refer to the number of edges joining the corresponding vertices in the graph. For example,

vertices 1 and 2 are joined by 1 edge, so 1 appears in row 1 column 2 and in row 2 column 1;

vertices 2 and 4 are joined by 2 edges, so 2 appears in row 2 column 4 and in row 4 column 2;
vertices 1 and 3 are joined by 0 edges, so 0 appears in row 1 column 3 and in row 3 column 1.

Note that every entry on the main (top-left to bottom-right) diagonal is 0, since the graph has no loops. Note also that the matrix is symmetrical about this main diagonal.

We can generalize this idea, as follows.

Definition *Let G be a graph without loops, with n vertices labelled 1, 2, 3, . . ., n. The* ***adjacency matrix M(G)*** *is the $n \times n$ matrix in which the entry in row i and column j is the number of edges joining the vertices i and j.*

Problem 2.13 Write down the adjacency matrix of the following graph:

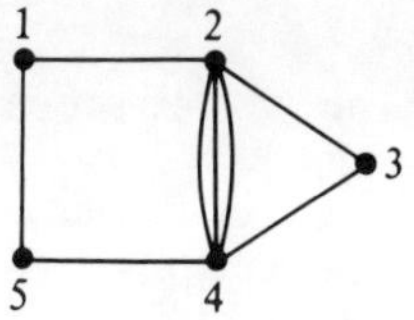

Problem 2.14 Draw the graph whose adjacency matrix is

$$\begin{bmatrix} 0 & 2 & 0 & 1 & 1 \\ 2 & 0 & 0 & 1 & 1 \\ 0 & 0 & 0 & 0 & 0 \\ 1 & 1 & 0 & 0 & 2 \\ 1 & 1 & 0 & 2 & 0 \end{bmatrix}$$

2.6 DIGRAPHS AND ISOMORPHISMS

In the previous chapter we described how digraphs can be used to represent one-way relationships between certain objects. We first obtained a pictorial representation in which the objects are represented by points (vertices) and the relationships are represented by directed lines (arcs) joining them. We choose to regard any pictures representing the same situation as the same digraph, and this leads to the following definitions.

Definitions *A* ***digraph*** *D consists of a set of elements, called* ***vertices****, and a list of ordered pairs of these elements, called* ***arcs****. The set of vertices is called the* ***vertex-set*** *of D, and the list of arcs is called the* ***arc-list*** *of D.*

Definitions *The* ***underlying graph*** *of D is the graph obtained by replacing each arc of D by the corresponding (undirected) edge. A* ***subdigraph*** *of D is a digraph all of whose vertices belong to the vertex-set of D and all of whose arcs belong to the arc-list of D.*

If v and w are vertices of D, then an arc of the form vw is said to be ***directed from v to w****, or to* ***join*** *v to w. Two or more arcs joining the same pair of vertices in the same direction are called* ***multiple arcs****, and an arc joining a vertex to itself is called a* ***loop****. A digraph with no loops or multiple arcs is called a* ***simple digraph****.*

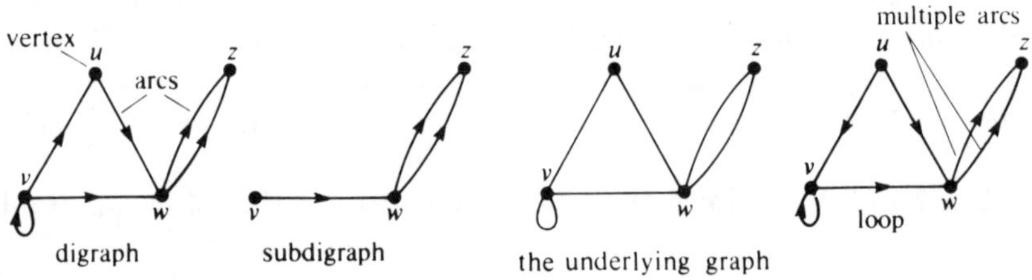

Problem 2.15 Write down the vertex-set and arc-list of each of the following digraphs:

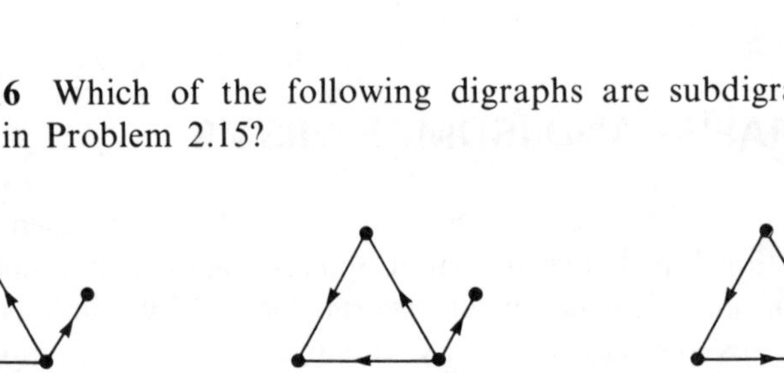

Problem 2.16 Which of the following digraphs are subdigraphs of the digraph (a) in Problem 2.15?

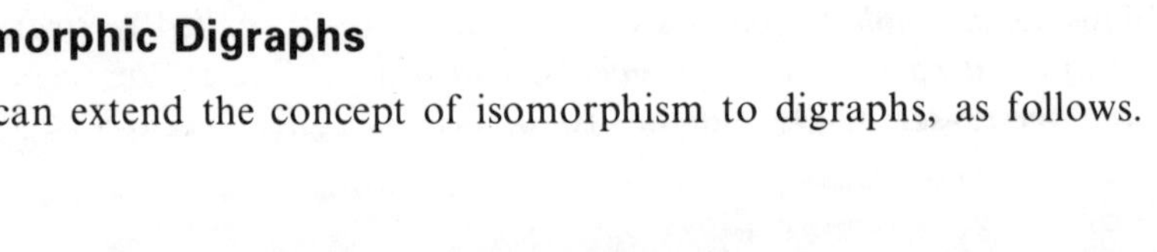

Isomorphic Digraphs

We can extend the concept of isomorphism to digraphs, as follows.

Definition *Two digraphs C and D are* **isomorphic** *if D can be obtained from C by relabelling the vertices—that is, if there is a one-to-one correspondence between the vertices of C and those of D, such that the number of arcs joining any pair of vertices in C is equal to the number of arcs joining the corresponding pair of vertices (in the same direction) in D.*

For example, the digraphs

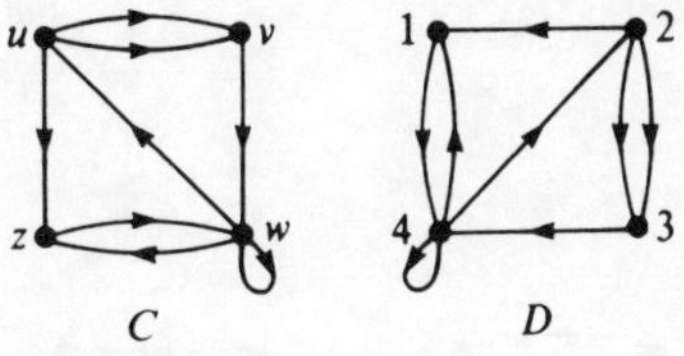

are isomorphic, as you can see by considering the one-to-one correspondence

C:	*u*	*v*	*w*	*z*
	↕	↕	↕	↕
D:	2	3	4	1

Note that

the two arcs *uv* in *C* correspond to the two arcs 23 in *D*;
the arcs *wz* and *zw* in *C* correspond to the arcs 41 and 14 in *D*;
the loop at *w* in *C* corresponds to the loop at 4 in *D*;
and so on.

As with graphs, when checking whether or not two digraphs are isomorphic, we can ignore the actual symbols used to label the vertices, since the vertices can be relabelled as necessary. In view of this, we often drop the labels altogether, and say (for example) that the following unlabelled digraphs are isomorphic.

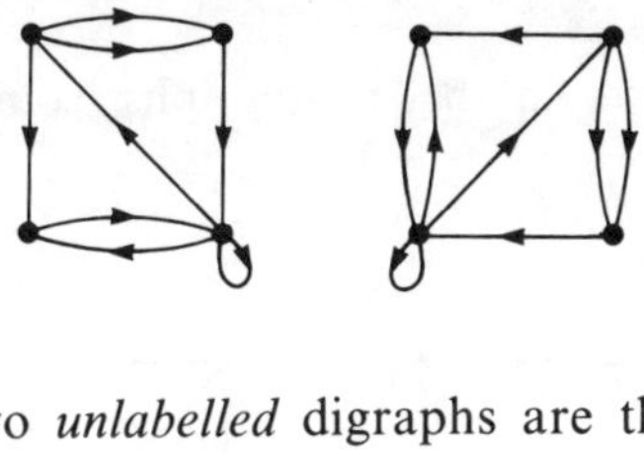

(When we say that two *unlabelled* digraphs are the same, we mean that they are isomorphic.)

We can summarize the above as follows.

Labelled digraphs

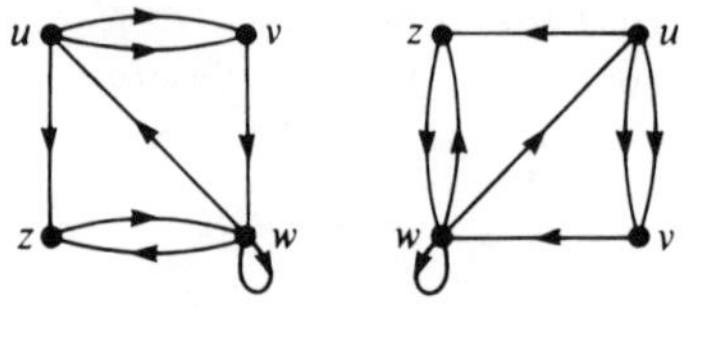

these labelled digraphs are the same
(and so are isomorphic)

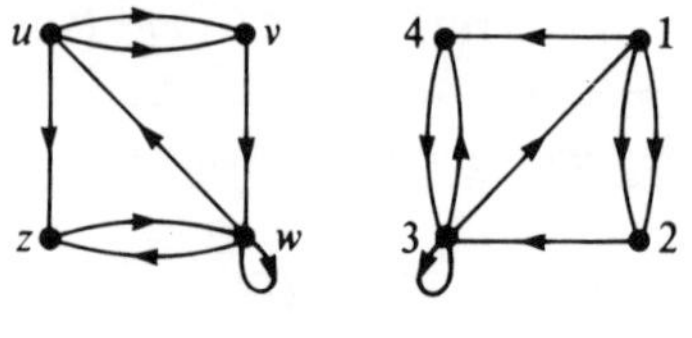

these labelled digraphs are not the same,
but are isomorphic

Unlabelled digraphs

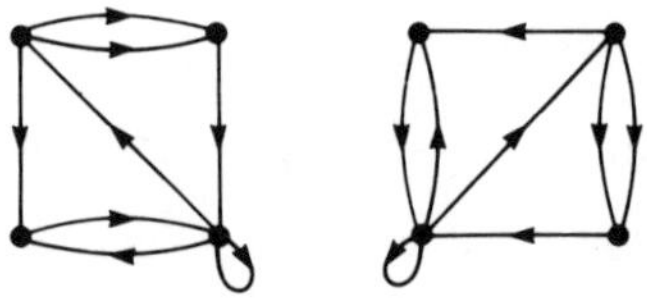

these unlabelled digraphs are isomorphic
(the same)

Problem 2.17

(i) Of the following four labelled digraphs, two are the same, one is isomorphic to these two, and the fourth is not isomorphic to any of the others. Which is which?

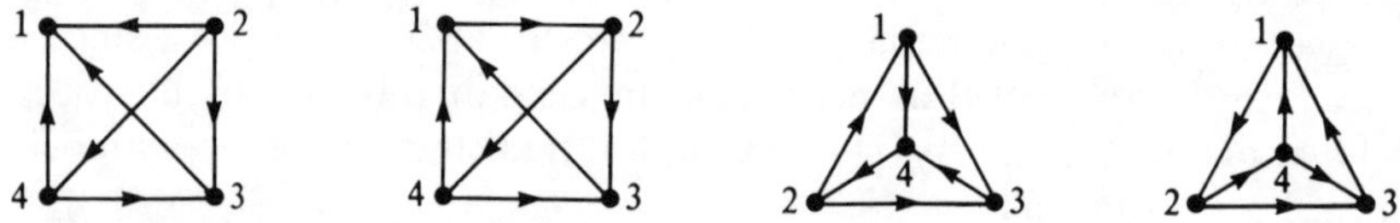

(a) (b) (c) (d)

(ii) Are the following two unlabelled digraphs isomorphic?

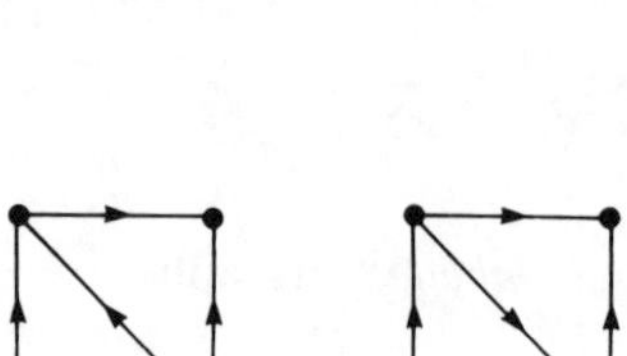

2.7 ADJACENCY AND INCIDENCE FOR DIGRAPHS

We start by defining the digraph analogues of *adjacency* and *incidence*. These are similar to the corresponding definitions for graphs, except that we need to take account of the directions of the arcs.

Definitions *Let v and w be vertices of a digraph. If v and w are joined by an arc e, then v and w are said to be* ***adjacent****. If the arc e is directed from v to w, then v is said to* ***dominate*** *w, and w is said to be* ***dominated by*** *v. In this case the arc is said to be* ***incident from*** *v and* ***incident to w****.*

v and *w* are *adjacent*;
v dominates w;
w is *dominated by v*;
e is *incident from v*
and *incident to w*.

Problem 2.18 In the following digraph, *u*, *v*, *w* and *z* are vertices, and *a*, *b*, *c*, *d*, *e* and *f* are arcs:

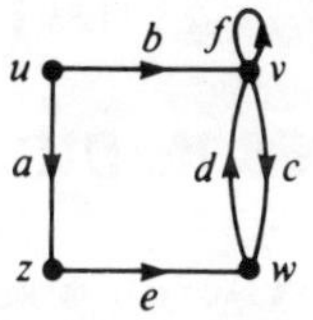

Classify each of the following statements as TRUE or FALSE:

(a) *u* and *z* are adjacent; (b) *v* and *z* are adjacent;
(c) *z* dominates *w*; (d) *v* is dominated by *w*;
(e) *b* is incident from *z*; (f) *f* is incident from *v*;
(g) *a* is incident to *u*; (h) *e* is incident to *z*.

Using the concept of incidence, we can give the digraph analogues of the degree of a vertex in a graph.

Definitions *Let D be a digraph, and let v be a vertex of D. The* ***out-degree of v*** *is the number of arcs incident from v, and is denoted by* ***outdeg v****. Similarly, the* ***in-degree of v*** *is the number of arcs incident to v, and is denoted by* ***indeg v****. The* ***out-degree sequence*** *of D is the sequence obtained by listing the out-degrees in non-decreasing order, and the* ***in-degree sequence*** *of D is defined analogously.*

(Note that, if the digraph contains loops, then each loop contributes 1 to both the out-degree and the in-degree of the corresponding vertex.)

For example, the digraph on the right has the following out-degrees and in-degrees:

outdeg $u = 1$, outdeg $v = 3$, outdeg $w = 2$,
outdeg $x = 0$, outdeg $y = 2$, outdeg $z = 2$,
indeg $u = 0$, indeg $v = 1$, indeg $w = 1$,
indeg $x = 0$, indeg $y = 6$, indeg $z = 2$,

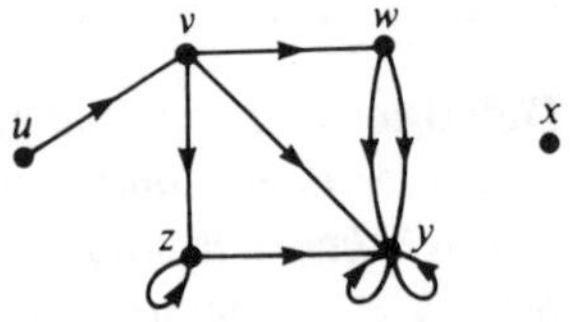

and thus the out-degree and in-degree sequences are (0, 1, 2, 2, 2, 3) and (0, 0, 1, 1, 2, 6), respectively.

Problem 2.19 Write down the out-degree and in-degree sequences of each of the following digraphs:

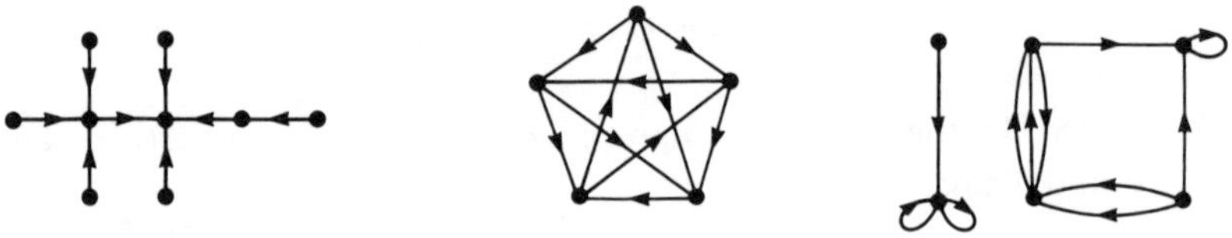

(a) (b) (c)

2.8 PATHS AND CYCLES IN DIGRAPHS

Just as you can 'get from one vertex of a graph to another' by tracing the edges of a walk, trail or path, so you can 'get from one vertex of a digraph to another' by tracing the arcs of a 'directed' walk, trail or path. This means that you have to follow the directions of the arcs as you go, just as if you were driving around a one-way system in town.

We can make this idea precise, as follows.

Definitions *A **walk of length k** in a digraph D is a succession of k arcs of D of the form*

$$uv, \ vw, \ wx, \ldots, \ yz.$$

*We denote this walk by uvwx . . . yz, and refer to it as a **walk from u to z**. If all the arcs (but not necessarily all the vertices) of a walk are different, then the walk is called a **trail**. If, in addition, all the vertices are different, then the trail is called a **path**.*

In the following digraph, the walk *vwxyvwyzzu* is a walk of length 9 from *v* to *u*, which includes the arc *vw* twice, and the vertices *v*, *w*, *y* and *z* twice. The walk *uvwyvz* is a trail which is not a path (since the vertex *v* occurs twice), whereas the walk *vwxyz* has no repeated vertices, and is therefore a path.

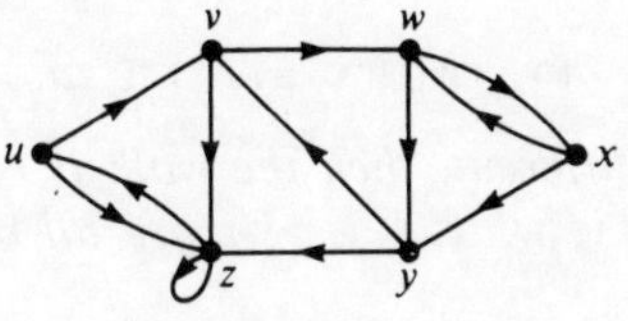

As with graphs, we can use the concept of a path to tell us whether or not a digraph is connected. Recall that a graph is connected if it is 'in one piece', and this means that there is a path between any given pair of vertices. For digraphs *these two ideas are not the same*, and this leads us to two different definitions of the word *connected*.

Definitions *A digraph D is* ***connected*** *if its underlying graph is a connected graph, and* ***disconnected*** *otherwise. It is* ***strongly-connected*** *if there is a path in D from any given vertex to any other.*

The difference between these types of connected digraph is illustrated below:

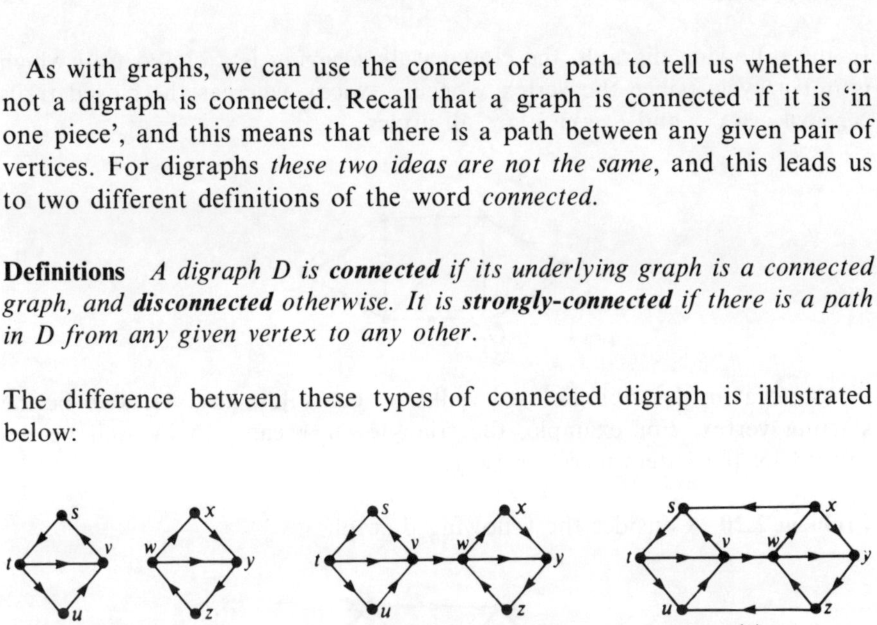

(a) (b) (c)

Digraph (a) is disconnected, since its underlying graph is a disconnected graph. Digraph (b) is connected but is not strongly-connected, since there is no path from *w* to *v*. Digraph (c) is strongly-connected, since there are paths joining all pairs of vertices. Alternatively, you can think of driving around a one-way system in a town. If the town is strongly-connected then you can drive from any part of the town to any other, following the directions of the one-way streets as you go, whereas if the town is merely connected then you can still drive from any part of the town to any other, but you may have to ignore the directions of the one-way streets! Note that every strongly-connected digraph is connected, but not every connected digraph is strongly-connected.

The terms *closed walk*, *closed trail* and *cycle* can also be defined for digraphs.

Definitions *A* ***closed walk*** *in a digraph D is a succession of arcs of D of the form*

$$uv,\ vw,\ wx,\ \ldots,\ yz,\ zu.$$

If all of these arcs are different, then the walk is called a ***closed trail****. If, in addition, the vertices u, v, w, x, . . ., y, z are all different, then the trail is called a* ***cycle****.*

In the following digraph, the closed walk *uvwyvzu* is a closed trail which is not a cycle (since the vertex *v* occurs twice), whereas the closed trails *zz*, *wxw*, *vwxyv* and *uvwxyzu* are all cycles.

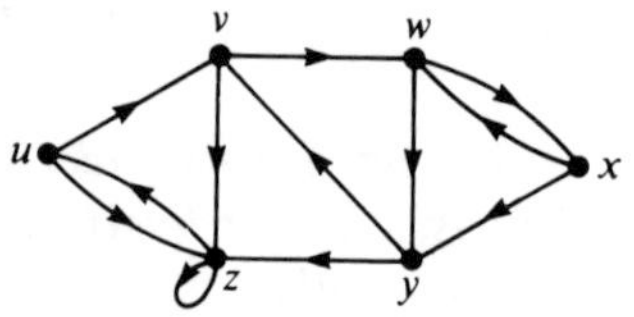

Note that in describing closed walks we can allow any vertex to be the starting vertex. For example, the triangle *vwyv* can equally well be described by the letters *wyvw* or *yvwy*.

Problem 2.20 Consider the following digraph:

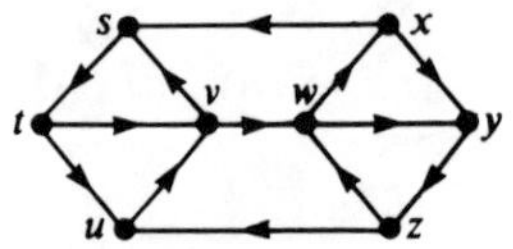

Find

(a) all the paths from *s* to *z*;
(b) all the paths from *z* to *s*;
(c) a closed trail of length 8 containing *s* and *z*.

Are there any cycles containing both *s* and *z*?

Problem 2.21 Classify each of the following digraphs as disconnected, connected but not strongly-connected, or strongly-connected.

(a) (b) (c) (d)

2.9 REPRESENTING A DIGRAPH

Earlier we saw two ways of representing a graph—as an adjacency matrix and as an incidence matrix. These representations have digraph analogues, which are frequently used when storing large digraphs in a computer. For convenience, we restrict our attention to digraphs without loops.

The Incidence Matrix

The incidence matrix of a digraph involves the incidence of vertices and arcs. Since an arc can be either incident from a vertex or incident to a vertex, we have to take account of this when defining the matrix. To see what is involved, consider the following example. In this digraph we have circled the labels of the vertices to distinguish them from the labels of the arcs.

Example 3

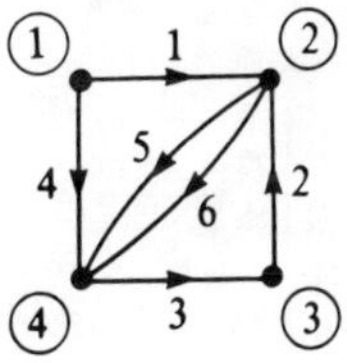

	col. 1	col. 2	col. 3	col. 4	col. 5	col. 6
row ① →	1	0	0	1	0	0
row ② →	−1	−1	0	0	1	1
row ③ →	0	1	−1	0	0	0
row ④ →	0	0	1	−1	−1	−1

On the left we have a digraph with four vertices and six arcs, and on the right we have a 4×6 matrix. Each of the numbers appearing in the matrix is 1, -1 or 0, depending on whether the corresponding arc is incident from, or incident to, or not incident with, the corresponding vertex. For example,

arc 4 is incident from vertex ①, so 1 appears in row 1 column 4;

arc 5 is incident to vertex ④, so −1 appears in row 4 column 5;
arc 4 is not incident with vertex ②, so 0 appears in row 2 column 4.

We can generalize this idea, as follows.

Definition *Let D be a digraph without loops, with n vertices labelled ①, ②, ③, . . . , ⓝ, and m arcs labelled 1, 2, 3, . . ., m. The* ***incidence matrix*** ***A(D)*** *is the* $n \times m$ *matrix in which the entry in row i and column j is*

1 *if arc j is incident from vertex* ⓘ;
−1 *if arc j is incident to vertex* ⓘ;
0 *otherwise.*

Note that the incidence matrix depends on the particular way in which the vertices and arcs are labelled, and we obtain one incidence matrix from another by interchanging rows (corresponding to relabelling the vertices) and columns (corresponding to relabelling the arcs).

Problem 2.22 Write down the incidence matrix of the following digraph:

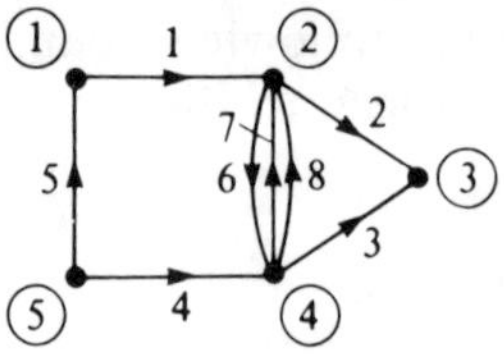

Problem 2.23 Draw the digraph whose incidence matrix is

$$\begin{bmatrix} 1 & -1 & 1 & -1 & 0 & 0 & 0 & 0 \\ -1 & 1 & 0 & 0 & -1 & 1 & 0 & 0 \\ 0 & 0 & 0 & 0 & 0 & 0 & 0 & 0 \\ 0 & 0 & 0 & 1 & 0 & -1 & -1 & -1 \\ 0 & 0 & -1 & 0 & 1 & 0 & 1 & 1 \end{bmatrix}$$

The Adjacency Matrix

The adjacency matrix of a digraph involves the adjacency of the vertices. When defining the adjacency matrix, we have to take into account the fact that each arc is directed. The following example shows how we deal with this.

Example 4

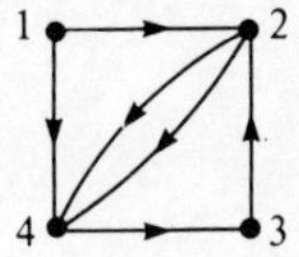

	col. 1 ↓	col. 2 ↓	col. 3 ↓	col. 4 ↓
row 1 →	0	1	0	1
row 2 →	0	0	0	2
row 3 →	0	1	0	0
row 4 →	0	0	1	0

On the left we have a digraph with four vertices, and on the right we have a 4×4 matrix. The numbers appearing in the matrix refer to the number of arcs joining the corresponding vertices in the digraph. For example,

vertices 1 and 2 are joined (in that order) by 1 arc, so 1 appears in row 1 column 2;
vertices 2 and 4 are joined (in that order) by 2 arcs, so 2 appears in row 2 column 4;
vertices 4 and 1 are joined (in that order) by 0 arcs, so 0 appears in row 4 column 1.

We can generalize this idea, as follows.

Definition *Let D be a digraph without loops, with n vertices labelled 1, 2, 3, . . ., n. The* ***adjacency matrix M(D)*** *is the $n \times n$ matrix in which the entry in row i and column j is the number of arcs from vertex i to vertex j.*

Note that the adjacency matrix depends on the particular way in which the vertices are labelled, and we obtain one adjacency matrix from another by interchanging a number of rows and the corresponding columns.

Problem 2.24 Write down the adjacency matrix of the following digraph:

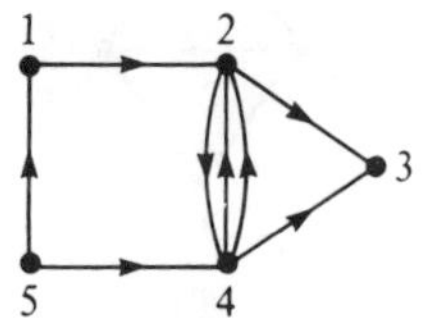

Problem 2.25 Draw the digraph whose adjacency matrix is

$$\begin{bmatrix} 0 & 1 & 0 & 0 & 1 \\ 1 & 0 & 0 & 1 & 0 \\ 0 & 0 & 0 & 0 & 0 \\ 1 & 0 & 0 & 0 & 0 \\ 0 & 1 & 0 & 2 & 0 \end{bmatrix}$$

EXERCISES 2

Graphs and Isomorphisms

2.1. Write down the vertex-set and edge-list of each of the following graphs:

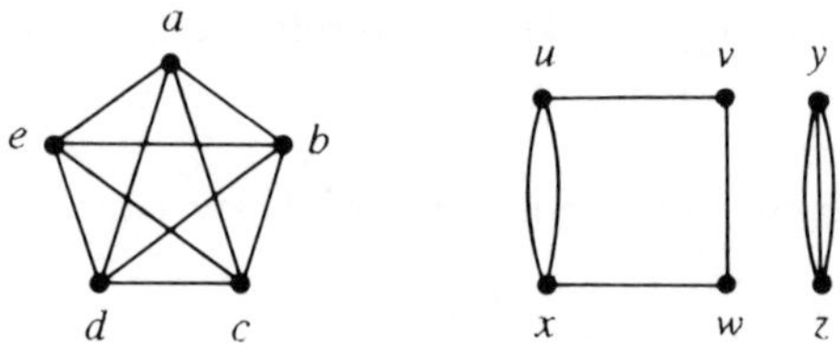

(a) (b)

2.2. Which of the following graphs are subgraphs of the graph (b) in Exercise 2.1?

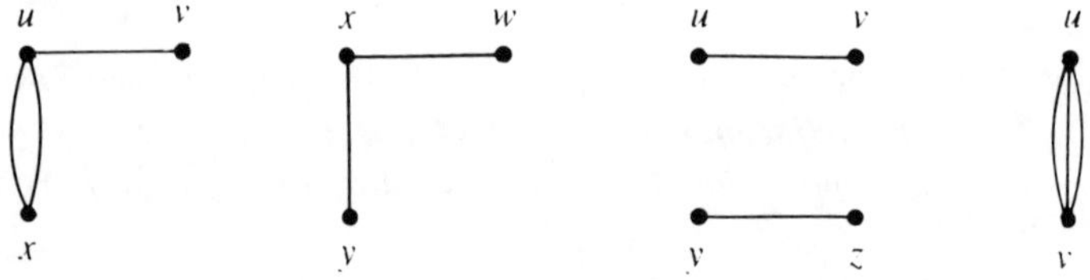

(a) (b) (c) (d)

2.3. Of the following three graphs, two are the same and the third is different. Which is the odd one out?

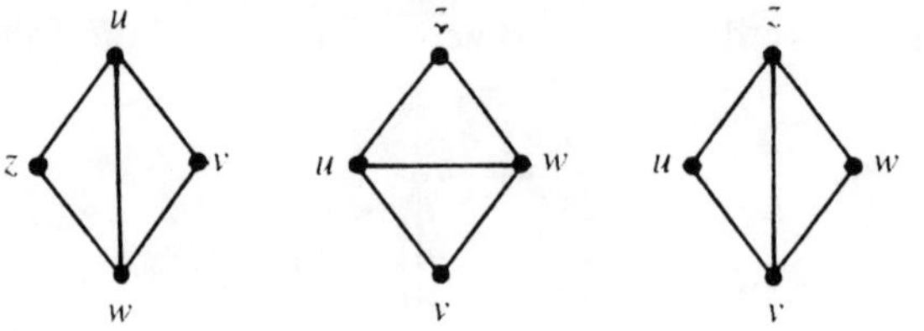

(a) (b) (c)

2.4. Are the following two graphs isomorphic? Give reasons for your answer.

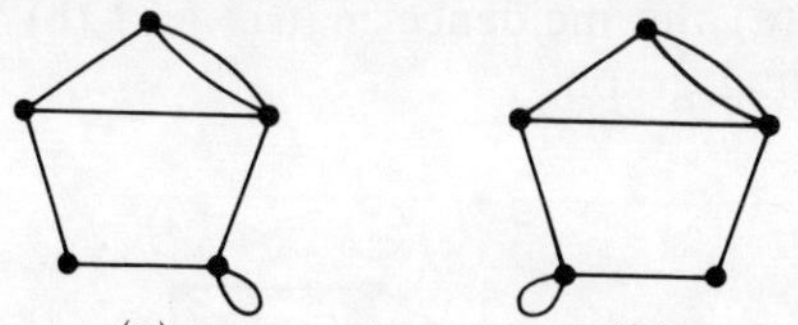

(a) (b)

2.5. Are the following two graphs isomorphic? Show that your conclusion is correct.

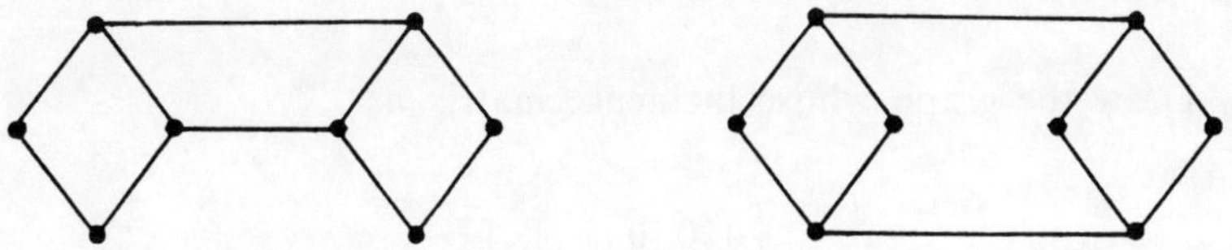

(a) (b)

Adjacency and Incidence for Graphs

2.6. For the graph (a) of Exercise 2.3, write down:

(a) all pairs of vertices which are adjacent;
(b) for each vertex, all the edges incident to it.

2.7. For the graph (a) of Exercise 2.3, write down:

(a) the degree of each vertex;
(b) the degree-sequence.

2.8. For the graph (b) of Exercise 2.1, write down:

(a) the degree of each vertex;
(b) the degree-sequence.

Paths and Cycles in Graphs

2.9. In the graph below find:

(a) a walk of length 7 between u and w;
(b) cycles of lengths 1, 2, 3 and 4;
(c) a path of maximum length.

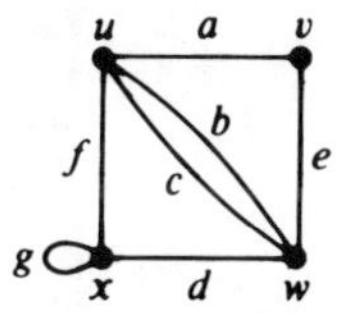

Representing a Graph

2.10. Write down (a) the incidence matrix and (b) the adjacency matrix of the following graph.

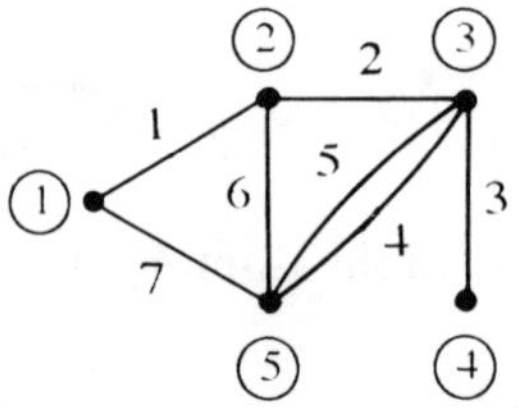

2.11. Draw the graph whose incidence matrix is:

$$\begin{bmatrix} 0 & 0 & 0 & 1 & 1 & 1 \\ 0 & 0 & 1 & 0 & 0 & 1 \\ 0 & 1 & 0 & 0 & 1 & 0 \\ 1 & 0 & 0 & 1 & 0 & 0 \\ 1 & 1 & 1 & 0 & 0 & 0 \end{bmatrix}$$

2.12. Draw the graph who adjacency matrix is:

$$\begin{bmatrix} 1 & 1 & 0 & 1 & 0 \\ 1 & 1 & 0 & 0 & 1 \\ 0 & 0 & 0 & 0 & 0 \\ 1 & 0 & 0 & 1 & 1 \\ 0 & 1 & 0 & 1 & 1 \end{bmatrix}$$

Digraphs and Isomorphisms

2.13. Write down the vertex-set and arc-list of each of the following digraphs:

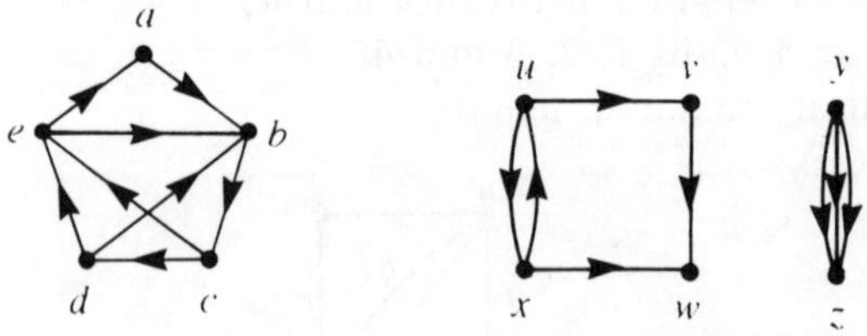

(a) (b)

Which of the following digraphs are subdigraphs of the digraph (b) at the foot of page 48?

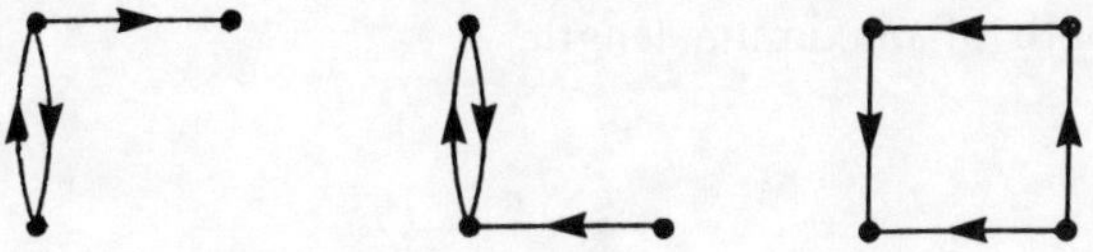

(a) (b) (c)

2.14. Which two of the following digraphs are isomorphic? Give reasons for your answer.

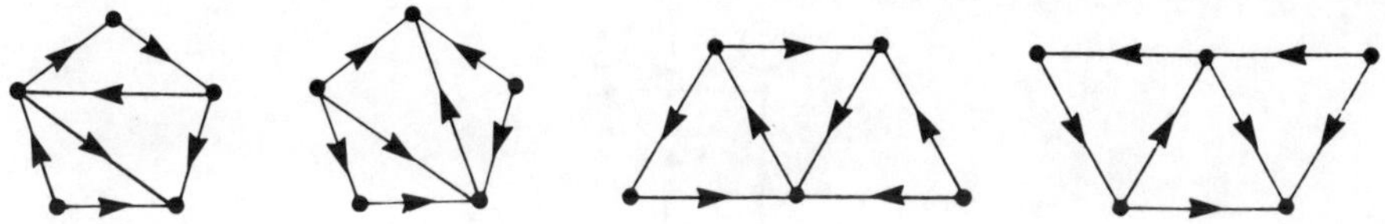

(a) (b) (c) (d)

Adjacency and Incidence for Digraphs

2.15. For the digraph (b) of Exercise 2.13, write down:

(a) all pairs of vertices which are adjacent;
(b) for each vertex, all other vertices which are dominated by it;
(c) for each vertex, all arcs incident from that vertex;
(d) for each vertex, all arcs incident to that vertex.

2.16. For the digraph (b) of Exercise 2.13, write down:

(a) the out-degree of each vertex;
(b) the in-degree of each vertex;
(c) the out-degree sequence;
(d) the in-degree sequence.

Paths and Cycles in Digraphs

2.17. In the following digraph, find (if possible):

(a) a walk of length 7 from u to w;
(b) cycles of length 1, 2, 3 and 4;
(c) a path of maximum length.

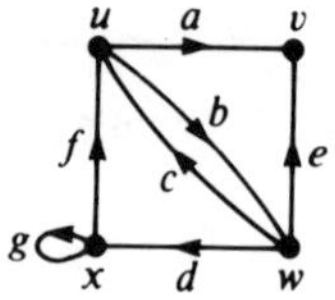

Representing a Digraph

2.18. Write down (a) the incidence matrix and (b) the adjacency matrix of the following digraph.

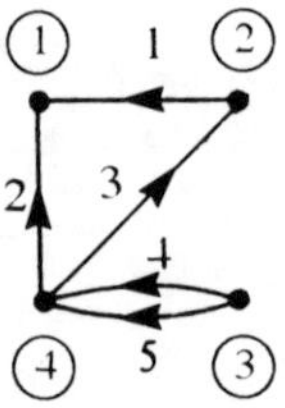

2.19. Draw the digraph whose incidence matrix is:

$$\begin{bmatrix} 1 & 1 & 0 & -1 & 0 \\ -1 & 0 & 0 & 0 & 1 \\ 0 & 0 & -1 & 0 & -1 \\ 0 & -1 & 1 & 1 & 0 \end{bmatrix}$$

2.20. Draw the digraph whose adjacency matrix is:

$$\begin{bmatrix} 0 & 1 & 0 & 1 \\ 0 & 0 & 1 & 1 \\ 0 & 0 & 0 & 0 \\ 0 & 0 & 1 & 0 \end{bmatrix}$$

CHAPTER 3

FLOWS IN BASIC NETWORKS

After reading this chapter, you should be able to:

- explain the terms *basic network*, *flow*, *saturated arc*, *value of flow*, *maximum flow*, and *flow-augmenting path*;
- use the maximum flow algorithm to find a maximum flow in a given basic network;
- explain the terms *cut*, *capacity of a cut*, and *minimum cut*;
- explain why the value of any flow cannot exceed the capacity of any cut;
- state and prove the max-flow min-cut theorem, and verify it for a given network.

Introduction

How much gas can flow through a given pipeline network in a day? How many vehicles can pass through a busy town in a given period of time? How can a manufacturer send the maximum number of items from factories to retailers at minimum cost?

Each of these questions involves finding the maximum amount of a commodity (gas, traffic, passengers) that can pass through a network in a given time. In each case, there are a number of restrictions (such as pipe diameter, road width, cost) which restrict the amount of flow through the various channels of the network. This chapter, and the following two chapters, are concerned with the problem of finding these 'maximum flows'.

We start by restricting our attention to a simplified type of network, often called a *basic network*. For such a network, the problem of finding a maximum flow can be solved completely, and we describe a step-by-step method which can be used to construct a maximum flow in any basic network, however large or complex. Next, we introduce the idea of a *minimum cut* or *bottle-neck* in a network, and we show how the maximum amount of flow in a network is related to the size of the smallest bottle-neck restricting the flow. The material covered in this chapter is fundamental to any discussion of network flow problems.

3.1 BASIC NETWORKS

Definition *A* ***basic network*** *is a digraph N satisfying the following conditions:*

(i) *N has exactly one source and one sink;*
(ii) *to each arc e of N there is assigned a positive number c(e), called the* ***capacity*** *of e, which represents the maximum amount of the commodity that can flow through the arc in a given time.*

A typical example of a basic network is the following:

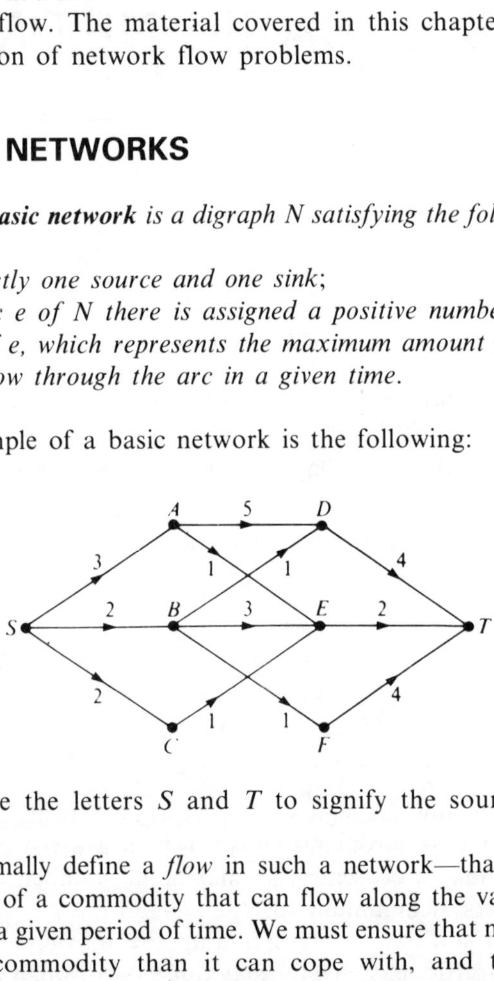

We usually use the letters S and T to signify the source and the sink, respectively.

We now formally define a *flow* in such a network—that is, a description of the amount of a commodity that can flow along the various channels of the network in a given period of time. We must ensure that no channel receives more of the commodity than it can cope with, and that none of the commodity is lost along the way. We therefore make the following definition.

Definition *A* ***flow*** *in a basic network N with source S and sink T is an assignment, to each arc e of N, of a non-negative number f(e), called the* ***flow along the arc e***, satisfying the following conditions:

(i) the feasibility condition: *the flow along each arc does not exceed the capacity of the arc—that is, $f(e) \leqslant c(e)$, for each arc e of N*;
(ii) the flow conservation law: *for each vertex V other than S and T, the sum of the flows along the arcs into V is equal to the sum of the flows along the arcs out of V.*

Such a flow is sometimes called *a flow from S to T* or an *(S,T)-flow*.

An example of a flow in a basic network is shown below; the first number (in bold type) next to each arc *e* is the flow $f(e)$, and the second number is the capacity $c(e)$:

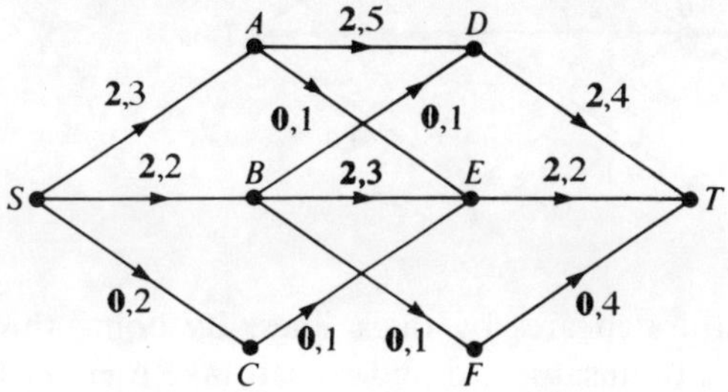

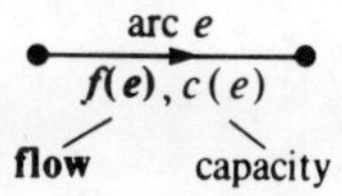

Definition *An arc e is said to be* ***saturated*** *if $f(e) = c(e)$, and* ***unsaturated*** *if $f(e) < c(e)$.*

In the above example, the arcs *SB* and *ET* are saturated, and the remaining arcs are unsaturated.

By the flow conservation law, none of the flow is lost at any vertex, and so the total flow out of the source *S* is always equal to the total flow into the sink *T*. This common value is called the **value of the flow**, and represents the amount of the commodity flowing through the network. (In the example given above, the value of the flow is 4.) Since our primary concern is with the *maximum* amount of the commodity flowing from *S* to *T*, it is natural to make the following definition.

Definition *A* ***maximum flow*** *is a flow of largest possible value.*

Note that we say '*a* flow of largest possible value' because there may be several different flows with the same largest value.

The rest of this chapter is devoted to the problem of finding a maximum flow in a given basic network.

3.2 FLOW-AUGMENTING PATHS

The basic idea we use when looking for maximum flows is to find a flow by inspection and then to increase its value step by step until we can increase it no further. Finding an initial flow by inspection is not difficult. For example, we can always take the *zero flow* in which the flow in each arc is 0.

To illustrate the procedure, we return to our example, in which we have a flow of value 4:

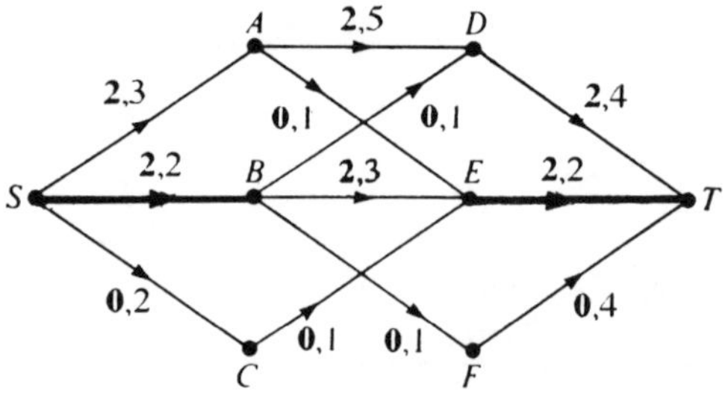

We have represented the saturated arcs by thick lines. By doing this, we can see at a glance which arcs are unsaturated and so can take a larger flow.

A path from S to T consisting entirely of unsaturated arcs is an example of a *flow-augmenting path*. By increasing the flow along the arcs of such a path, we can increase the value of the flow.

For example, the path $SADT$ consists entirely of unsaturated arcs, and we can increase the flow along each of these arcs by 1 (so that SA becomes saturated with a flow of 3, AD has a flow of 3, and DT has a flow of 3); this increases the value of the flow from 4 to 5:

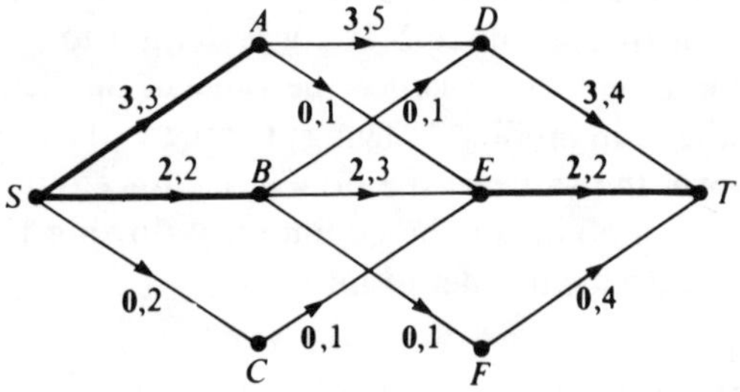

There are now no paths from S to T consisting entirely of unsaturated arcs (since any such path must start with SCE, and there are no unsaturated arcs out of E), and so the process cannot be repeated. However, the flow of value 5 is not a maximum flow. The following diagram illustrates a flow of value 6:

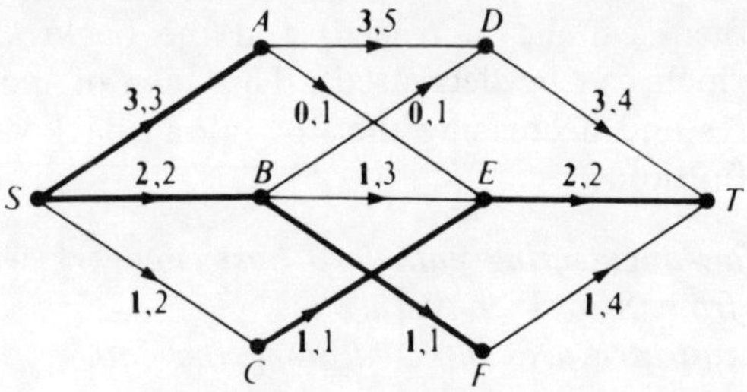

Since our primary aim is to construct maximum flows, this is rather discouraging. Either we must abandon the method or we must find some way of modifying it so that it works in all cases.

Fortunately the situation is not nearly as bad as it seems. All we need to do is to specify more carefully what we mean by a 'flow-augmenting path'. Up to now this term has meant a path from S to T consisting entirely of unsaturated arcs, but we now give it a less restricted meaning. To see what is involved, compare the above two diagrams. To get the flow with value 6 from the one with value 5 we *increase* the flow along SC, CE, BF and FT, and *decrease* the flow along BE:

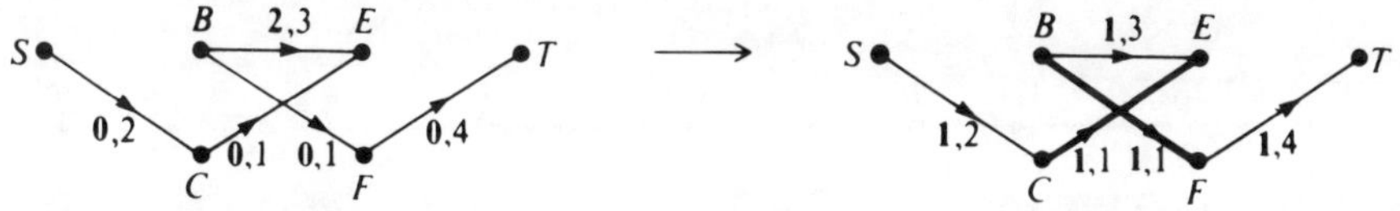

(a) (b)

We can think of diagram (a) as a 'path' containing four *forward* arcs, SC, CE, BF and FT, and one *backward* arc EB:

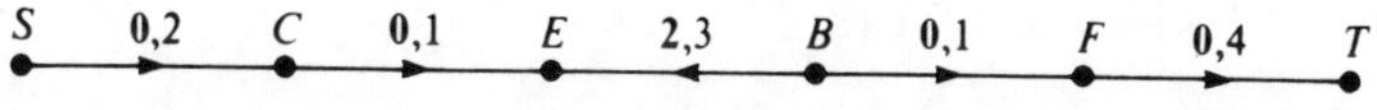

Note that we write the backward arc as EB, and not BE, since the flow is directed from S to T.

To obtain diagram (b), we increase the flow by 1 along the forward arcs of this path and decrease the flow by 1 along the backward arcs:

This is possible, since the forward arcs are all unsaturated (so that the flow along them can be increased) and the backward arc carries a non-zero flow (which can be decreased). This idea of increasing the flow along forward arcs and decreasing the flow along backward arcs leads to the following definition.

Definitions *A **flow-augmenting path** in a basic network with source S and sink T is a path from S to T consisting of:*
***forward arcs**: unsaturated arcs directed along the path*
and
***backward arcs**: arcs directed against the direction of the path and carrying a non-zero flow.*

Another example of a flow-augmenting path is the following:

The arcs *SA*, *BC* and *ET* are forward arcs, and the arcs *AB*, *CD* and *DE* are backward arcs.

We can increase the value of the flow from *S* to *T* by 2, by increasing the flow along the forward arcs by 2 and decreasing the flow along the backward arcs by 2:

Since the backward arc *CD* now carries a zero flow (which cannot be decreased), we can increase the flow no further. We have indicated this by representing the arc *CD* by a thick line.

Remark Our convention of using thick lines for saturated forward arcs and for backward arcs carrying zero flow is often helpful, but it is not useful in complicated networks, where it may not be clear which arcs are backward and which are forward.

Problem 3.1 For each of the following flow-augmenting paths, find the largest amount by which the flow from *S* to *T* can be increased, and draw a diagram showing the resulting flow:

(a)

(b)

(c)
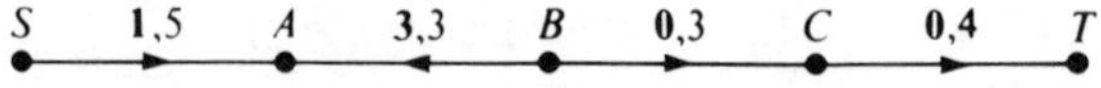

Problem 3.2

(a) Give a general rule for calculating the largest amount by which the flow can be increased along a flow-augmenting path.
(b) Verify that your rule works by applying it to the flow-augmenting path:

Problem 3.3 The following diagram shows a network with a flow of value 124. By finding a flow-augmenting path, obtain a flow of value 136.

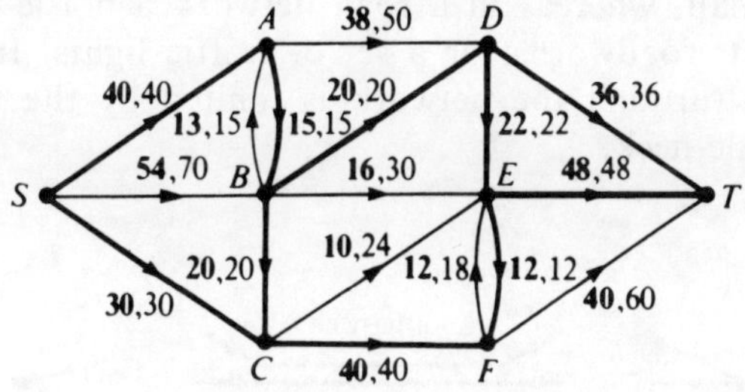

Are there any more flow-augmenting paths in this network?

Finding a Maximum Flow

The method described above is just what we need for dealing with small networks. In order to obtain a maximum flow, we find a succession of flow-augmenting paths by inspection and increase the flow along them as much as we can. When there are no more flow-augmenting paths, the flow attained is a maximum flow.

For large and complicated networks, the method soon becomes unmanageable, and a more systematic method is needed. At the end of this chapter, we discuss an algorithm known as the *maximum flow algorithm*. It can easily be applied to a particular network either by hand (in the case of small networks) or by using a computer (in the case of large ones). It is based on the idea of systematically locating flow-augmenting paths along which the flow can be increased. At each stage of the algorithm we increase the value of the flow along such a path, and the process stops when no more paths of this kind can be found. The resulting flow is then a maximum flow.

The maximum flow algorithm is a reliable method, but it has several disadvantages. The main disadvantage is that it can be very tedious, and may require a large number of iterations to produce a maximum flow. Furthermore, we cannot tell whether or not the flow we have produced actually is a maximum flow until the process stops and no more flow-augmenting paths can be found. It would be very useful to have an alternative method for determining whether or not a given flow is a maximum flow. In the next section we describe such a method.

3.3 MAXIMUM FLOWS AND MINIMUM CUTS

Let us consider a network with a bottle-neck in it. In a water or gas network such a bottle-neck can occur if the pipe diameter at some point becomes very small, whereas in a road network a bottle-neck may result from an accident, roadworks, or a set of traffic lights. In each case, the amount of flow through the network is limited by the amount of flow through the bottle-neck.

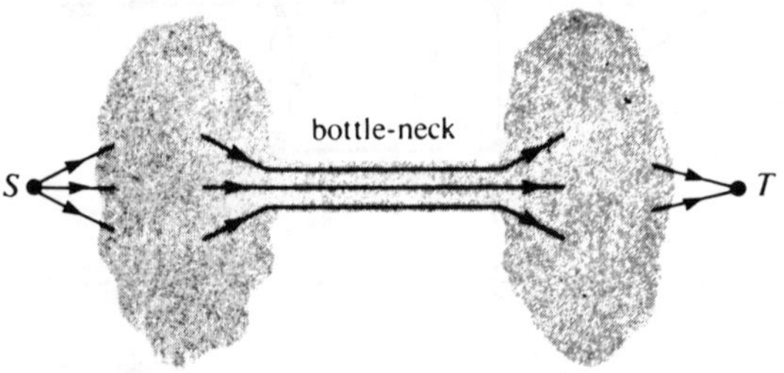

It seems probable that information about the amount of flow through the 'worst bottle-neck' in the system may give useful information about the maximum amount of flow in the network. This is indeed the case, and forms the underlying idea for the whole of this section.

In each part of the following problem, we ask you to consider a bottle-neck consisting of a few arcs of small capacity through which the commodity in question has to flow. We then make this idea more precise by introducing the idea of a *cut*, and replacing the intuitive term 'bottle-neck' by *minimum cut*.

Problem 3.4 For each of the following networks (in which the numbers next to the arcs are capacities), find:

the value of a maximum flow;
the worst bottle-neck.

(Note that a bottle-neck may not necessarily consist of just one arc.)

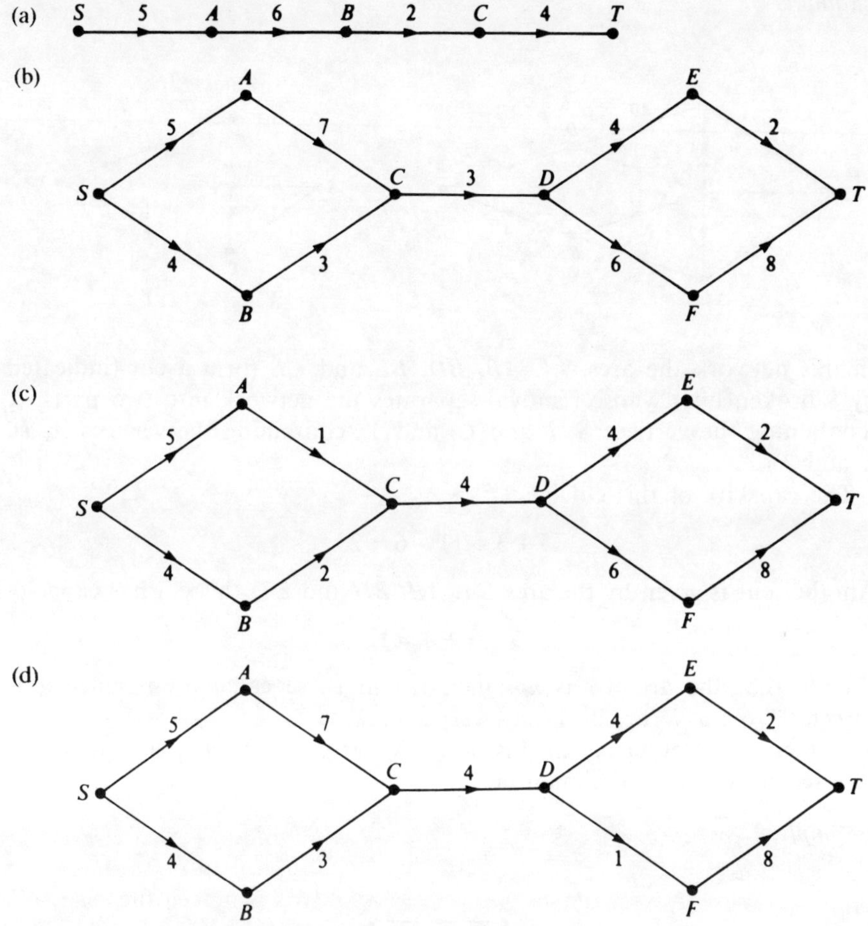

Definitions *A **cut** in a network with source S and sink T is a set of arcs whose removal separates the network into two parts X and Y, one containing S and the other containing T.*

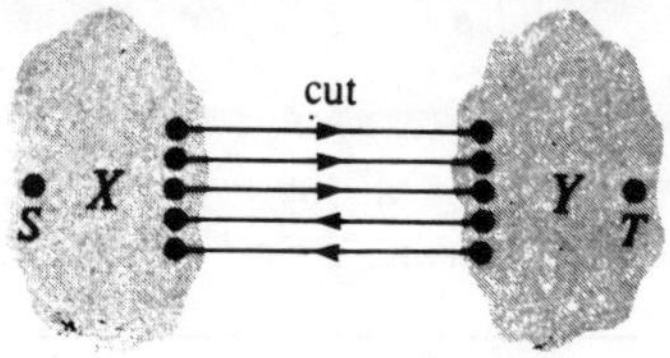

*The **capacity of a cut** is the sum of the capacities of those arcs in the cut which are directed from X (the part containing S) to Y (the part containing T). A **minimum cut** is a cut of smallest possible capacity.*

Example 1

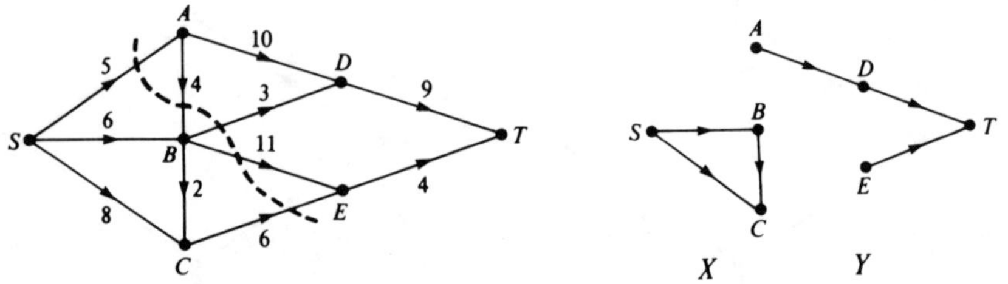

In this network the arcs *SA*, *AB*, *BD*, *BE* and *CE* form a cut (indicated by a broken line) whose removal separates the network into two parts, *X* (containing the vertices *S*, *B* and *C*) and *Y* (containing the vertices *A*, *D*, *E* and *T*).

The capacity of this cut is

$$5 + 3 + 11 + 6 = 25.$$

Another cut is given by the arcs *SA*, *AB*, *BD* and *ET*; this cut has capacity

$$5 + 3 + 4 = 12.$$

Note that the arc *AB* is *not* included in these calculations since it is directed *from* a vertex in *Y* *to* a vertex in *X*.

In fact, every other cut in this network has capacity greater than 12, so this second cut is a minimum cut.

Example 2

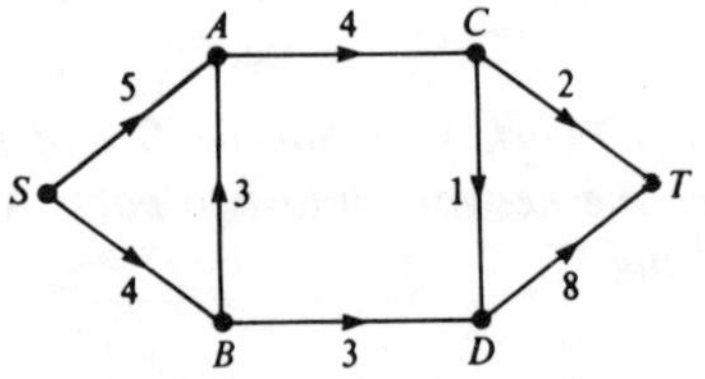

For the above network, we can list all the cuts. In the following table we give the arcs in each cut, the vertices in *X* and *Y*, and the capacity of the cut:

	arcs in cut	vertices in *X*	vertices in *Y*	capacity of cut
1.	*SA*, *SB*	*S*	*A*, *B*, *C*, *D*, *T*	$5 + 4 = 9$
2.	*SB*, *BA*, *AC*	*S*, *A*	*B*, *C*, *D*, *T*	$4 + 4 = 8$
3.	*SA*, *BA*, *BD*	*S*, *B*	*A*, *C*, *D*, *T*	$5 + 3 + 3 = 11$

	arcs in cut	vertices in X	vertices in Y	capacity of cut
4.	AC, BD	S, A, B	C, D, T	$4+3=7$
5.	SB, BA, CD, CT	S, A, C	B, D, T	$4+1+2=7$
6.	BD, CD, CT	S, A, B, C	D, T	$3+1+2=6$
7.	SA, BA, CD, DT	S, B, D	A, C, T	$5+3+8=16$
8.	AC, CD, DT	S, A, B, D	C, T	$4+8=12$
9.	CT, DT	S, A, B, C, D	T	$2+8=10$

Note that in each of the cuts (2), (5), (7) and (8), one of the arcs is directed *from* a vertex in Y *to* a vertex in X, and so is not included when calculating the capacity.

In this example there is only one minimum cut: $\{BD, CD, CT\}$, that is, cut (6), with capacity 6.

Problem 3.5 For each of the following networks (in which the numbers next to the arcs are capacities), write down the minimum cut, its capacity, and the corresponding sets X and Y:

(a)

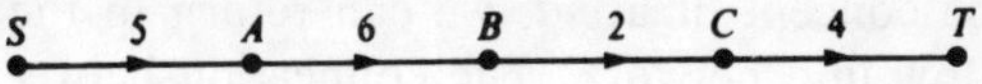

(b)

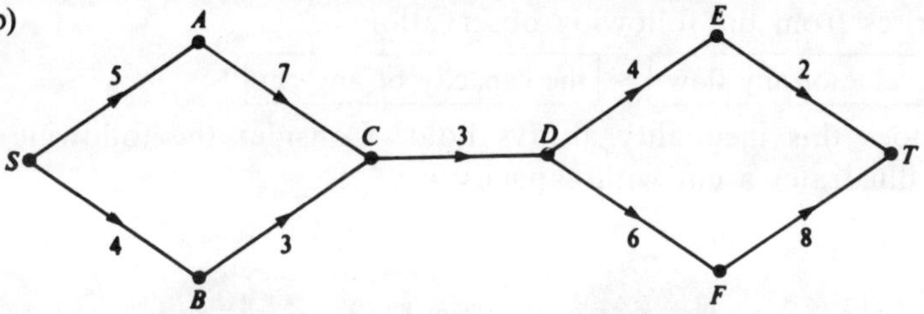

(c)

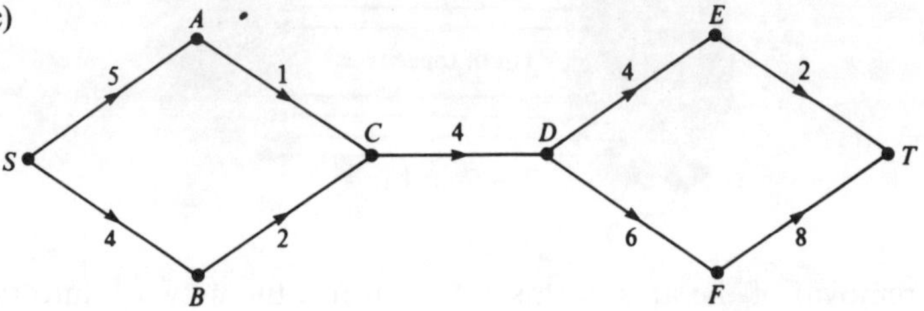

(d)

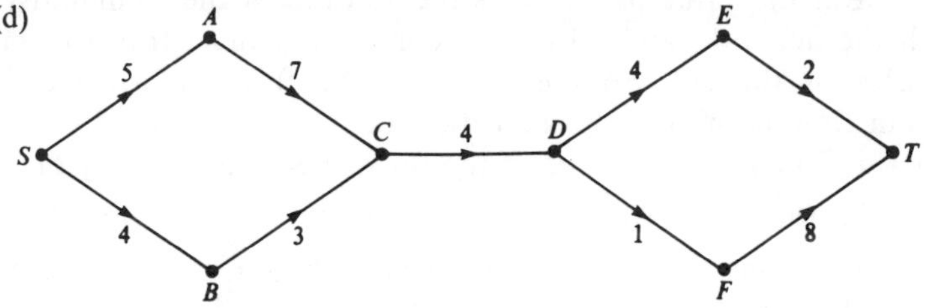

Problem 3.6

(a) Construct an example of a network with two minimum cuts.
(b) The following network has eight cuts. Draw up a table (similar to that given in Example 2) listing these cuts, the vertices in X and Y, and the capacity of each cut. Which are the minimum cuts?

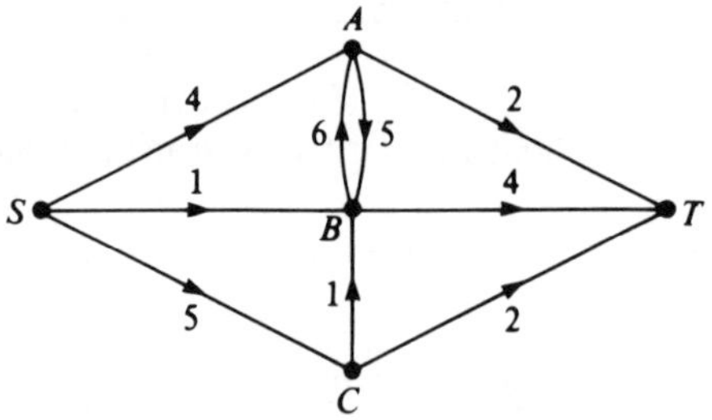

3.4 THE MAX-FLOW MIN-CUT THEOREM

Now that we have the concept of a *cut*, we can return to the problem of finding a maximum flow in a network. The connection between flows and cuts arises from the following observation:

| the value of any flow | $\leqslant$ | the capacity of any cut |

Why does this inequality always hold? Consider the following diagram which illustrates a cut with capacity k.

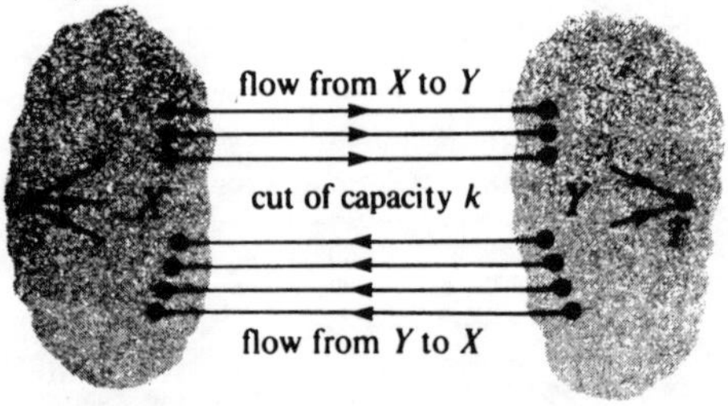

The removal of the arcs in this cut separates the network into two parts, X and Y. But the value of a flow is the amount of the commodity flowing through the network, and all of this commodity must travel from X to Y along arcs of the cut. So the value of the flow cannot exceed k, the maximum flow permitted through the cut.

In fact, if F denotes the value of the flow, then F is given by the equation

$$F = (\text{the total flow from } X \text{ to } Y) - (\text{the total flow from } Y \text{ to } X).$$

Since the total flow from X to Y is at most k, and the total flow from Y to X is at least 0, we have $F \leqslant k - 0$; that is, $F \leqslant k$.

But we can go further:

Since

the value of *any* flow	≤	the capacity of *any* cut

this inequality must be true for a *maximum* flow in particular. So

the value of any *maximum* flow	≤	the capacity of *any* cut

But this is true for *any* cut, and so it must be true for a *minimum* cut in particular. So

the value of any *maximum* flow	≤	the capacity of any *minimum* cut

This last inequality is very important in practice. In particular, *if we can find a flow with value k, and a cut with capacity equal to this value of k, then*:

- *the flow is a maximum flow* (since any larger flow would violate the inequality);
- *the cut is a minimum cut* (since any smaller cut would violate the inequality).

To see what can be deduced from the above inequalities, try the following problems.

Problem 3.7

(a) If you find a cut with capacity 7 in a network, what can you deduce about a maximum flow?
(b) If you find a flow with value 11 in a network, what can you deduce about a minimum cut?
(c) If you find a flow with value 4 and a cut with capacity 6, what can you deduce?
(d) If you find a flow with value 6 and a cut with capacity 4, what can you deduce?

Problem 3.8 The following diagram illustrates a flow of value 136 in a network. Find a cut of capacity 136, and deduce that the given flow is a maximum flow.

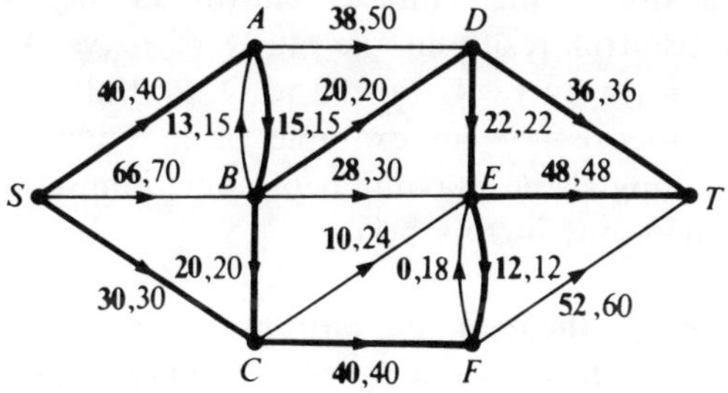

Problem 3.9 For each of the following networks, find a flow and a cut for which the value of the flow is equal to the capacity of the cut. What can you deduce about the flow in each case?

(a)

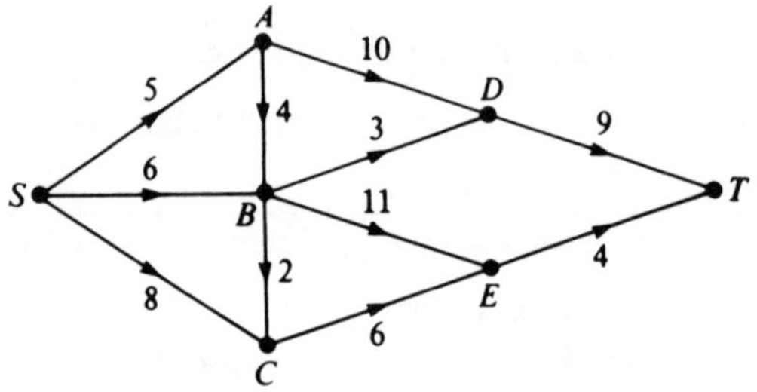

(b)

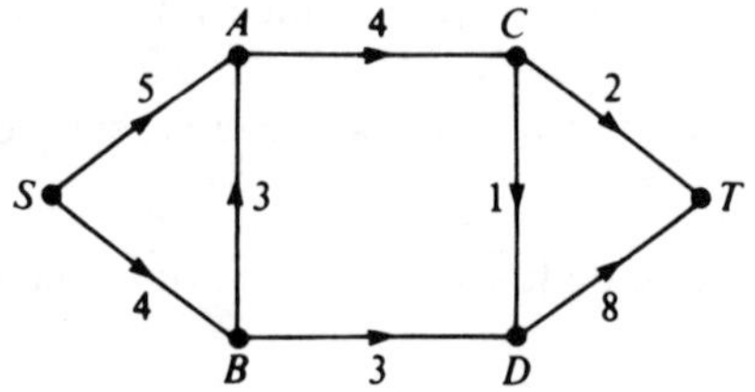

(c)

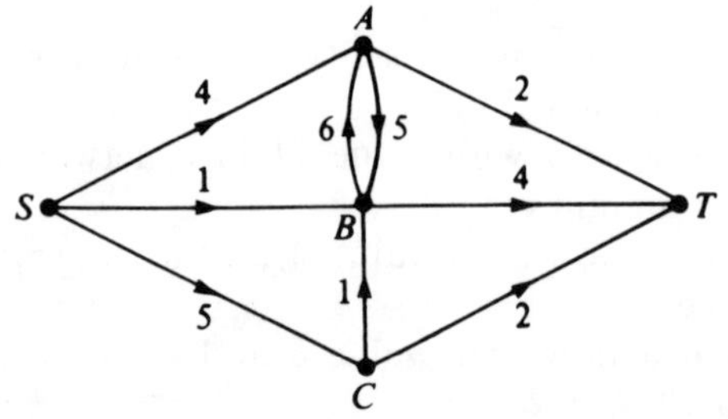

In your solution to the above problem, you may have observed that for each network the value of a maximum flow is actually equal to the capacity of a minimum cut. This is not a coincidence, and the statement that these two numbers are always the same is known as the *max-flow min-cut theorem*. It is the central result in the study of network flows, and was first proved in this form by L. R. Ford and D. R. Fulkerson in 1956. (The max-flow min-cut theorem is an example of a 'minimax theorem'—the minimum of something = the maximum of something else. You will meet several minimax theorems in this book.)

The max-flow min-cut theorem *In any basic network, the value of a maximum flow is equal to the capacity of a minimum cut.*

Proof We have already shown that the value of any maximum flow can never exceed the capacity of any minimum cut. To show that these two numbers are actually equal, we need to find a cut whose capacity is equal to the value of any maximum flow.

Consider any maximum flow. Let X be the set of all vertices which can be reached from S by a flow-augmenting path, and let Y be the remaining vertices. Then T must lie in the set Y, because if T were in X then we could increase the flow from S to T, which is impossible since we started with a maximum flow.

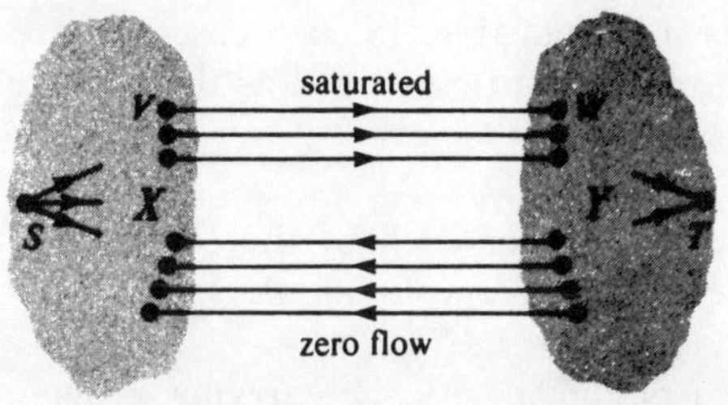

We complete the proof by looking at the cut consisting of those arcs which have one end in X and the other end in Y. Any arc which is directed *from* a vertex V in X *to* a vertex W in Y must be saturated, since if it were not saturated there would be a flow-augmenting path from S to W, and W would have to be put into X instead of Y. By a similar argument, any arc which is directed *from* a vertex in Y *to* a vertex in X must carry a zero flow. So the capacity of the cut is equal to the total flow from X to Y, and this is equal to the value of the flow in the network. Since the flow is a maximum flow, this completes the proof. □

We illustrate the proof by the following example of a maximum flow:

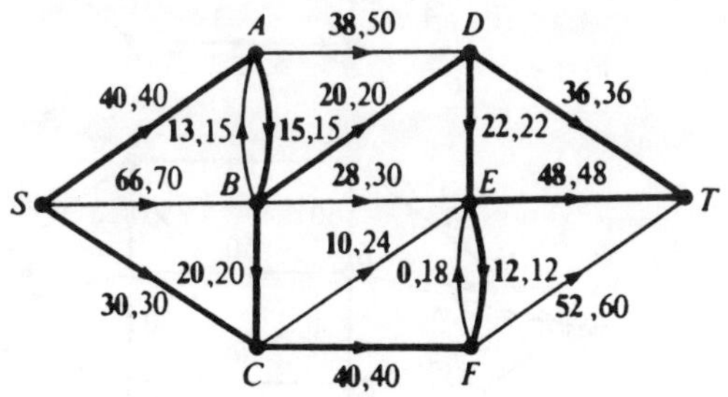

The vertices A, B, C, D and E can all be reached from S by flow-augmenting paths, whereas F and T cannot. So $X = \{S, A, B, C, D, E\}$ and $Y = \{F, T\}$.

The cut in the above example consists of the saturated arcs *CF*, *EF*, *DT* and *ET*, directed from *X* to *Y*, and the arc *FE*, carrying a zero flow and directed from *Y* to *X*. The capacity of the above cut is $40 + 12 + 36 + 48 = 136$, which is the value of the above flow.

An important consequence of this theorem is that, if you can spot a cut of minimum capacity in a network, then you can immediately deduce the value of a maximum flow. In many examples, such as the ones occurring in some of the problems in this section, a minimum cut is easily found by inspection, and a maximum flow can then be found without much difficulty. However, for large and complicated networks, finding a minimum cut by inspection is impracticable. In such cases we use a more systematic method for finding maximum flows, such as the algorithm described in the next section.

Problem 3.10 Decide whether each of the following statements is TRUE or FALSE, and give reasons for your answers.

(a) Every minimum cut in a network carrying a maximum flow consists entirely of saturated arcs.
(b) If the capacity of every arc of a network is an integer, then the value of a maximum flow is also an integer.

Problem 3.11 By finding a maximum flow and a minimum cut in each case, verify the max-flow min-cut theorem for each of the following networks.

(a)

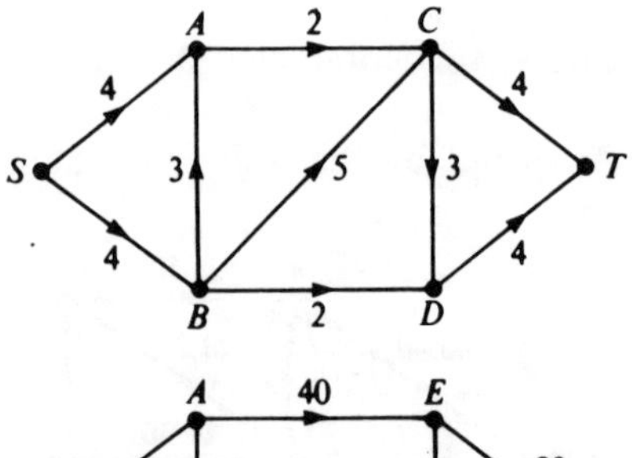

(b)

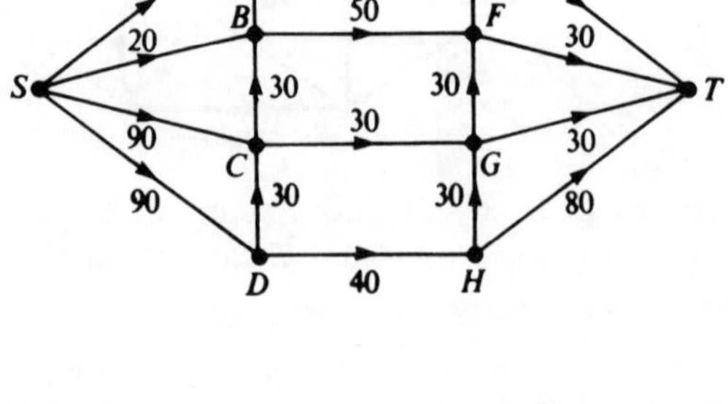

3.5 THE MAXIMUM FLOW ALGORITHM

In this section we describe the algorithm, referred to earlier in the chapter, for finding a maximum flow in a basic network. The idea of the algorithm is to start with an initial flow found by inspection, and then to increase the value of the flow systematically by finding flow-augmenting paths from the source *S* to the sink *T*. We can always find a suitable initial flow, since if necessary we can start with zero flow along each arc.

The algorithm involves the alternate use of two parts:

Part A: The Labelling Procedure, which labels appropriate vertices in order to find flow-augmenting paths.

Part B: The Flow-augmenting Procedure, which increases the flow along flow-augmenting paths found by using Part A.

When Part A fails to find a flow-augmenting path, the algorithm terminates and a maximum flow has been found. (The algorithm will always terminate if the capacities have integral values; it may fail to terminate if the capacities are irrational.)

We start by giving each arc a flow of zero.

Next, we try to find a flow-augmenting path from *S* to *T* by proceeding from vertex to vertex along unsaturated arcs. We keep track of our progress by labelling each vertex we come to with a letter and a number. The letter corresponds to the previous vertex on the flow-augmenting path, and the number is the increase in flow which is possible along the flow-augmenting path from *S* to the vertex we are considering. If we manage to reach the sink *T* by this method, we proceed as follows:

A: we use the labels to trace the flow-augmenting path back to *S*;

B: we increase the flow along this path by the appropriate amount, and we then erase the labels.

We repeat this procedure until we obtain a maximum flow.

We illustrate the algorithm by an example.

Worked problem using the maximum flow algorithm Find a maximum flow in the following basic network:

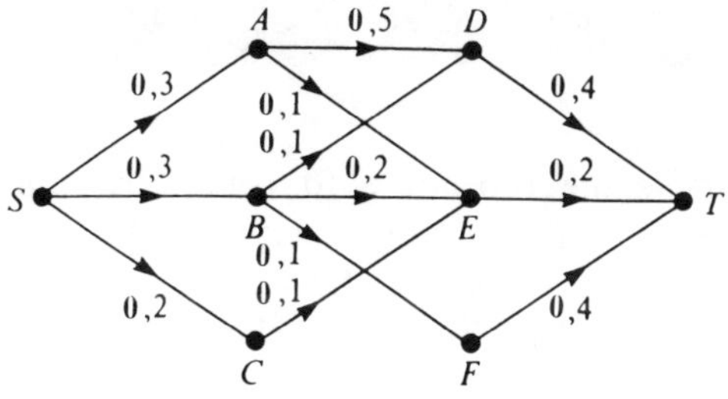

Solution We start by giving each arc a zero flow, as indicated on the diagram.

We start from S and proceed along an unsaturated arc to any unlabelled vertex. We shall be systematic and choose vertices in alphabetical order—the first vertex is therefore A. The arc SA has capacity 3, so the flow along SA can be increased by 3. So we give A the label (S, 3) to show that A can accept a flow of 3 along the path from the previous vertex, S.

From A we continue to D, which is the first possible vertex alphabetically. The flow along AD can be increased by 5, but the flow into A from S can be increased by only 3. So we give D the label (A, 3) to indicate that D can accept a flow of 3 from S via A.

We proceed on to the next vertex T in a similar way, and give T the label (D, 3).

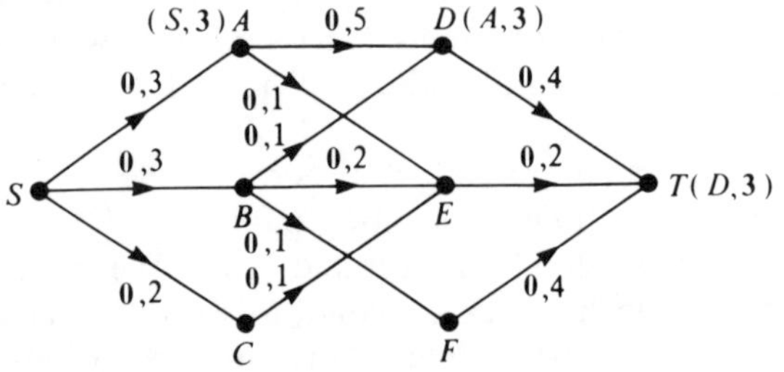

We have now reached the sink T, and so we have found a flow-augmenting path. Backtracking through the labels, we identify the path as $SADT$. We now increase the flow along this path by 3, as indicated on the following diagram, and we erase the labels.

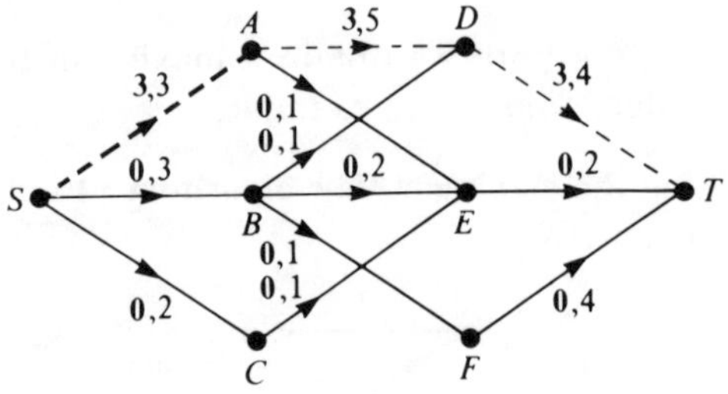

We now use the same procedure to find a second flow-augmenting path, via vertex B. We label vertex B with (S, 3), vertex D with (B, 1), and vertex T with (D, 1). We send a flow of 1 along the flow-augmenting path $SBDT$ and erase the labels. We thus obtain the following flow pattern.

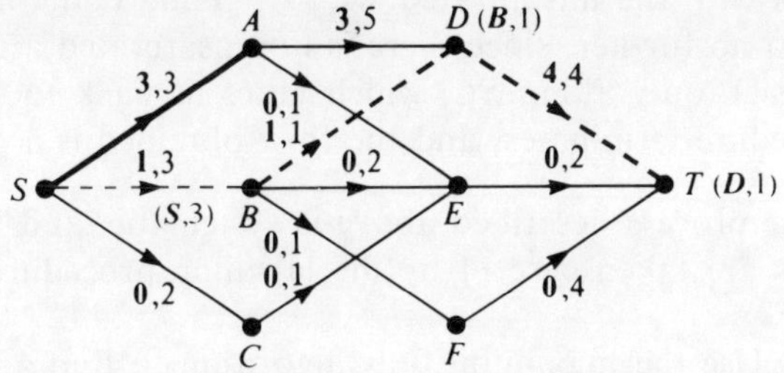

There is still an unsaturated arc leading into B, so the next flow-augmenting path also proceeds from S to B. We label B with (S, 2), then E with (B, 2) and finally T with (E, 2). So we have found a flow-augmenting path $SBET$, and we increase the flow along it by 2, and erase the labels.

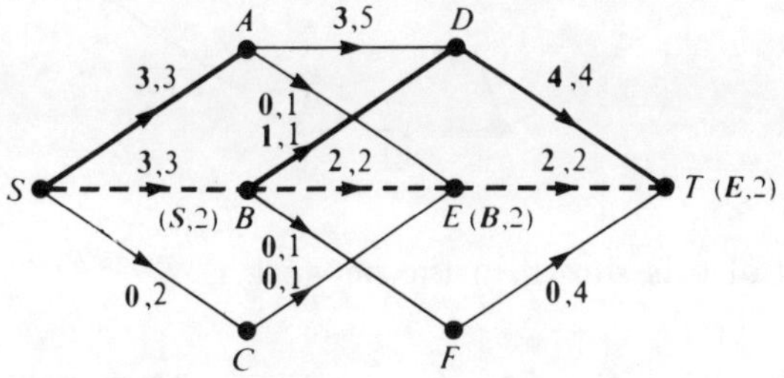

Following the procedure again, we label C with (S, 2) and E with (C, 1). However, we now seem to be stuck, because there is no unsaturated arc leaving E. But there is an arc coming into E which has a non-zero flow—namely BE with a flow of 2—so we can send a flow out of E in the direction EB by decreasing the flow along BE. We give B the label (E^-, 1), where the minus sign indicates that we have used a *backward* arc.

Continuing, we give F the label (B, 1) and T the label (F, 1), so our flow-augmenting path is $SCEBFT$, and we can increase the flow along it by 1, as shown below. Again, we erase the labels.

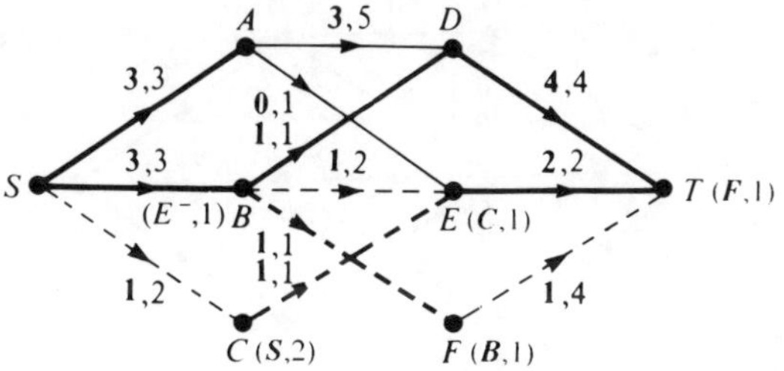

We start again with the unsaturated arc *SC*, giving *C* the label (*S*, 1). But now we can go no further, since there is no unsaturated arc out of *C*, and we can backtrack only along *SC*, which takes us back to *S*.

So the procedure terminates, and the flow obtained is a maximum flow.

Notice that the process described above is systematic, and that false starts and dead ends are taken care of by the labelling procedure.

Problem 3.12 Use the maximum flow algorithm to find a maximum flow in the following basic network:

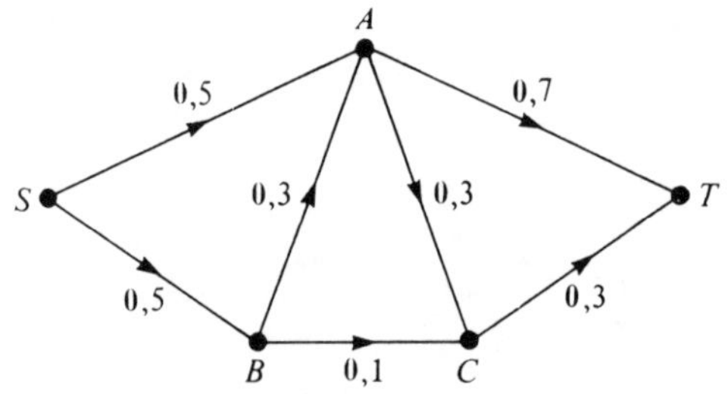

(An initial flow of 0 is shown in the network.)

Tabular Method for the Maximum Flow Algorithm

An alternative way of applying the algorithm without having to draw diagrams is to use a tabular method. This method is not so easy to understand at first as the method described earlier, but lends itself more easily to conversion to a computer program. Also, once mastered, this method is probably quicker and easier to work through by hand. We illustrate the tabular procedure by using it to solve the earlier worked example.

Worked problem using the tabular method Find a maximum flow in the following basic network:

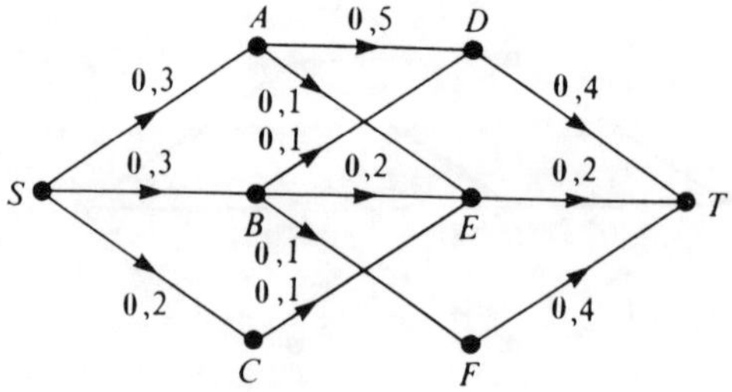

Solution We draw up a table showing the capacities of all the arcs, and a zero flow in each arc.

	S	A	B	C	D	E	F	T
S		(0,3)	(0,3)	(0,2)				
A					(0,5)	(0,1)		
B					(0,1)	(0,2)	(0,1)	
C						(0,1)		
D								(0,4)
E								(0,2)
F								(0,4)
T								

Each arc corresponds to a particular cell in the above table. So arc *SA*, for example, corresponds to row *S* and column *A*. (An arc *AS*, directed in the reverse sense from *A* to *S* would correspond to row *A* and column *S*.)

An entry in the table gives the flow and capacity of the corresponding arc. As before, all arcs are initially given zero flow. For example, arc *AD* has the entry (0,5) in the table, indicating zero flow and a capacity of 5. Thus the above table contains exactly the same amount of information as the corresponding network diagram.

We now start the labelling procedure. Looking along the *S* row, we see that *SA* can take a flow of 3 units. We give *A* the label (*S*, 3) to indicate that a flow of 3 can be sent from *S*. Now, looking along the *A* row, we find that we can send a flow of up to 5 units to vertex *D*, so we label *D* with (*A*, 3), since the flow of 3 on the previous label is less than 5. Continuing in the same way, we label *T* with (*D*, 3). We record the labels above the appropriate vertices, as shown below:

		(*S*, 3)			(*A*, 3)			(*D*, 3)
	S	A	B	C	D	E	F	T
S		(3,3)	(0,3)	(0,2)				
A					(3,5)	(0,1)		
B					(0,1)	(0,2)	(0,1)	
C						(0,1)		
D								(3,4)
E								(0,2)
F								(0,4)
T								

Backtracking through the labels from *T*, we find the flow-augmenting path *SADT*, along which we can send a flow of 3 (the minimum flow value in the labels). This flow is indicated in the above table by a flow value of 3 inserted for each arc of the path *SADT*.

Looking at the above table, we see that *SA* has a flow of 3 and a capacity of 3; that is, it is saturated. So in the next labelling iteration we go on to vertex *B* and label it (*S*, 3). We then label *D* with (*B*, 1) and *T* with (*D*, 1), and insert the flow of 1 along the path *SBDT* as shown in the following table, in which the previous labels have been erased.

			(*S*, 3)		(*B*, 1)			(*D*, 1)
	S	*A*	*B*	*C*	*D*	*E*	*F*	*T*
S		(3,3)	(1,3)	(0,2)				
A					(3,5)	(0,1)		
B					(1,1)	(0,2)	(0,1)	
C						(0,1)		
D								(4,4)
E								(0,2)
F								(0,4)
T								

Next we label *B*, *E* and *T* and send a flow of 2 along the path *SBET*, as shown in the following table:

			(*S*, 2)			(*B*, 2)		(*E*, 2)
	S	*A*	*B*	*C*	*D*	*E*	*F*	*T*
S		(3,3)	(3,3)	(0,2)				
A					(3,5)	(0,1)		
B					(1,1)	(2,2)	(0,1)	
C						(0,1)		
D								(4,4)
E								(0,2)
F								(0,4)
T								

In the next iteration, we label *C* with (*S*, 2) and then *E* with (*C*, 1). We are then apparently stuck, since if we look along the *E*-row we find no unsaturated arc leading from E. However, if we look down the E-column, we see that there is a flow of 2 from B to E. We can therefore send a flow of up to 2 units in the reverse direction from E to B by reducing the flow from B to E. In a case like this, when we have a backward flow, we label the appropriate vertex on the vertical axis instead of above the horizontal axis. So we label B on the vertical axis with $(E^-, 1)$ to indicate a flow of 1 from E to B. The minus sign is inserted to show that we have a backward flow along the arc. Next, we label F with $(B, 1)$ and T with $(F, 1)$. We can increase the flow along $SCEBFT$ by 1 unit, as indicated in the table below. (Note that the flow along the backward arc EB is shown by reducing the value of the flow along BE from 2 to 1.)

					(S, 2)		(C, 1)	(B, 1)	(F, 1)
		S	*A*	*B*	*C*	*D*	*E*	*F*	*T*
	S		(3,3)	(3,3)	(1,2)				
	A					(3,5)	(0,1)		
$(E^-, 1)$	*B*					(1,1)	(1,2)	(1,1)	
	C						(1,1)		
	D								(4,4)
	E								(2,2)
	F								(1,4)
	T								

Next we label C with $(S, 1)$. However, there are now no unsaturated arcs from C, and we have no backward arcs except CS, so no more labelling is possible, and the algorithm terminates. The maximum flow is $4 + 2 + 1 = 7$.

For reference, a formal statement of the maximum flow algorithm is given below:

ALGORITHM FOR FINDING A MAXIMUM FLOW IN A BASIC NETWORK

Part A: The Labelling Procedure

START with a flow found by inspection, and with all the vertices unlabelled.

STEP 1 Choose any unsaturated arc SV_1 and assign to the vertex V_1 the label (S, a_1), where a_1 is the amount of flow needed to saturate the arc SV_1.

STEP 2 Choose any unlabelled vertex V_2 such that

either (i) $V_1 V_2$ is an unsaturated arc
or (ii) $V_2 V_1$ is an arc with a non-zero flow.

In case (i), assign to the vertex V_2 the label (V_1, a_2), where a_2 is the amount needed to saturate $V_1 V_2$, or a_1, whichever is the smaller.

In case (ii), assign to the vertex V_2 the label (V_1^-, a_2), where a_2 is the amount of flow in $V_2 V_1$, or a_1, whichever is the smaller.

STEP 3 Continue in this way, choosing vertices $V_3, V_4, \ldots$

At each stage, choose the vertex V_{k+1} to be any unlabelled vertex such that

either (i) $V_k V_{k+1}$ is an unsaturated arc
or (ii) $V_{k+1} V_k$ is an arc with a non-zero flow.

In case (i), assign to the vertex V_{k+1} the label (V_k, a_{k+1}), where a_{k+1} is the amount needed to saturate $V_k V_{k+1}$, or a_k, whichever is the smaller.

In case (ii), assign to the vertex V_{k+1} the label (V_k^-, a_{k+1}), where a_{k+1} is the amount of flow in $V_{k+1} V_k$, or a_k, whichever is the smaller.

STEP 4 Continue in this way until

either (a) the vertex T becomes labelled
or (b) there is no vertex V_{k+1} with property (i) or (ii).

In case (a), a flow-augmenting path has been found, so proceed to Part B.

In case (b), backtrack along the path until vertex V_i is reached with the property that either there is an unsaturated arc $V_i W$ or there is an arc $W V_i$ carrying a non-zero flow, where W is as yet unlabelled. If you find such a vertex V_i, regard W as V_{i+1} and label it as in Step 3.

STEP 5 Continue in this way, labelling vertices where possible and back-tracking where necessary until

either the vertex T becomes labelled
or no further labelling is possible.

In the former case, proceed directly to Part B.

In the latter case, no increase in flow is possible, and a maximum flow has already been obtained so STOP.

Part B: The Flow-augmenting Procedure

The following procedure is used when the sink T has been labelled.

STEP 6 Look at the label of T—it will be of the form (V_k, a), for some vertex V_k and some positive number a. Increase the flow along V_kT by a.

STEP 7 Look at the label of V_k—it will be of the form (V_{k-1}, a) or (V_{k-1}^{-}, a) for some vertex V_{k-1}. In the former case, increase the flow along $V_{k-1}V_k$ by a, and in the latter case decrease the flow along V_kV_{k-1} by a.

STEP 8 Repeat Step 7 successively for the vertices $V_{k-1}, V_{k-2}, \ldots, V_1$, and conclude by increasing the flow along SV_1 by a.

STEP 9 Erase all the labels, and start the labelling procedure (Step 1) using the flow obtained in Steps 6–8.

EXERCISES 3

Flows in Basic Networks

3.1. In the following basic network, the arcs AD, BE, CD and ET are saturated, and the flows along SA, SB and SC are all the same. Find all the missing flows.

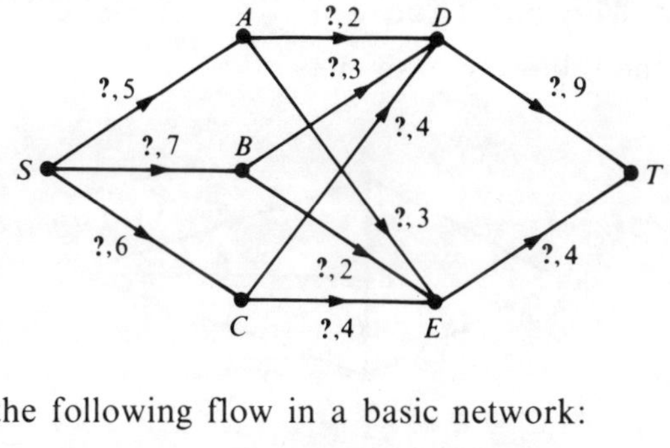

3.2. Consider the following flow in a basic network:

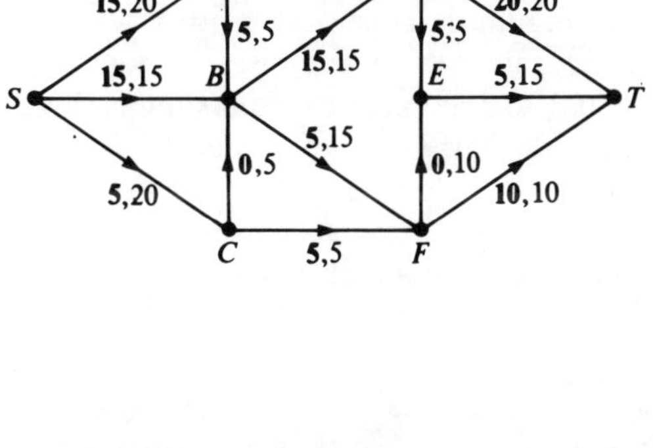

(a) Which of the arcs are saturated?
(b) What is the value of the flow from S to T?
(c) There is just one flow-augmenting path involving only forward arcs; find it, and determine the amount by which the flow along it can be increased.
What is the value of the new flow?
(d) With this new flow, there is just one flow-augmenting path (involving backward arcs); find it, and hence determine the value of a maximum flow in the network.

3.3. Consider the following flow in a basic network:

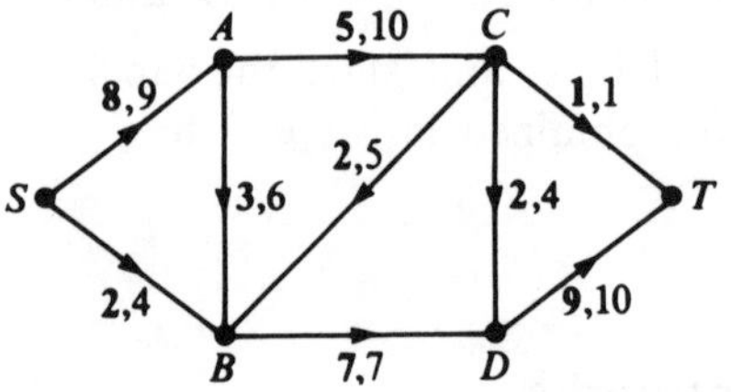

(a) Find all the flow-augmenting paths.
(b) What is the value of a maximum flow?

The Max-flow Min-cut Theorem

3.4. Consider the following basic network:

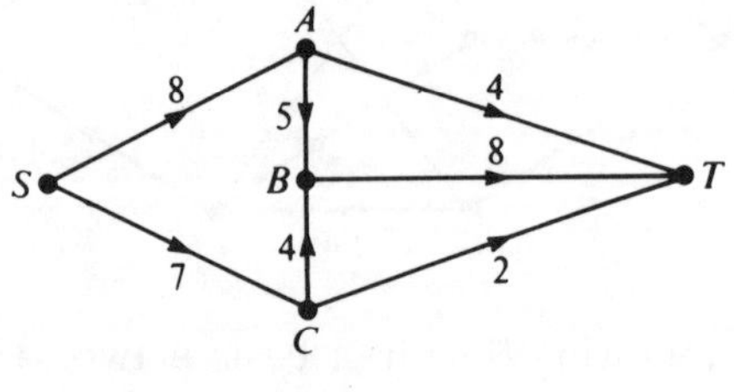

Draw up a table listing all the cuts, the vertices in the sets X and Y, and the capacity of each cut.
Which are the minimum cuts?

3.5. For each of the networks in Exercises 3.2 and 3.3, find a cut whose capacity is equal to the value of a maximum flow.
Are these cuts necessarily minimum cuts?

3.6. Classify each of the following statements as TRUE or FALSE.
(a) If a cut in a basic network consists entirely of saturated arcs, then it must be a minimum cut.

(b) If a cut in a basic network is a minimum cut, then it must consist entirely of saturated arcs.

The Maximum Flow Algorithm

3.7. Use the algorithm to find a maximum flow in the following basic network:

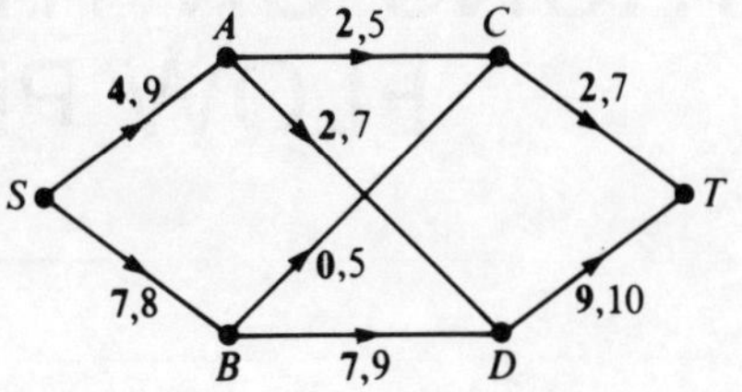

3.8. Use the algorithm to find a maximum flow in the following basic network:

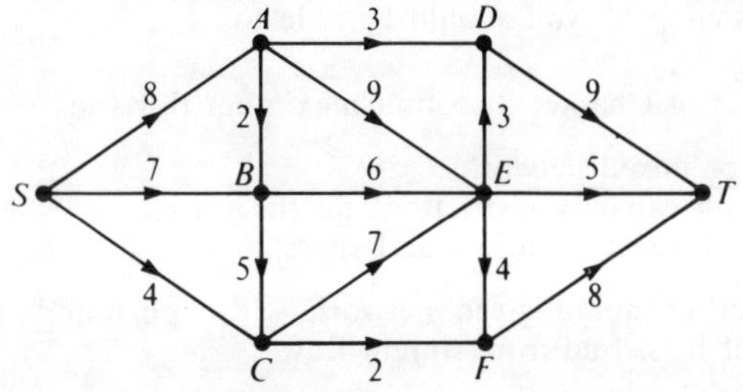

3.9. Use the algorithm to find a flow of 20 in the following basic network. Is this a maximum flow?

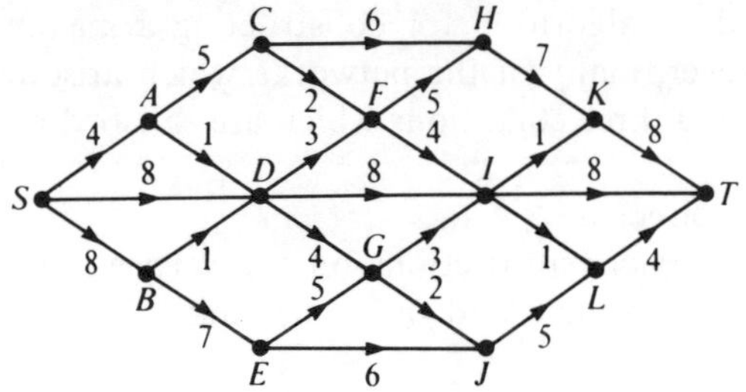

CHAPTER 4

VARIATIONS ON THE BASIC FLOW PROBLEM

After reading this chapter, you should be able to:

- adapt the ideas of Chapter 3 to find maximum flows in

  undirected or mixed networks,
  networks with capacity restrictions on the vertices,
  networks with several sources and sinks;

- decide whether or not a given network with upper and lower capacities is feasible, and if it is, find a maximum flow;
- state the max-flow min-cut theorem for feasible networks.

Introduction

In the previous chapter we considered the flow problem for basic networks, and we presented an algorithm for constructing a maximum flow in such a network. However, many of the networks which arise in practice do not satisfy the following three conditions which are satisfied by a basic network:

- all edges are directed;
- there are no capacity restrictions on the vertices;
- there is only one source and only one sink.

For example, the network of streets in a busy town does not usually satisfy these conditions, since it generally involves both one-way and two-way streets, has road junctions at which the traffic flow is restricted by traffic signals, and contains several points at which traffic can enter the system. In this chapter we show how the ideas developed for basic networks can easily be adapted to find maximum flows in networks which do not satisfy the above conditions.

We also show how the ideas in the previous chapter can be applied to networks with both lower and upper capacities. These are networks in which it is necessary to specify both a maximum and a minimum flow through each arc. The minimum flow constraint may arise in various ways—for example, it may be due to physical reasons (an oil pipeline in Alaska may need a specified minimum flow to prevent the oil from freezing), or to economic reasons (a company cannot function efficiently unless there is a certain minimum of trade).

4.1 THREE TRANSFORMATIONS

Undirected or Mixed Networks

An **undirected network** is one in which flow is allowed in either direction along an edge. A network containing both directed and undirected edges (such as a road network with both one-way and two-way streets) is called a **mixed network**. We can transform a mixed or undirected network into a basic network by regarding each two-way channel as two one-way channels, one in each direction. In other words, we replace each two-way edge in the network by two one-way arcs with the same capacity:

For example, the following mixed network gives rise to the basic network on the right:

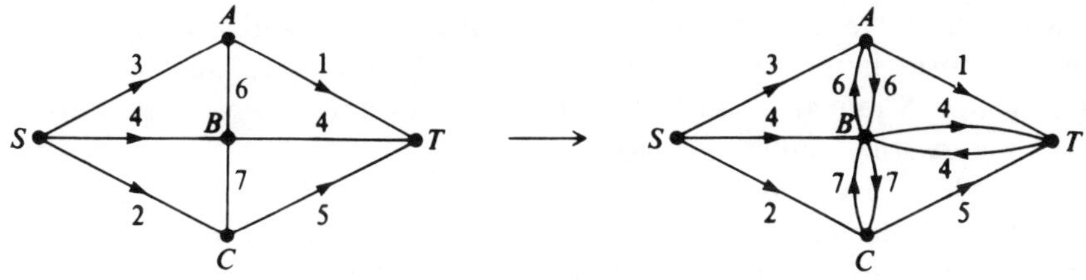

Networks with Capacity Restrictions on the Vertices

If a network has one or more vertices with capacity restrictions (such as the flow of traffic through a particular junction, or the flow of mail through a particular sorting office), we can transform it into a basic network by replacing each such vertex by two vertices joined by an arc. More precisely, we replace each vertex V of capacity k by two vertices V_1 and V_2 joined by an arc from V_1 to V_2 of capacity k. All arcs directed towards V in the original network are directed towards V_1 in the new network, and all arcs directed away from V in the original network are directed away from V_2 in the new network, as follows:

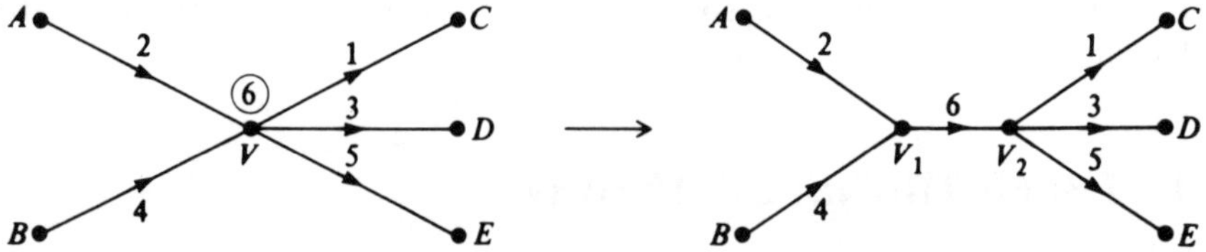

For example, the following network with capacity restrictions on the vertices gives rise to the basic network on the right:

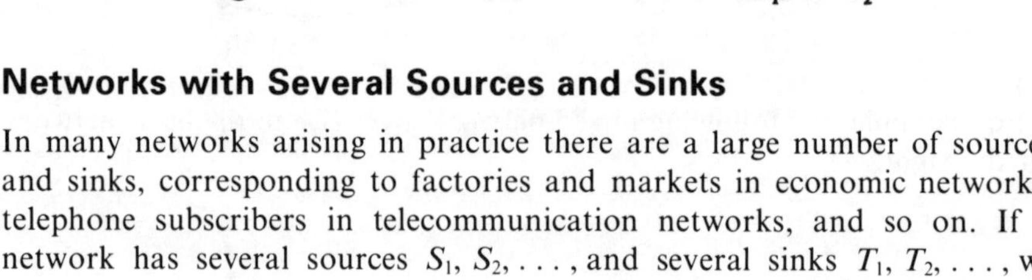

Networks with Several Sources and Sinks

In many networks arising in practice there are a large number of sources and sinks, corresponding to factories and markets in economic networks, telephone subscribers in telecommunication networks, and so on. If a network has several sources $S_1, S_2, \ldots$, and several sinks $T_1, T_2, \ldots$, we can transform it into a basic network by adjoining two new vertices—a *super-source* S joined to all the existing sources and a *super-sink* T joined to all the existing sinks, as follows:

Each added arc SS_i is assigned a capacity equal to the sum of the capacities of the arcs out of S_i, and each added arc T_iT is assigned a capacity equal to the sum of the capacities of the arcs into T_i.

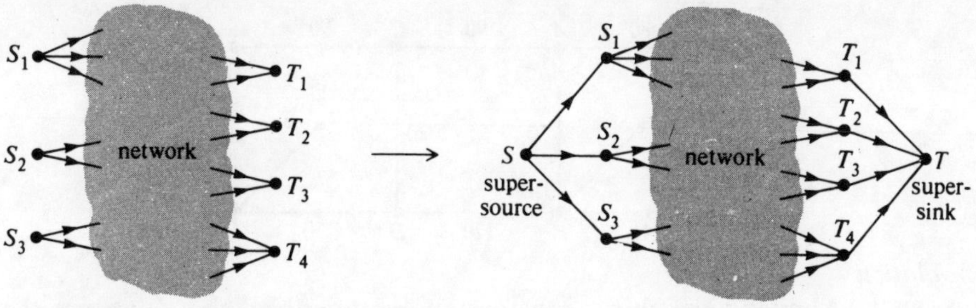

For example, in the following network the arc SS_1 is given capacity $3 + 4 + 2 = 9$, and the arc T_3T is given capacity $2 + 5 = 7$:

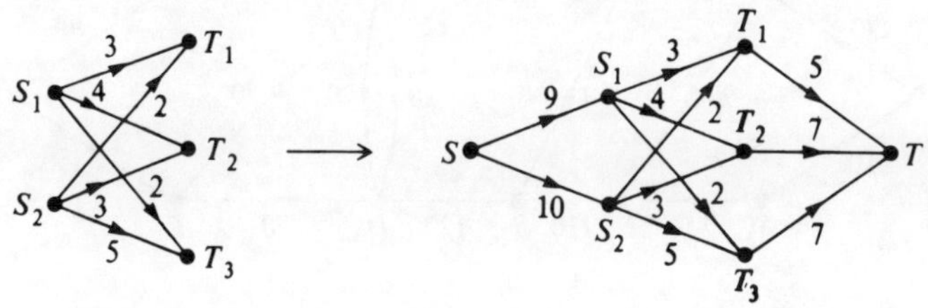

(Some authors assign infinite capacity to the arcs SS_i and T_iT to indicate that any amount of flow is allowed through these arcs. If fact, any capacities can be assigned to the arcs SS_i and T_iT, as long as they are not less than the capacities given here.)

Note that the maximum flow from S to T in this basic network is equal to the maximum flow from all the original sources to the original sinks.

Using the Three Transformations

Using these three transformations, we can solve a variety of network flow problems by the following procedure:

STEP 1 turn the problem into one involving a basic network;
STEP 2 solve this basic network problem by the methods of Chapter 3;
STEP 3 interpret the solution in terms of the original network.

The following worked problem illustrates this procedure.

Worked problem A manufacturer of farm machinery is faced with the problem of sending as many tractors as he can from two factories S_1 and S_2 to two markets T_1 and T_2 along the channels of the following network. How many can he send?

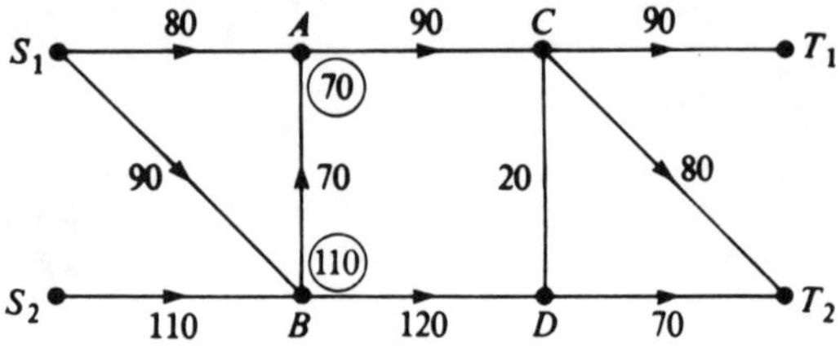

Solution

STEP 1 Applying the three transformations described above, we get the following basic network:

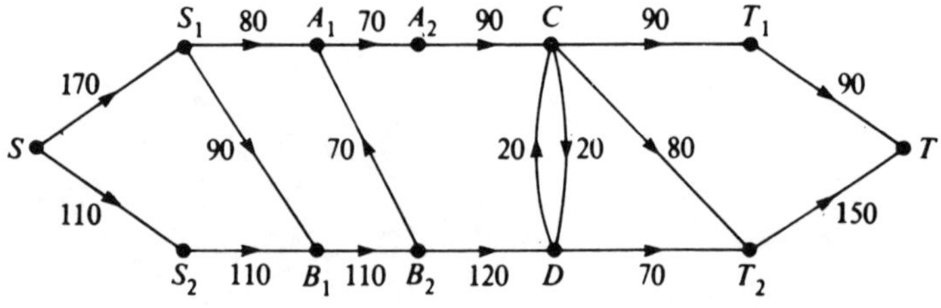

STEP 2 Using the methods of Chapter 3, we can easily check that this basic network has a maximum flow of value 160 corresponding to the minimum cut $\{A_1A_2, CD, DC, DT_2\}$. Such a maximum flow is shown in the following diagram:

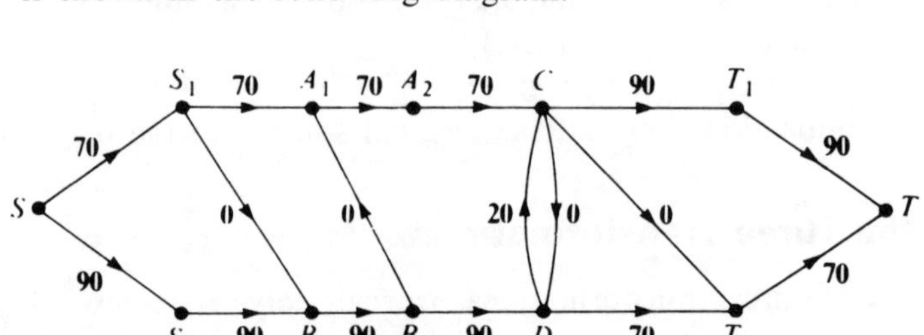

STEP 3 It follows that the manufacturer can send 160 tractors as indicated on the following diagram:

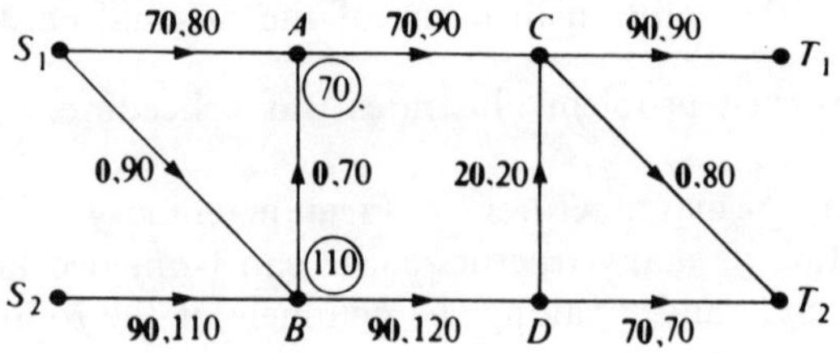

Problem 4.1 By reducing the following network to a basic network, find a maximum flow from the sources S_1 and S_2 to the sinks T_1 and T_2:

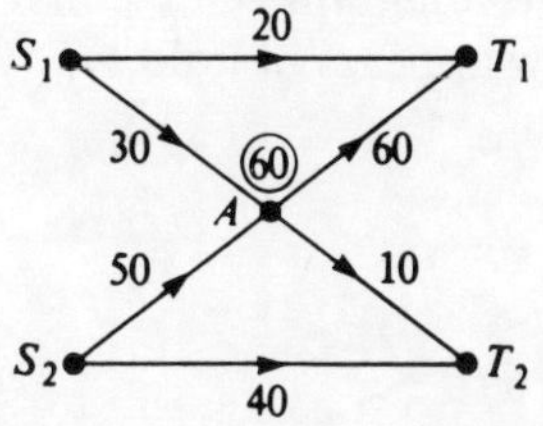

4.2 NETWORKS WITH LOWER AND UPPER CAPACITIES

In many practical problems (for example, the oil pipe-line example mentioned at the beginning of this chapter), it is necessary to specify both the maximum and the minimum allowable flow along each arc. We refer to the minimum allowable flow along an arc as the **lower capacity** of the arc, and the maximum allowable flow as the **upper capacity** of the arc. (An ordinary basic network can be regarded as a network of this type in which all lower capacities are zero.)

The following diagram depicts a network with lower and upper capacities on its arcs; for each arc, the first number represents the lower capacity of the arc, and the second number represents the upper capacity:

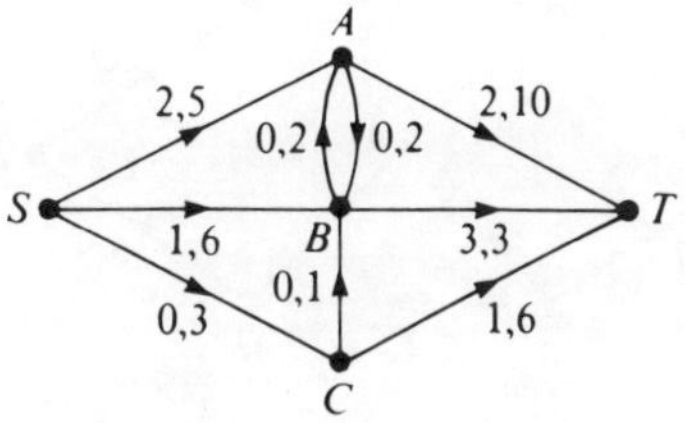

Definition *A **flow** in a network with lower and upper capacities is an assignment of non-negative numbers to the arcs of the network in such a way that the following conditions are satisfied:*

(i) (the feasibility condition): *the flow along each arc is not less than the lower capacity and not more than the upper capacity*:

$$\text{lower capacity} \leqslant \textbf{flow} \leqslant \text{upper capacity};$$

(ii) (the flow conservation law): *the flow into each vertex (other than S or T) is equal to the flow out of it.*

Terms such as the **value of a flow** and a **maximum flow** are defined in exactly the same way as before.

As with the basic flow problem, it is convenient to represent both flows and capacities on the same diagram. Our convention is illustrated in the following diagram: the flow appears in bold type *between* the lower capacity and the upper capacity.

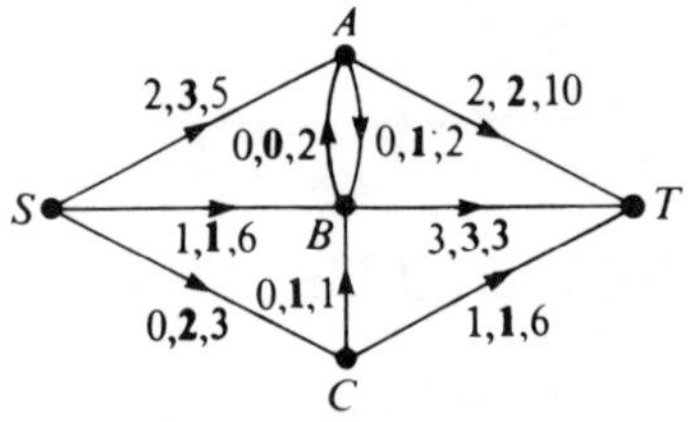

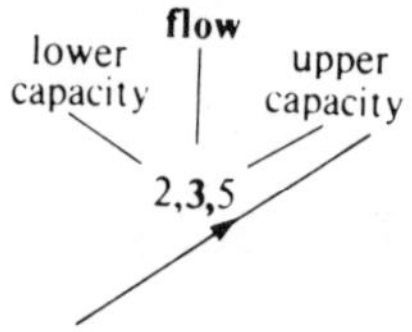

In this example, the value of the flow is equal to the flow out of S (or the flow into T), which is 6.

The following worked problem illustrates how to find a maximum flow in a network with lower and upper capacities.

Worked problem Find a maximum flow in the following network with lower and upper capacities:

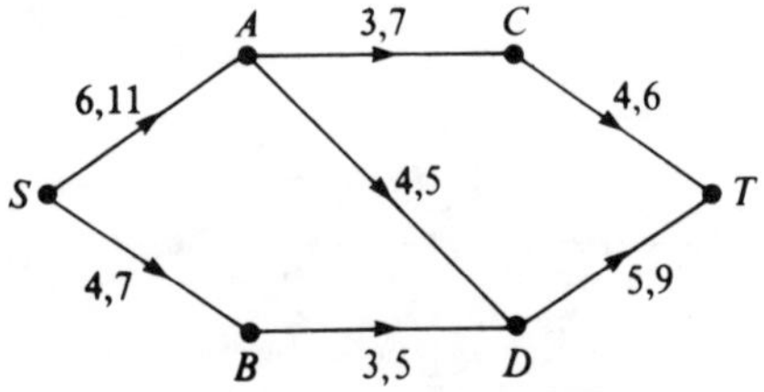

Solution We begin by finding any possible flow. In this example we can do this by inspection. There are several possible flows in this network—for example, the following diagram shows a flow of value 12:

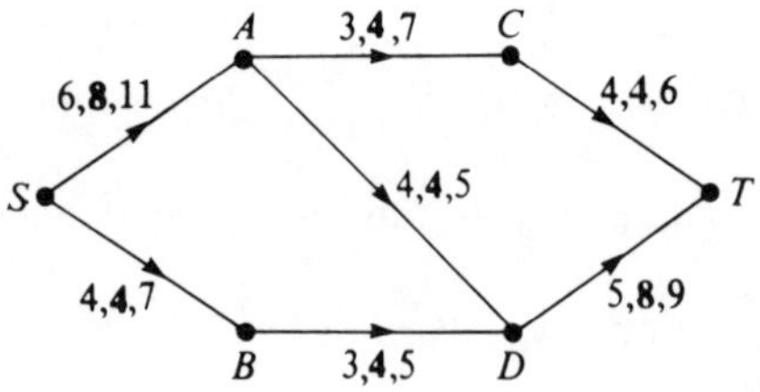

Next, by considering flow-augmenting paths we find a maximum flow in the network.

We can increase the flow by 2 along the path *SACT*, and by 1 along either of the paths *SADT* and *SBDT*. This gives the two maximum flows of value 15 shown in the following diagrams:

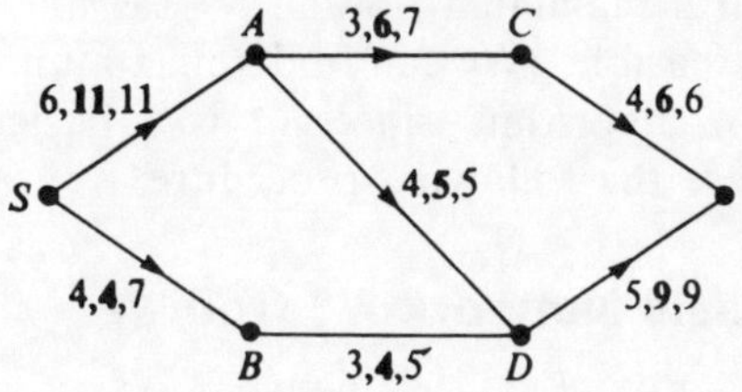

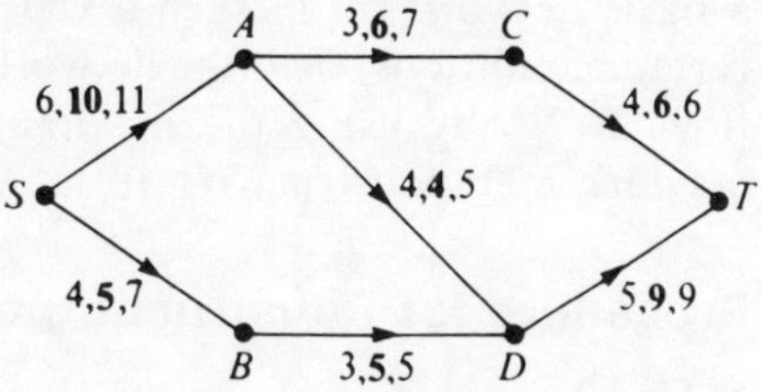

The above problem illustrates the fact that, if we can find a flow in a network with lower and upper capacities, then we can increase this flow to a maximum flow by means of the techniques described in Chapter 3. In particular, we can use the labelling procedure to find flow-augmenting paths, provided that *when decreasing the flow in an arc we do not produce a flow less than the lower capacity of the arc.*

However, there is a snag! Actually finding an initial flow is not always easy. (We can no longer choose the zero flow as the initial flow.) Indeed, there may not even be such a flow, as shown by the following network:

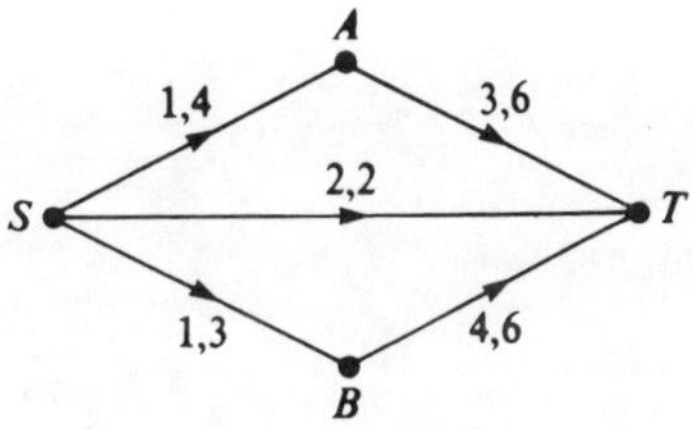

It is impossible to find a flow in this network since the flow along *SB* cannot exceed 3, and the flow along *BT* must be at least 4. In such a network, the capacity restrictions are incompatible. This leads us to make the following definition.

Definition *A network with lower and upper capacities is **feasible** if there exists a flow satisfying all the capacity restrictions. If no such flow exists, then the network is **infeasible.***

In view of the above remarks, it would be useful to have a systematic method for testing whether or not a given network is feasible, and if it is, for finding a valid flow. We now present such a method, which we call the *feasibility algorithm*; it is useful when we cannot find a flow by inspection. The method reduces the problem to that of finding a maximum flow in a related basic network—something we already know how to do.

The idea behind the algorithm is to construct from the given network N a basic network N^* in such a way that if a maximum flow in N^* satisfies certain conditions, then the network N is feasible. (We can find a maximum flow in N^* by using the maximum flow algorithm, since N^* is a basic network.) To construct N^* from N we use the following procedure.

Procedure for Constructing the Basic Network N^* from N

network N — network N^*

STEP 1
Take the vertices of N^* to be the vertices of N and two new vertices: a new source S^* and a new sink T^*.

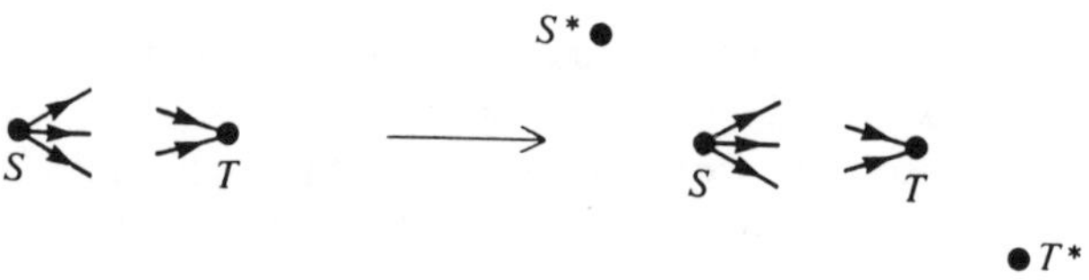

STEP 2
Replace lower and upper capacities l, k on each arc VW of N by a single capacity $k - l$ on the corresponding arc of N^*.

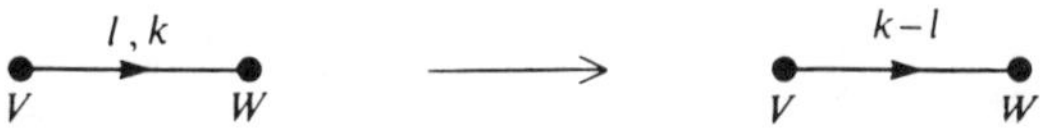

STEP 3
For each arc VW with lower capacity l in N, add in N^* two arcs S^*W and VT^*, each with capacity l.

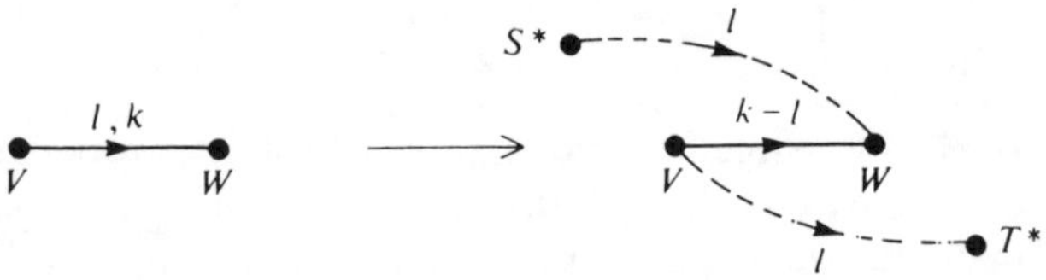

STEP 4
In N^*, add a new arc from T to S with infinite capacity.

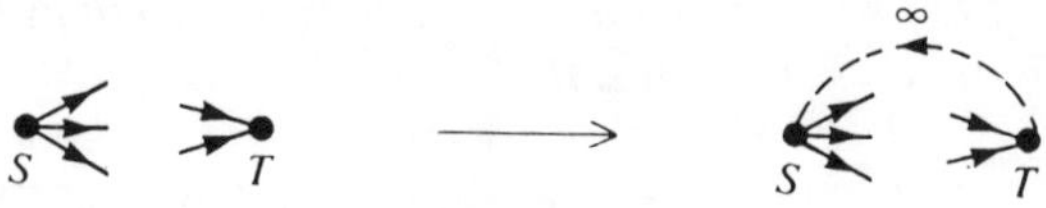

To understand how this procedure works, consider the diagram in Step 2 above. In replacing the upper and lower capacities l and k in N by a single capacity $l - k$, we are effectively sending a flow of value l through the arc VW, thus eliminating the need for a lower capacity, and saying that VW can take a further flow of value $k - l$. We thus reduce the problem to a basic flow problem. However, to maintain the flow in the arc at its original value, we must add a flow of value l in the opposite direction, from W to V. We arrange this by introducing an extra sources S^* and an extra arc from S^* to W with capacity l to supply this extra flow at W, and an extra sink T^* and an extra arc from V to T^* with capacity l to remove it at V (Steps 1 and 3). We do this for each arc VW in the network N. However, some arcs begin at S and some arcs end at T so, to avoid the extra flow draining from S and building up at T, we include an extra arc of infinite capacity from T to S.

Determining Feasibility

If, in the new network N^*, we can find a flow in which all the arcs out of S^* and all the arcs into T^* are saturated, then the balance of flow will have been restored, and the network problems in N and N^* are equivalent. These ideas lead to the following theorem for determining the feasibility of a given network.

Feasibility theorem *The original network N is feasible if and only if in the related basic network N^* there is a flow from S^* to T^* such that all the arcs out of S^* and all the arcs into T^* are saturated.*

Note that if a flow in N^* can be found such that all the arcs out of S^* and all the arcs into T^* are saturated, then it must be a maximum flow. So we test for feasibility of N by trying to find a maximum flow in the basic network N^*. We already know how to do this—we use the maximum flow algorithm, described in Chapter 3.

Finding a Maximum Flow in *N*

Having established that a given network N with lower and upper capacities is feasible, we may want to find a maximum flow in N. To do this, we first need to find some valid flow which we can then improve—but we cannot start with a zero flow as we can for a basic network. However, we can obtain a valid flow in N from the maximum flow found in N^* by 'undoing' the steps in the construction. We remove the vertices S^* and T^* and all the extra arcs, and add to the flow in each original arc VW the original lower capacity l. We thus obtain a flow in N which is compatible with all the lower and upper capacities. We can now increase this flow to a maximum flow using the algorithm described in Chapter 3—provided that, *when augmenting a backward arc, we do not reduce the flow below the lower capacity of the arc.*

We illustrate the procedure in the following worked problem.

Worked problem Determine whether the given network N is feasible, and if it is, find a maximum flow.

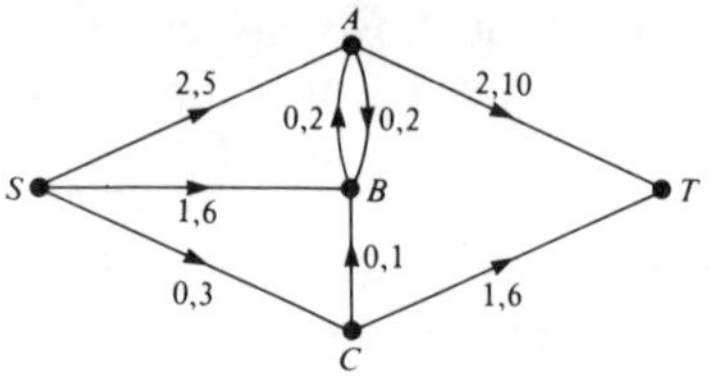

Solution First we construct the basic network N^*.

In Steps 1 and 2 we add the vertices S^* and T^*, and on each arc we replace the lower and upper capacities by a single capacity.

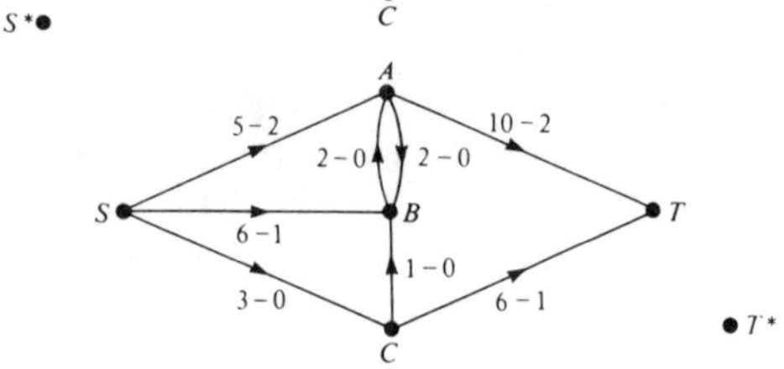

In Step 3, for each arc VW, we add arcs S^*W and VT^*; these arcs are shown by dashed lines.

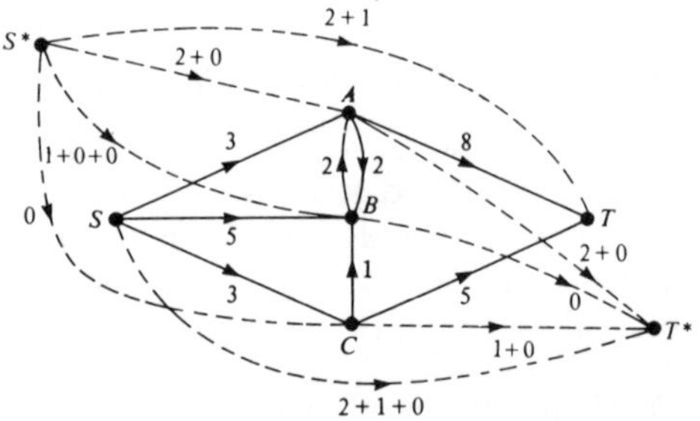

Finally, in Step 4 we add an arc of infinite capacity from T to S to obtain the corresponding basic network N^*.

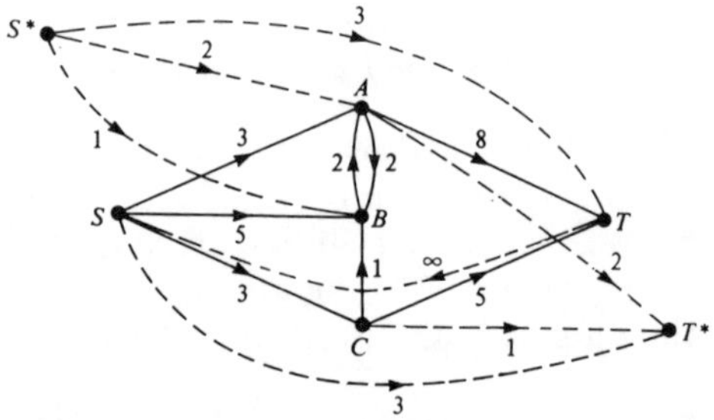

the network N^*

To determine whether N is feasible, we find a maximum flow in N^* either by applying the maximum flow algorithm of Chapter 3 or by inspection. (In this case, we can find this maximum flow by inspection.)

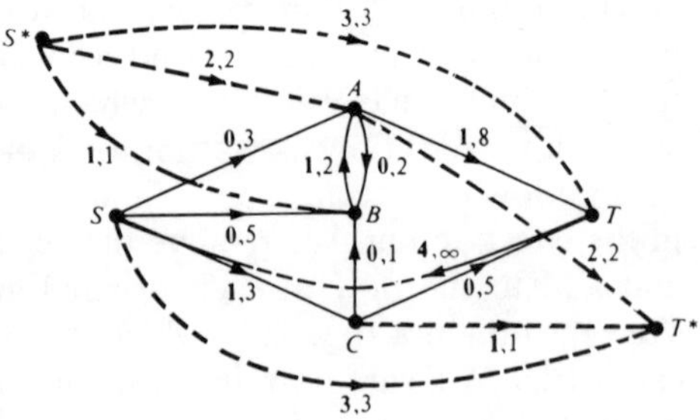

All arcs out of S^* and all arcs into T^* are saturated so, by the Feasibility Theorem, the network N is feasible.

We can find a valid flow in N by converting the network N^* with its maximum flow back to a network with lower and upper capacities.

First, we remove the vertices S^* and T^* and all the extra arcs added in Steps 1, 3 and 4 of the construction.

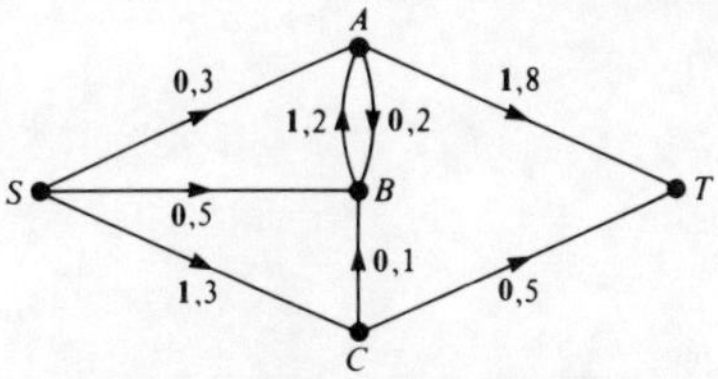

Next, we 'undo' Step 4—we restore the lower and upper capacities and increase the flow on each arc by the lower capacity of the arc. For arc AT, for example, the lower capacity is 2; so we increase the flow by 2 to obtain a flow of 3. The upper capacity of AT is 10, so we label this arc 2, **3**, 10. We thus obtain the flow in N shown on the right:

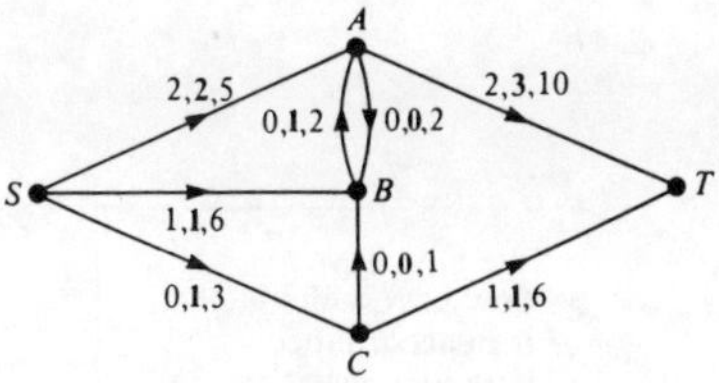

This is a valid flow, but it is not a maximum flow. However, we can easily find a maximum flow, either by inspection or by using the algorithm of Chapter 3. Such a maximum flow has an extra 3 units of flow along the path SAT, an extra 1 unit of flow along the path $SBAT$, and an extra 2 units of flow along the path SCT, making a flow of value 10.

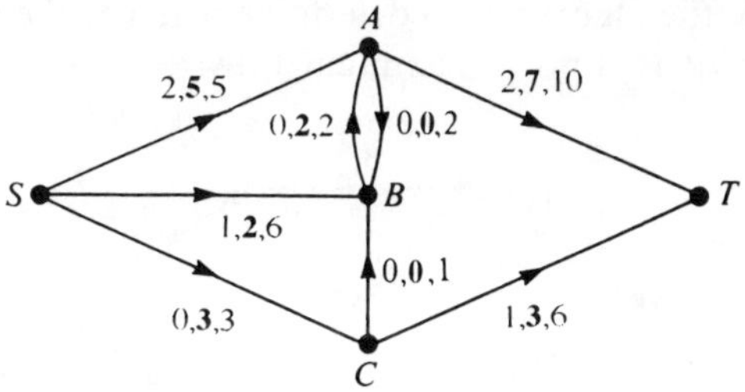

We give a flow chart of the algorithm for reference.

The Feasibility Algorithm

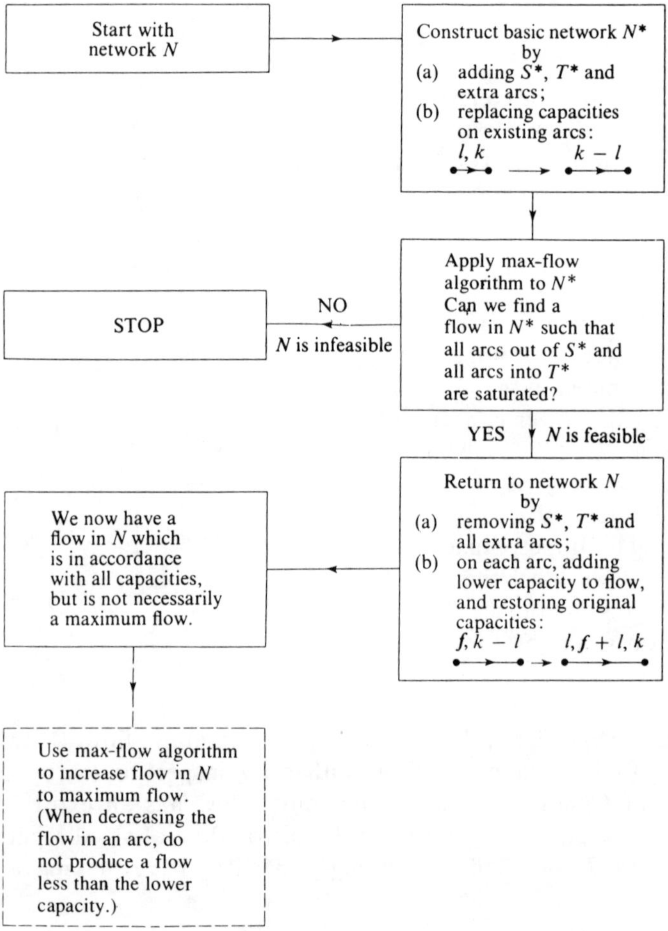

Problem 4.2 Use the algorithm to determine whether the following network is feasible, and if it is, find a maximum flow.

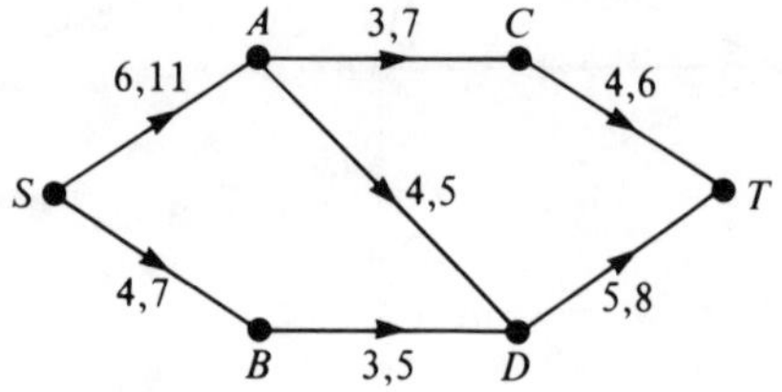

The Generalized Max-Flow Min-cut Theorem

We conclude this chapter by stating an analogue of the max-flow min-cut theorem for networks with lower and upper capacities. First, we need to redefine the capacity of a cut.

Definition *Let N be a network with lower and upper capacities, and let C be a cut which separates N into two parts: X (containing S) and Y (containing T). Then the* ***capacity of the cut C*** *is*

(*the sum of the upper capacities of the arcs of C which are directed from X to Y*)
– (*the sum of the lower capacities of the arcs of C which are directed from Y to X*).

(Note that if the lower capacity of each arc is zero, then this definition coincides with the definition given in Chapter 3.)

For example, for the network N shown below,

the capacity of the cut $\{SA, SB\}$ is $9 + 6 = 15$;
the capacity of the cut $\{AC, DA, BD\}$ is $10 + 9 - 3 = 16$;
the capacity of the cut $\{SB, DA, CT\}$ is $6 + 7 - 3 = 10$.

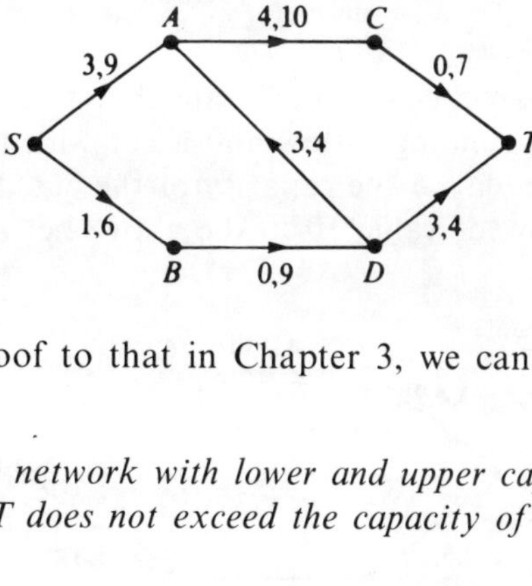

By an analogous proof to that in Chapter 3, we can prove the following theorem.

Theorem 4.1 *In any network with lower and upper capacities, the value of any flow from S to T does not exceed the capacity of any cut.*

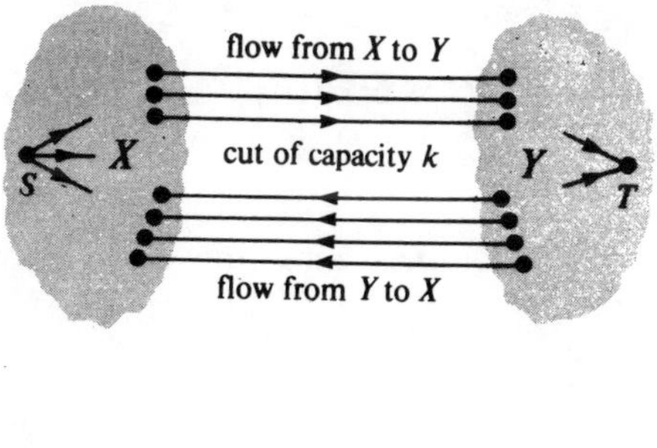

Proof Consider a cut C with capacity k (say). The removal of the arcs of C separates the network into two parts, X and Y. If the sum of the upper capacities of the arcs of C directed from X to Y is k_1, and the sum of the lower capacities of the arcs of C directed from Y to X is k_2, then $k_1 - k_2 = k$. Let F denote the value of the flow. Then F is given by the equation

$$F = \text{(the total flow from } X \text{ to } Y) - \text{(the total flow from } Y \text{ to } X).$$

Since the total flow from X to Y is at most k_1, and the total flow from Y to X is at least k_2, we have $F \leqslant k_1 - k_2$; that is, $F \leqslant k$. □

Just as before, it is not difficult to show that the value of any maximum flow from S to T is equal to the capacity of any minimum cut. (The proof of this theorem is very similar to that of the max-flow min-cut theorem, and we omit it.)

The Generalized Max-Flow Min-Cut Theorem *In any feasible network with lower and upper capacities, the value of a maximum flow is equal to the capacity of a minimum cut.*

Just as in Chapter 3, we can demonstrate that this theorem holds for a particular network by finding a flow and a cut with the property that the value of the flow is equal to the capacity of the cut. For example, we can easily construct a flow of value 10 in the above network:

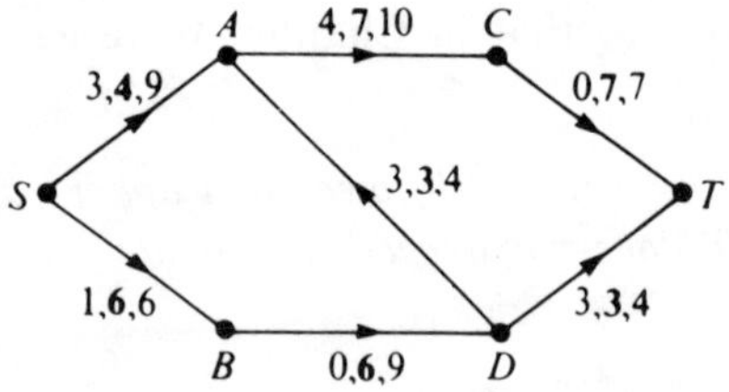

Since $\{SB, DA, CT\}$ is a cut of capacity 10, it follows that this cut is a minimum cut, and that the flow just given is a maximum flow.

Problem 4.3 Verify the generalized max-flow min-cut theorem for each of the following networks:

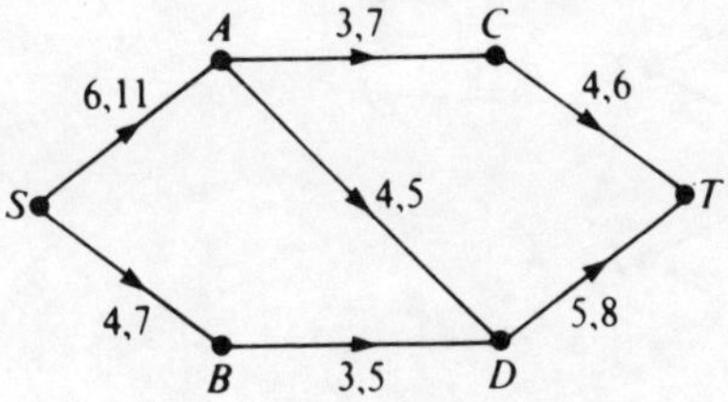

(a)

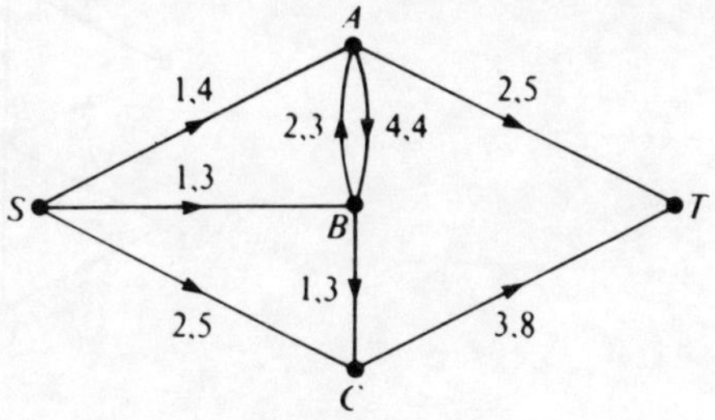

(b)

EXERCISES 4

Variations on the Basic Flow Problem

4.1. Transform the following network into a basic network:

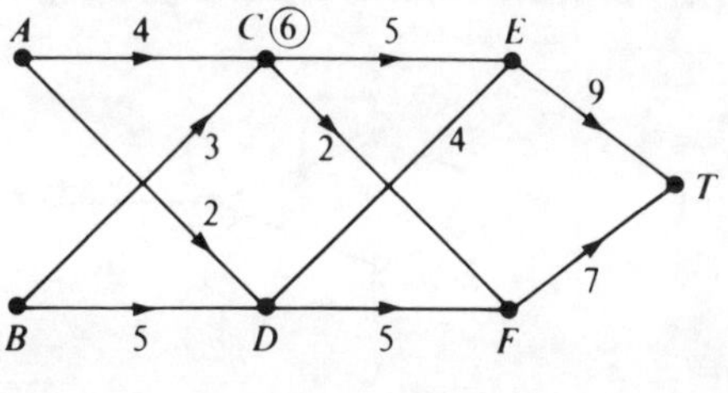

4.2. Transform the following network into a basic network:

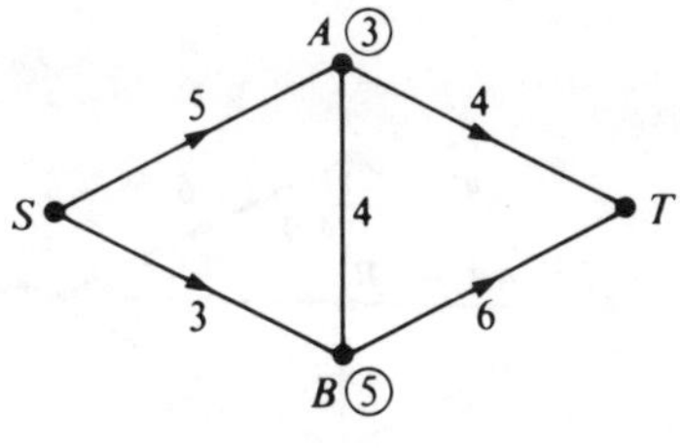

4.3. Transform the following network into a basic network. Find a maximum flow in the basic network and use it to find a maximum flow in the original network.

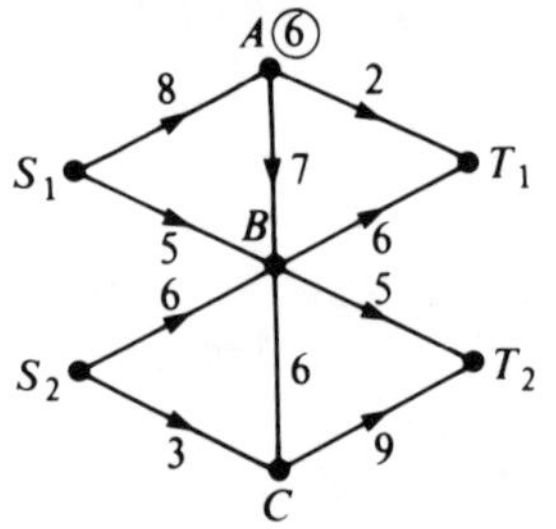

Networks with Lower and Upper Capacities

4.4. Determine by inspection which of the following networks are feasible:

(a) 

(b) 

(c) 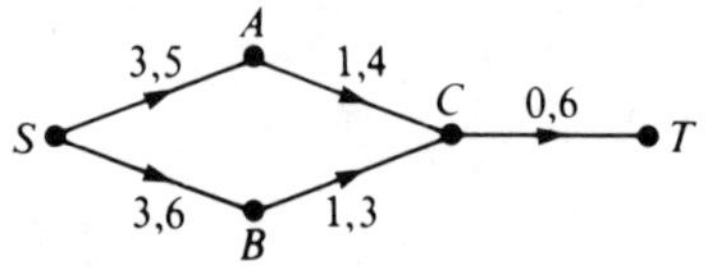

4.5. Use the Feasibility Theorem to determine whether network (c) of Exercise 4.4 is feasible.

4.6. Use the Feasibility Theorem to show that the following network is feasible, and hence find a maximum flow from S to T:

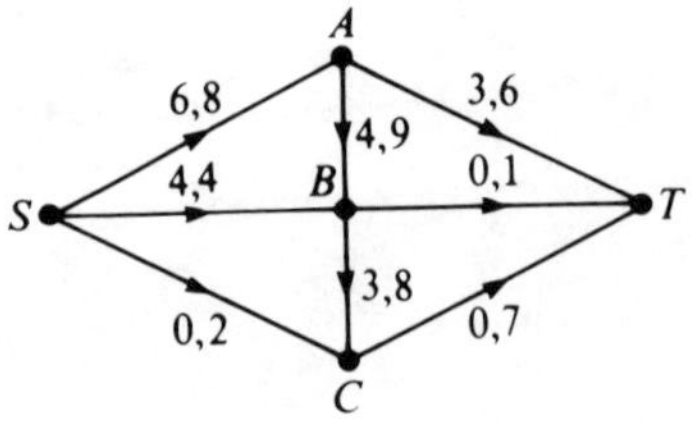

4.7. Consider the following network:

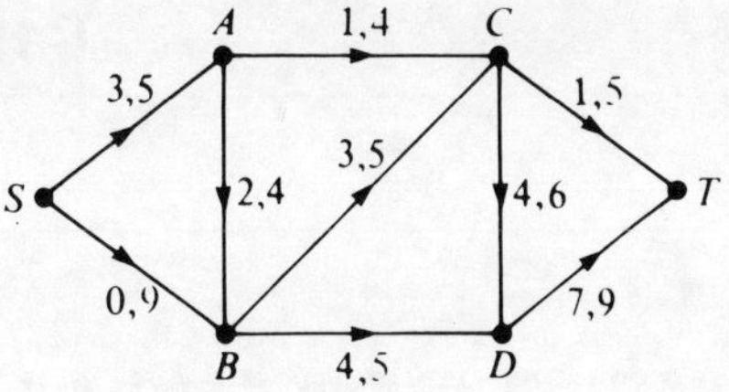

(a) Find by inspection a feasible flow in the network.

(b) Check that the generalized max-flow min-cut theorem holds for this network.

(c) Use the algorithm in the text to find a maximum flow in this network.

4.8. Suppose that the flow in each arc of a basic network is bounded below (that is, the value must be greater than or equal to some minimum value) instead of being bounded above (by the capacity). Design an algorithm to find a minimum flow in such a network, and formulate and prove the corresponding min-flow max-cut theorem.

CHAPTER 5

MULTI-TERMINAL FLOWS

After reading this chapter, you should be able to:

- explain the term *multi-terminal flow*;
- explain the connection between a network N and the corresponding complete network $C(N)$;
- obtain all the capacities in $C(N)$ from those in a maximum tree;
- use the method of Gomory and Hu to construct a maximum tree, and hence find the flow between each pair of vertices.

Introduction

Up to this point we have primarily been concerned with problems involving the flow of a commodity from a single source to a single sink. However, many of the network problems arising in practice involve the flow of commodities between many different pairs of locations. For example, an airline network involves flights between a number of cities; a telephone network links many pairs of subscribers or exchanges; and a road network involves the flow of traffic between a number of different locations. Problems of this type are known as **multi-terminal flow problems**.

Multi-terminal flow problems of the types just described are, in general, too complicated to be solved here. We therefore restrict our attention to one particular type of multi-terminal flow problem:

find the maximum flow between any two locations, assuming that only a single pair of locations can communicate with each other at any one time.

Although this simplified version of the problem may seem too restrictive to be of any practical use, it is important in certain applications. For example, the technique we describe is very useful in problems involving the expansion or improvement of an existing system such as a road network (widening a road or building an over-pass) or a telephone network (constructing a new exchange or increasing the capacity of certain links).

In order to solve this simplified multi-terminal flow problem, we could proceed by considering separately each possible source and sink and using the techniques of the previous chapters to find the maximum flow in each case. If the number of vertices in the network is n, then the number of ways of choosing the source and sink is $n(n-1)$, since the source can be chosen in n ways, and the sink can then be chosen in $n-1$ ways. It follows that we should have to do $n(n-1)$ separate maximum flow calculations—which is a lot of work!

We can halve the number of calculations by restricting our attention to *undirected* networks, since any flow from a vertex V to a vertex W in an undirected network immediately gives rise to a flow from W to V. (Just reverse it!) So for undirected networks we should have to do $\frac{1}{2}n(n-1)$ separate maximum flow calculations. But this is still an enormous number—even for networks with only 20 vertices we should need to carry out 190 different network flow calculations.

Fortunately, we can get away with much less work than this. Provided that we restrict our attention to undirected networks, we can use an algorithm due to R. E. Gomory and T. C. Hu to reduce the number of network flow calculations from $\frac{1}{2}n(n-1)$ to just $n-1$. (For example, a network with 20 vertices would require only 19 flow calculations instead of the 190 calculations mentioned above.) The approach that Gomory and Hu used in solving multi-terminal problems is entirely different from those described earlier, and involves the construction of a particular type of tree.

5.1 THE CAPACITY RULE

Before describing the method of Gomory and Hu, we consider a simple example which leads us to formulate a result known as the *capacity rule*.

Example 1 Consider the following undirected network N:

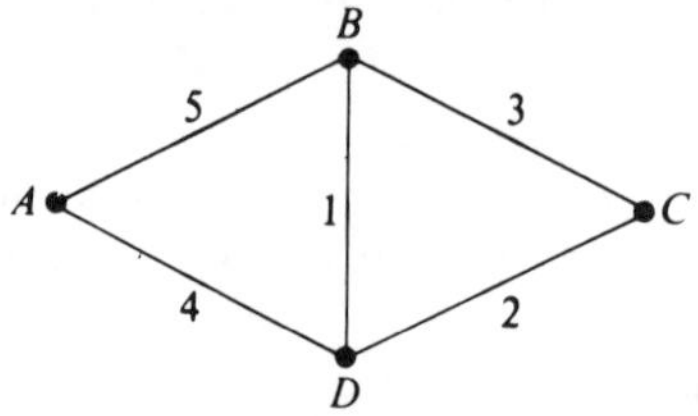

the network N

We calculate $f(VW)$, the value of a maximum flow between each pair of vertices V and W of N. (When calculating $f(VW)$, it is often simplest to look at the various cuts separating the vertices V and W and to find a cut with smallest capacity. By the max-flow min-cut theorem, this capacity gives the required value.) For example, $f(AC)$, the value of a maximum flow between vertices A and C (or C and A), is 5, and $f(AB)$, the value of a maximum flow between vertices A and B (or B and A), is 8.

We now construct a new undirected network by taking the complete graph with vertices A, B, C and D, and assigning to each edge VW a capacity equal to $f(VW)$. For example, the capacity of the edge AC is 5, and the capacity of the edge AB is 8. We call this new network the **complete network**

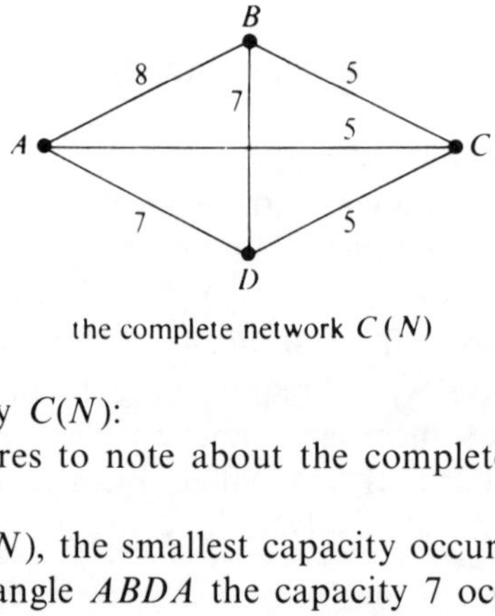

the complete network $C(N)$

of N, and denote it by $C(N)$:

There are two features to note about the complete network $C(N)$:

1. in each cycle of $C(N)$, the smallest capacity occurs more than once (for example, in the triangle $ABDA$ the capacity 7 occurs twice, and in the cycle $ABCDA$ the capacity 5 occurs twice);
2. $C(N)$ has four vertices but only three different capacities (5, 7 and 8).

Problem 5.1 Consider the undirected network N at the top of page 99.
(a) Find $f(VW)$ for each pair of vertices V and W.
(b) Construct the complete network $C(N)$, and write down statements analogous to statements 1 and 2 in the above example.

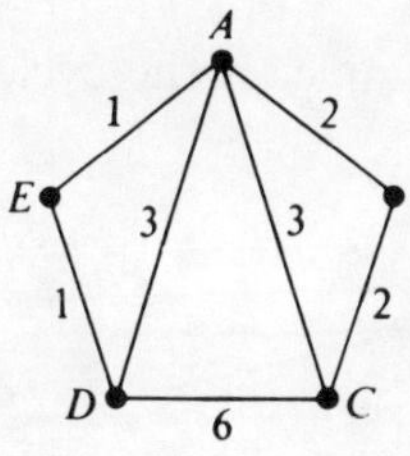

The statements 1 and 2 given above are special cases of the following results for an undirected network N and its complete network $C(N)$:
1. in each cycle of $C(N)$, the smallest capacity occurs more than once;
2. if N has n vertices, then $C(N)$ has at most $n - 1$ different capacities.

It follows from statement 2 that at most $n - 1$ of the $\frac{1}{2}n(n-1)$ possible flow values $f(VW)$ are different. Before describing how to calculate these $n - 1$ flow values, we prove statements 1 and 2.

The proof of statement 1 is similar to proofs we have seen earlier (in Chapter 3).

Theorem 5.1 *Let N be an undirected network, and let $C(N)$ be the corresponding complete network. Then, in any cycle of $C(N)$, the smallest capacity occurs more than once.*

Proof Let $f(VW)$ be the smallest capacity in some cycle of $C(N)$. Since $f(VW)$ is the value of a maximum flow from V to W in the network N, it follows from the max-flow min-cut theorem that there is a minimum cut in N with capacity $f(VW)$. The removal of the edges in this cut splits the network N into two parts, X (containing V) and Y (containing W).

Now let $V'W'$ be any other edge of the cycle with one end in X and the other end in Y. (There must be such an edge, because in going from V to W we cross from X to Y, and we eventually need to return to X in order to complete the cycle. In fact, we may need to cross several times from X to Y in tracing the cycle.)

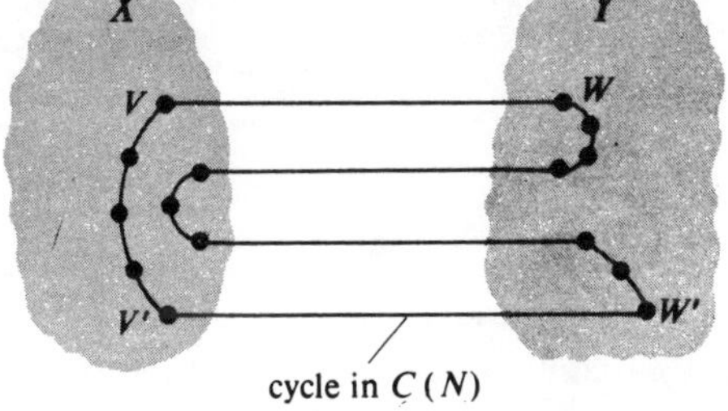

Note that it is possible for V and V', or W and W', to be the same vertex. This happens, for example, if the cycle is a triangle:

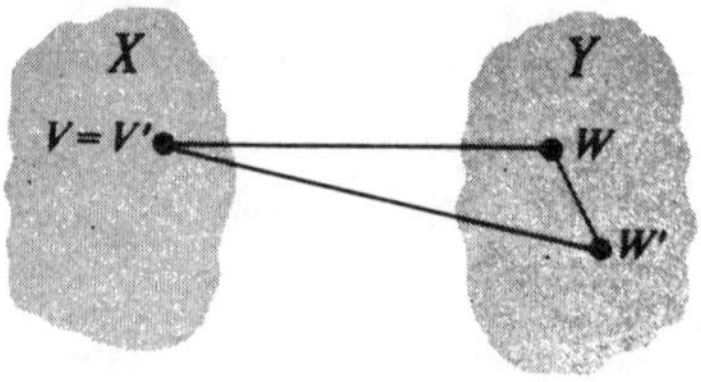

Since the minimum cut separating V from W also separates the vertices V' and W', any flow between V' and W' must have value not exceeding $f(VW)$, so

$$f(V'W') \leqslant f(VW).$$

But $f(VW)$ is the *smallest* capacity occurring in the cycle, so

$$f(VW) \leqslant f(V'W').$$

It follows from these two inequalities that $f(VW) = f(V'W')$, so the smallest capacity occurs at least twice. □

To prove statement 2, we construct in $C(N)$ a tree such that the sum of the capacities of all its edges is a maximum—such a tree is called a **maximum tree**. The construction is done by successively choosing edges of largest capacity in such a way that no cycles are created. (This procedure is a form of the greedy algorithm discussed in Chapter 8. You are asked to prove that it works in Exercise 8.15.)

The following example illustrates the method.

Example 1 continued Let N and $C(N)$ be as before:

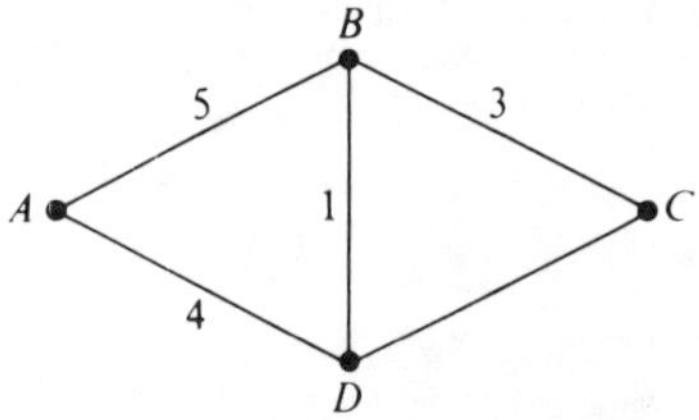

the network N

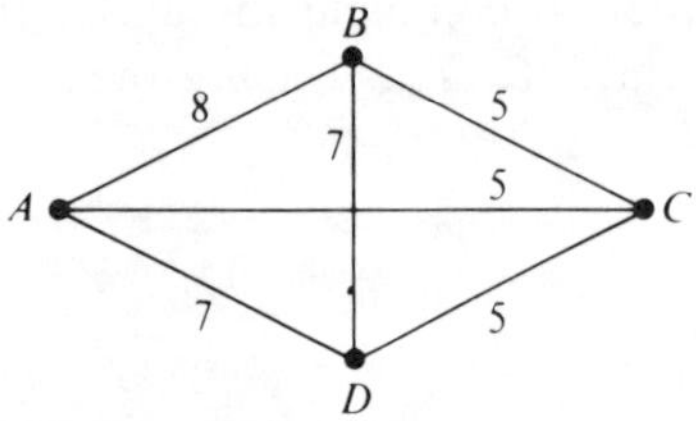

the complete network $C(N)$

First choice We choose an edge of largest capacity in $C(N)$; this is AB, with capacity $f(AB) = 8$.

Second choice We choose an edge of next largest capacity; this is either AD or BD, with capacity $f(AD) = f(BD) = 7$.

Third choice We cannot include both AD and BD in the tree because a tree cannot contain a cycle ($ABDA$). So we choose one of the three edges AC, BC, CD, with capacity

$$f(AC) = f(BC) = f(CD) = 5.$$

Since every tree with n vertices has exactly $n - 1$ edges, this completes the tree.

Note that there are two possible choices at the second stage and three possible choices at the third stage. These possibilities give rise to six different maximum trees, each with total capacity $8 + 7 + 5 = 20$:

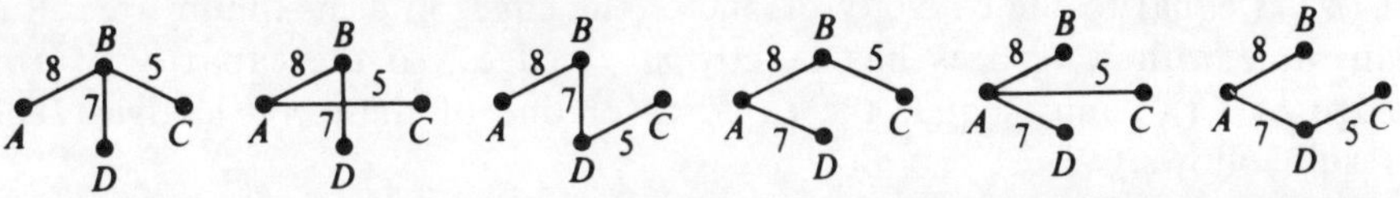

The importance of these trees lies in the fact that we can use them to reconstruct all the capacities in $C(N)$. To do this, we use the following rule.

The capacity rule *To find $f(VW)$, look at the path joining the vertices V and W in a maximum tree, and take the smallest of the capacities along this path.*

For example, applying the rule to the first of the six trees pictured above, we get

$$f(AC) = \min\,(f(AB), f(BC)) = \min\,(8, 5) = 5,$$
$$f(AD) = \min\,(f(AB), f(BD)) = \min\,(8, 7) = 7,$$
$$f(CD) = \min\,(f(CB), f(BD)) = \min\,(5, 7) = 5,$$

which are the correct values.

Problem 5.2 Check the values of $f(AD)$, $f(BC)$ and $f(CD)$ by applying the capacity rule to the second of the six trees pictured above.

Problem 5.3 The following diagram represents a maximum tree in a complete network:

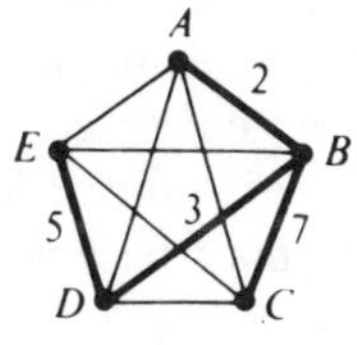

Use the capacity rule to write down the values of

$f(AC)$, $f(AD)$, $f(AE)$, $f(BE)$, $f(CD)$, $f(CE)$.

Problem 5.4 Prove the capacity rule by applying Theorem 5.1 to the cycle obtained when the edge VW is added to a maximum tree.

We are now in a position to prove statement 2.

Theorem 5.2 *Let N be an undirected network with n vertices, and let $C(N)$ be the corresponding complete network. Then $C(N)$ has at most $n - 1$ different capacities.*

Proof It follows from the capacity rule that the capacity of any edge of $C(N)$ is equal to the capacity of one of the edges in a maximum tree. But any tree with n vertices has exactly $n - 1$ edges, so the capacity of any edge of $C(N)$ must equal the capacity of one of these $n - 1$ edges. The result follows. □

As yet we have not accomplished much as far as the solution of the original multi-terminal flow problem is concerned. We have merely shown that there are at most $n - 1$ different flow values $f(VW)$, and that these can easily be found if we know the capacities of the edges in a maximum tree. However, we cannot consider the original problem solved until we have a method for constructing such a maximum tree directly from the original network. We now present such a method, known as the *method of Gomory and Hu*. Note that their method works only for *undirected* networks. (For directed networks, the maximum flow from a vertex V to a vertex W may be different from the maximum flow from W to V, and there is no known general method which can efficiently take account of this in calculating the $n(n - 1)$ different maximum flows.)

5.2 THE ALGORITHM OF GOMORY AND HU

In this section we describe the method of Gomory and Hu for finding the maximum flows between all pairs of vertices in a given undirected network. As we have seen, for an undirected network with n vertices, there are $\frac{1}{2}n(n - 1)$ maximum flows to be found. The method of Gomory and Hu uses the properties of minimum cuts to construct a particular type of tree, called a *cut tree*, by doing only $n - 1$ maximum flow calculations; some of these calculations may be simplified by condensing certain vertices of the given network into a single vertex. Once a cut tree has been constructed, the remaining maximum flows may be found by using the capacity rule described in the previous section.

Underlying Ideas—flow-equivalent Networks and Cut Trees

The algorithm is based on the idea of *flow-equivalent networks.*

Definition *Two undirected networks are* ***flow-equivalent*** *if*
(i) *they have the same number of vertices;*
(ii) *the maximum flows between pairs of corresponding vertices in the two networks are the same.*

Example 1 revisited In the previous section we saw that the network N shown below on the left and the two trees shown on the right are all flow-equivalent.

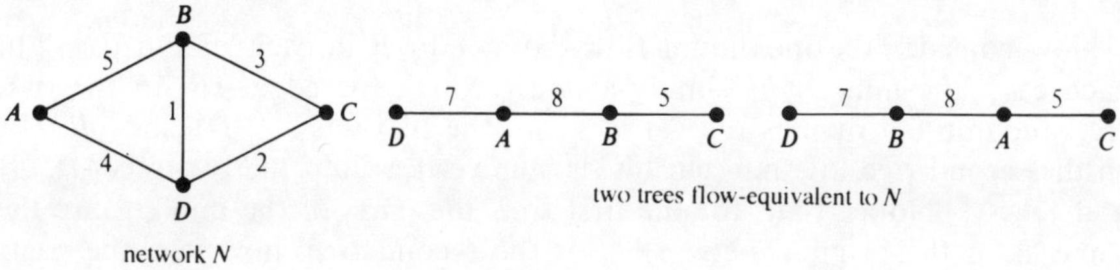

This example illustrates two things. First, for any undirected network, we can find a flow-equivalent network which is a tree. Secondly, in general there will be many such trees.

The algorithm of Gomory and Hu constructs a particular type of flow-equivalent tree; such trees have a special property involving minimum cuts.

Let us look at a minimum cut in our example. Suppose we find the maximum flow between A and B. The corresponding min cut is shown by the dashed line. It divides the vertices of the network into two sets, one set X containing A, and the other set Y containing B.

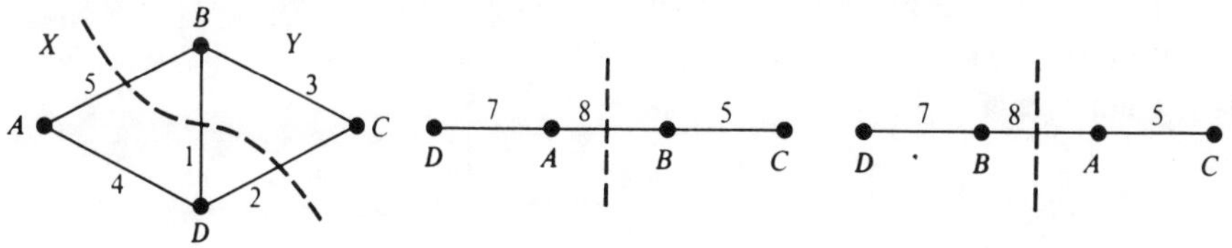

min cut separates N into 2 parts: $\{A, D\}$, $\{B, C\}$

min cut separates tree into 2 parts: $\{A, D\}$, $\{B, C\}$

min cut separates tree into 2 parts: $\{D, B\}$, $\{A, C\}$

In this example, X is the set of vertices $\{A, D\}$, and Y is the set $\{B, C\}$.

An important property of a minimum cut (see Chapter 3) is that the capacity of a minimum cut is equal to the value of a maximum flow. In this example, the capacity of the cut is simply the sum of the capacities of the arcs forming the cut—since the flow along each arc must go from X to Y—so the value of the maximum flow is 8.

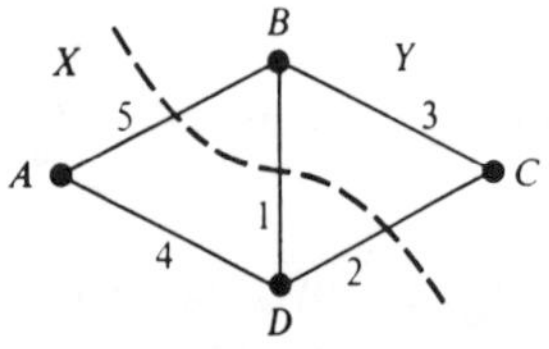

min cut separates
N into 2 parts:
$\{A, D\}$, $\{B, C\}$

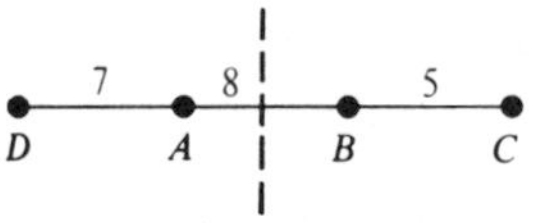

min cut separates
tree into 2 parts:
$\{A, D\}$, $\{B, C\}$

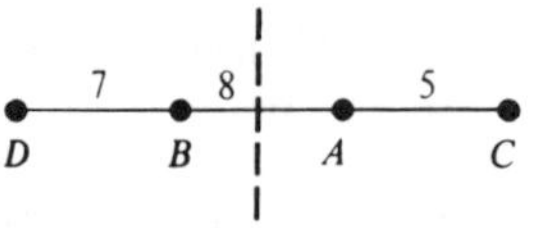

min cut separates
tree into 2 parts:
$\{D, B\}$, $\{A, C\}$

Now consider the maximum flow from A to B in each of the trees. In each case, the min cut is simply a line through the edge AB. In the first tree, the min cut divides the vertices into the two sets $\{A, D\}$ and $\{B, C\}$. In the second tree, the min cut divides the vertices into the two sets $\{D, B\}$ and $\{A, C\}$. Notice that, for the first tree, the parts of the min cut are the same as in the original network. For the second tree, however, the parts of the min cut are not the same. If you find the min cuts between the other pairs of vertices, you will see that the first tree has this property for *all* the min cuts.

Clearly, the first flow-equivalent tree is rather special. We call it a *cut tree* for the network N.

Definition *A flow-equivalent tree is called a* **cut tree** *if each branch of the tree represents a minimum cut of the original network—that is, each branch, when removed, separates the vertices into two sets X and Y corresponding to a minimum cut of the original network.*

Problem 5.5 Determine whether each of the following trees is

(a) flow-equivalent to N;
(b) a cut tree for N.

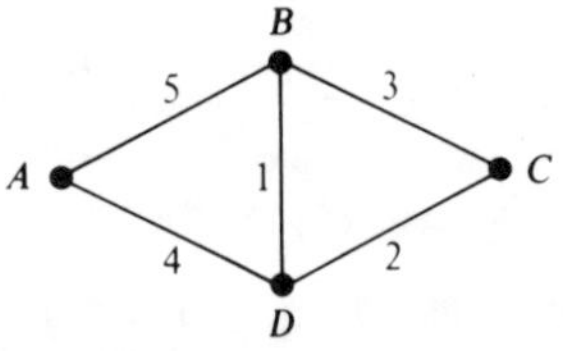

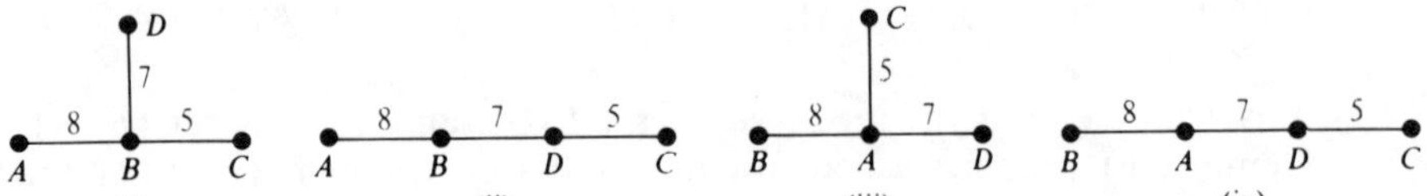

(i) (ii) (iii) (iv)

5.3 THE BASIC ALGORITHM

Before giving the full algorithm of Gomory and Hu we describe a basic form; we then refine it to incorporate a technique called *condensing the vertices.*

The algorithm constructs a cut tree, branch by branch, by doing a maximum flow calculation corresponding to each branch. If the original network has n vertices, then our cut tree has $n - 1$ edges, so we need do only $n - 1$ max flow calculations. We illustrate the algorithm in the following worked problem.

Worked problem Use the basic algorithm of Gomory and Hu to find a cut tree for the following network:

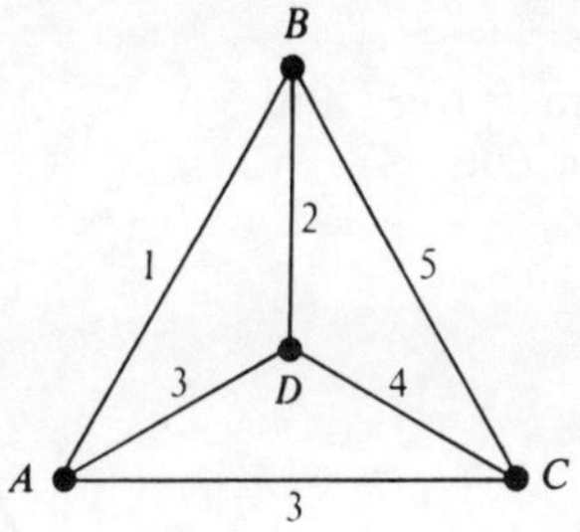

Solution This network has 4 vertices, so we can find a cut tree in 3 steps. Note that in the diagrams of the network given in the solution *the numbers on the arcs are **flows**, not capacities.*

STEP 1
Choose any two vertices, say A and B.

(i) Find a max flow from A to B and a corresponding min cut:

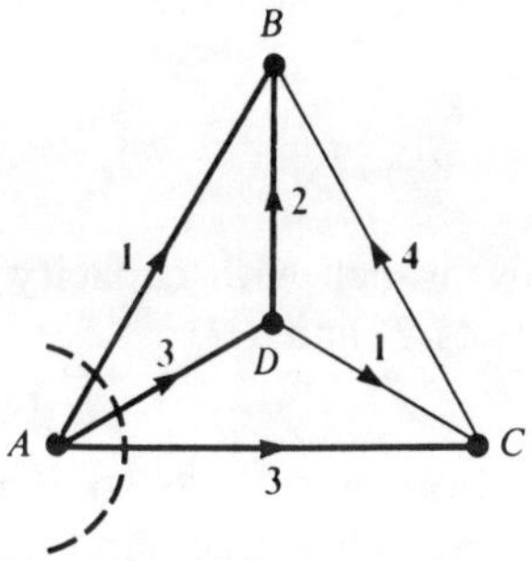

(ii) Write down the effect of the min cut: min cut has capacity 7 and separates $\{A\}$ and $\{B, C, D\}$

(iii) Draw branch with capacity 7 and vertices A and BCD:

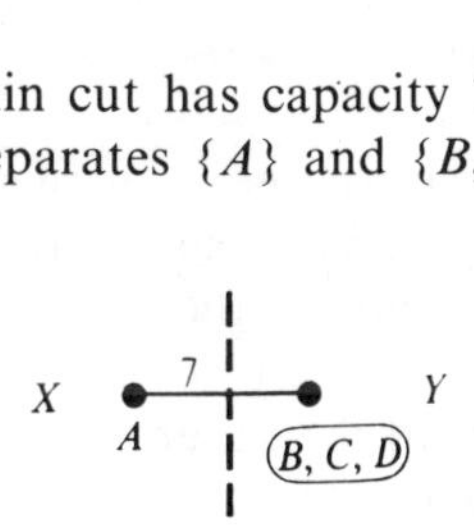

The two vertices of the branch correspond to the two parts of the min cut, so we label one vertex A and the other $\boxed{BCD}$. The vertex $\boxed{BCD}$ is called a **generalized vertex**.

Notice that the statement 'a min cut of capacity 7 separates the vertex A from the vertices B, C and D' applies *both* to the network diagram and to the first branch of the tree.

We now move on to STEP 2.

We split up the generalized vertex $\boxed{BCD}$. We choose any two vertices, say B and C, and we do a max flow calculation from B to C.

STEP 2 Choose any 2 vertices from the generalized vertex $\boxed{BCD}$, say B and C.

(i) find a max flow from B to C and a corresponding min cut:

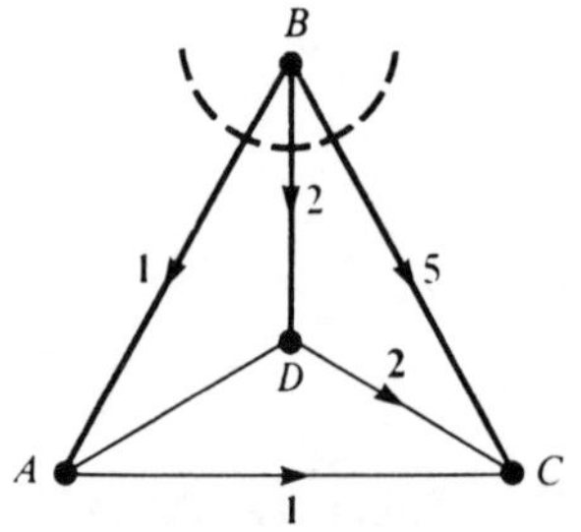

(ii) write down the effect of the min cut.

min cut has capacity 8 and separates $\{B\}$ and $\{A, C, D\}$

(iii) draw branch with capacity 8 and vertices B and $\boxed{CD}$:

(iv) join branch to tree so that min cuts in the tree and in the network have the same effect.

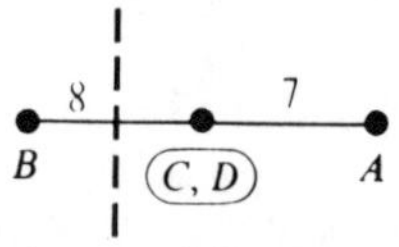

Note that, in the min cut in the network, the vertices C and D belong to the same part as A, so we add the branch containing A to the generalized vertex $\boxed{CD}$, not to the vertex B.

We now move on to Step 3.

We split up the generalized vertex $\boxed{CD}$, so we do a max flow calculation from C to D.

STEP 3

Consider the remaining 2 vertices, C and D.

(i) find a max flow from C to D and a corresponding min cut:

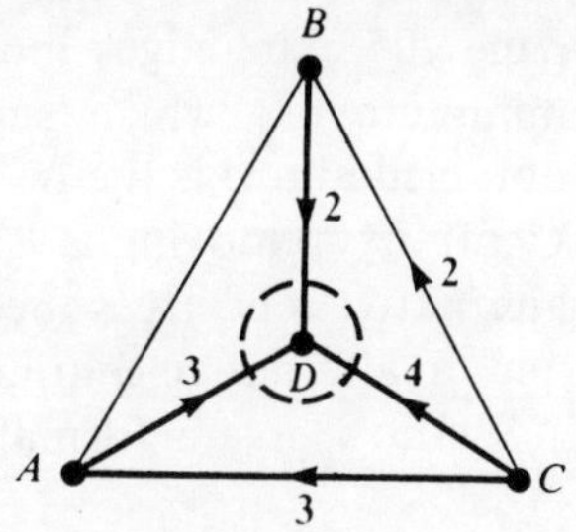

(ii) write down the effect of the min cut:

min cut has capacity 9 and separates $\{D\}$ and $\{A, B, C\}$.

(iii) draw branch with capacity 9 and vertices D and C:

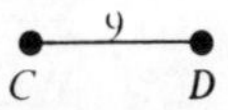

(iv) join branch to tree so that min cuts in the tree and in the network have the same effect:

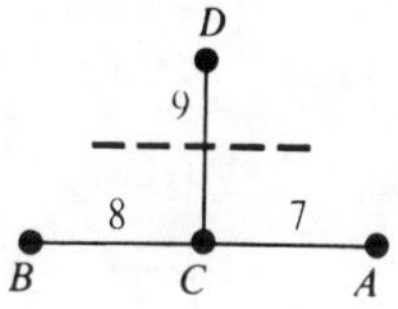

Notice that for the final diagram to be a cut tree, we must be able to draw a cut separating D from the other vertices. Thus the tree cannot have all the vertices strung out in a line $BCDA$ or in a line $BDCA$: the only possibility is the tree shown above on the right. This is the required cut tree.

Remarks You may be wondering why in the above algorithm we first find a max-flow between vertices when it would suffice to simply find a min-cut. In this simple case, we can find a min-cut by inspection, but in general we cannot. However, we can find a min-cut by using the max-flow algorithm. The final iteration of the labelling procedure gives us a min-cut which separates the labelled vertices plus the START vertex from the other vertices.

In the above example, in Step 1 we could have obtained a different maximum flow pattern from A to B, by again saturating the arcs AB, AD and AC, but sending a flow of 1 along DB, 2 along DC, and 5 along CB. In this case, we would still have obtained the *same* minimum cut. At Step 2, we could have chosen any two vertices from B, C and D. You might like to try working through this example making some different choices on the way. You should end up with the same cut tree.

You can see why this must be so if you look back at the network N on page 105. The edges incident to C have total capacity 12—so there is no minimum cut which separates C from the other vertices: the cut tree obtained above is the only flow-equivalent tree in which we cannot separate C off by removing a single branch. (In other examples, there may be alternative cut trees because there may be alternative maximum flows and/or alternative minimum cuts at one or more steps in the calculation.)

We now give a formal statement of the basic algorithm.

THE BASIC ALGORITHM OF GOMORY AND HU

START with a network with n vertices.

STEP 1 Choose any 2 vertices, say A and B.

(i) Find a maximum flow f_1 from A to B.

(ii) Write down the two parts X and Y of a corresponding minimum cut.

(iii) Draw 2 generalized vertices, one corresponding to X and one corresponding to Y; join them with an edge having capacity f_1.

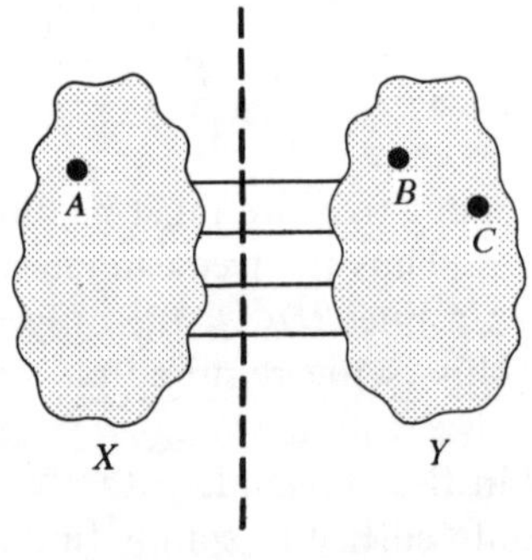

removal of arcs of min cut splits network into 2 parts, X and Y

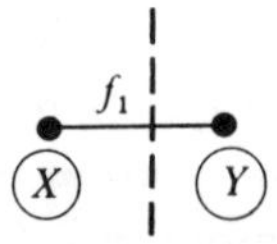

removal of edge splits tree into 2 parts, corresponding to X and Y

(In both the original network and the tree, the removal of the mincut separates the vertices into the same two parts X and Y.)

STEP 2 Choose any 2 vertices from one of the generalized vertices (both from X or both from Y), say B and C in Y.

(i) Find a maximum flow f_2 from B to C.

(ii) Write down the two parts of a corresponding minimum cut.

(iii) Draw two generalized vertices, corresponding to the two parts Y_1 and Y_2 into which the minimum cut divides Y; join them with an edge having capacity f_2.

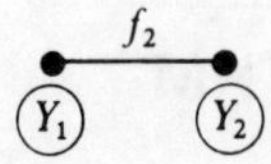

(iv) Attach the branch found in Step 1:

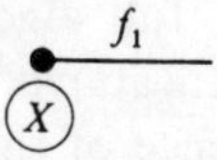

to Y_1 if X is in the same part of the cut as Y_1

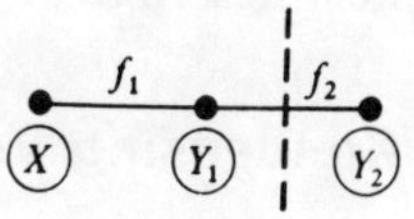

to Y_2 if X is in the same part of the cut as Y_2

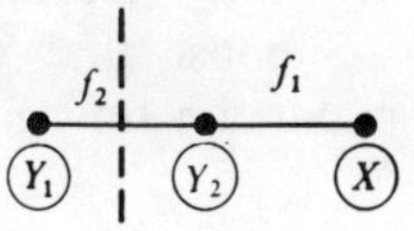

ONE OF THESE CASES MUST OCCUR.

Now continue splitting up the generalized vertices and adding branches in this way until no generalized vertex remains—this will happen after $n - 1$ steps. STOP

Points to Note when Applying the Algorithm

(a) When splitting up a generalized vertex containing more than two vertices, it does not matter which two vertices you choose.
(b) There may be more than one way of achieving a maximum flow—choose any.
(c) For a particular maximum flow, there may be more than one minimum cut—choose any.
(d) For a given undirected network, there may be more than one cut tree.

Problem 5.6 Use the algorithm to find a cut tree for the following network by first finding a maximum flow from A to B:

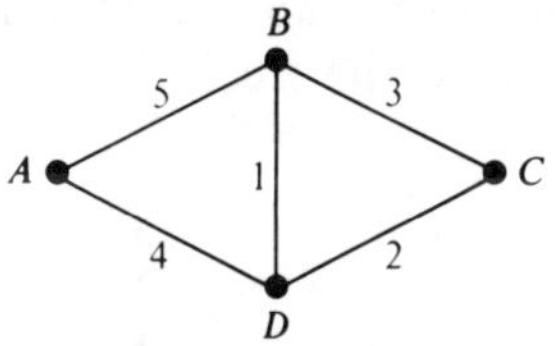

(Remember that, at each stage, when a new branch is added to the tree, the branch MUST SEPARATE the two sets of vertices of the corresponding minimum cut.)

5.4 THE FULL ALGORITHM

The full algorithm of Gomory and Hu improves on the basic algorithm given above by simplifying some of the maximum flow calculations. The essence of their technique is a process called *condensing vertices*, which we now describe.

Condensing Vertices

When finding a cut tree for a large network, we can simplify some of the max-flow calculations by condensing certain vertices into a single vertex.

Suppose that when applying the algorithm we have a min cut for a maximum flow between two vertices A and B.

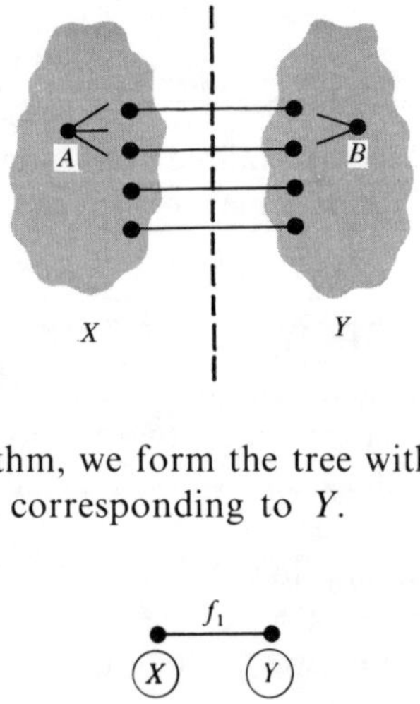

In Step 1 of the algorithm, we form the tree with one vertex corresponding to X and one vertex corresponding to Y.

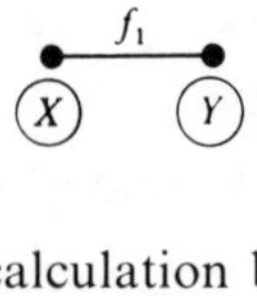

In Step 2, we do a max flow calculation between 2 vertices in Y, and use the corresponding min cut to split up the set Y into two parts, Y_1 and Y_2. When we add our branch to the tree,

EITHER X gets tacked on to Y_1

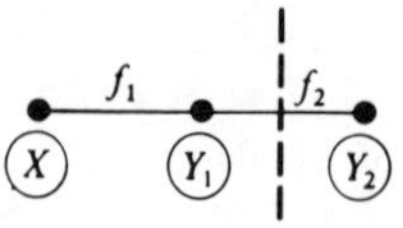

OR X gets tacked on to Y_2

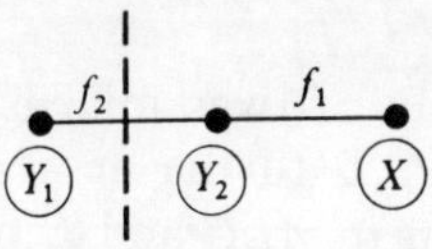

The essential property which allows us to condense vertices is that *all the vertices in X are in the same part of this second min cut*. This is the statement that Gomory and Hu proved; it leads directly to the following theorem.

Condensing theorem *In a max-flow calculation between any two vertices in the part Y of a min cut X, Y, all the vertices in X may be condensed to a single vertex to which all arcs of the cut are joined, provided that the capacities of the arcs are modified so that the capacity of the min cut is unchanged.*

Example 2 Consider the following example in which the five thick edges in the middle form a min cut with capacity 11 for flow from A to B:

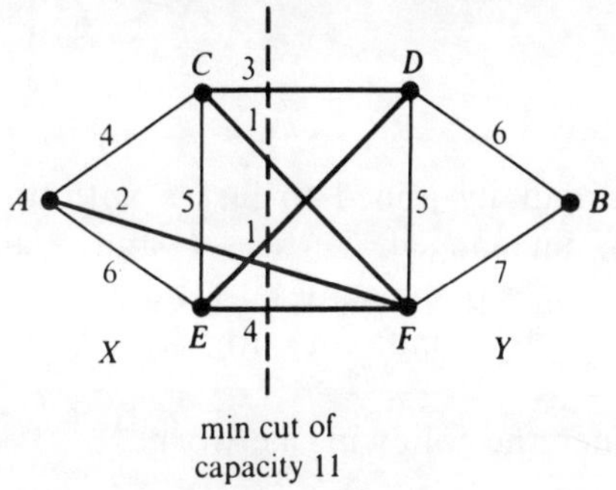

The cut separates the graph into 2 parts: the set X containing the vertices A, C and E, and the set Y containing the vertices B, D and F.

The theorem tells us that, if we want to find a max flow between any 2 vertices in Y, then we may condense the vertices in X to a single vertex $\boxed{ACE}$, as shown in the following diagram.

Condense vertices in X:

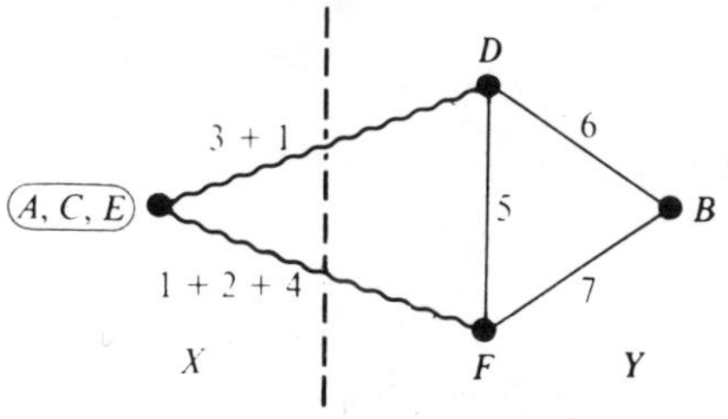

Now the vertex B was not originally joined to any vertex in X, so B is not joined to $\boxed{ACE}$.

Originally, D was joined to C and to E by edges of capacities 3 and 1, so we join D to the new vertex with an edge of capacity $3 + 1$. Also, F was joined to A, C and E by edges of capacities 1, 2 and 4, so F is joined to the new vertex by an edge of capacity $1 + 2 + 4$. Notice that X and Y are still separated by a min cut of capacity 11.

Similarly, the theorem tells us that, if we want to find a max flow between any 2 vertices in X, we may condense the vertices in Y to a single vertex $\boxed{BDF}$, as in the following diagram:

Condense vertices in Y:

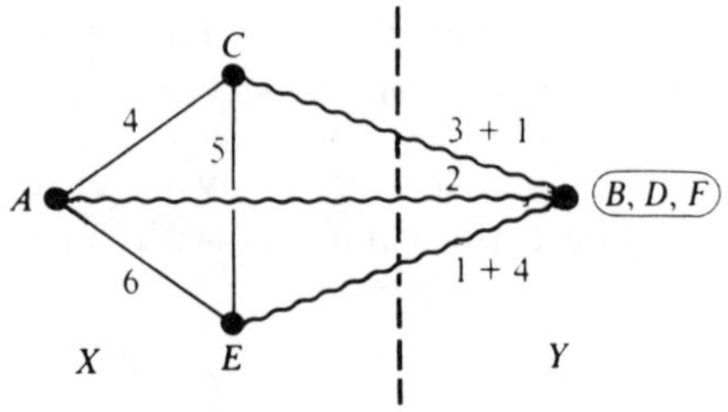

The vertex A was originally joined to just F with an edge of capacity 2, so this edge remains unchanged. Vertices C and E are joined to the new condensed vertex by edges of capacities $3 + 1$ and $4 + 1$ respectively. Again, X and Y are separated by a min cut with capacity 11.

Problem 5.7 Consider the following network:

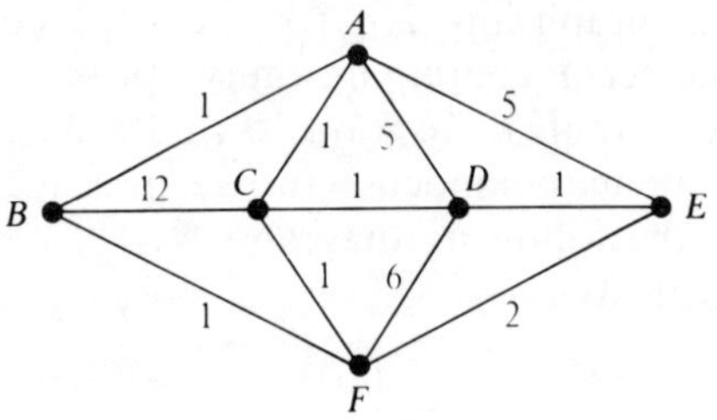

(a) Condense the vertices A, D, E and F.
(b) Condense the vertices B and C.
(c) Condense the vertices B, C, D and F.
(d) Condense the vertices B and C to a single vertex and then condense the vertices A and E to a single vertex.

The Full Algorithm

The full algorithm of Gomory and Hu uses the technique of condensing vertices to simplify the calculations of maximum flow. Thus the statement of the full algorithm is obtained from that of the basic algorithm by changing Step 2, part (i) to the following:

(i) Condense the vertices in X and then find a maximum flow from B to C. (If the vertices chosen are both in X, we condense the vertices in Y.)

We now illustrate the use of the full algorithm with the following worked problem.

Worked problem Use the full algorithm of Gomory and Hu to find a cut tree for the following network:

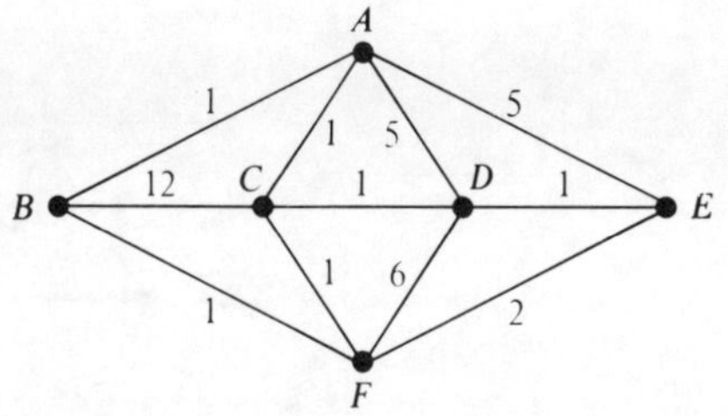

Solution The given network is the same as the network in Problem 5.7; the results of condensing the vertices obtained in that problem are needed in the application of the algorithm. The steps of the algorithm are shown on the right-hand page of the following two-page spread. On the left-hand page, we give the solution to Problem 5.7; each part appears opposite to the place where it is used in the algorithm. On the right-hand page, we list the steps of the algorithm on the left, and build up the cut tree down the right-hand side.

Problem 5.8 Use the full algorithm of Gomory and Hu to find a cut tree for the following network:

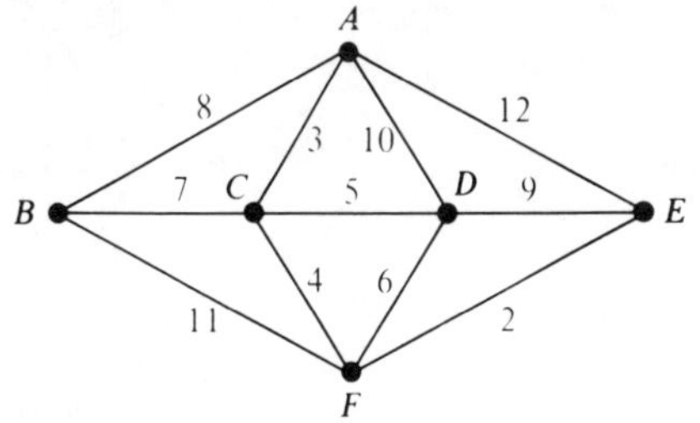

Solution to Problem 5.7

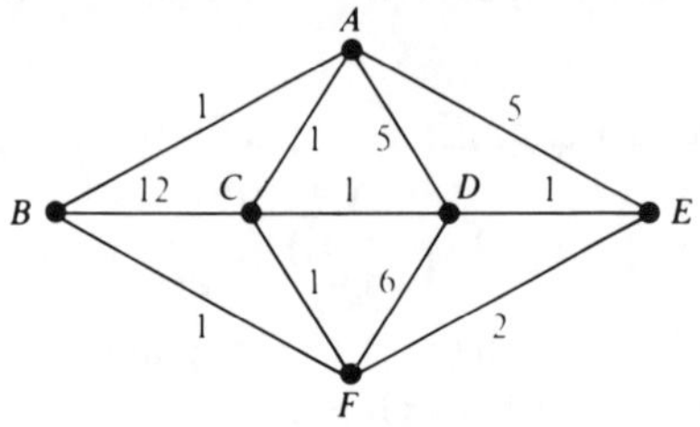

(i) Condense
A, *D*, *E*, *F*.

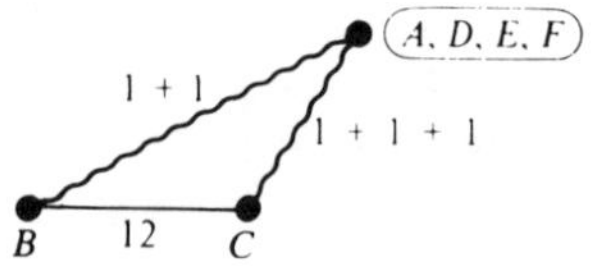

(ii) Condense
B and *C*.

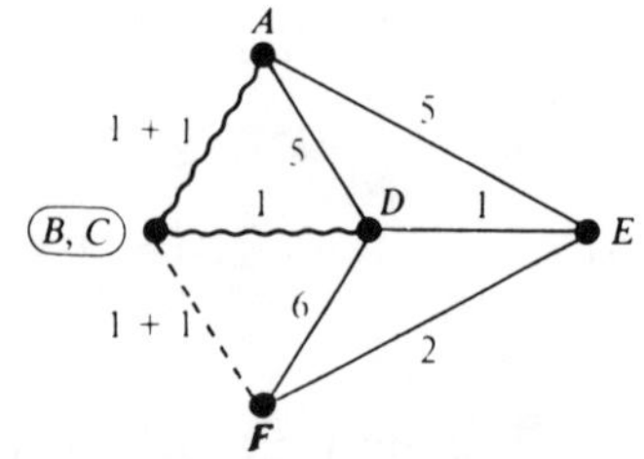

(iii) Condense
B, *C*, *D*, *F*.

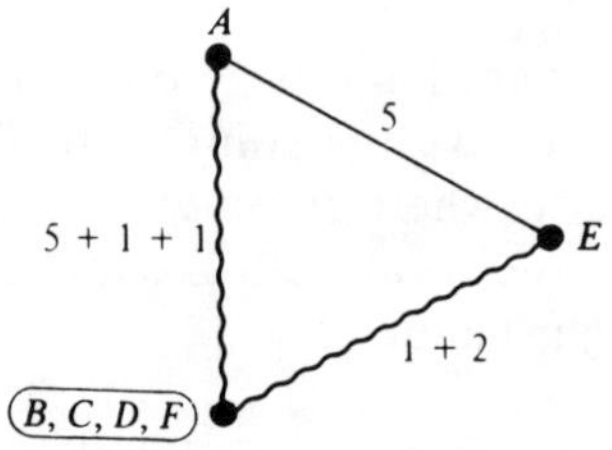

(iv) Condense
B and *C*
and
A and *E*.

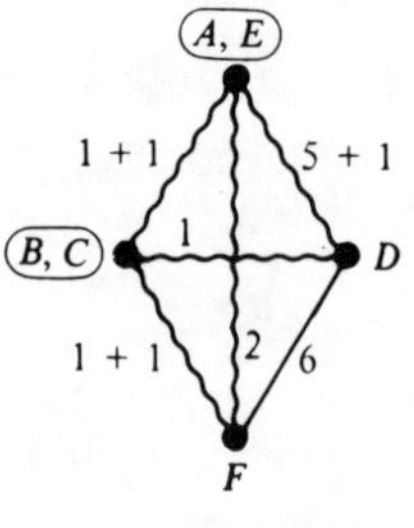

Solution to Worked Problem

STEP 1 Find max flow from A to B.

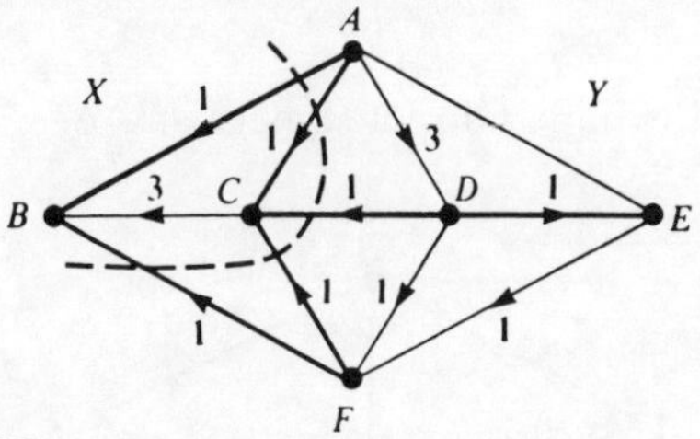

max flow is 5; min cut separates $\{B, C\}$, $\{A, D, E, F\}$.

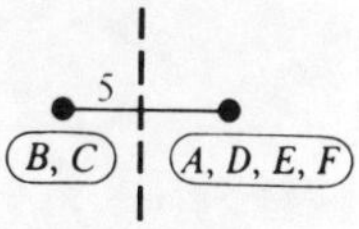

STEP 2 Find max flow from B to C.

Condense A, D, E, F.

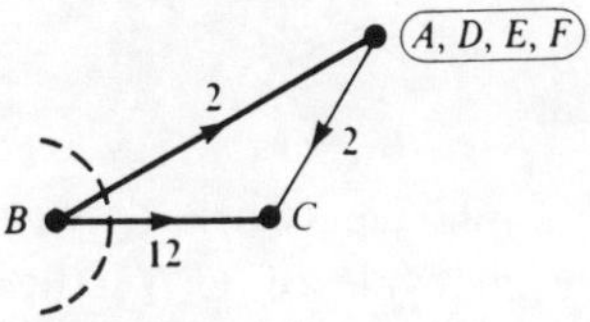

max flow is 14; min cut separates $\{B\}$, $\{A, C, D, E, F\}$.

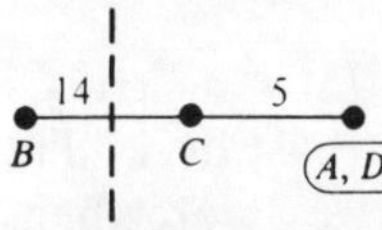

STEP 3 Find max flow from A to D.

Condense B, C.

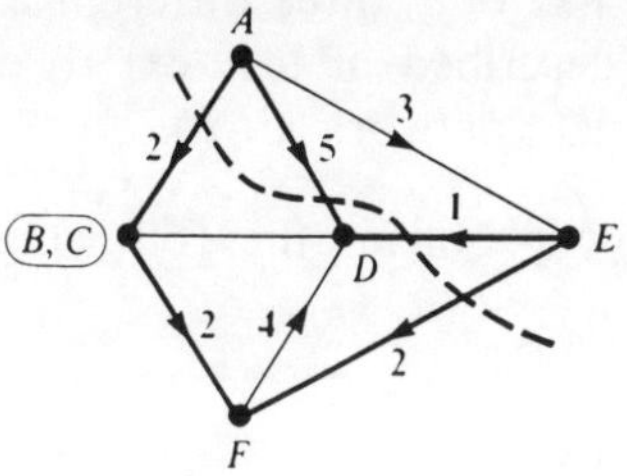

max flow is 10; min cut separates $\{A, E\}$, $\{B, C, D, F\}$.

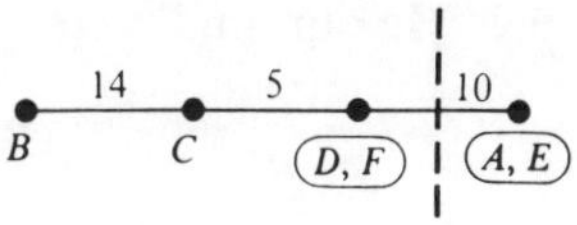

STEP 4 Find max flow from A to E.

Condense B, C, D, F.

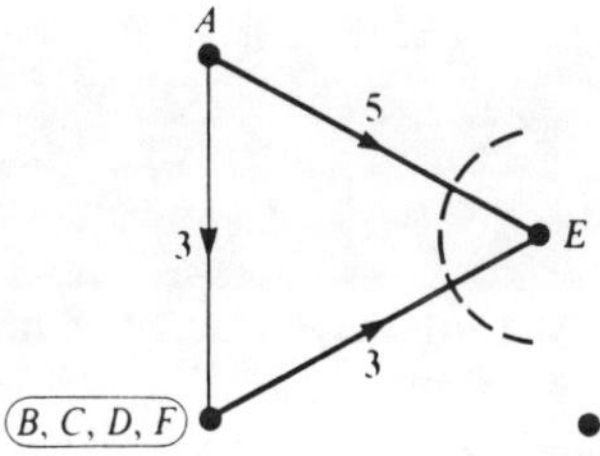

max flow is 8; min cut separates $\{E\}$, $\{A, B, C, D, F\}$.

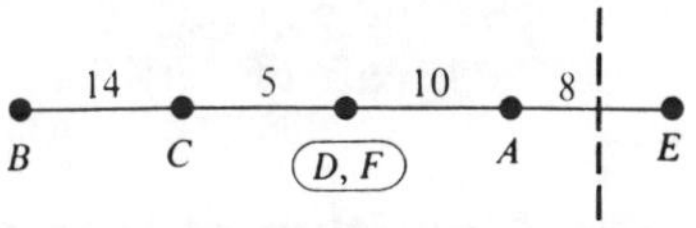

STEP 5 Find max flow from D to F.

Condense A, E.
Condense B, C.

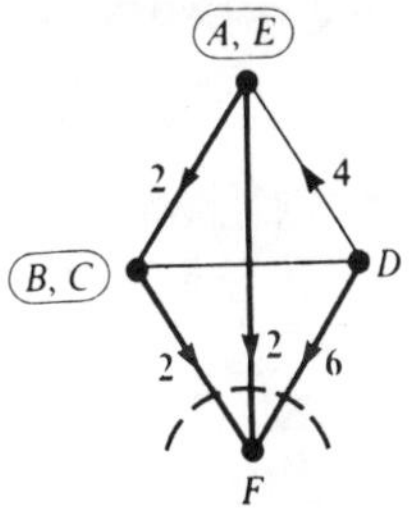

max flow is 10; min cut separates $\{F\}$, $\{A, B, C, D, E\}$.

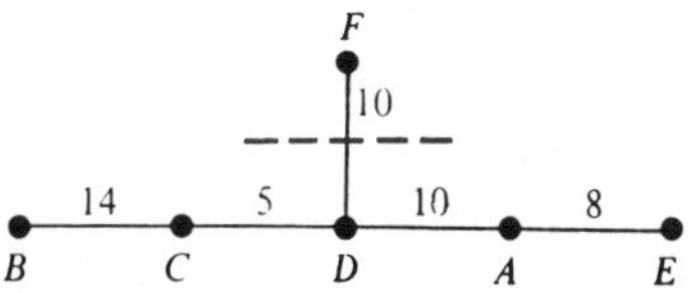

EXERCISES 5

The Capacity Rule

5.1. Consider the following undirected network N:

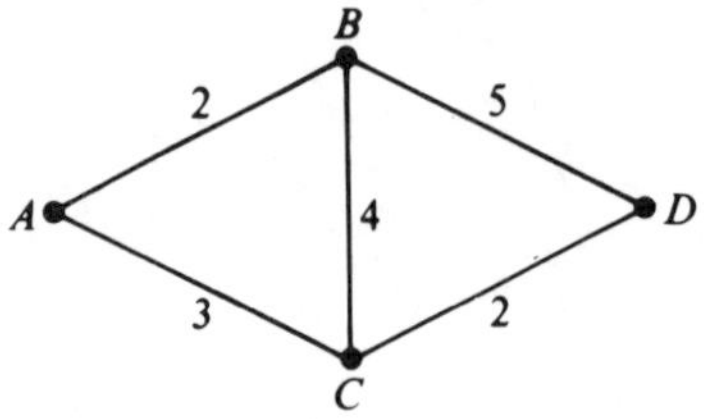

(a) Construct the complete network $C(N)$.
(b) Check that in each cycle of $C(N)$ the smallest capacity occurs more than once.
(c) Check that $C(N)$ has only three different capacities.
(d) Use the method described in the text to construct a maximum tree for the complete network $C(N)$.

5.2. In an undirected flow problem, the following maximum tree was obtained:

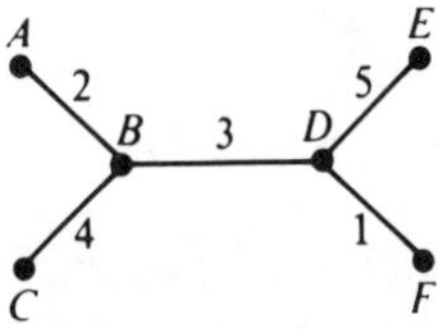

Use the capacity rule to find the value of a maximum flow between each pair of vertices.

The Algorithm of Gomory and Hu

5.3. Consider the following undirected network N:

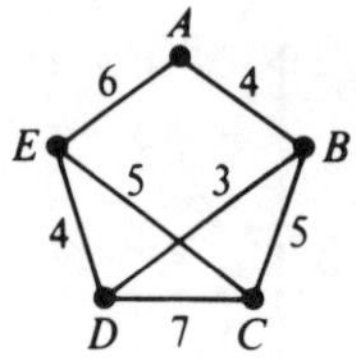

(a) Use the basic algorithm of Gomory and Hu to find a cut tree for N.
(b) Use the capacity rule to determine the value of a maximum flow between each pair of vertices.

5.4. Consider the following undirected network:

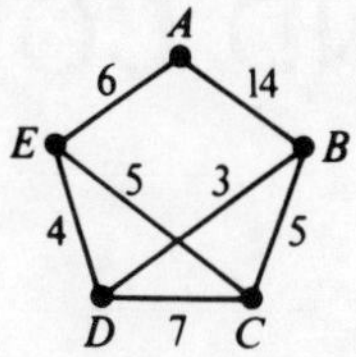

Find the network obtained by:

(a) condensing the vertices C, D and E;
(b) condensing the vertices A and B.

5.5. Use the full algorithm of Gomory and Hu to find a cut tree for the network in Exercise 5.4.

5.6. (a) Use the full algorithm of Gomory and Hu to find a cut tree for the following undirected network:

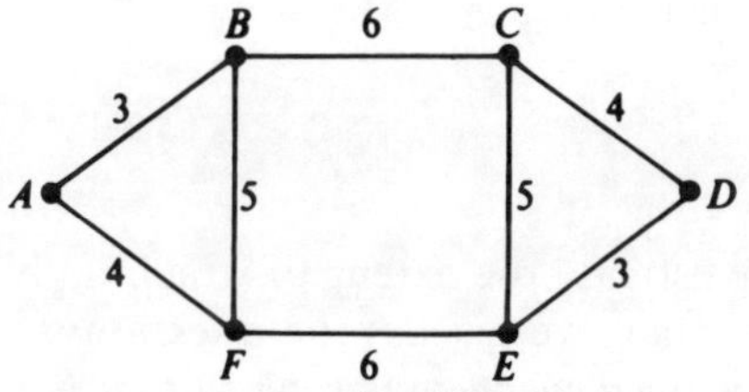

(b) Use the cut tree to write down the values of the maximum flows between each pair of vertices.

CHAPTER 6

PATHS AND CONNECTIVITY

After reading this chapter, you should be able to:

- explain the terms *edge-connectivity*, *vertex-connectivity*, *k-edge-connected*, *k-connected*, *bridge*, *cut-vertex*, *cutset* and *vertex-cutset*;
- understand the various forms of Menger's theorem for graphs and digraphs, and explain how the edge-form for digraphs is related to the max-flow min-cut theorem for networks;
- discuss the reliability of a telecommunication network in terms of the concepts introduced in this chapter.

Introduction

In this chapter we investigate the extent to which a given graph or digraph is connected. In particular, we discuss the question: *how many edges do we need to remove from a given connected graph so that it becomes disconnected?* This, and other similar questions related to connectivity, are important ones to consider when designing telecommunication networks, road systems and other networks—for example, in a telecommunication network it is essential that the network should still be operable if some of the links between the exchanges become damaged, or are blocked by other calls. After discussing the theory of connectivity in graphs and digraphs in general, we describe its relevance to such networks.

6.1 CONNECTED GRAPHS AND DIGRAPHS

In Chapter 2 we introduced the idea of a *connected graph*—that is, a graph which is 'in one piece'. We noted that in any connected graph there is a path between any given pair of vertices, and this led to the following definitions.

Definitions *A graph G is **connected** if there is a path in G between any given pair of vertices, and **disconnected** otherwise. Every disconnected graph can be split up into a number of connected subgraphs, called **components**.*

For example:

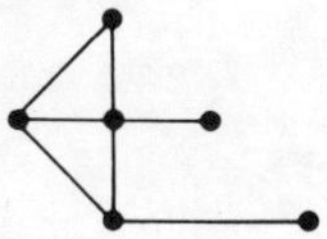

a connected graph

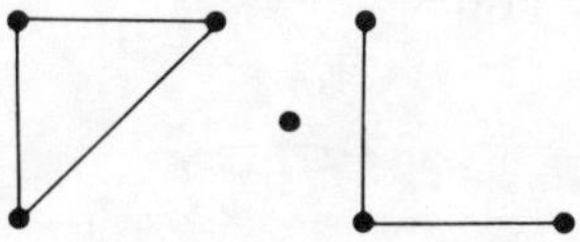

a disconnected graph with 3 components

In the case of digraphs, it is not true in general that a digraph which is 'in one piece' must have a (directed) path between any given pair of vertices, and this observation led us to define two types of connected digraph, as follows.

Definitions *A digraph D is **connected** if its underlying graph is connected, and **disconnected** otherwise. It is **strongly-connected** if there is a path in D from any given vertex to any other.*

For example:

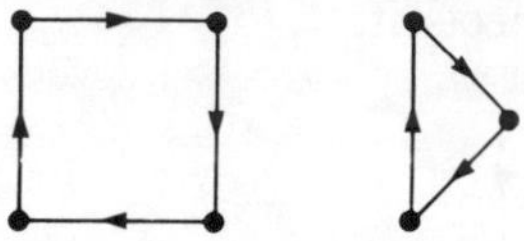

a disconnected digraph

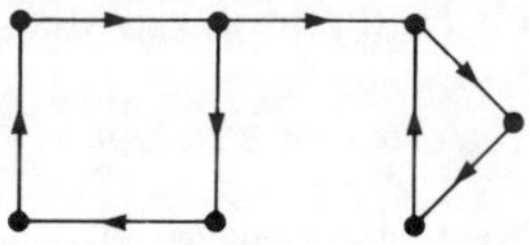

a connected digraph
(not strongly-connected)

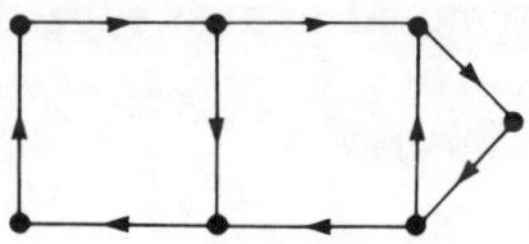

a strongly-connected
digraph

Edge-connectivity

For many applications it is necessary to know more about a graph than just whether or not it is connected. For example, in telecommunication networks there are usually several different paths between a given pair of subscribers (vertices). In such a situation, it is important to know how many links (edges) can be blocked or broken without preventing a call being made between the two subscribers. In order to answer this and similar questions, we need to investigate connected graphs a little further.

Consider the following graphs:

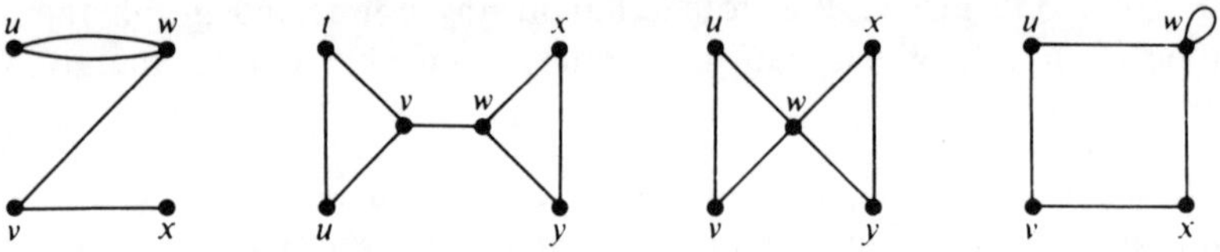

(a) (b) (c) (d)

Graph (a) can be split into two components by removing one of the edges *vw* and *vx*. We say that the removal of either of these edges *disconnects* the graph.

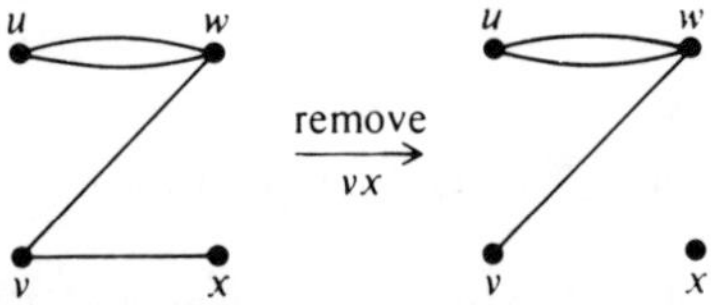

Graph (b) can also be disconnected by the removal of a single edge—the edge *vw*.

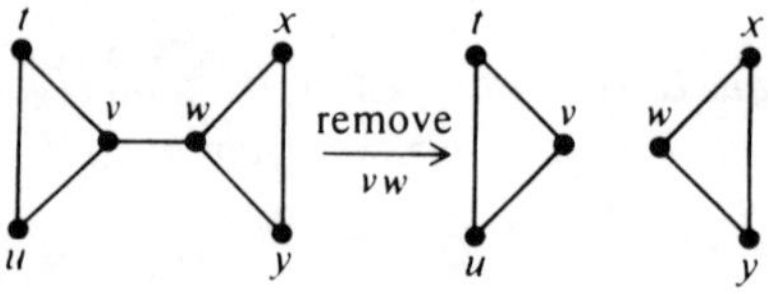

Graph (c) cannot be disconnected by removing a single edge, but the removal of two edges (such as *uw* and *vw*) disconnects it.

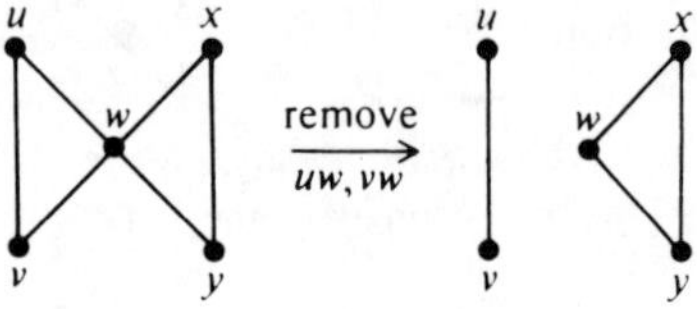

Similarly, graph (d) can be disconnected by removing the two edges *uw* and *wx*.

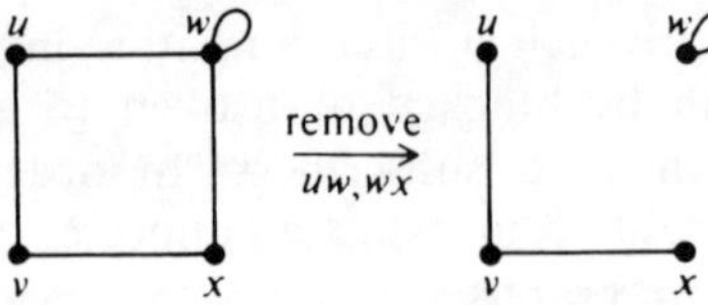

A single edge whose removal disconnects a graph, such as *vw* or *vx* in graph (a), or *vw* in graph (b), is called a **bridge**.

With these examples in mind, we define the *edge-connectivity* of a graph as follows.

Definitions *The* ***edge-connectivity*** $\lambda(G)$ *of a connected graph G is the smallest number of edges whose removal disconnects G. When* $\lambda(G) \geq k$, *the graph G is said to be* ***k-edge-connected***.

Thus, graphs (a) and (b) have edge-connectivity 1, and graphs (c) and (d) have edge-connectivity 2. All four graphs are 1-edge-connected; graphs (c) and (d) are 2-edge-connected, but not 3-edge-connected.

If we wish to disconnect a graph by removing edges from it, we usually have a choice of edges to delete. In view of this, it seems natural to consider ways of disconnecting a graph which do not involve 'redundant edges'.

Consider the following graph *G*.

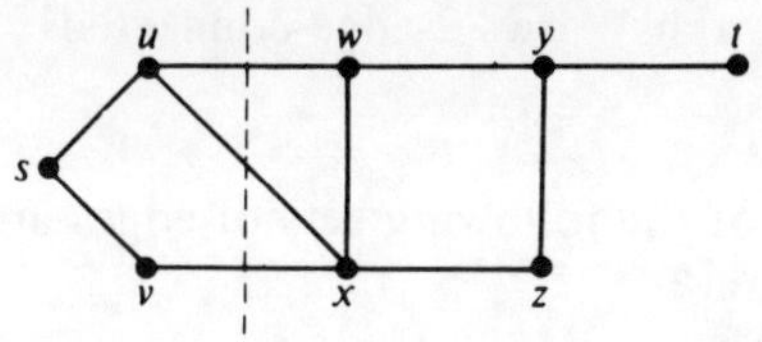

We can disconnect *G* by removing the three edges *uw*, *ux* and *vx*, but we cannot disconnect it by removing just two of these edges. We can also disconnect *G* by removing the edges *uw*, *wx*, *xz* and *yz*, but the edge *yz* is redundant here since we need remove only the edges *uw*, *wx* and *xz* to disconnect *G*. A set of edges in which no edge is redundant—for example, {*uw*, *ux*, *vx*} or {*uw*, *wx*, *xz*}—is called a *cutset*.

Definition *A* ***cutset*** *of a connected graph G is a set S of edges with the following properties*:

(i) *removal of all the edges in S disconnects G*;
(ii) *removal of some (but not all) of the edges in S does not disconnect G.*

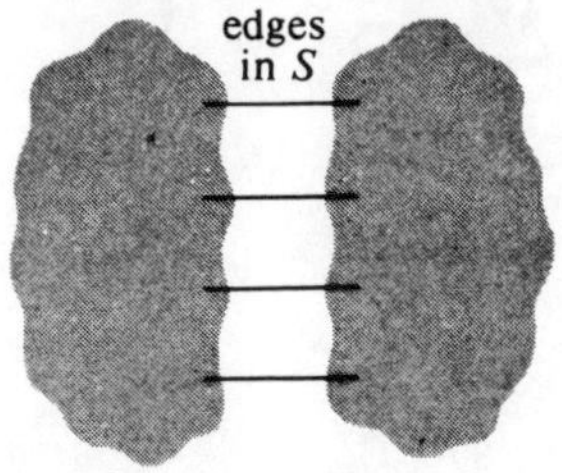

Note that *two cutsets of a graph need not necessarily have the same number of edges*. For example, in the above graph, the sets $\{uw, ux, vx\}$ and $\{wy, xz\}$ are both cutsets. Note also that *the edge-connectivity* $\lambda(G)$ *of a graph G is simply the size of the smallest cutset of G*.

Problem 6.1 Write down the value of $\lambda(G)$ for each of the following graphs G:

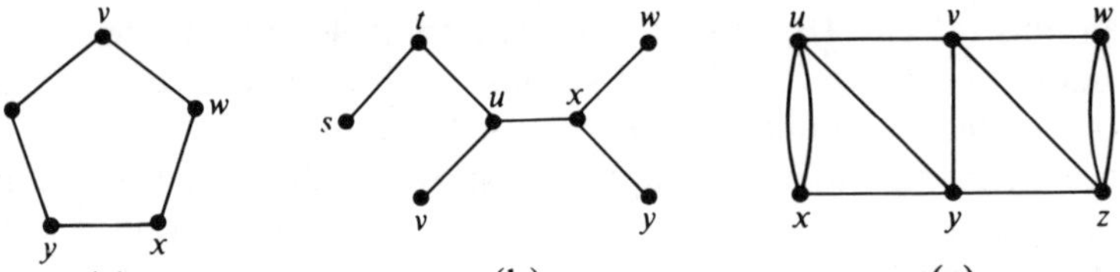

(a) (b) (c)

Which of these graphs are 2-edge-connected? Which are 3-edge-connected?

Problem 6.2 Which of the following sets of edges are cutsets of the graph G on the previous page?

(a) $\{su, sv\}$; (b) $\{ux, wx, yz\}$; (c) $\{ux, vx, wx, yz\}$;
(d) $\{yt\}$; (e) $\{wx, xz, yz\}$; (f) $\{uw, wx, wy\}$.

Vertex-connectivity

We can also think of connectivity in terms of the minimum number of *vertices* which need to be removed in order to disconnect a graph. When we remove a vertex, we must also remove the edges incident to it:

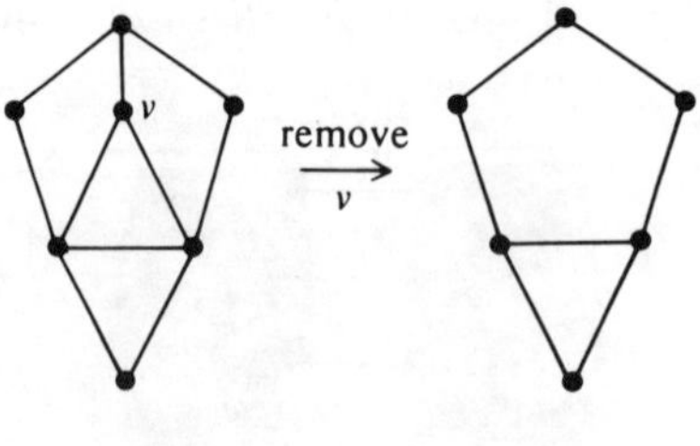

Consider again the graphs (a)–(d).

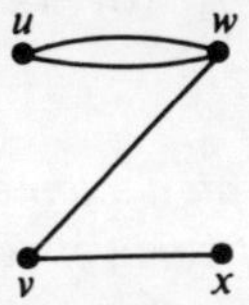

(a)

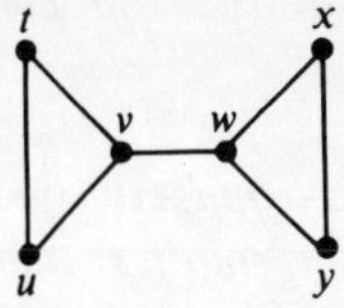

(b)

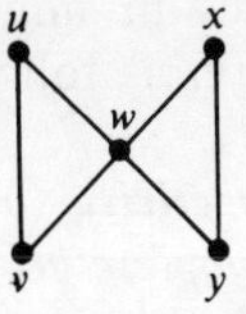

(c)

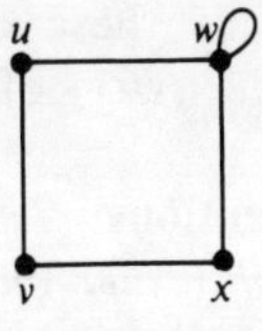

(d)

Graphs (a) and (b) can be disconnected by the removal of a single vertex (either *v* or *w*).

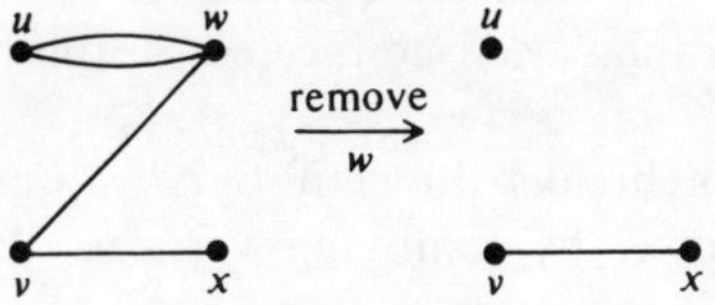

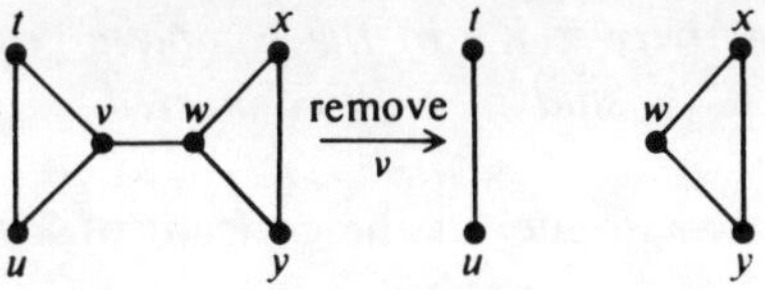

Graph (c) can also be disconnected by removing just one vertex (the vertex *w*).

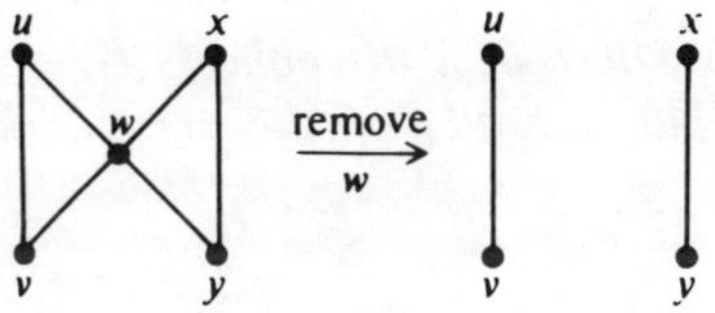

Graph (d) cannot be disconnected by removing a single vertex, but the removal of two non-adjacent vertices (such as *v* and *w*) disconnects it.

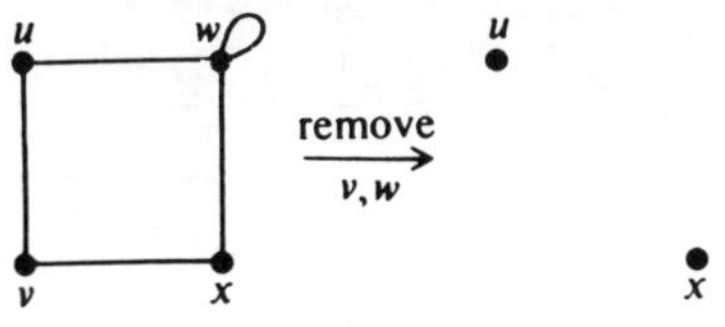

A single vertex whose removal disconnects a graph, such as *v* or *w* in graph (b), or *w* in graph (c), is called a **cut-vertex**.

With these examples in mind, we define the *connectivity* (or *vertex-connectivity*) of a graph as follows.

Definitions *The* ***connectivity*** *(or* ***vertex-connectivity****)* $\kappa(G)$ *of a connected graph (other than a complete graph) is the smallest number of vertices whose removal disconnects G. When* $\kappa(G) \geqslant k$*, the graph is said to be* ***k-connected*** *(or* ***k-vertex-connected****).*

(We use the simpler terms *connectivity* and *k-connected* when there is no possibility of confusion with *edge-connectivity* and *k-edge-connected.*)

Thus graphs (a), (b) and (c) have connectivity 1, and graph (d) has connectivity 2. All four graphs are 1-connected; graph (d) is 2-connected, but not 3-connected.

The above definition breaks down if G is a complete graph, since we cannot then disconnect G by removing vertices. We therefore make the following definition.

Definition *The* ***connectivity*** $\kappa(K_n)$ *of the complete graph* K_n *is* $n - 1$*. When* $n - 1 \geqslant k$*, the graph* K_n *is said to be* ***k-connected****.*

There is also a 'vertex analogue' of the concept of a cutset. This is defined as follows.

Definition *A* ***vertex-cutset*** *of a connected graph G is a set S of vertices with the following properties:*

(i) *removal of all the vertices in S disconnects G;*
(ii) *removal of some (but not all) of the vertices in S does not disconnect G.*

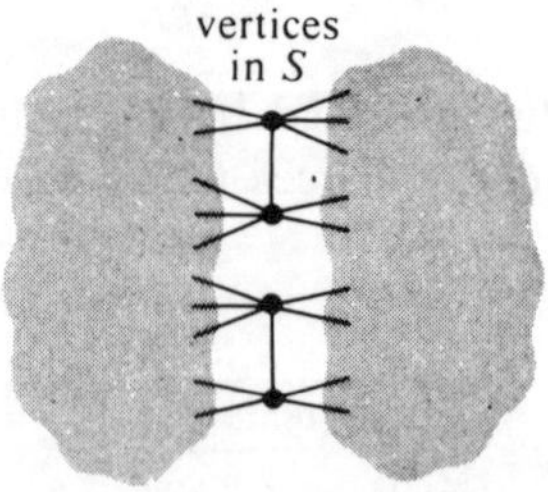

For example, we can disconnect the following graph by removing the two vertices u and x, but we cannot disconnect it by removing just one of these vertices. It follows that $\{u, x\}$ is a vertex-cutset.

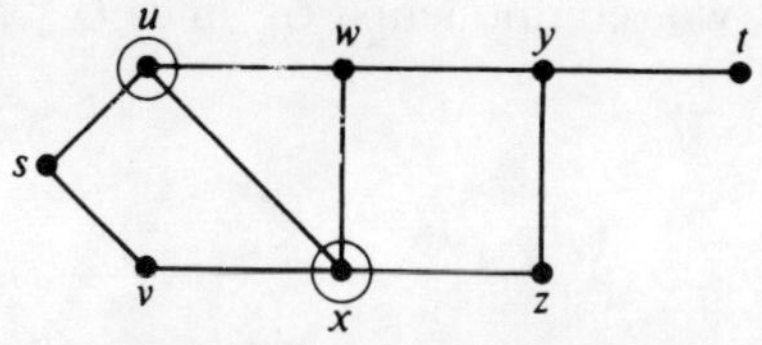

Note that *two vertex-cutsets of a graph need not necessarily have the same number of vertices.* For example, in the above graph, the sets $\{u, x\}$ and $\{y\}$ are both vertex-cutsets. Note also that *the connectivity $\kappa(G)$ of a graph G is simply the size of the smallest vertex-cutset of G.*

Problem 6.3 Write down the value of $\kappa(G)$ for each of the following graphs G:

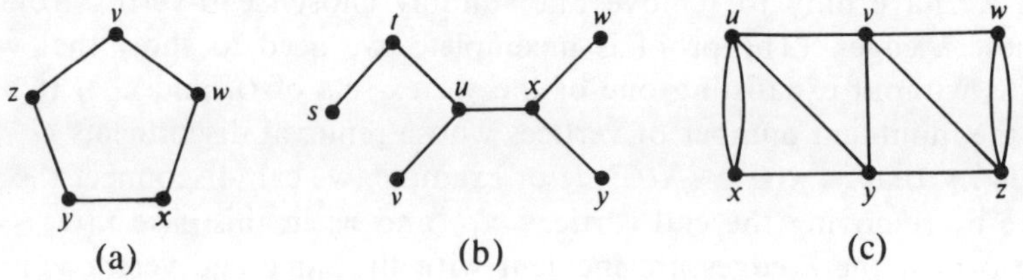

Which of these graphs are 2-connected? Which are 3-connected?

Problem 6.4 Which of the following sets of vertices are vertex-cutsets of the graph at the top of the page?

(a) $\{u, v\}$; (b) $\{v, w\}$; (c) $\{u, x, y\}$; (d) $\{w, z\}$.

In the above examples, you may have noticed that the connectivity $\kappa(G)$ does not exceed the edge-connectivity $\lambda(G)$. This inequality holds, in fact, for all connected graphs.

Theorem 6.1 *For any connected graph G,*

$$\kappa(G) \leqslant \lambda(G) \leqslant \delta(G),$$

where $\delta(G)$ is the smallest vertex-degree in G.

Outline of proof If G is K_n, then $\kappa(G) = \lambda(G) = \delta(G) = n - 1$.

If G is not K_n, and v is a vertex of degree $\delta(G)$, then G can be disconnected by removing all the $\delta(G)$ edges incident with v. It follows that $\lambda(G)$, the *minimum* number of edges whose removal disconnects G, cannot exceed $\delta(G)$. So $\lambda(G) \leqslant \delta(G)$.

It remains to show that $\kappa(G) \leqslant \lambda(G)$. Let G be a graph with edge-connectivity λ. Then there is at least one set of λ edges whose removal disconnects G into two components G_1 and G_2, as illustrated by the following diagram.

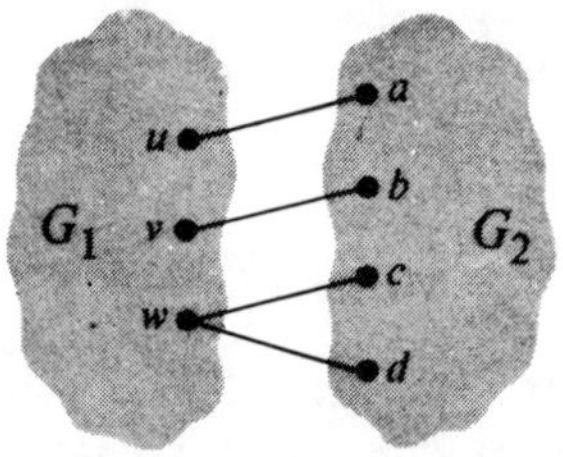

But we can also remove these edges by removing at most λ vertices, since we have only to remove one suitably chosen end-vertex from each of these λ edges. (The proof is incomplete: we need to show that we can do this without exhausting one of the vertex-sets of G_1 and G_2.) It follows that the minimum number of vertices whose removal disconnects G cannot exceed λ—that is, $\kappa(G) \leq \lambda(G)$. (For example, we can disconnect the graph above by removing the end-vertices a, b and w; in this case $\kappa(G) < \lambda(G)$, since two of the λ edges are incident with the same end-vertex w.) □

Note that it is possible for both inequalities in Theorem 6.1 to be strict inequalities (that is, $\kappa(G) < \lambda(G) < \delta(G)$). For example, in the following graph, $\kappa(G) = 1$, $\lambda(G) = 2$ and $\delta(G) = 3$:

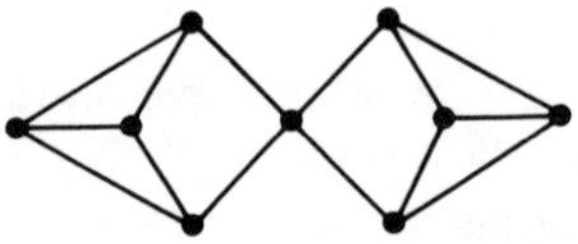

6.2 MENGER'S THEOREM FOR GRAPHS (EDGE-FORM)

In this section we discuss an important result which relates the above ideas to the number of 'disjoint paths' between two vertices in a graph. This result is known as *Menger's theorem*. Some of our discussion here will seem reminiscent of our treatment of the max-flow min-cut theorem in Chapter 3: the connection between these two theorems will emerge as we progress.

We start by defining *disjoint paths* in a graph.

Definitions *Let G be a connected graph, and let s and t be vertices of G. A path between s and t is called an* **st-path**. *Two or more st-paths are*

edge-disjoint *if they have no edges in common, and* ***vertex-disjoint*** *if they have no vertices in common (apart from s and t).*

For example, in the following diagram,

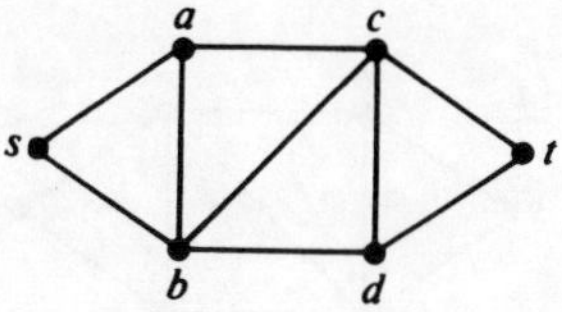

the paths *sact* and *sbdt* are both edge-disjoint and vertex-disjoint *st*-paths;
the paths *sact* and *sbct* are neither edge-disjoint nor vertex-disjoint (since they have the edge *ct* in common);
the paths *sact* and *sbcdt* are edge-disjoint, but not vertex-disjoint (since they have the vertex *c* in common).

Problem 6.5 Consider the following graph:

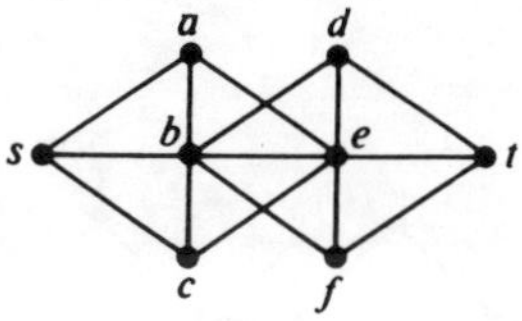

Write down

(a) three edge-disjoint *st*-paths;
(b) two *st*-paths which are edge-disjoint, but not vertex-disjoint;
(c) two vertex-disjoint *st*-paths.

Does this graph contain *three* vertex-disjoint *st*-paths?

Problem 6.6

(a) Prove that if two *st*-paths in a graph are vertex-disjoint, then they must also be edge-disjoint.
(b) Give an example of a graph in which no two edge-disjoint *st*-paths are vertex-disjoint.

We also need the following definitions.

Definitions *Let G be a connected graph, and let s and t be vertices of G. We say that certain* ***edges separate s from t*** *if the removal of these edges* *destroys all paths between s and t. Similarly, we say that certain* **vertices separate s from t** *if the removal of these vertices destroys all paths between s and t.*

For example, in the following graph,

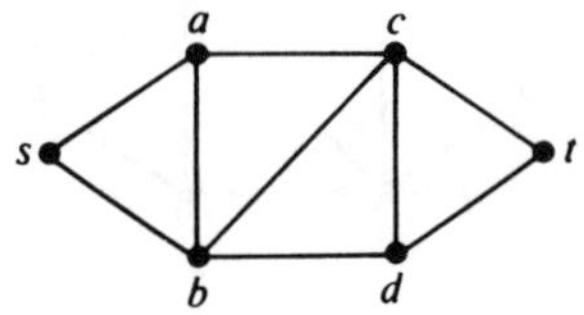

the edges *ac*, *bc* and *bd* separate *s* from *t*, as do the edges *sa*, *ac*, *bc*, *bd* and *dt*;
the vertices *b* and *c* separate *s* from *t*, as do the vertices *a*, *b* and *d*.

We now show how these ideas are related to those of edge-disjoint and vertex-disjoint *st*-paths. Before doing this in general, we motivate our discussion with three examples.

Example 1

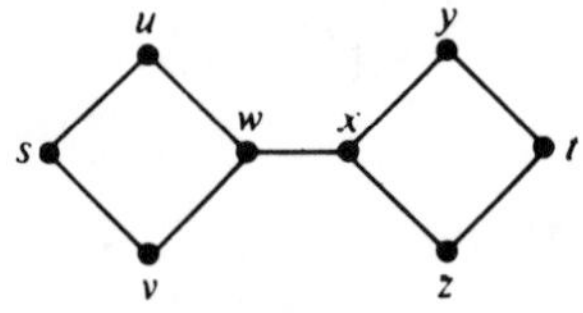

In this graph the single edge *wx* separates *s* from *t*. It follows that *there cannot be two edge-disjoint st-paths*, since all *st*-paths must include the edge *wx*.

Example 2

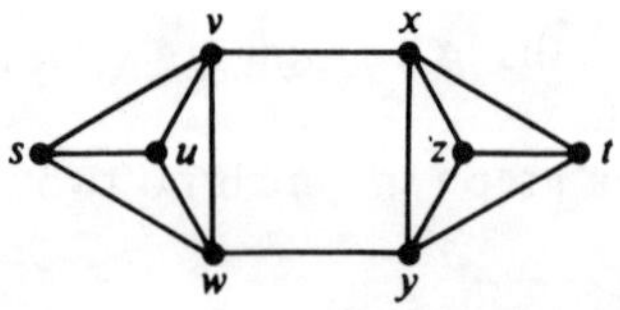

In this graph the two edges *vx* and *wy* separate *s* from *t*. It follows that *there are at most two edge-disjoint st-paths*, since all *st*-paths must include one of these edges.

Example 3

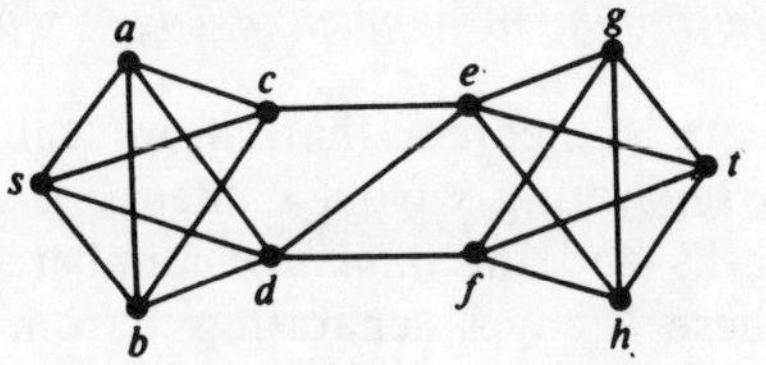

In this graph the three edges *ce*, *de* and *df* separate *s* from *t*. It follows that *there are at most three edge-disjoint st-paths*, since all *st*-paths must include one of these edges.

More generally, consider a set of edges separating *s* from *t* in an arbitrary connected graph.

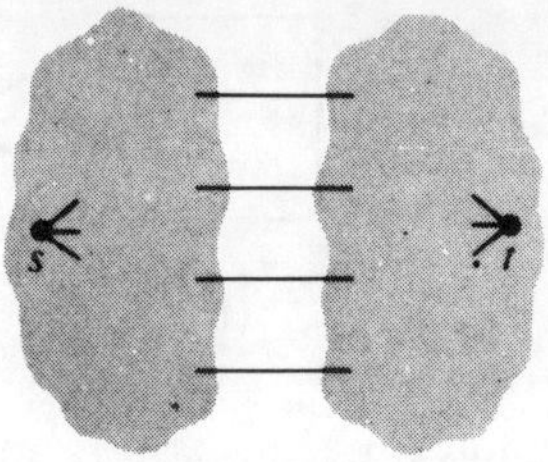

Since the removal of these edges destroys all paths between *s* and *t*, every *st*-path must include at least one of them. It follows that *the maximum number of edge-disjoint st-paths cannot exceed the number of edges in this set*. Since this applies to *any* set of edges separating *s* from *t*, we have

the *maximum* number of edge-disjoint *st*-paths	$\leqslant$	the number of edges in *any* set separating *s* from *t*

But this is true for *any* set of edges separating *s* from *t*, and so it must be true for a set with the smallest possible number of edges. So

the *maximum* number of edge-disjoint *st*-paths	$\leqslant$	the *minimum* number of edges separating *s* from *t*

The above argument should seem familiar, because it is similar to that preceding the max-flow min-cut theorem in Chapter 3. The two numbers in the above inequality are, in fact, always equal. This is the *edge-form of Menger's theorem for graphs*, which may be stated formally as follows.

Menger's theorem for graphs (edge-form) *Let G be a connected graph, and let s and t be vertices of G. Then the maximum number of edge-disjoint st-paths is equal to the minimum number of edges separating s from t.*

It follows from Menger's theorem that, if we can find k edge-disjoint st-paths and k edges separating s from t, then k is the *maximum* number of edge-disjoint st-paths and the *minimum* number of edges separating s from t. Note that these k edges separating s from t necessarily form a cutset. It follows that, when looking for them, we need consider only cutsets whose removal disconnects G into two components, one containing s and the other containing t.

Problem 6.7 By finding k edge-disjoint st-paths, and k edges separating s from t (for the same value of k), and using the edge-form of Menger's theorem, find the maximum number of edge-disjoint st-paths for each of the following graphs:

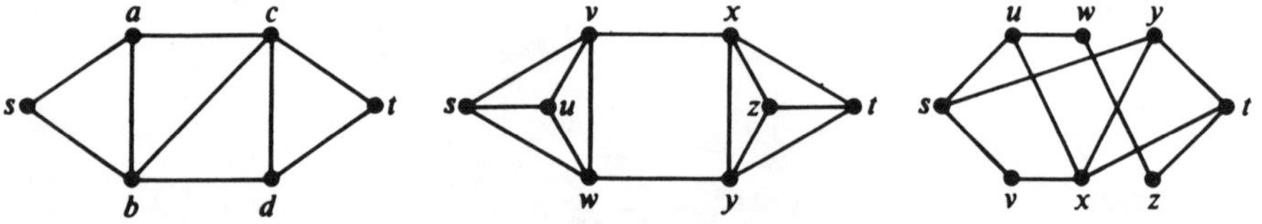

(a) (b) (c)

We can use Menger's theorem to obtain a result about edge-connectivity. Recall that the edge-connectivity $\lambda(G)$ of a connected graph G is the smallest number of edges whose removal disconnects G. So, by Menger's theorem, there are at least $\lambda(G)$ edge-disjoint paths between any given pair of vertices. Since G is k-edge-connected if and only if $\lambda(G) \geqslant k$, we can restate this result as follows.

Corollary of Menger's theorem for graphs (edge-form) *A connected graph G is k-edge-connected if and only if any two vertices of G are connected by at least k edge-disjoint paths.*

We shall find this corollary useful when we discuss telecommunication networks at the end of this chapter.

6.3 SOME ANALOGUES OF MENGER'S THEOREM

We now present some analogues of Menger's theorem, starting with Menger's theorem for digraphs (arc-form), and continuing with the vertex-forms for both graphs and digraphs. At this stage we state all these results without proof. In Section 6.4 we show how Menger's theorem for digraphs (arc-form) can be proved using the max-flow min-cut theorem, and we indicate how some variations of Menger's theorem can be deduced.

Menger's Theorem for Digraphs (Arc-form)

Many of the concepts introduced earlier for graphs have analogues for digraphs. For example, the following definitions are almost identical to those given for graphs.

Definitions *Let D be a connected digraph, and let s and t be vertices of D. A path from s to t is called an* ***st-path****. Two or more st-paths are* ***arc-disjoint*** *if they have no arcs in common, and* ***vertex-disjoint*** *if they have no vertices in common (apart from s and t).*

For example, in the following digraph,

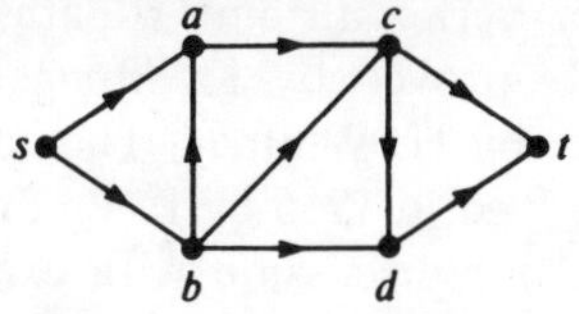

the paths *sact* and *sbdt* are both arc-disjoint and vertex-disjoint *st*-paths;
the paths *sact* and *sbct* are neither arc-disjoint nor vertex-disjoint;
the paths *sact* and *sbcdt* are arc-disjoint but not vertex-disjoint.

We also say that certain **arcs separate *s* from *t*** if the removal of these arcs destroys all paths from *s* to *t*. Similarly, we say that certain **vertices separate *s* from *t*** if the removal of these vertices destroys all paths from *s* to *t*.

For example, in the above digraph,

the arcs *ac*, *bc* and *bd* separate *s* from *t*, as do the arcs *sa*, *ac*, *bc*, *bd* and *dt*;
the vertices *b* and *c* separate *s* from *t*, as do the vertices *a*, *b* and *d*.

Using this terminology, we state Menger's theorem for digraphs:

Menger's theorem for digraphs (arc-form) *Let D be a connected digraph, and let s and t be vertices of D. Then the maximum number of arc-disjoint st-paths is equal to the minimum number of arcs separating s from t.*

As with Menger's theorem for graphs, if we can find *k* arc-disjoint *st*-paths and *k* arcs separating *s* from *t* then *k* is the *maximum* number of arc-disjoint *st*-paths and the *minimum* number of arcs separating *s* from *t*.

Problem 6.8 By finding k edge-disjoint st-paths, and k arcs separating s from t (for the same value of k), and using the arc-form of Menger's theorem, find the maximum number of arc-disjoint st-paths for each of the following digraphs:

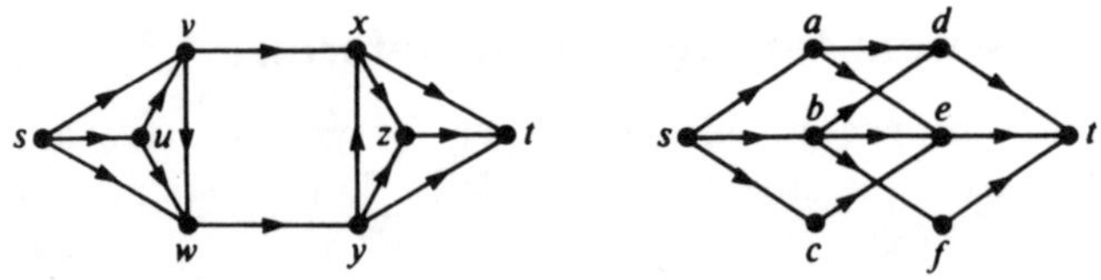

(a) (b)

Menger's Theorem for Graphs (Vertex-form)

We have seen how Menger's theorem (edge-form) relates the number of edge-disjoint st-paths in a graph to the smallest number of edges separating s from t, and how this result relates to edge-connectivity. We now state an analogous theorem for vertex-disjoint st-paths. This is the version of Menger's theorem actually proved by K. Menger in 1927. The corollary was proved five years later by H. Whitney. The edge-form and arc-form of Menger's theorem were proved in 1955 by L. R. Ford and D. R. Fulkerson.

As before, we motivate our discussion with examples.

Example 4

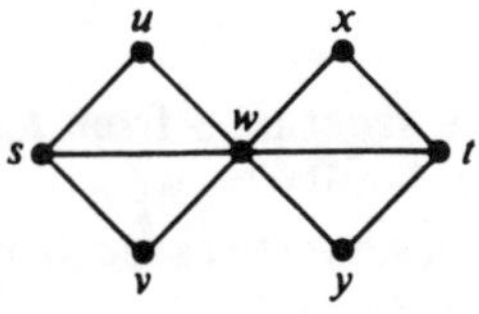

This graph has (vertex-)connectivity 1, and the vertex w separates s from t. It follows that *there cannot be two vertex-disjoint st-paths*, since all st-paths must include the vertex w.

Example 5

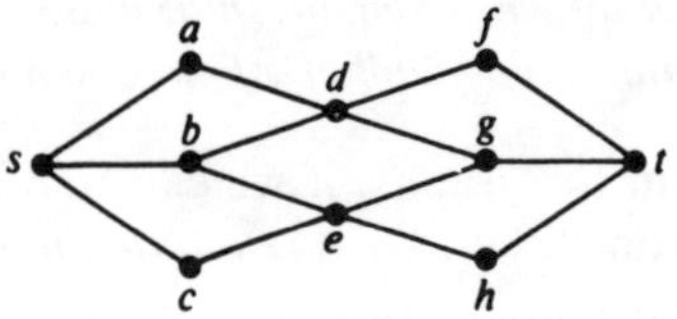

This graph has connectivity 2, and the vertices d and e separate s from t. It follows that *there are at most two vertex-disjoint st-paths*, since all st-paths must include one of these vertices.

More generally, consider a set of vertices separating non-adjacent vertices s and t in an arbitrary connected graph.

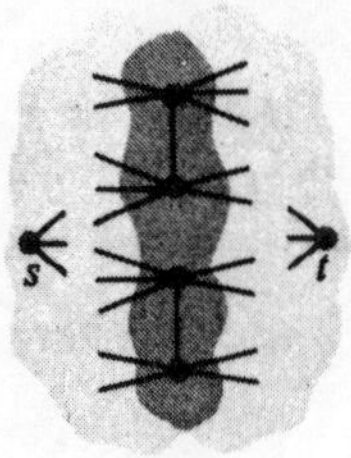

Since the removal of these vertices destroys all paths between s and t, every st-path must include at least one of them. It follows that *the maximum number of vertex-disjoint st-paths cannot exceed the number of vertices in this set*.

As with the edge-form of Menger's theorem, these numbers are, in fact, equal. This is the *vertex-form of Menger's theorem*, which we state formally as follows.

Menger's theorem for graphs (vertex-form) *Let G be a connected graph, and let s and t be non-adjacent vertices of G. Then the maximum number of vertex-disjoint st-paths is equal to the minimum number of vertices separating s from t.*

As before, it follows that, if we can find k vertex-disjoint st-paths and k vertices separating s from t, then k is the *maximum* number of vertex-disjoint st-paths and the *minimum* number of vertices in a vertex-cutset separating s from t. Note that these k vertices separating s from t necessarily form a vertex-cutset. It follows that, when looking for them, we need consider only vertex-cutsets whose removal disconnects G into two or more components, one containing s and another containing t.

We can also use this theorem to obtain a result about vertex-connectivity. Recall that the connectivity $\kappa(G)$ of a graph G (other than a complete graph) is the smallest number of vertices whose removal disconnects G. So, by the vertex-form of Menger's theorem, there are at least $\kappa(G)$ vertex-disjoint paths between any given pair of vertices. Since G is k-connected if and only if $\kappa(G) \geq k$, we can restate this result as follows.

Corollary of Menger's theorem for graphs (vertex-form) *A connected graph G is k-connected if and only if any two vertices of G are connected by at least k vertex-disjoint paths.*

Problem 6.9 By finding k vertex-disjoint st-paths, and k vertices separating s from t (for the same value of k), and using the above statement of Menger's theorem, find the maximum number of vertex-disjoint st-paths for the following graph:

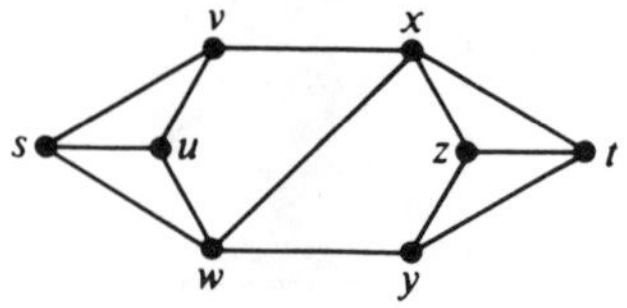

Menger's Theorem for Digraphs (Vertex-form)

Finally, for completeness, we present the vertex-form of Menger's theorem for digraphs. This is almost identical to the vertex-form for graphs.

Menger's theorem for digraphs (vertex-form) *Let D be a connected digraph, and let s and t be non-adjacent vertices of D. Then the maximum number of vertex-disjoint st-paths is equal to the minimum number of vertices separating s from t.*

Problem 6.10 By finding k vertex-disjoint st-paths, and k vertices separating s from t (for the same value of k), and using the above statement of Menger's theorem, find the maximum number of vertex-disjoint st-paths for the following digraph:

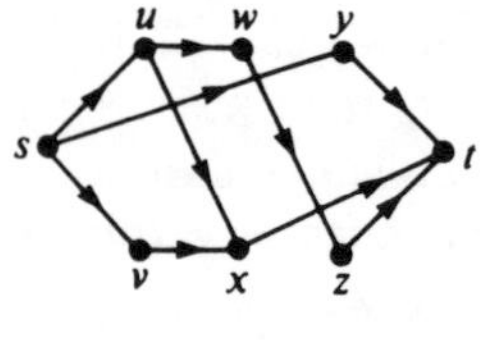

6.4 THE PROOF OF MENGER'S THEOREM

We now prove Menger's theorem. We start by deducing the arc-form for digraphs from the max-flow min-cut theorem for basic networks (proved in Chapter 3), and then show how the corresponding result for graphs follows immediately. Finally, we show how the vertex-form for both graphs and digraphs follows from these other versions.

We begin by restating the max-flow min-cut theorem.

The max-flow min-cut theorem *In any basic network, the value of a maximum flow is equal to the capacity of a minimum cut.*

We now show how to deduce the arc-form of Menger's theorem for digraphs. First, we restate the theorem.

Menger's theorem for digraphs (arc-form) *Let D be a connected digraph, and let s and t be vertices of D. Then the maximum number of arc-disjoint st-paths is equal to the minimum number of arcs separating s from t.*

Proof We associate with the digraph D a basic network which is formed from D by giving each arc a capacity of 1, as indicated, for example, in the following diagram:

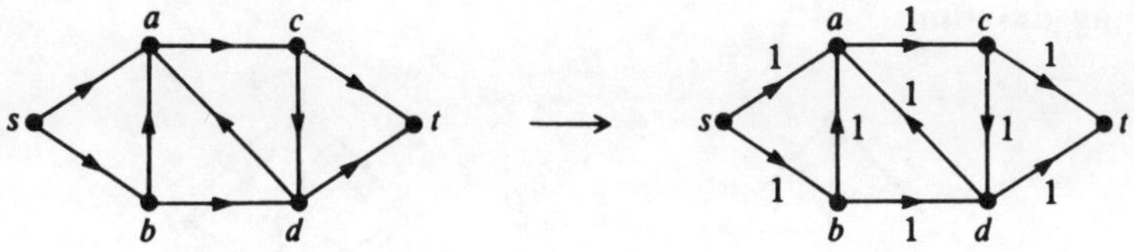

digraph **network**

Since each arc has capacity 1, a flow from s to t in the network must consist of a flow of value 1 along each of a number of arc-disjoint paths from s to t. Thus

the value of a maximum flow	=	the maximum number of arc-disjoint st-paths in D

Now consider a set of arcs which correspond to a cut in the network—for example, the set of arcs $\{ac, da, bd\}$ in the above diagram. Such a set of arcs separates s from t. Moreover, since the capacity of each arc of the network is 1, the capacity of the cut is simply the number of these arcs directed from the component containing s to that containing t. (For example, two of the arcs ac, da and bd are so directed, and thus the capacity of the corresponding cut is 2.) It follows that

the capacity of a minimum cut	=	the minimum number of arcs separating s from t in D

But

the value of a maximum flow	=	the capacity of a minimum cut

by the max-flow min-cut theorem, and so

the maximum number of arc-disjoint *st*-paths in *D*	=	the minimum number of arcs separating *s* from *t* in *D*

as required. □

We can now deduce the following theorem.

Menger's theorem for graphs (edge-form) *Let G be a connected graph, and let s and t be vertices of G. Then the maximum number of edge-disjoint st-paths is equal to the minimum number of edges separating s from t.*

Outline of proof We transform the graph *G* into a digraph *D*(*G*) by replacing each edge by two arcs, one in each direction, as indicated in the following diagram:

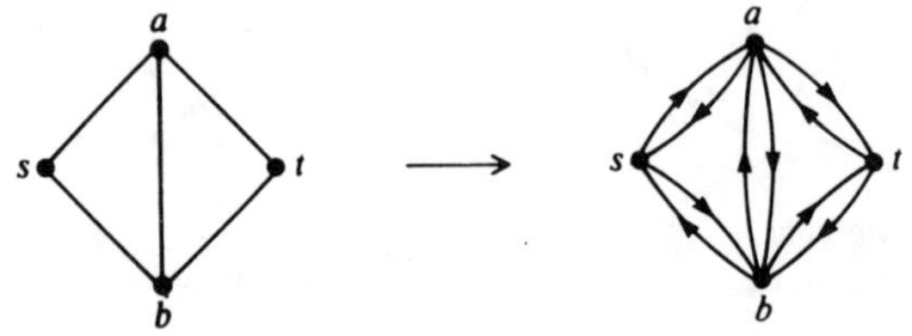

the graph *G* **the digraph *D*(*G*)**

It can be shown that

the maximum number of edge-disjoint *st*-paths in *G*	=	the maximum number of arc-disjoint *st*-paths in *D*(*G*)

and that

the minimum number of edges of *G* separating *s* from *t*	=	the minimum number of arcs of *D*(*G*) separating *s* from *t*

But, by Menger's theorem for digraphs (proved above),

the maximum number of arc-disjoint *st*-paths in *D*(*G*)	=	the minimum number of arcs of *D*(*G*) separating *s* from *t*

and so

the maximum number of edge-disjoint st-paths in G	=	the minimum number of edges of G separating s from t

as required. □

We can also deduce the following theorem.

Menger's theorem for digraphs (vertex-form) *Let D be a connected digraph, and let s and t be non-adjacent vertices of D. Then the maximum number of vertex-disjoint st-paths is equal to the minimum number of vertices separating s from t.*

Outline of proof We transform the digraph D into another digraph D' by replacing each vertex v of D (other than s and t) by two vertices v_1 and v_2 joined by an arc, as indicated by the following diagram:

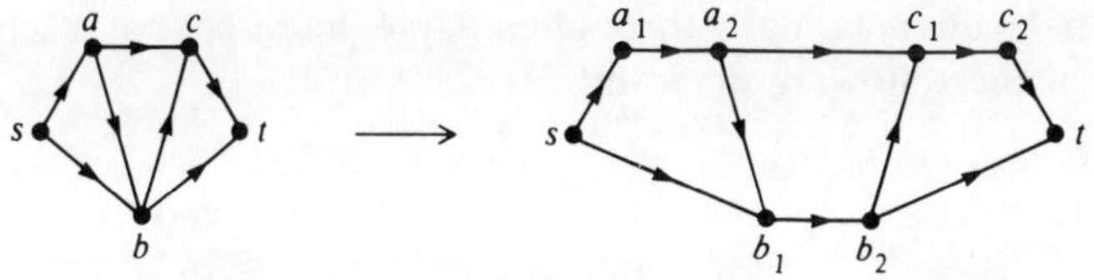

the digraph D the digraph D'

(This construction is essentially the same as the one we used to deal with networks involving vertex-capacities in Chapter 4.)

All arcs of D directed towards a vertex v become arcs of D' directed towards the vertex v_1, and all arcs of D directed away from v become arcs of D' directed away from vertex v_2.

It is not difficult to see that two or more st-paths in D are vertex-disjoint if and only if the corresponding st-paths in D' are arc-disjoint. Applying the arc-form of Menger's theorem to D', we obtain the vertex-form of Menger's theorem for D.

Finally, we can deduce the vertex-form of Menger's theorem for graphs:

Menger's theorem for graphs (vertex-form) *Let G be a connected graph, and let s and t be non-adjacent vertices of G. Then the maximum number of vertex-disjoint st-paths is equal to the minimum number of vertices separating s from t.*

Outline of proof This form of Menger's theorem is deduced from the vertex-form for digraphs in the same way as the edge-form for graphs is deduced from the arc-form for digraphs—namely, by considering the digraph $D(G)$.

6.5 TELECOMMUNICATION NETWORKS

In this section we look at the use of graph theory to represent telecommunication networks and show how it can be of value in the design of efficient networks. We shall see that the notion of connectivity is important in this context.

Optimal Connectivity

The international telephone system is one of the largest combinatorial structures built by man, and can connect a caller to over 400 million subscribers throughout the world. A local call may be routed simply from one subscriber via a local exchange to the called subscriber, but an international call may pass through a large hierarchy of exchanges as indicated in the following diagram:

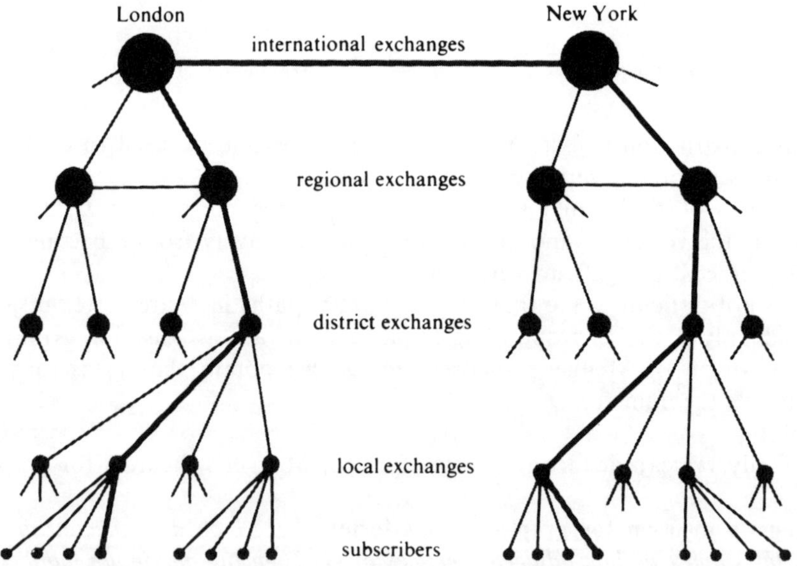

A graph of a telecommunication network may contain a very large number of vertices and edges. The vertices may represent telephone exchanges and subscribers, or they may represent the switches in an exchange. In each case the edges represent links between them. It is important that such telephone networks should be reliable. One aspect of reliability is that calls between subscribers should be possible even if a few exchanges or links are damaged. Consider the following graph, which represents a possible interconnection of exchanges:

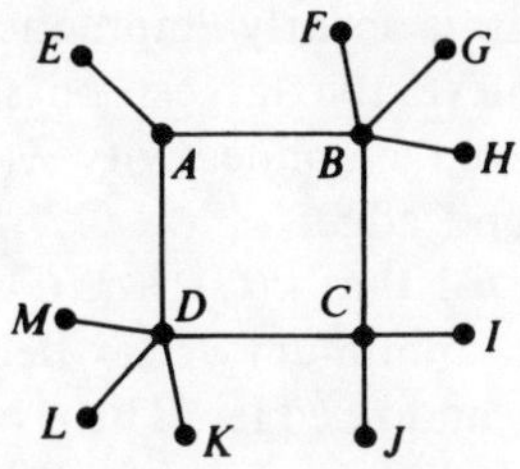

If the link *AB* is damaged, then communication between exchanges *A* and *B* is still possible, but if the link *EA* is damaged, then exchange *E* cannot communicate with any other exchange. The minimum number of links which when damaged will prevent the system from functioning fully is clearly equal to the *edge-connectivity* of the corresponding graph. Similarly, the *vertex-connectivity* tells us how many exchanges must suffer damage before there is a breakdown in communications between the remaining exchanges.

So one reason for providing alternative paths between exchanges (or other types of vertex) is so that communication between the exchanges will still be possible if one path is damaged. Another important reason for providing alternative paths is that a particular link or exchange in the path between two exchanges may also form part of the path between another pair of exchanges. The link or the exchange may therefore be already used to capacity by other calls when a new call is attempted, thus preventing the new call from being made. In such a case we say that the new call is *blocked*.

For any two particular exchanges, the maximum number of alternative paths, no two of which pass through the same intermediate exchange, is (in the terminology of graph theory) the maximum number of vertex-disjoint paths between the corresponding vertices of the graph. By the corollary to the vertex-form of Menger's theorem, the smallest number of such paths between any two exchanges is equal to the *vertex-connectivity*. Similarly, the maximum number of alternative paths, no two of which use the same link between exchanges, is the maximum number of edge-disjoint paths between the corresponding vertices of the graph. By the corollary to the edge-form of Menger's theorem, the smallest number of such paths between any two exchanges is equal to the *edge-connectivity*.

If reliability were the only consideration, a telecommunication system would have as many alternative paths as possible between exchanges—that is, it would have the largest possible vertex-connectivity and the largest possible edge-connectivity. To achieve this, every exchange would need to be connected to every other exchange, and the corresponding graph would be a complete graph. This is clearly impracticable. What can be done, however, is to try to achieve the largest possible values of the vertex-connectivity $\kappa(G)$ and the edge-connectivity $\lambda(G)$ for a graph G with a given number of vertices and edges.

We know from Theorem 6.1 that $\kappa(G) \leqslant \lambda(G) \leqslant \delta(G)$ for any connected graph G, where $\delta(G)$ is the minimum vertex-degree in G.

Suppose G has n vertices and m edges. Then, by the handshaking lemma, the sum of all the vertex-degrees is $2m$. It follows that the *average* of the vertex-degrees is $2m/n$, so the *minimum* degree of the vertices cannot be greater than this. Combining these results, we obtain the following inequalities:

$$\kappa(G) \leqslant \lambda(G) \leqslant \delta(G) \leqslant 2m/n$$

A graph G for which these inequalities are all equalities has the maximum vertex-connectivity and the maximum edge-connectivity possible for any graph with the same number of vertices and edges. Such a graph is said to have **optimal connectivity**.

All graphs with optimal connectivity are regular graphs (since the minimum vertex-degree equals the average of the vertex-degrees), but not every regular graph has optimal connectivity. For example, the graph G below is regular of degree 3, but $\kappa(G) = \lambda(G) = 2$, so G does not have optimal connectivity.

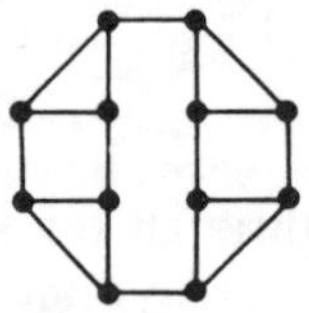

Problem 6.11 Show that the following regular graphs (defined in Chapter 2) all have optimal connectivity:

(a) the cycle graph C_n;
(b) the complete graph K_n;
(c) the complete bipartite graph $K_{r,r}$.

Problem 6.12

(a) There are two non-isomorphic simple graphs with 6 vertices and 9 edges which have optimal connectivity. Draw them.
(b) Draw a regular graph with 7 vertices and 14 edges which does not have optimal connectivity.

We have seen that the values of κ and λ for a graph give us some information about the reliability of the corresponding telecommunication system, or part of such a system. However, the values of κ and λ do not tell us the whole story: two graphs with the same number of vertices, the same number of edges, and the same values of κ and λ, may not correspond to equally reliable systems.

Consider, for example, the following two graphs, each of which has 10 vertices and 12 edges, and satisfies $\kappa = \lambda = 2$.

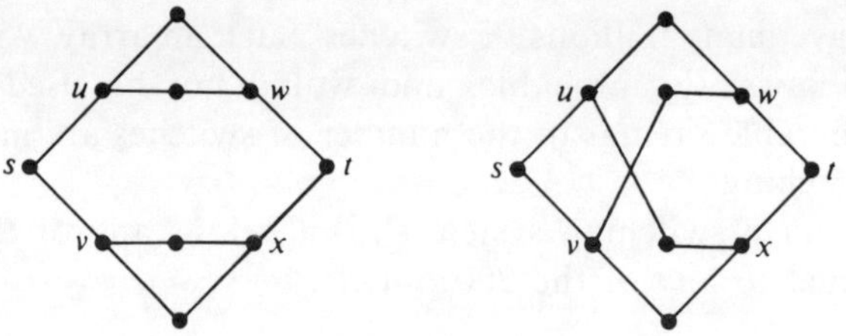

(a) (b)

In the first graph, all paths from the vertex s to the vertex t can be destroyed by removing the edges of any one of the four cutsets with 2 edges—namely, $\{su, sv\}$, $\{wt, xt\}$, $\{su, xt\}$ or $\{sv, wt\}$. However, the second graph has only two cutsets with 2 edges separating s from t—namely, $\{su, sv\}$ and $\{wt, xt\}$. So if all the edges contained in these cutsets have equal likelihood of being blocked or damaged, we would expect that the second graph would be more reliable than the first, and it can be shown that this is indeed the case.

In order to have full information about the reliability of the network represented by a graph, we need not only to know the edge-connectivity and vertex-connectivity, but also to have a full list of all the cutsets of the graph.

Switching in Exchanges

We now look at switching in exchanges, and show how graphs such as the two above may arise in practice.

In an exchange, incoming calls must be routed to the required destinations. This may be done using arrays of switches which may be reed-relay switches or semiconductor switches. A rectangular array as illustrated below allows any inlet row to be connected to any outlet column.

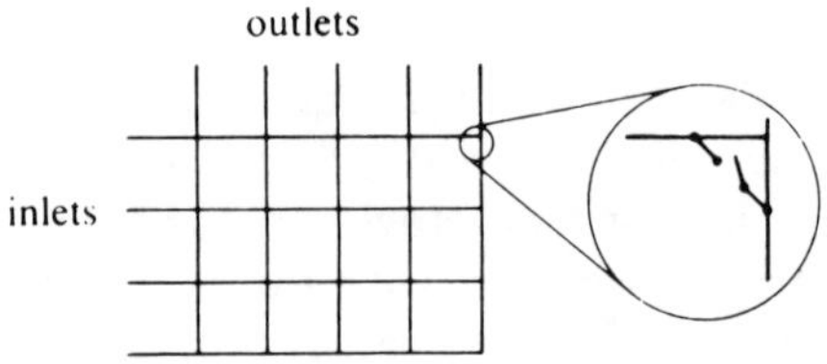

Such an array may be used for a small exchange with, say, ten subscribers, and involving about 100 switches. However, a typical local telephone exchange has many thousands of subscribers, and the corresponding array would have many millions of switches. Such an array would be large, expensive and possibly unreliable, and would not be used in practice. Instead, considerable savings in the number of switches are made by using multi-stage switching.

In the three-stage switching system shown below, any of the 200 inlets can be connected to any of the 200 outlets.

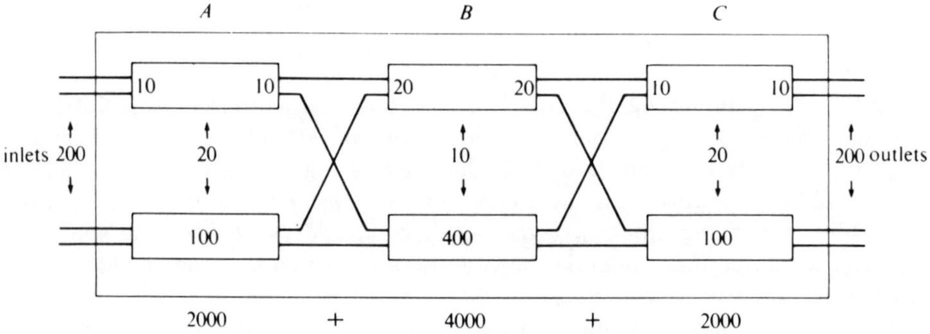

The 200 inlets are connected to twenty 10×10 arrays of switches in stage A. So stage A has $20 \times 10 \times 10 = 2000$ individual switches or crosspoints. Each of the ten outlets from any of these arrays must go to a different array in the next stage. So stage B needs ten 20×20 arrays which amounts to $10 \times 400 = 4000$ crosspoints. Finally, stage C needs twenty 10×10 arrays and 2000 crosspoints to connect a call to any outlet.

The total number of crosspoints required for the three-stage system is therefore 2000 + 4000 + 2000 = 8000 crosspoints. A single 200 × 200 array would have 200 × 200 = 40 000 crosspoints. So the multi-stage system gives us a great saving in switches—only 8000 instead of the 40 000 required for the single array, a saving of 80%. In addition, the three-stage system uses manageable sizes of arrays, only 10 × 10 and 20 × 20. This system in fact uses the smallest number of crosspoints for any three-stage 200 inlet and 200 outlet multi-stage switch.

Even greater economies can be made by using five-stage or even seven-stage switching systems. However, in these systems, unlike the three-stage system, each array is not linked to every array at the next stage. This gives us a choice of how we connect the arrays. To investigate different possibilities, it is convenient to use a graphical representation.

To illustrate the approach, consider first the following simple three-stage switching system with four inlets and four outlets:

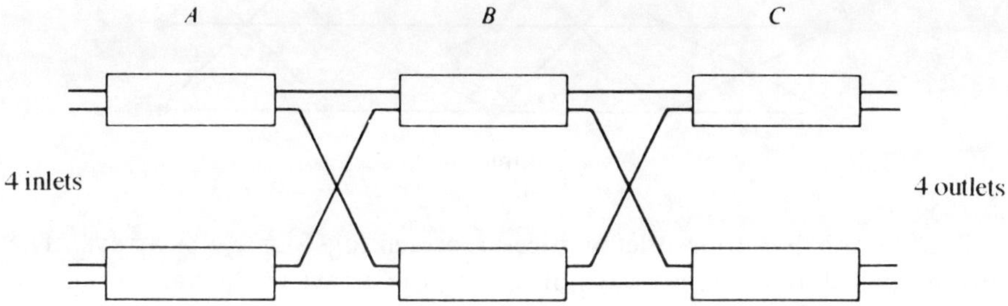

This can be represented by the following graph in which each 2 × 2 array is represented by a vertex, and the links between arrays are represented by edges.

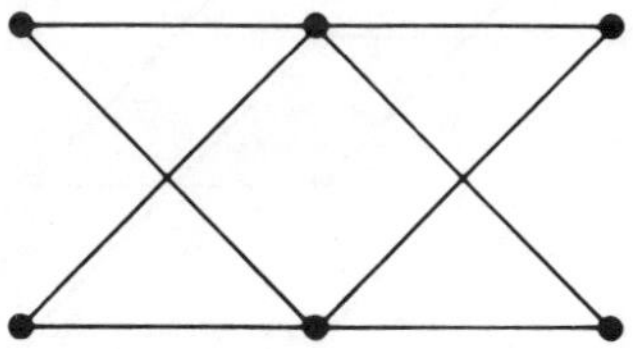

Notice that for simplicity the edges entering the inlets and leaving the outlets have been omitted.

Now let us look at two more elaborate graphs of a five-stage network.

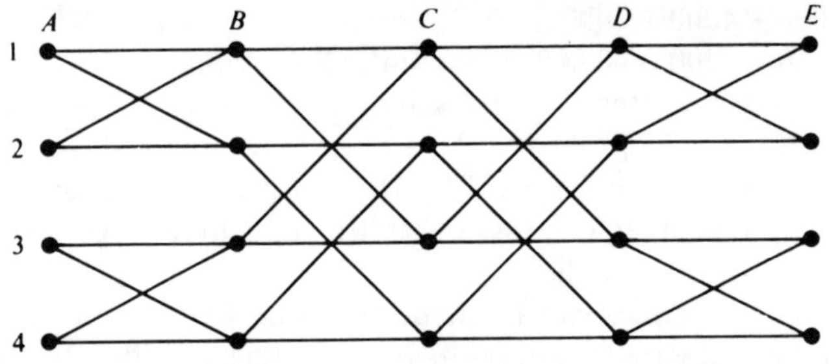

graph (a)

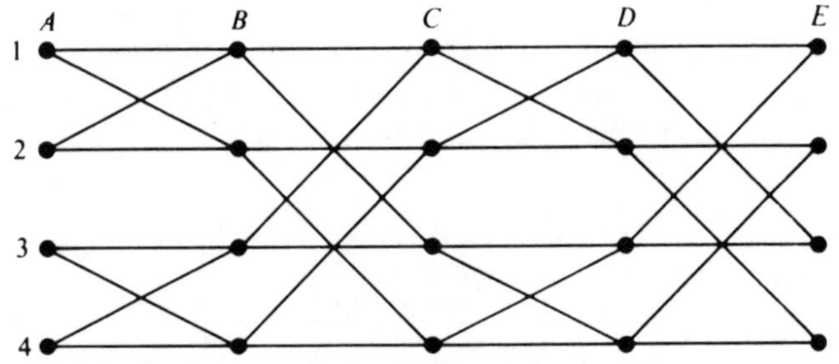

graph (b)

Each graph has four inlet vertices representing four 2×2 arrays. The two graphs show different ways of linking the arrays in the various stages. Which graph is preferable? To answer this question, consider all possible paths from inlet $A1$ to outlet $E3$ on graph (a).

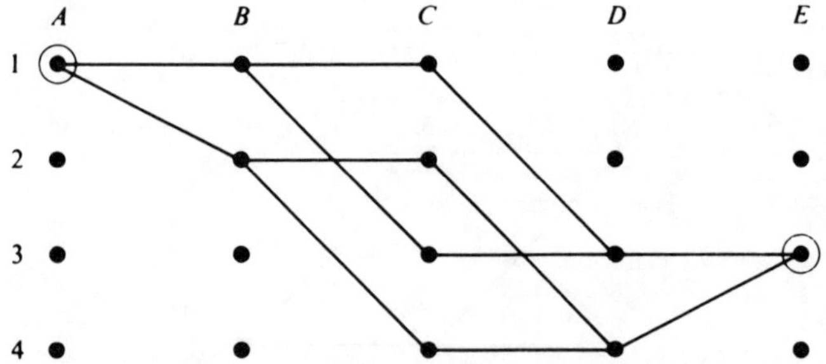

channel graph from $A1$ to $E3$ in graph (a)

Taken together, the four paths form the *channel graph* from $A1$ to $E3$. For any inlet-outlet pair on graph (a) we find essentially the same graph:

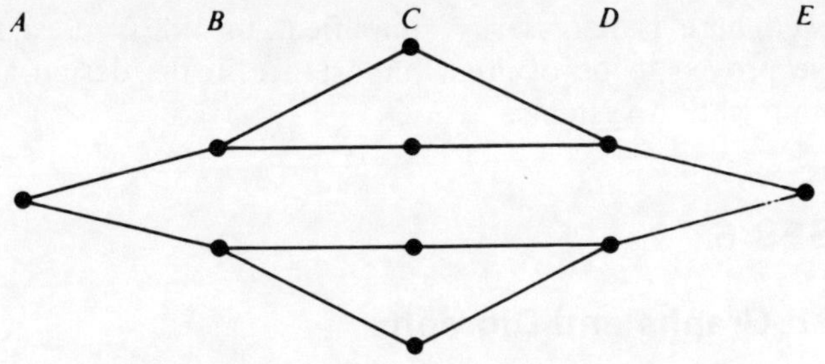

typical channel graph in graph (a)

Because all the channel graphs are isomorphic to this graph, we say that the first multi-stage graph is *channel-regular*.

Similarly, all the channel graphs in graph (b) are isomorphic to the following graph:

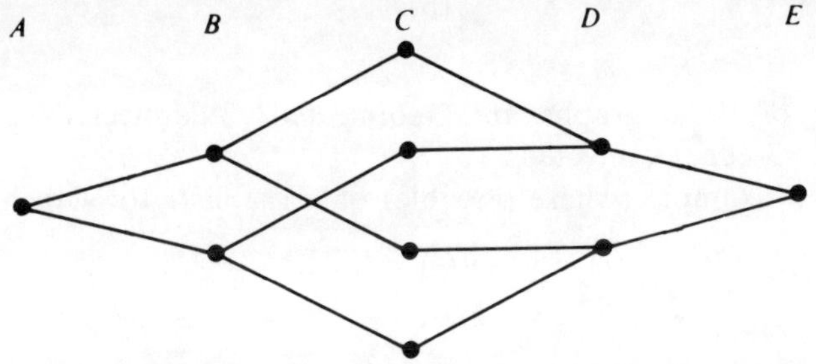

typical channel graph in graph (b)

We have already met these graphs (just after Problem 6.12 on page 141).

Both have the same values of edge-connectivity and vertex-connectivity, namely $\kappa = \lambda = 2$. However, in the first channel graph, all paths from A to E can be destroyed by removing the edges of any one of four cutsets with 2 edges, but in the second channel graph there are only two cutsets with the same property. So the first channel graph has more cutsets of size 2 which separate A from E, and in fact it also has more such cutsets of size 3. We would therefore expect the second five-stage switching system to be less prone to blocking or failure than the first. In fact, the second system is the least prone to blocking of all five-stage systems with the same number of vertices and edges.

This illustrates again the result we demonstrated earlier, namely that, in order to have full information about the reliability of a network represented by a graph, we need to know not only the edge-connectivity and vertex-connectivity, but also all the cutsets of the graph. Although the account given here is necessarily simplified, the ideas presented in this section have proved to be of great importance in the design and analysis of telecommunication systems.

EXERCISES 6

Connected Graphs and Digraphs

6.1. Write down the values of $\kappa(G)$ and $\lambda(G)$ for each of the following graphs G:

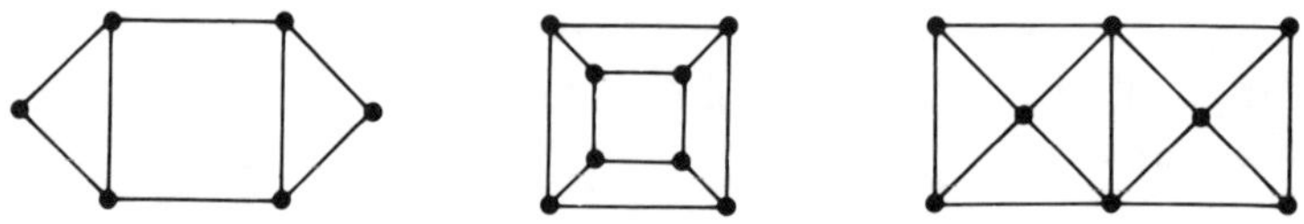

(a) (b) (c)

Which of these graphs are 2-connected? 3-connected? 2-edge-connected? 3-edge-connected?

6.2. Give an example (where possible) of a graph G for which:

(a) $\kappa(G) = 2, \quad \lambda(G) = 3, \quad \delta(G) = 4;$

(a) $\kappa(G) = 3, \quad \lambda(G) = 2, \quad \delta(G) = 4;$

(a) $\kappa(G) = 2, \quad \lambda(G) = 2, \quad \delta(G) = 4.$

6.3. Find a 4-connected graph with 8 vertices and 16 edges.

Menger's Theorem and its Analogues

6.4. In the complete bipartite graph $K_{10,13}$, let v be any vertex in the set with 10 vertices, and w be any vertex in the other set.

(a) Find the maximum number of edge-disjoint paths between v and w.

(b) Find the maximum number of vertex-disjoint paths between v and w.

6.5. (a) By finding k arc-disjoint st-paths and k arcs separating s from t (for the same value of k), find the maximum number of arc-disjoint st-paths in the following digraph:

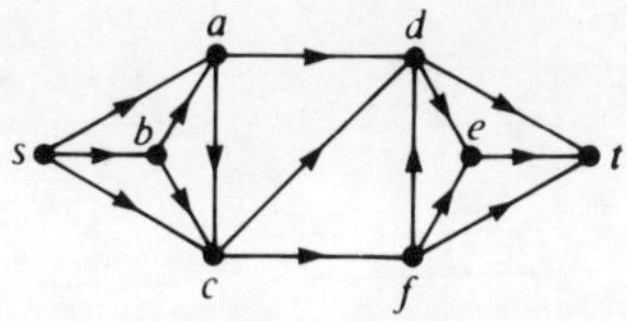

(b) Similarly, find the maximum number of vertex-disjoint *st*-paths in the above digraph.

Telecommunication Networks

6.6. Which of the graphs in Exercise 6.1 have optimal connectivity?

6.7. A small private inter-office telephone system has 6 extensions connected by means of an exchange using reed-relay switches.

(a) Draw the graph representing this system.
(b) Find the size of the switching unit array and the number of crosspoints.

6.8. A 3-stage switching system is to be used to cope with 150 inlets and 150 outlets. The first stage consists of 10 switching units.

(a) How many switching units are needed at stages 2 and 3, and what is the size of each array at each stage?
(b) How many crosspoints are there altogether?
(c) How many crosspoints would there be in a single-stage switch?

6.9. Draw a diagram of a 5-stage switching system with 200 inlets and 200 outlets, using only 10×10 arrays in the system. How many crosspoints are used?

6.10. Design a 3-stage switching system with 72 inlets and 72 outlets with the minimum number of crosspoints.

6.11. The following graphs are the channel graphs of two channel-regular 5-stage switching systems. Which system is more prone to blocking?

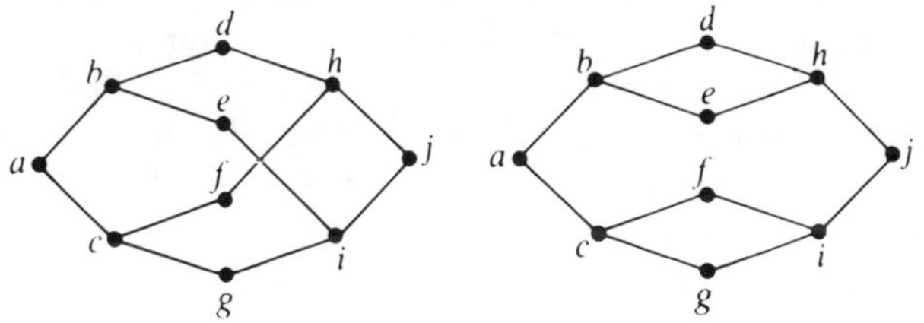

CHAPTER 7

LONGEST AND SHORTEST PATH ALGORITHMS

After reading this chapter, you should be able to:

- use the shortest path algorithm to find the shortest path between two vertices of a weighted graph or digraph;
- use a modified form of the algorithm to find the longest path between two vertices of a weighted graph or digraph without cycles;
- find a maximum flow at minimum-cost in a weighted network by combining the shortest path algorithm with the maximum flow algorithm.

Introduction

In this chapter we consider a number of problems which involve finding the shortest or longest paths in a network. First we describe the shortest path algorithm, and then show how it may be modified to find the longest path in a network without cycles. This longest path algorithm forms the basis of the algorithm used in Chapter 15 to find a critical path in an activity network for scheduling problems. Finally, we give an algorithm for finding a maximum flow at minimum cost in a network; this algorithm is obtained by combining the shortest path algorithm with the maximum flow algorithm.

7.1 THE SHORTEST PATH ALGORITHM

The shortest path algorithm successively assigns potentials to all the vertices of a network. To see how it works, consider the following example:

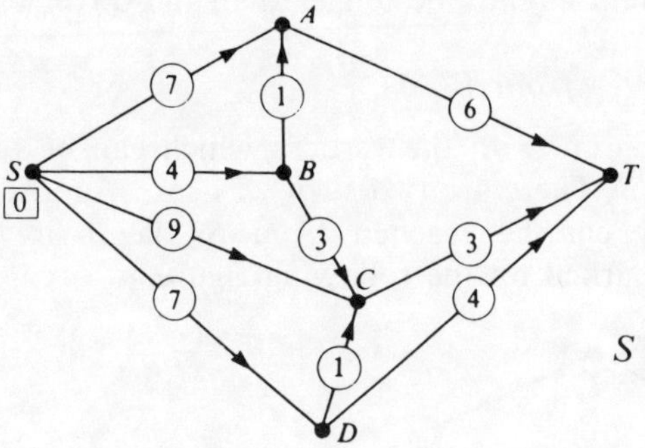

S has potential 0

The problem is to find the shortest path from *S* to *T* in the network, where the length of each arc is represented by the number in the circle on the arc. We wish to assign a potential to each vertex; this represents the shortest distance along any path from *S* to that vertex. We denote a potential by a number inside a square. We start by assigning zero potential to *S*—this is indicated by a zero in a square on the above figure.

Vertices Reached Directly from S

The next step is to consider each vertex which can be reached from *S* by a single arc, that is, vertices *A*, *B*, *C* and *D*. We label each of these vertices with a number equal to the distance along the corresponding arc from *S*, as shown below.

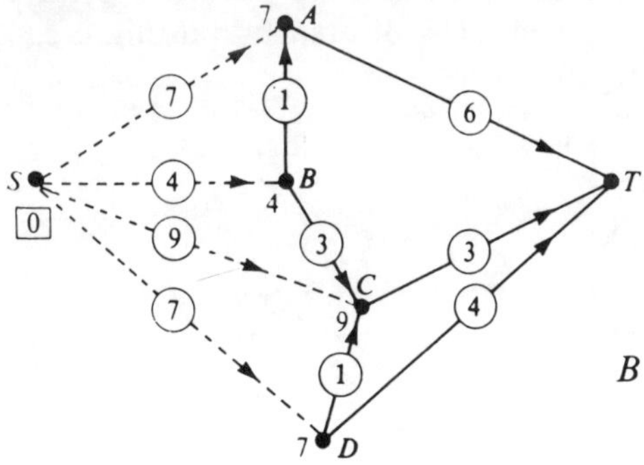

B has potential 4

Now we take the *lowest* number on these vertices (that is, the shortest distance from *S*) and mark it as a potential. The shortest of the four distances *SA*, *SB*, *SC* or *SD* is *SB*, which is a distance of 4, so this is marked as a potential by putting a square round it in the diagram below. We know this must be the shortest distance to *B* along any path from *S*, since any other path would have to go via *A*, *C* or *D*, and the first stage alone of such a path would already be longer than the direct path.

Vertices Reached Directly from B

We now repeat the procedure with the vertices which can be reached by a single arc from vertex *B*. There are two such vertices, *A* and *C*, and the distances in which they can be reached from *S* via *B* are 5 and 7 respectively. These are marked on the following figure.

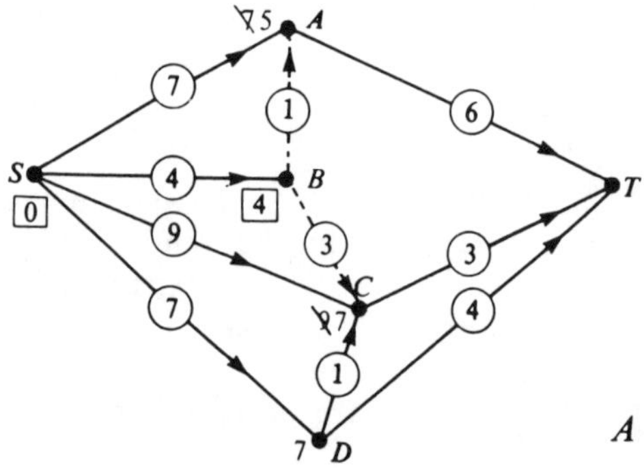

A has potential 5

Note that the distance to *A* via *B* of 5 is less than the previously marked label of 7, so we cross out the 7 and replace it by 5. The same applies with vertex *C*. If, however, the existing label were lower than the new distance, we would leave it in place since we are trying to assign potentials equal to the shortest distances to vertices.

The shortest marked distance which is not a potential is the distance of 5 at *A*, so we make this a potential, indicated by putting a square round it, as shown in the next figure.

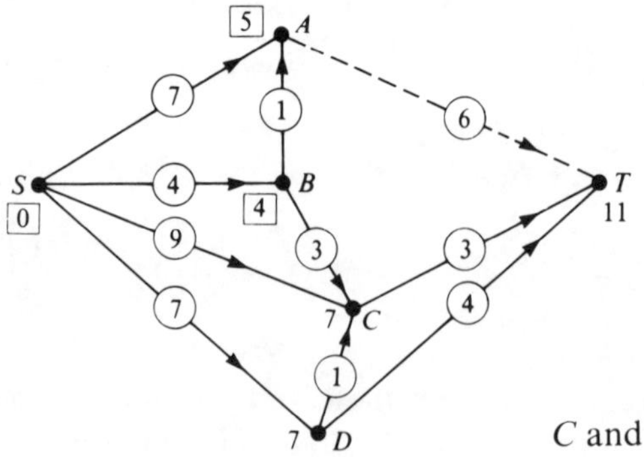

C and *D* have potential 7

Vertices Reached Directly from A

Only one vertex (vertex T) can be reached from the vertex just assigned a potential (vertex A) by a single arc, so we label T with the distance from S via A. This can be obtained by adding the potential at A to the distance along arc AT, that is, $5 + 6 = 11$.

There are now two vertices (C and D) with a lowest labelled distance of 7, and these distances therefore become potentials. We must now look at all the vertices which can be reached from these two vertices by a single arc.

Vertices Reached Directly from C or D

There is now only one vertex not assigned a potential, namely T. The path via D has a total distance of $7 + 4 = 11$, which is the same value as the current label on T, corresponding to the path via A. However, the path via C has a distance of 10, which is clearly the shortest of the three paths, so the label at T, and hence its potential, is 10, as shown on the figure below.

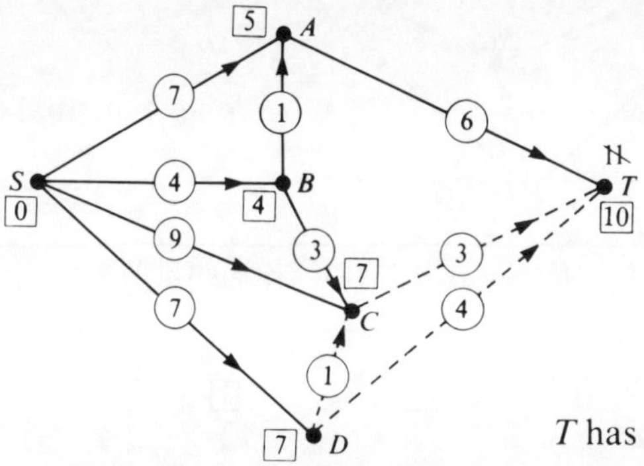

T has potential 10

We have found the shortest distance to T, namely 10, and it remains only to identify the corresponding shortest path. We do this by backtracking from T, choosing at each stage the vertex whose potential is exactly

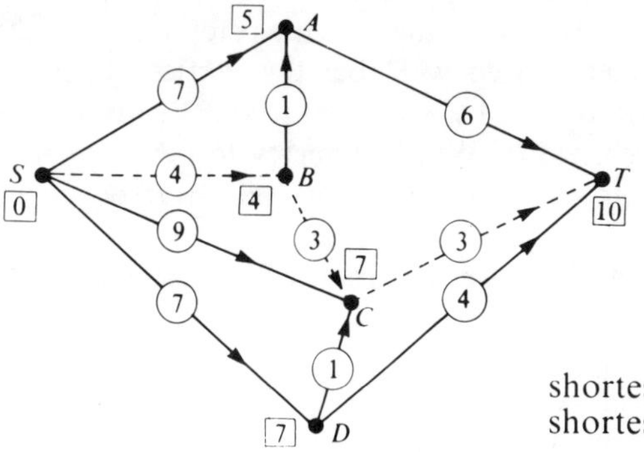

shortest distance is 10
shortest path is *SBCT*

equal to the potential at the previous vertex less the distance along the arc joining them. Thus we reach *T* from *C*, then *C* from *B*, and *B* from *S*. Hence the shortest path is *SBCT*.

Tabular Method for Finding a Shortest Path

It is sometimes convenient to use a tabular method, rather than to annotate a diagram. The tabular method enables the steps of the algorithm to be clearly displayed, and this permits more accurate working and easier checking, particularly for large-scale examples. The method of laying out the table is illustrated below for our example.

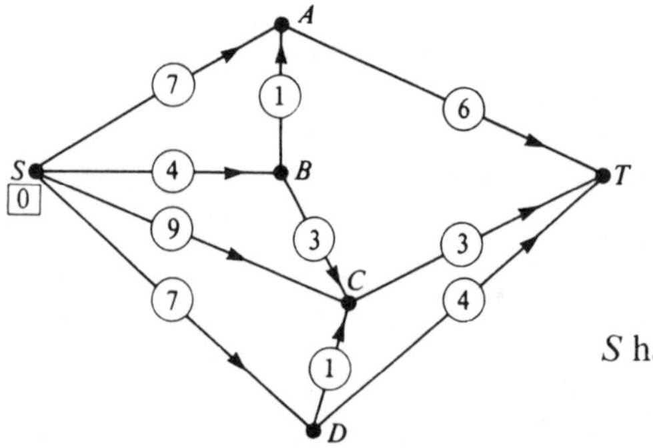

S has potential 0

step	vertex started from	vertices assigned labels A	B	C	D	T
STEP 1	*S*	7	[4]	9	[7]	
STEP 2	*B*	[5]		[7]	↑	
STEP 3	*A*			↑ (•)	•	11
STEP 4	*C*					[10]
	D					11

We start by assigning zero potential to *S*. Then, at each step, we consider only those vertices which can be reached directly from the vertex (or vertices) just assigned a potential in the previous step (called 'vertex started from' in the table). At each step we label the vertices being considered, and assign a potential (indicated by a square) to one or more of them—that is, those with the smallest value. Where a vertex to be assigned a potential was labelled in an earlier step, we use a dot and an arrow to indicate the step at which the potential was assigned.

When the table is complete, we can read off the shortest path(s) from the table. For our example, we find that

T has potential 10 assigned from *C*,
C has potential 7 assigned from *B*,
B has potential 4 assigned from *S*,

so

the shortest path is *SBCT* with length 10.

We can see from the table that in this case there is only one shortest path.

Note that there may be more than one shortest path, and this possibility will be clear from the table, since two or more vertices will be assigned the same potential at some step. (This is illustrated in the following problem.)

Problem 7.1 Use the tabular method to find the shortest path from S to T in the following network:

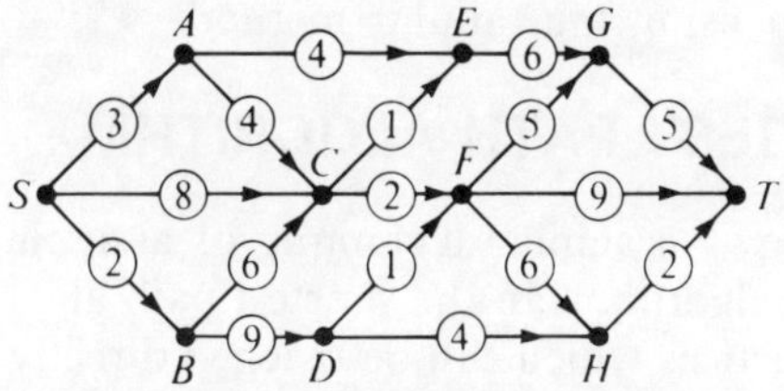

DESCRIPTION OF THE SHORTEST PATH ALGORITHM

START Assign potential 0 to S.

GENERAL STEP Consider the vertex (or vertices) just assigned a potential. For each such vertex V, consider each vertex W which can be reached from V along an arc VW, and assign W the label

$$(\text{potential of } V) + (\text{distance } VW)$$

unless W already has a *smaller* label.

When all such vertices W have been labelled, choose the smallest vertex-label in the network which is not already a potential, and make it a potential at each vertex where it occurs.

REPEAT the GENERAL STEP with the new potential(s).

STOP when T has been assigned a potential. The shortest distance from S to T is the potential of T.

To find the shortest path, work backwards from T and include an arc VW whenever

$$(\text{potential of } W) - (\text{potential of } V) = \text{distance } VW.$$

Problem 7.2 Use the algorithm to find the shortest path from S to T in the following network.

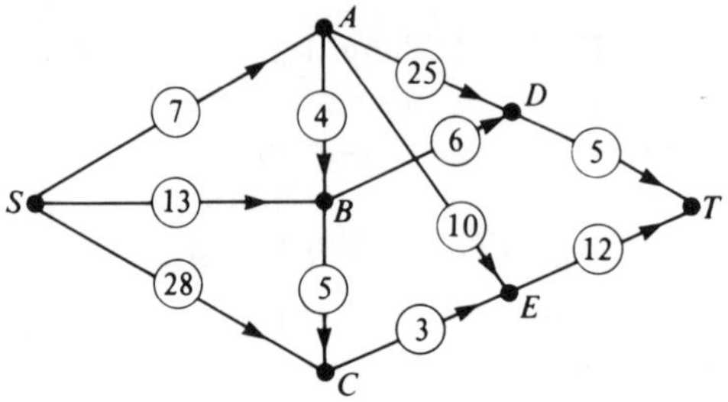

Repeat the problem using the tabular method.

7.2 THE LONGEST PATH ALGORITHM

This algorithm follows a similar technique of assigning potentials to that of the shortest path algorithm. In the shortest path algorithm, we consider at each stage all the vertices which can be reached directly from vertices which have just been assigned potentials. In the longest path algorithm, however, we consider at each stage only those vertices which can be reached directly only from vertices of known potential (and not from any other vertices).

Note that the longest path algorithm is restricted to networks with no cycles, since if a network contains a cycle then there are paths of indeterminate length and the algorithm will not terminate.

As an example, let us go through the procedures of the algorithm in order to find the longest path from S to T in the following network, which we considered earlier in this chapter:

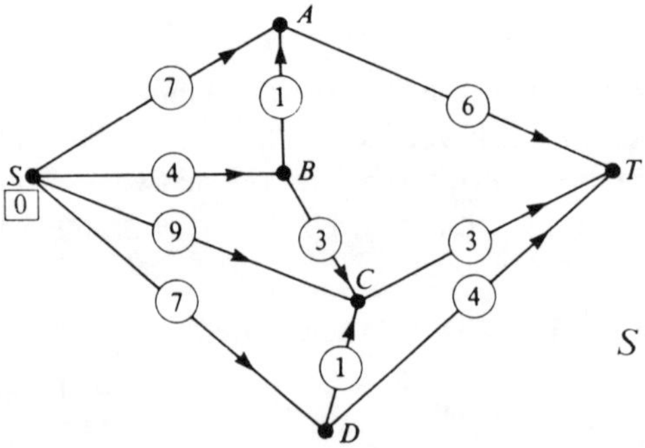

S has potential 0

Vertices Reachable Only from S

At each stage we look at those vertices which can be reached only from vertices of known potential, and not from any other vertex. We start by assigning zero potential to S, and then look at vertices reachable only from S. There are two such vertices, vertex B and vertex D. No other vertex is reachable directly only from vertex S. Vertex A, for example, can also be reached from *B*, whose potential we do not yet know, and so *A* is not considered at this stage.

Since *SB* is the only path to *B*, it must be the longest path, so we assign to *B* a potential equal to the length of *SB*, which is 4 units. In a similar way we assign a potential of 7 to *D*.

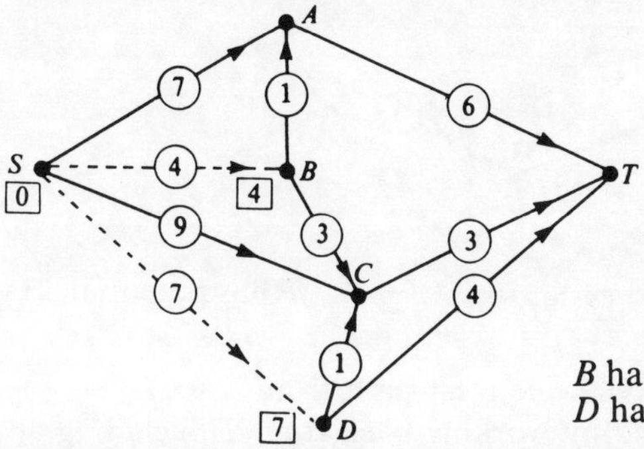

B has potential 4
D has potential 7

Vertices Reachable Only from S, B and D

Now that we know the potentials of *B* and *D*, as well as *S*, we can proceed to look at vertices *A* and *C*.

Vertex *A* can be reached directly only from *S* or *B*, both of which already have potentials assigned to them. The path from *S* direct to *A* has length 7, while the path from *S* to *A* via *B* has length 4 + 1 = 5. So the length of the longest path from *S* to *A* is 7 units, which we mark on the diagram as the value of the potential at *A*.

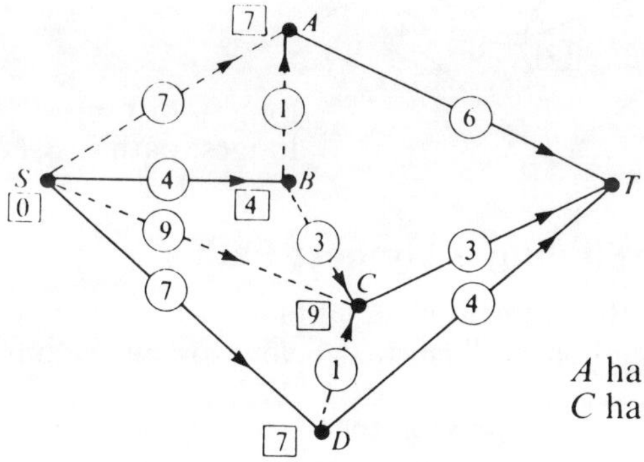

A has potential 7
C has potential 9

There are three paths to *C*, namely *SC*, *SBC* and *SDC* of respective lengths 9, 7 and 8 units. So the potential of *C* is 9 units.

Vertices Reachable Only from S, A, B, C and D

There remains only vertex *T*, which can be reached directly from *A*, *C* and *D*. The path from *S* to *T* via *A* has total length equal to the potential at A plus the length AT, which is $7 + 6 = 13$. The path via C has total length $9 + 3 = 12$, and that via D has length $7 + 4 = 11$. So the potential at T is 13.

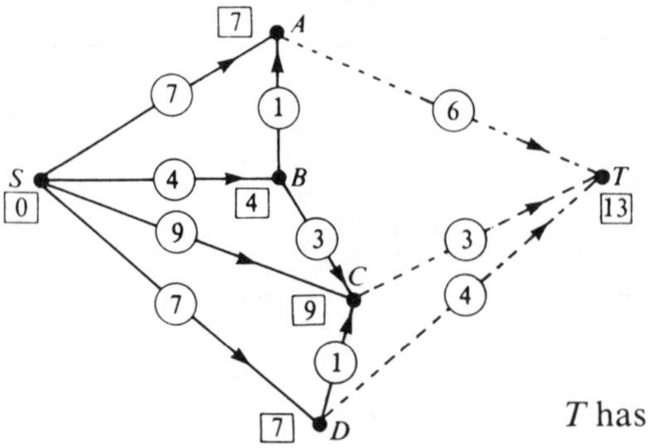

T has potential 13

Thus 13 is the length of the longest path from S to T.

The actual path is found by working back from T, including an arc only if its length is equal to the difference in the potentials of the end vertices; thus the longest path is SAT.

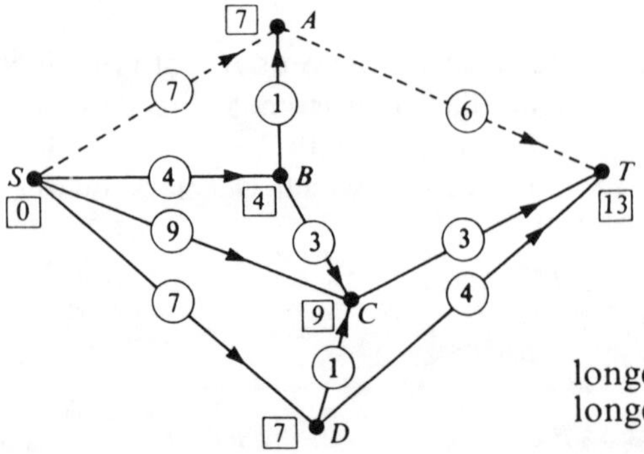

longest distance is 13
longest path is SAT

Tabular Method for Finding a Longest Path

As with the shortest path algorithm, a tabular method can be used to record the steps. The method is illustrated below for our example.

step	vertices assigned labels	origin vertices S	A	B	C	D
STEP 1 (S)	B	[4]				
	D	[7]				
STEP 2 (S, B, D)	A	[7]		5		
	C	[9]		7		8
STEP 3 (S, A, B, C, D)	T		[13]		12	11

We start by assigning zero potential to *S*. Then, at each step, we consider those vertices which can be reached directly only from vertices of known potential. In the table, these vertices of known potential are indicated in parentheses beside the step number. Vertex labels are inserted in the appropriate columns under the heading 'origin vertices' to indicate from which vertex each currently labelled vertex is reached. At each step, one or more of these labels becomes a potential, indicated by a square.

When the table is complete, we can read off the longest path(s) from the table. For our example, we find that

T has potential 13 assigned from *A*,

A has potential 7 assigned from *S*,

so

the longest path is *SAT* with length 13.

We can see from the table that in this case there is only one longest path.

DESCRIPTION OF THE LONGEST PATH ALGORITHM

START Assign potential 0 to *S*.

GENERAL STEP Consider all the vertices which can be reached directly only from vertices with known potentials.

For each such vertex *W*, consider each arc *VW* directed into *W* and assign *W* the label

(potential of *V*) + (distance *VW*)

unless *W* already has a *larger* label.

When all such arcs *VW* have been considered, make the vertex label at *W* a potential.

REPEAT the GENERAL STEP with the new potential(s).

STOP when *T* has been assigned a potential. The longest distance from *S* to *T* is the potential of *T*.

To find the longest path, work backwards from *T* and include an arc *VW* whenever

(potential of *W*) − (potential of *V*) = distance *VW*.

Problem 7.3 Use the longest path algorithm to find the longest path from *S* to *T* in the following network.

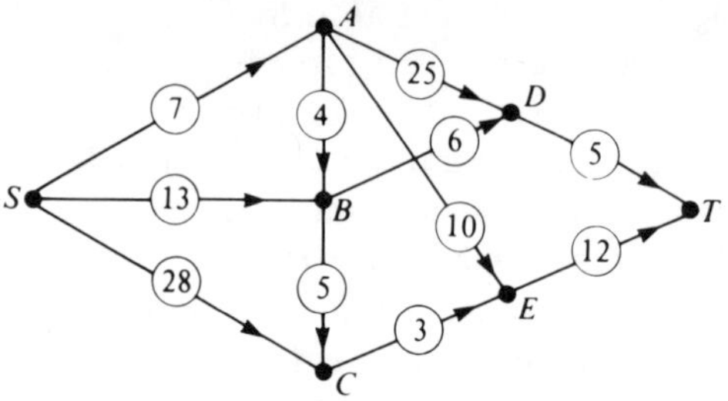

Repeat the problem using the tabular method.

7.3 THE MAXIMUM-FLOW MINIMUM-COST ALGORITHM

Once again, we illustrate the algorithm by working through an example before giving a formal statement of it. Consider the following network:

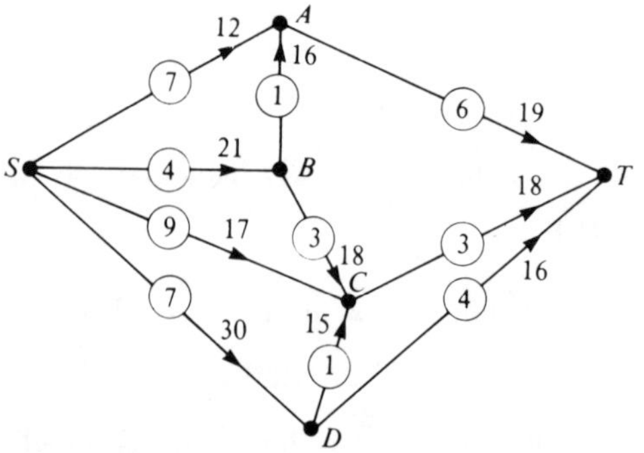

Each arc has a capacity, indicated by the number written next to the arc. Each arc also has a cost associated with it, and this is shown by the circled number on the arc. The problem is to find a maximum flow from S to T at minimum total cost.

The procedure used by the algorithm begins by finding the shortest path through the network, and then sends as much flow as possible through it. In our example, the shortest path becomes the path with the minimum total cost. The next step is to find the cheapest flow-augmenting path, and to send as much flow as possible through it. This step is repeated until the maximum possible flow is reached.

First Flow-augmenting Path SBCT

We use the shortest path algorithm to find the shortest (that is, the cheapest) path from S to T. In fact, this is the same network as the one we used as an example for the shortest path algorithm, so we have already done this.

The shortest path we obtained was *SBCT*, with a cost of 10 units. So we send the maximum flow possible along this path. Since the minimum capacity of any arc along *SBCT* is 18, we send a maximum flow of 18 along this path. The saturated arcs *BC* and *CT* are shown now as thick lines.

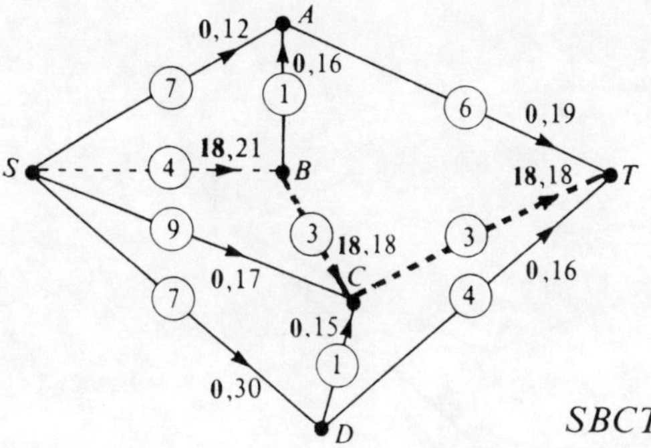

SBCT has flow 18

Second Flow-augmenting Path SDT

We now have to recalculate the potentials for the network without the two saturated arcs *BC* and *CT*, so that we can find the cheapest flow-augmenting path (that is, the shortest path through the network without the arcs *BC* and *CT*). These new potentials are indicated by the numbers in squares in the following diagram.

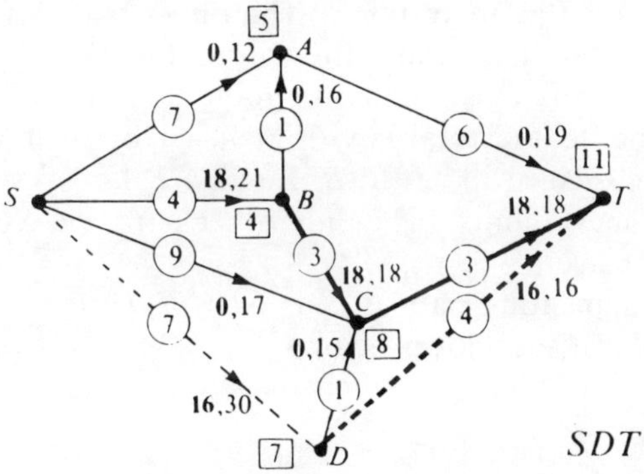

SDT has flow 16

There are in fact two cheapest paths, *SBAT* and *SDT*, both of which have a total cost of 11. However, *SBAT* is limited to a flow of 3 because of the remaining capacity of arc *SB*, whereas *SDT* can take a flow of 16. We choose the path with the highest capacity, and so send a flow of 16 along *SDT* as indicated on the diagram above.

Third Flow-augmenting Path SBAT

We now recalculate the potentials to find the next flow-augmenting path which has the minimum cost, leaving out the saturated arcs *BC*, *CT* and *DT*. The potentials are the same as in the previous diagram, since the cheapest path remaining is the path *SBAT*. So this is the third flow-augmenting path, and we send the maximum possible flow of 3 along this path. The current flow pattern is shown below.

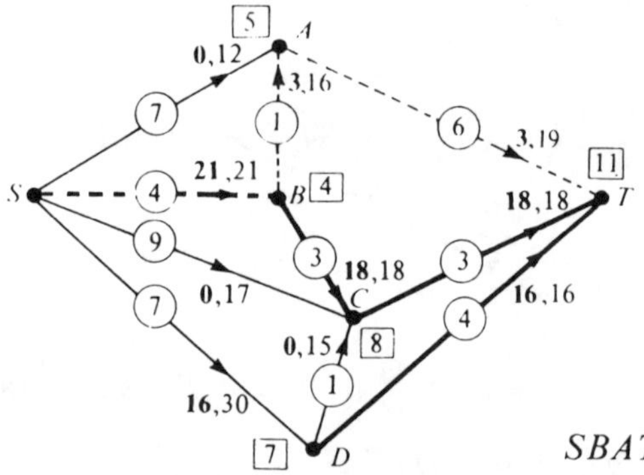

SBAT has flow 3

Fourth Flow-augmenting Path SDCBAT

In recalculating potentials we now meet the need to consider backward flows along arcs, as we did in Chapter 3. This arises with the saturated arc *BC*, since reducing the flow along *BC* has the same effect as sending a backward flow from *C* to *B*.

Consider the available paths to vertex *A*. The direct path *SA* has a cost of 7 units. However, we can also send a flow along the path *SDCBA*. The cost of going from *S* to *C* via *D* is 8 units. In going from *C* to *B* we reduce the flow along *BC*, and so the total cost is reduced by 3 units for each unit of flow, giving a cost of 5 to get to *B*. Finally, *BA* costs one unit, so the potential at *A* becomes 6 units. (The only other path, via *SC*, is clearly going to be more expensive.)

The cheapest flow-augmenting path is thus *SDCBAT*, and we can send a flow of 13 units along it, as shown below.

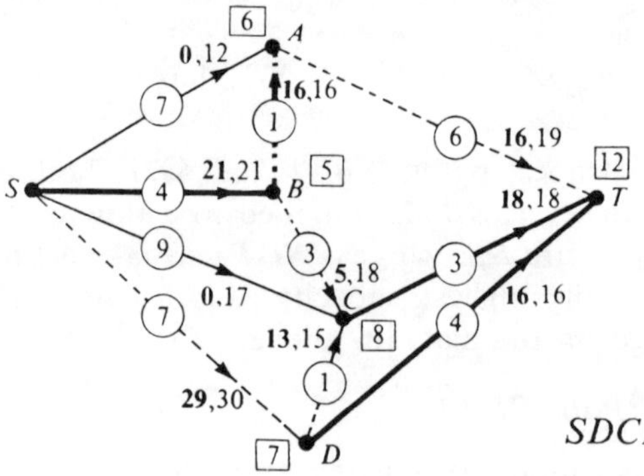

SDCBAT has flow 13

Fifth Flow-augmenting Path SAT

With so many arcs now saturated, there remains only one flow-augmenting path, *SAT*, so we send the maximum possible flow of 3 units along this path, arriving at the final flow pattern shown below.

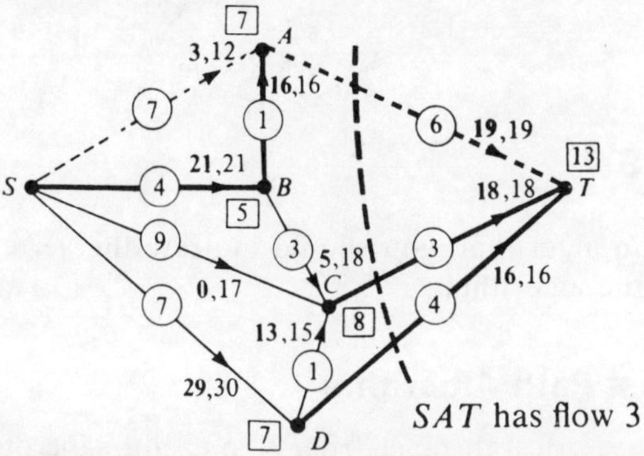

value of flow is 53
total cost is 584

min cut is {*AT*, *CT*, *DT*}

The value of the maximum flow is 53, which can be checked by finding the minimum cut *AT*, *CT*, *DT* which also has value 53. The total cost is 584.

DESCRIPTION OF THE MAXIMUM-FLOW MINIMUM-COST ALGORITHM

START — Use the shortest path algorithm to find the cheapest path from *S* to *T*, and send as much flow as possible along it.

GENERAL STEP — Use the shortest path algorithm (modified to cover the possibility of a backward flow) to find the cheapest flow-augmenting path from *S* to *T* and send as much flow as possible along it.

(If there is more than one flow-augmenting path, choose the one which can take the largest increase in flow.)

REPEAT the GENERAL STEP until a maximum flow is reached.

Problem 7.4 Use the algorithm to find a maximum flow at minimum cost from *S* to *T* in the following network. What is the value of this flow and what is the minimum cost?

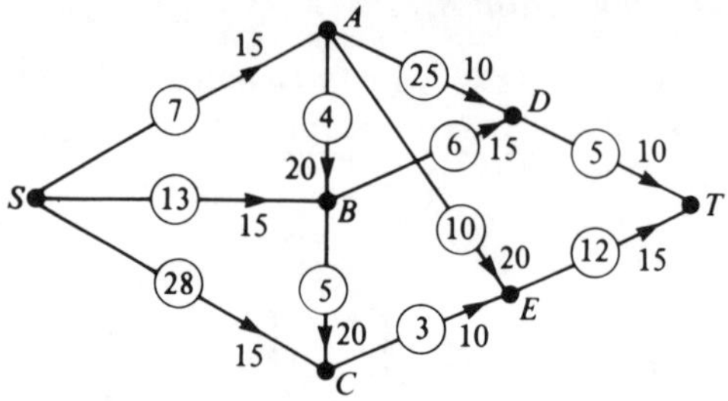

EXERCISES 7

When using an algorithm, you should indicate the vertex labels clearly at each step of the algorithm.

The Shortest Path Algorithm

7.1. Find the shortest distance from S to each of the other vertices in the following weighted graph, by using the algorithm.

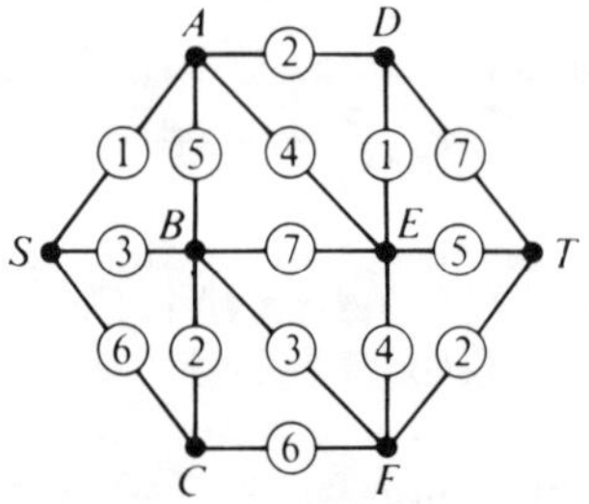

7.2. Find the shortest path from S to T in the following weighted digraph, by using the algorithm.

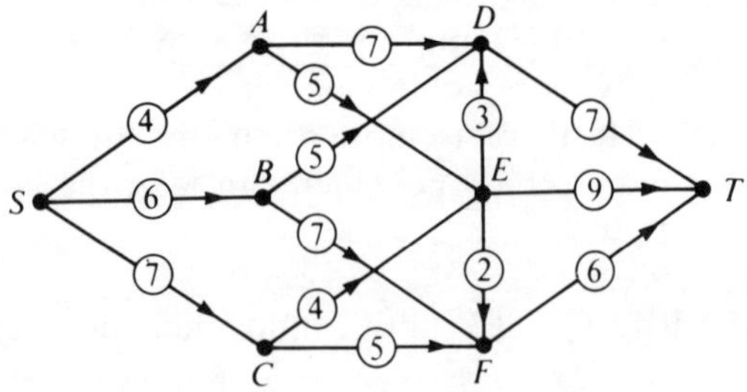

7.3. Find the shortest path from S to T in the following weighted digraph, by using the algorithm.

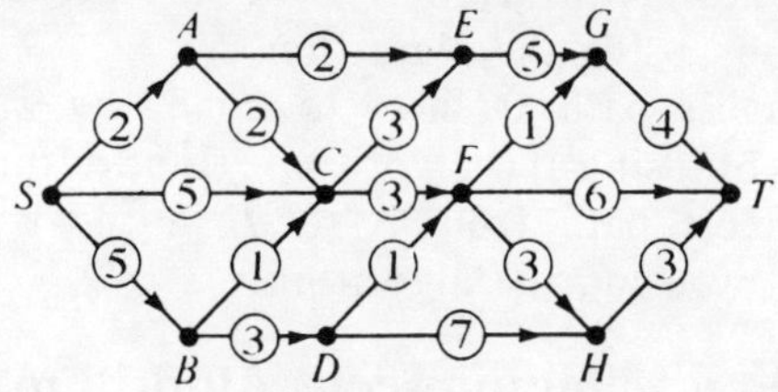

7.4. Consider the following network, in which the circled numbers represent lengths, and the uncircled numbers represent capacities of the arcs.

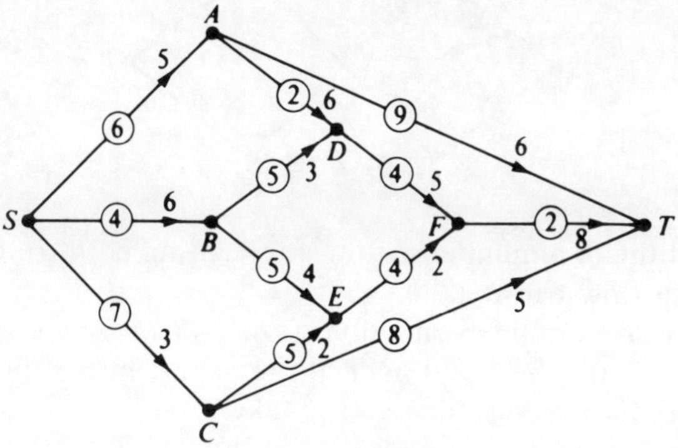

Ignoring the capacities, use the algorithm to find the shortest path from S to T.

7.5. Consider the following weighted digraph, in which the circled numbers on the arcs represent lengths.

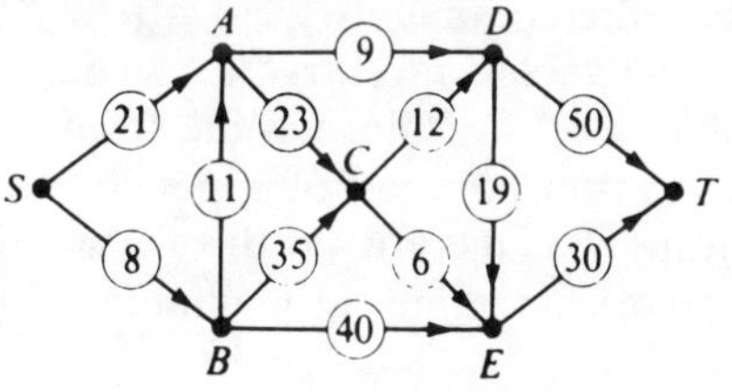

Using the algorithm, find the shortest path from S to T. Show clearly how the algorithm is being applied.

The Longest Path Algorithm

7.6. Find the longest path from S to T in the weighted digraph of Exercise 7.2, by using the algorithm.

7.7. Find the longest path from S to T in the weighted digraph of Exercise 7.3, by using the algorithm.

7.8. Find the longest path from S to T in the weighted digraph of Exercise 7.4, by using the algorithm. (Ignore the capacities.)

7.9. Find the longest path from S to T in the weighted digraph of Exercise 7.5, by using the algorithm.

The Maximum-flow Minimum-cost Algorithm

7.10. Consider the following network:

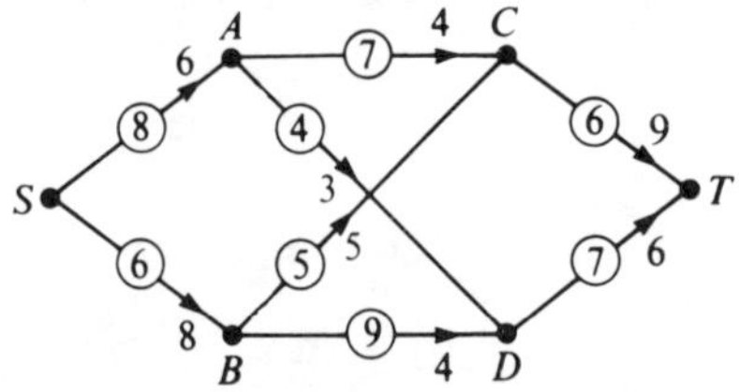

(a) Find the minimum-cost flow-augmenting path from S to T. How much flow can it take?

(b) If the maximum permissible flow is sent down the path found in part (a), which path is then the minimum-cost flow-augmenting path? How much flow can it take?

(c) By repeating this process, find a flow with value 10 at minimum cost.

(d) Find a maximum flow at minimum cost. What is the value of this minimum cost?

7.11. A fleet of express delivery vehicles leaves a depot S, collects packages from four branch offices A, B, C and D of a company, and delivers them to the company's head office T.

In the following table, an entry (m, n) in row X and column Y indicates the existence of a direct route from X to Y with a capacity of n vehicles and a transport cost of m pounds per vehicle. (The absence of an entry means that there is no direct route from X to Y.)

Using the algorithm, find a maximum flow from S to T at minimum cost.

	A	B	C	D	T
S	(5,3)	(3,5)	—	—	—
A	—	—	(4,10)	(6,10)	—
B	—	—	(5,10)	(9,10)	—
C	—	—	—	—	(1,2)
D	—	—	—	—	(1,5)

7.12. In the following network, the circled numbers represent costs, and the uncircled numbers represent arc capacities.

Use the algorithm to find a maximum flow from S to T at minimum cost.

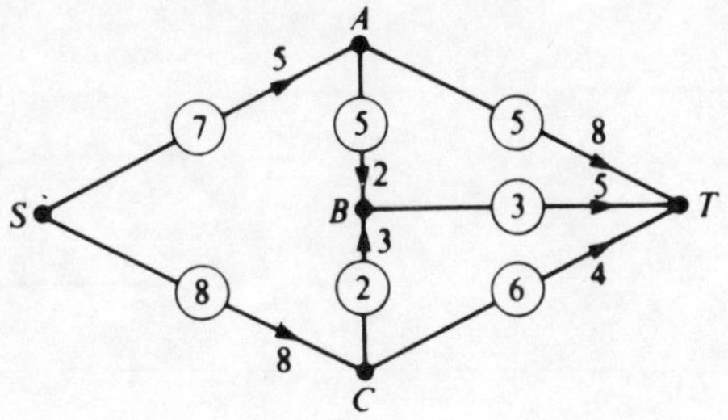

7.13. In the following network, the circled numbers represent costs and the uncircled numbers represent arc capacities.

Use the algorithm to find a maximum flow from S to T at minimum cost.

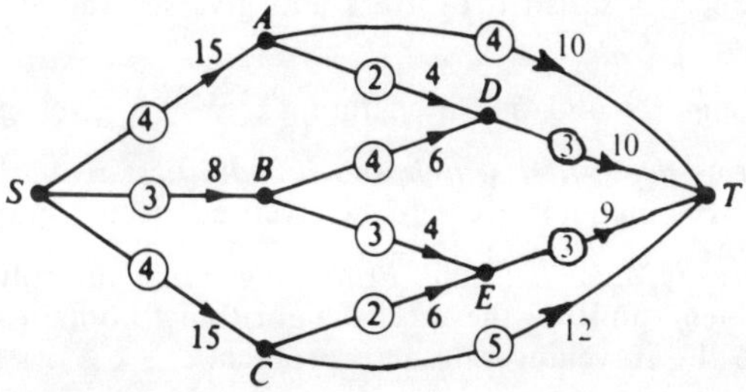

7.14. Consider the network of Exercise 7.4, and let the circled number on each arc be the cost of sending one unit of a commodity along the arc. Find a minimum-cost flow from S to T with a flow of value 10 units. Calculate the total cost, and determine whether this flow is a maximum flow.

CHAPTER 8

TREES

After reading this chapter, you should be able to:

- state some properties satisfied by trees and give several equivalent definitions of a tree;
- define a *spanning tree* and find a spanning tree of a given graph;
- explain the terms *depth-first search* and *breadth-first search* for trees and find the DFS and BFS spanning trees for a given connected graph;
- describe the *greedy algorithm* and *Prim's algorithm* for solving the minimum connector problem, and use the greedy algorithm to obtain a lower bound for the solution to the travelling salesman problem;
- state Cayley's theorem for counting labelled trees;
- explain how tree algorithms have been used to generate efficient gas pipe-line networks.

Introduction

The tree is in many ways the simplest non-trivial type of graph, and has applications in many areas of graph theory. The concept arose historically in connection with Kirchhoff's work on electrical network theory and with Cayley's work on the enumeration of chemical molecules. More recently, it has found applications in such areas as computer science, decision theory and linguistics. Our main interest is in algorithms associated with tree structures. In particular, we look at algorithms for constructing minimum spanning trees and their application to the travelling salesman problem and at algorithms for searching trees and labelling parts of them. We also look at the problem of counting trees. We end this chapter by looking at an application of trees to gas pipe-line networks. Before discussing these topics, we consider the general mathematical properties of trees.

8.1 MATHEMATICAL PROPERTIES OF TREES

In Chapter 2 we defined a tree as follows:

Definition *A **tree** is a connected graph which contains no cycles.*

For example, the graph in the following diagram is a tree with 6 vertices.

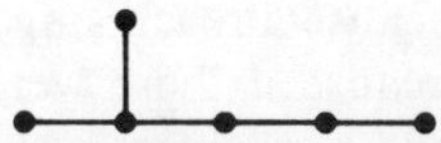

Problem 8.1 Write down as many properties of trees as you can.

If in the tree shown above we add an edge joining one of the vertices to a new vertex, we create a tree with 7 vertices. This illustrates a general procedure for increasing the size of a tree, which works because adding such an edge creates no cycles. It can be carried out systematically by joining the edge to each existing vertex in turn. If we do this with the above tree, we obtain the following trees with 7 vertices:

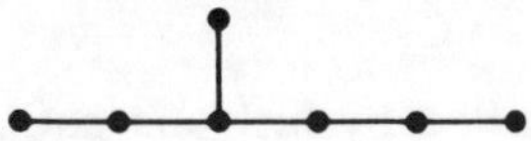

(a)

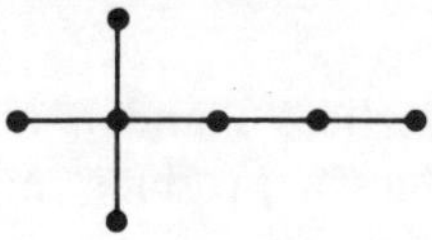

(b)

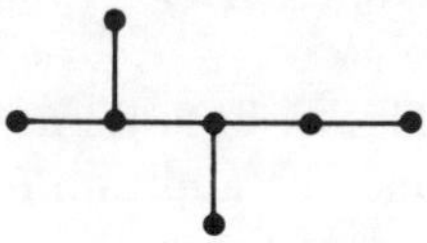

(c)

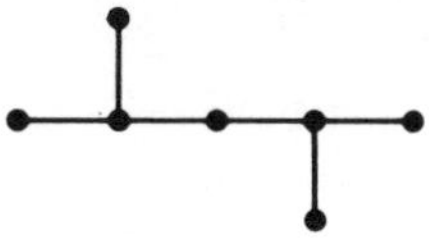

(d)

(e)

(f)

Notice that we can omit tree (f) from this list since it is isomorphic to tree (a), and so we get five trees with 7 vertices from our original tree with 6 vertices.

The difficulty of generating trees by this procedure is in recognizing duplicates, but at least we can be certain that each tree with 7 vertices can be obtained in this way at least once from some tree with 6 vertices.

Starting with the tree with just one vertex, we can build up any tree we wish by successively adding a new edge and a new vertex. At each stage, the number of vertices exceeds the number of edges by 1, so it follows that:

every tree with n vertices has exactly $n - 1$ edges.

At no stage is a cycle created, since each added edge joins an old vertex to a new one. It follows that:

any two vertices in a tree are connected by exactly one path.

There is at least one path, because at each stage the tree remains connected, and there is at most one, because if there were two or more paths joining two given vertices, they would contain a cycle.

In particular, any two adjacent vertices are connected by only one path, so if the edge forming this path is removed, there is then no path between the two vertices. It follows that:

each edge of a tree is a bridge.

Moreover,

the addition of an edge between any two vertices of a tree creates exactly one cycle.

This is because any two vertices v and w are connected by just one path, and the addition of the edge vw produces a single cycle—the cycle consisting of the path and the edge vw.

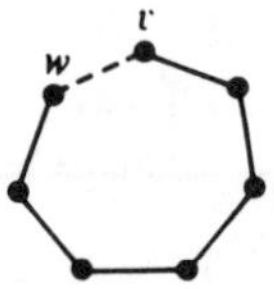

Several of the properties just obtained can be used as alternative definitions of a tree. In the following theorem we state six of these alternative definitions; they are all *equivalent*: any one of them can be taken as the definition of a tree, and the other five can then be deduced.

Theorem 8.1 *Let T be a graph with n vertices. Then the following statements are equivalent*:

(i) *T is connected and contains no cycles*;
(ii) *T is connected and has $n - 1$ edges*;
(iii) *T has $n - 1$ edges and contains no cycles*;
(iv) *T is connected and each edge is a bridge*;
(v) *any two vertices of T are connected by exactly one path*;
(vi) *T contains no cycles, but the addition of any new edge creates exactly one cycle.*

Problem 8.2

(a) Draw the three trees with five vertices.
(b) By adding a new edge and a new vertex in turn to each vertex of the trees with five vertices, generate the six different trees with six vertices.

8.2 CONSTRUCTING SPANNING TREES

We have already seen an example of one method for constructing trees, namely the process of constructing larger trees from smaller ones by repeatedly adding an extra vertex and edge.

Often we require to find a tree which is a subgraph of a given graph. Usually we require that the tree *spans* all the vertices; such a tree is called a *spanning tree*.

Definition *Let G be a connected graph. A* **spanning tree** *in G is a subgraph of G which includes all the vertices of G and is a tree.*

For example, the following diagram illustrates a graph and three of its spanning trees.

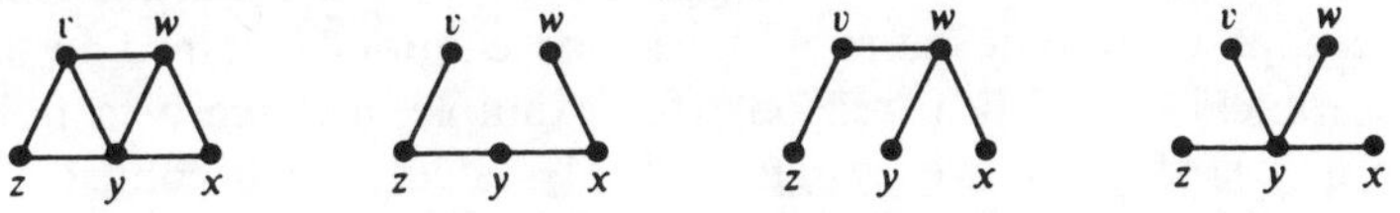

a graph G spanning tree spanning tree spanning tree

We shall see in Chapter 10 that spanning trees play an important part in the theory of electrical networks. A spanning tree can be constructed from a given connected graph G by either of the following methods.

Building-up method *Select edges of G one at a time in such a way that no cycles are created; repeat this procedure until all the vertices are included.*

For example, in the above graph *G*, choose the edges *vz*, *wx*, *xy* and *yz*; then no cycles are created, and we obtain the first of the spanning trees shown above.

Cutting-down method *Choose any cycle, and remove any one of its edges. Repeat this procedure until there are no cycles left.*

For example, from the graph *G*, remove the edges

vy (destroying the cycle *vwyv*)
yz (destroying the cycle *vwyzv*)
xy (destroying the cycle *wxyw*).

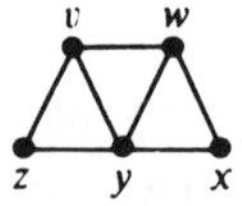

a graph *G*

In this way we obtain the second of the spanning trees shown above.

Problem 8.3 Use the building-up or the cutting-down method to construct all possible spanning trees of the above graph *G*. (There are 21 in total, including the three shown above.)

If the edges of the graph have weights associated with them (representing distances, for example) we may be required to find a spanning tree of minimum total length. This is known as the *minimum connector problem*.

The Minimum Connector Problem

Consider the following example:

Example 1 An irrigation canal system is to be built, connecting a number of given locations. The cost of digging and maintaining each canal is known, except that some pairs of locations cannot be joined by a canal for geographical or political reasons (for example, a gorge or a politically inviolate area). How do we design a canal system which connects all the locations at minimum possible total cost?

This problem can be interpreted in two different ways, depending on whether or not we allow extra 'locations' where canals may intersect. For example, in the case of the canal system shown below, we may be able to reduce the total cost by creating an extra location at the point *E*, and linking it to *A*.

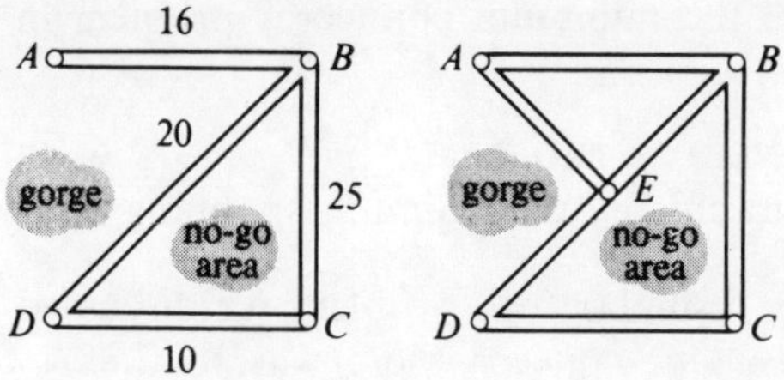

canal system system with extra location E

Unfortunately, for many minimum connector-type problems, the cost of inserting an extra location (which may be a telephone exchange or power station) can greatly exceed the possible saving in cost, and the resulting mathematical analysis becomes rather complicated. In view of this, we adopt the second interpretation of the problem and assume that each connection joins two existing locations.

We can represent the minimum connector problem graphically by taking the locations as vertices and the canals as edges, giving us a weighted graph. The problem is then to find a subgraph of minimum total weight (the sum of the weights on the edges), passing through each vertex. Note that *a subgraph of minimum weight must always be a spanning tree*, because if there is a cycle we can lower the total cost by removing any one of its edges.

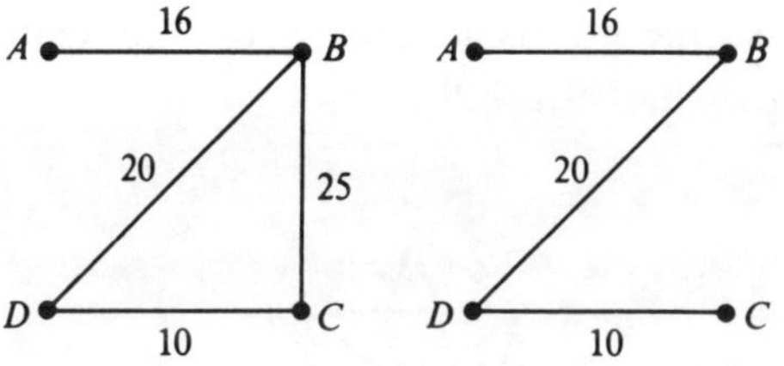

graph minimum spanning tree

In our example, the graph has total weight 71. Removal of any one of the edges in the cycle BCD lowers the total weight, and gives a spanning tree. Clearly, the spanning tree of minimum total weight is obtained by removing the edge BC, and the minimum total cost is therefore

$$16 + 20 + 10 = 46.$$

Definition *Let T be a spanning tree of minimum total weight in a connected weighted graph G. Then T is a* ***minimum spanning tree*** *(or a* ***minimum connector****) of G.*

We can now restate the minimum connector problem in graphical terms, as follows.

THE MINIMUM CONNECTOR PROBLEM
Given a weighted graph, find a minimum spanning tree in it.

Note that this is very similar to a problem we discussed in Chapter 5 on multi-terminal flows. Our concern there was to find a tree of *maximum* total capacity (which we called a *maximum tree*), and we described how to do this by successively choosing edges of largest capacity in such a way that no cycles are created. An analogous construction works for the minimum connector problem; it is known as the *greedy algorithm*, or *Kruskal's algorithm*, and was published in 1956.

The greedy algorithm *To construct a minimum spanning tree in a connected weighted graph G, successively choose edges of G of minimum weight in such a way that no cycles are created.*

The name 'greedy algorithm' arises from the fact that at each stage we make the greediest choice available (that is, the edge involving the smallest weight) with no concern for what is happening elsewhere in the graph. Algorithms of this kind do not usually succeed in practice, but this is one algorithm which does.

We illustrate the use of this algorithm by the following worked problem.

Worked problem Use the greedy algorithm to find a minimum spanning tree in the following weighted graph:

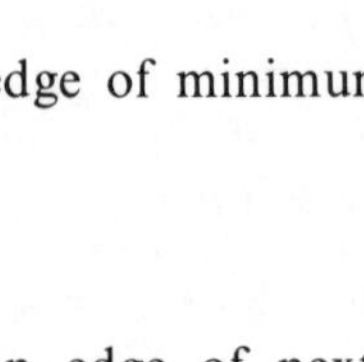

Solution
First choice We choose an edge of minimum weight; this is AE, with weight 2.

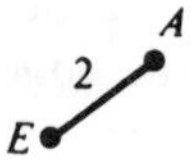

Second choice We choose an edge of next smallest weight; this is either AC or CE, with weight 4. Let us choose CE.

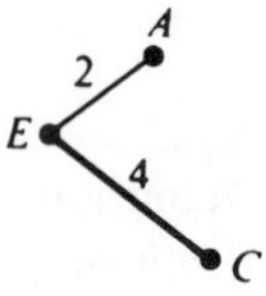

Third choice We cannot now include AC in the tree, since it would create a cycle ($ACEA$), so we choose an edge of next smallest weight; this is BC, with weight 5.

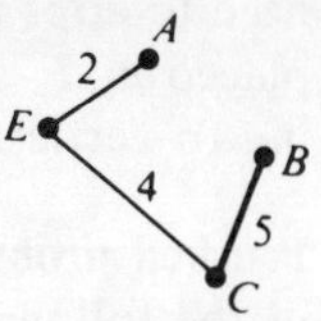

Fourth choice The edges of next smallest weight are AB and BE, each with weight 6. Since either of these would create a cycle ($ABCEA$ or $BCEB$), we choose instead the edge DE, with weight 7.

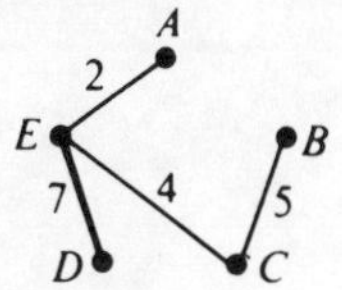

This completes the spanning tree, which is a minimum spanning tree of weight 18. (Note that if we had chosen the edge AC at the second stage, rather than the edge CE, then we should have obtained a different spanning tree, but its weight would still be 18.)

Problem 8.4 Use the greedy algorithm to find a minimum spanning tree in the following weighted graph:

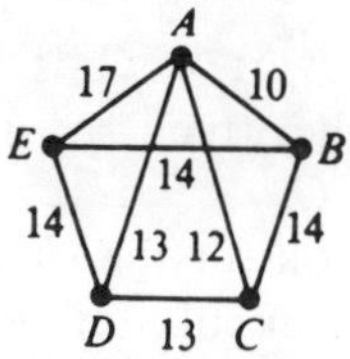

Although the greedy algorithm can easily be applied by hand when the graph is small, it is not particularly well suited for efficient computer implementation, owing to the need to arrange the edges in order of ascending weight, and the need to recognize cycles as they are created. Both of these difficulties can easily be overcome by a slight modification of the algorithm; the result is known as *Prim's algorithm.*

Prim's algorithm *To construct a minimum spanning tree T in a connected weighted graph G, build up T step by step, as follows*:

(i) *put an arbitrary vertex into T*;
(ii) *successively add edges of minimum weight joining a vertex already in T to a vertex not in T.*

The great advantage of Prim's algorithm is that we can operate directly on the table of weights rather than on the graph itself. If the graph is large, this makes the method more suitable for computer implementation. All we need to do is to delete a row of the table whenever the corresponding vertex is placed in T, and then choose the smallest entry in the column corresponding to vertices in T. The following worked problem illustrates the method.

Worked problem Use Prim's algorithm to find a minimum spanning tree in the following weighted graph:

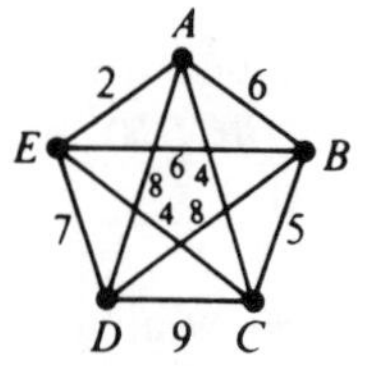

	A	B	C	D	E
A	—	6	4	8	2
B	6	—	5	8	6
C	4	5	—	9	4
D	8	8	9	—	7
E	2	6	4	7	—

Solution

First choice We choose any vertex, say B, and put it into T.

Delete row B.

Look for the smallest entry in column B; it is 5 in row C.

	A	B	C	D	E
A	—	6	4	8	2
C	4	⑤	—	9	4
D	8	8	9	—	7
E	2	6	4	7	—

Second choice BC is the edge of minimum weight joining B to other vertices. Put the edge BC and the vertex C into T.

The vertex C is now in the tree, so delete row C.

Look for the smallest entry in columns B and C; it is 4 in rows A and E.

	A	B	C	D	E
A	—	6	④	8	2
D	8	8	9	—	7
E	2	6	④	7	—

Third choice CA and CE are the edges of minimum weight joining B and C to other vertices. Choose one of these, say CA, and put the edge CA and the vertex A into T.

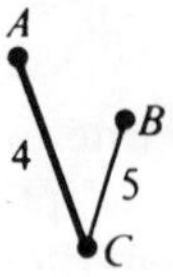

The vertex A is now in the tree, so delete row A.

Look for the smallest entry in columns A, B and C; it is 2 in row E.

	A	B	C	D	E
D	8	8	9	—	7
E	②	6	4	7	—

Fourth choice AE is the edge of minimum weight joining A, B and C to other vertices. Put the edge AE and the vertex E into T.

The vertex E is now in the tree, so delete row E.

Look for the smallest entry in columns A, B, C and E; it is 7 in row D.

	A	B	C	D	E
D	8	8	9	—	⑦

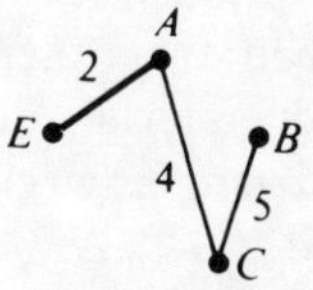

Fifth choice ED is the edge of minimum weight joining A, B, C and E to D. Put the edge ED and the vertex D into T.

All the vertices are now in the tree.

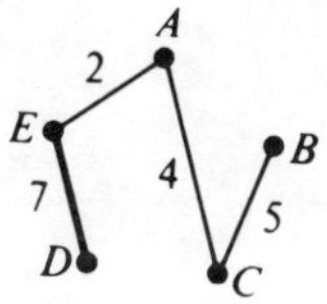

This completes the spanning tree, which is a minimum spanning tree of weight

$$5 + 4 + 2 + 7 = 18.$$

Problem 8.5 Use Prim's algorithm to find a minimum spanning tree in the following weighted graph:

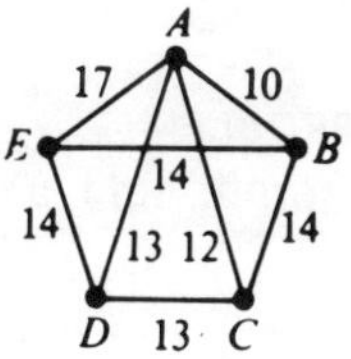

Application to the Travelling Salesman Problem

In Chapter 1 we discussed the travelling salesman problem, in which a salesman wishes to visit a number of cities and return to his starting-point, covering the minimum possible total distance on the way. This problem is an important one in practice, since it can appear in a number of different guises. An example is the *job-sequencing problem*, which can be stated as follows.

THE JOB-SEQUENCING PROBLEM

A number of independent jobs are to be carried out on a single machine. These jobs are complicated, and the machine must be set up for each new job after completing the previous one. The machine is initially set up for one of the jobs, and it must be reset for that job when the other jobs are completed. If the set-up costs (involving labour and material) depend on the job just completed and the job about to be started, how can we order the jobs so that the total set-up cost is minimized?

The link between the two problems may be seen by considering the problems graphically. In the travelling salesman problem we draw a weighted complete graph in which the vertices correspond to the cities visited, the edges correspond to the routes joining them, and the weights correspond to the distances between pairs of cities.

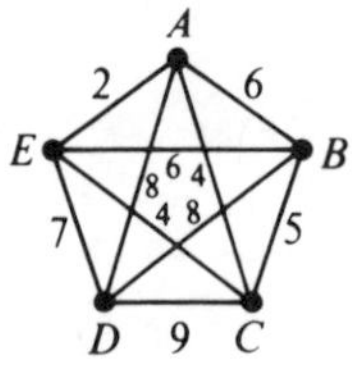

In the job-sequencing problem we draw a weighted complete graph in which the vertices correspond to jobs, the edges link these jobs, and the weights correspond to the set-up costs associated with pairs of jobs. In each case, our aim is to find a cycle of minimum total weight which passes through every vertex—in other words, a *minimum-weight Hamiltonian cycle*. The travelling salesman problem may therefore be restated in graphical terms as follows.

THE TRAVELLING SALESMAN PROBLEM

Given a weighted complete graph, find a minimum-weight Hamiltonian cycle in it.

In view of the simple nature of the greedy algorithm for solving the minimum connector problem, we might hope that there is a simple algorithm here as well. Unfortunately, as we remarked in Chapter 1, no such algorithm is known. We could, of course, try all possible Hamiltonian cycles, but this is a hopeless task, even on a computer, unless the number of vertices is very small. For a job-sequencing problem involving (say) 100 jobs, there would be 100! ($\approx 93 \times 10^{157}$) sequences to be considered, and no method along these lines would be worth attempting.

In view of this, we are forced to look for approximate solutions to the problem. One method, which often works well in practice, is to find a *lower bound* for the minimum-weight Hamiltonian cycle by solving the minimum connector problem instead. To justify this method, we argue as follows.

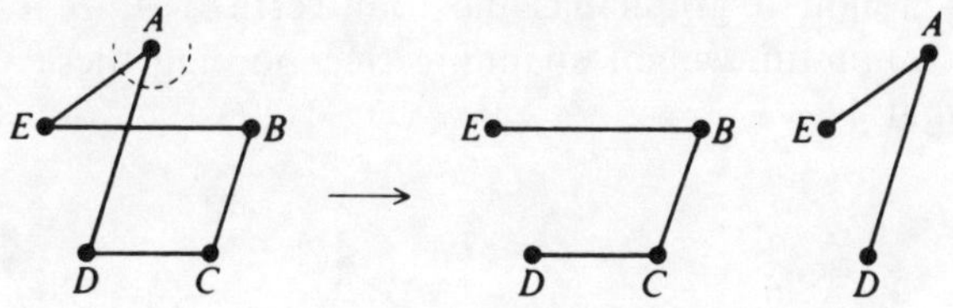

If we take a minimum-weight Hamiltonian cycle in a weighted complete graph, and remove a vertex A and its incident edges, we get a path passing through the remaining vertices. Such a path must be a spanning tree for the complete graph formed by these remaining vertices, and the weight of the Hamiltonian cycle is obtained by adding the weight of this spanning tree to the weights of the two edges incident to A. We can therefore obtain a *lower bound* for the solution of the travelling salesman problem by adding the weight of a minimum spanning tree joining those vertices to the two smallest weights of edges incident to A.

Worked problem Find a lower bound for the travelling salesman problem for the following weighted graph:

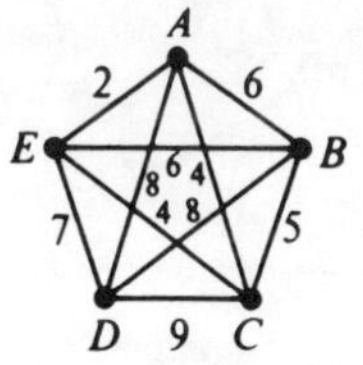

Solution If we remove the vertex A, then the remaining weighted graph has the four vertices B, C, D and E.

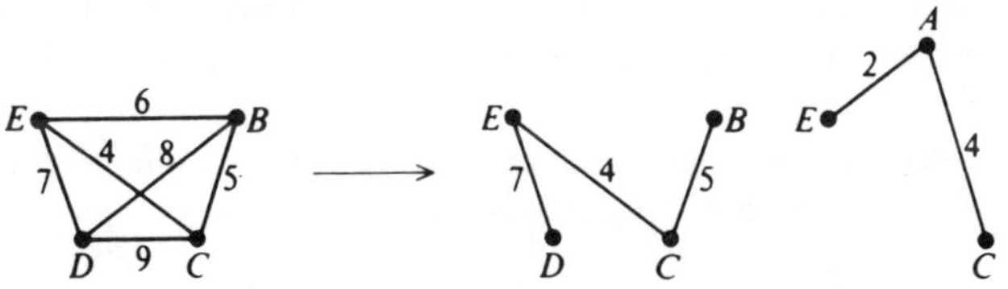

The minimum-weight spanning tree joining these vertices is the tree whose edges are BC, CE and ED, with total weight 16. The two edges of smallest weight incident to A are AE and AC, with total weight 6. The required lower bound for the travelling salesman problem is therefore

$$16 + 6 = 22.$$

A better lower bound is obtained by removing the vertex D. In this case, the remaining weighted graph has the four vertices A, B, C and E, and there are two minimum-weight spanning trees joining these vertices, each with total weight 11.

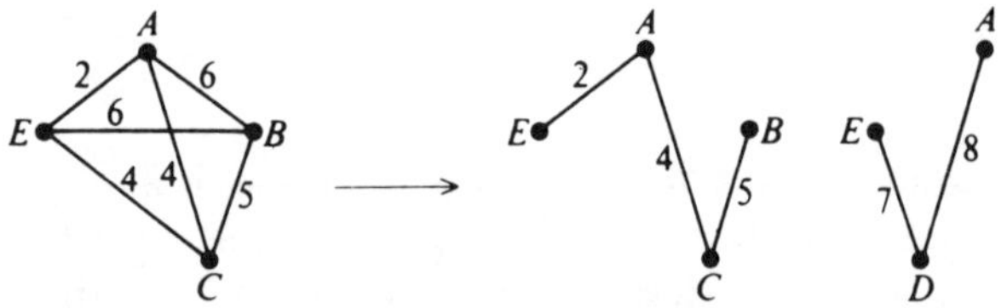

The two edges of smallest weight incident to D are DE, and DA or DB, with total weight 15. The required lower bound for the travelling salesman problem is therefore $11 + 15 = 26$. This second lower bound is in fact equal to the actual solution of the travelling salesman problem, which is given by the cycle $ACBDEA$ with total weight 26.

Problem 8.6 In the above example, find the lower bounds obtained by removing

(a) the vertex B; (b) the vertex E.

8.3 SEARCHING TREES

A problem which frequently arises in practice is that of searching through some given tree structure in a systematic way. For example, a computer file is often organized as a tree-like data-structure in some form of random-access memory (RAM), and a systematic tree search is necessary whenever a particular piece of information is required. In practice, this usually involves examining every part of the tree until the desired vertex or edge is found; in order to avoid unnecessary wastage of time and processing resources, we need a search technique which is guaranteed to visit all parts of the tree eventually without visiting any particular vertex too often.

There are two well-known search methods, which differ in the pattern of search they employ. They are usually known as **depth-first search** (DFS) and **breadth-first search** (BFS). Each of these methods lists the vertices as they are encountered, and indicates the direction in which each edge is first traversed. The methods differ only in the way in which the vertex-lists are constructed. Although our discussion here is introduced in the context of trees, both methods can easily be applied to more general types of graph. In such applications, they effectively search the graph by searching through all the vertices of an appropriate spanning tree.

Depth-first Search (DFS)

The basic idea of depth-first search is to penetrate as deeply as possible into a tree before fanning out to other vertices.

Example 2 Consider the following tree:

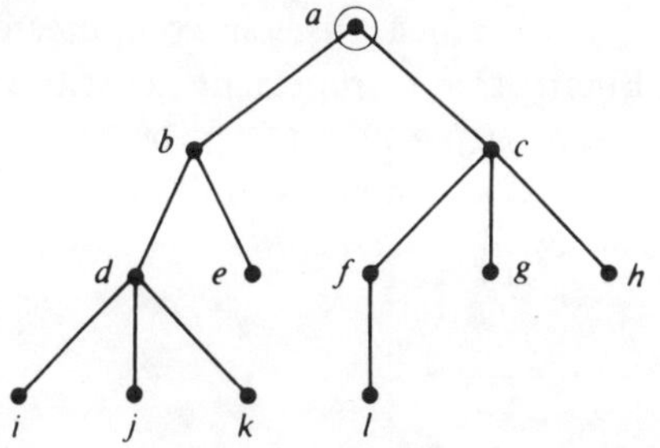

In order to perform a depth-first search starting at the vertex a, we start by choosing any vertex adjacent to a (b, say), then any vertex adjacent to b (d, say), and so on. At each stage, we choose a vertex not previously used, if it is possible to do so. We number the vertices as we go. Thus, we may start by assigning to a, b, d and i the labels 1, 2, 3 and 4, respectively. Once we have reached i we are stuck, since there are no new vertices to go to. We must therefore backtrack to d, from which we can go to j, assigning it the label 5. Backtracking to d again takes us to k, to which we assign the label 6. We now have to backtrack, via d, to b, from which we can go to e, assigning it the label 7. Backtracking to a takes us to c, f and l, and (eventually) to g and h. This gives us the following labelled tree.

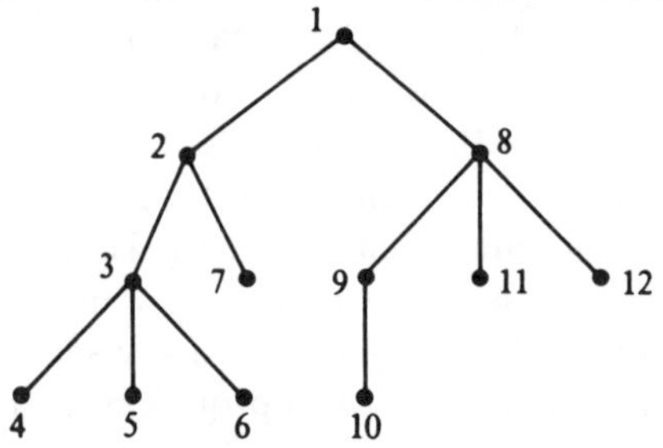

labelled tree

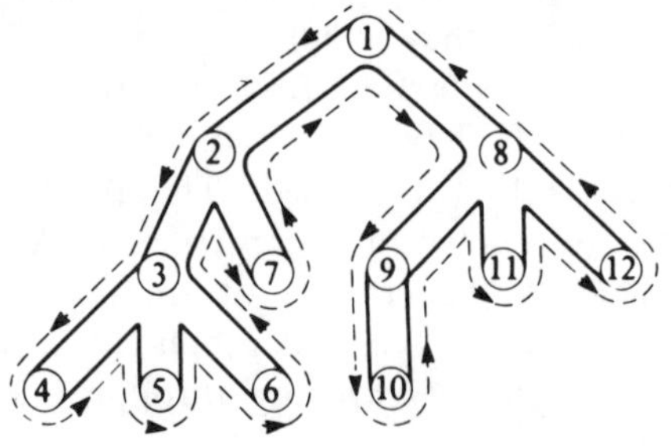

wall diagram

Because we need to backtrack whenever we get stuck, DFS is sometimes called *backtrack search*.

In choosing which adjacent vertex to take at each stage, we have selected the left-hand one in every case, although it is not necessary to be so systematic.

Another way of obtaining this labelling is to regard the tree as a solid wall, and to go for a walk around it, always keeping the wall on our left. Whenever we reach a new vertex, we give it the next label, as shown in the above diagram. A similar idea arises in computer programming, where we sometimes use a binary tree to represent an arithmetic expression. For example, the expression $a + ((b - c) \times d)$ can be represented by the following tree; we retrieve this expression by walking around the tree, writing down each vertex as we walk *underneath* it.

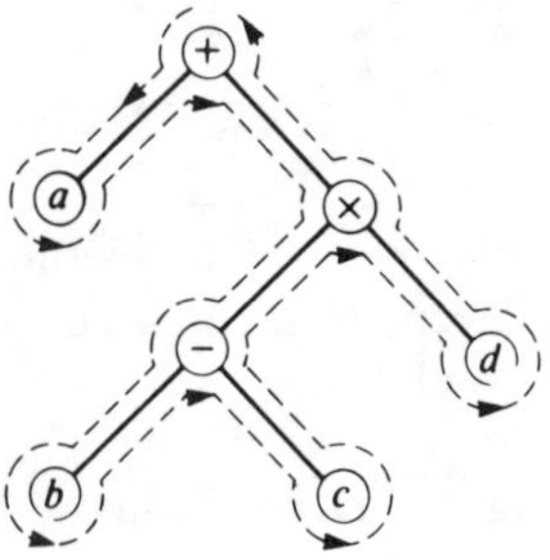

It is often useful, especially in applications to computer science, to describe a particular depth-first search by means of a list. Each 'word' in this list corresponds to the path from the initial vertex to the current vertex (the vertex being considered at that time). To form the next word in the list, we apply the following rules.

1. If the last vertex in the word is adjacent to a new vertex, add this new vertex to the word;
2. If the last vertex in the word is not adjacent to a new vertex, delete the last vertex from the word.

Worked problem

Apply the rules to the following tree:

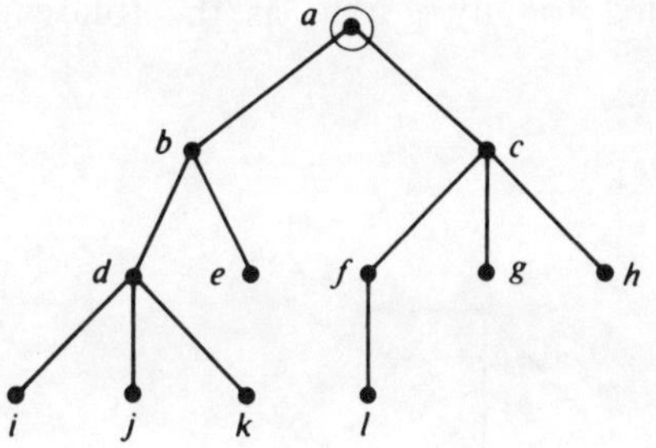

Solution The complete list corresponding to the above depth-first search, together with the label assigned at each stage, is given on the right. Note that the word *abd* becomes *abdi* (by rule 1), and the word *abdi* becomes *abd* (by rule 2).

List	Label
a	$a = 1$
ab	$b = 2$
abd	$d = 3$
$abdi$	$i = 4$
abd	
$abdj$	$j = 5$
abd	
$abdk$	$k = 6$
abd	
ab	
abe	$e = 7$
ab	
a	
ac	$c = 8$
acf	$f = 9$
$acfl$	$l = 10$
acf	
ac	
acg	$g = 11$
ac	
ach	$h = 12$
ac	
a	

In forming this list, only the *last* letter in each word is relevant. Because of this, the list is sometimes called **last-in first-out**, often abbreviated to LIFO. (In computer science, such a list is referred to as a *stack*.)

Finally, we can abbreviate this list by writing down the vertices in the order in which they are labelled. For example, we can represent the above depth-first search by the list *abdijkecflgh*.

Problem 8.7 Perform a depth-first search on the following tree, starting at the vertex *a*. Write down the corresponding list, the label of each vertex, and the abbreviated list.

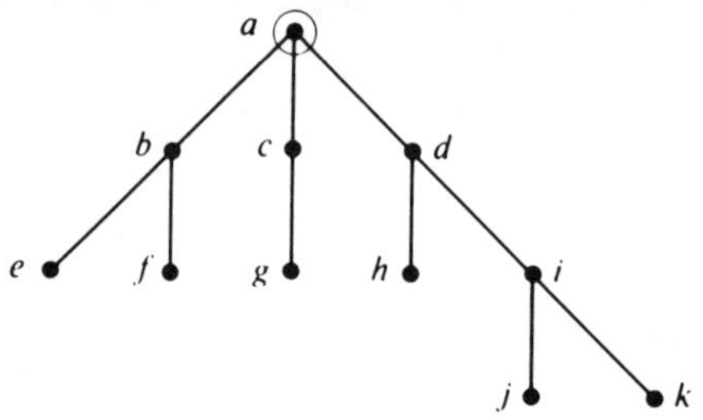

The above procedures can be extended to any graph, as the following example shows:

Example 3 Consider the following graph:

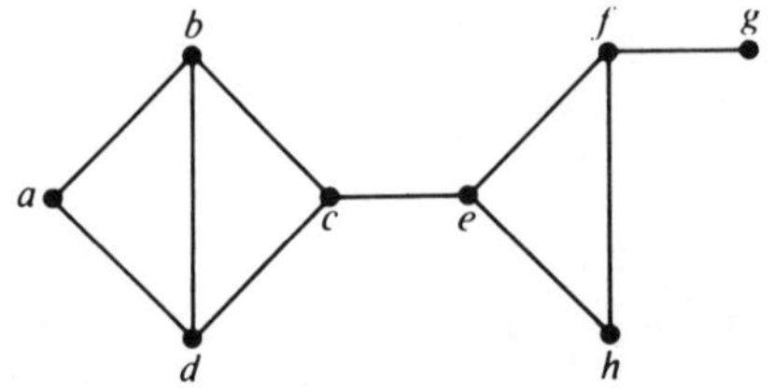

We can perform a depth-first search on this graph by starting at *a*, going to *b*, *c* and *d*, backtracking to *c*, going to *e*, *f* and *g*, backtracking to *f*, going to *h*, and returning to *a*. This gives us the following labelling:

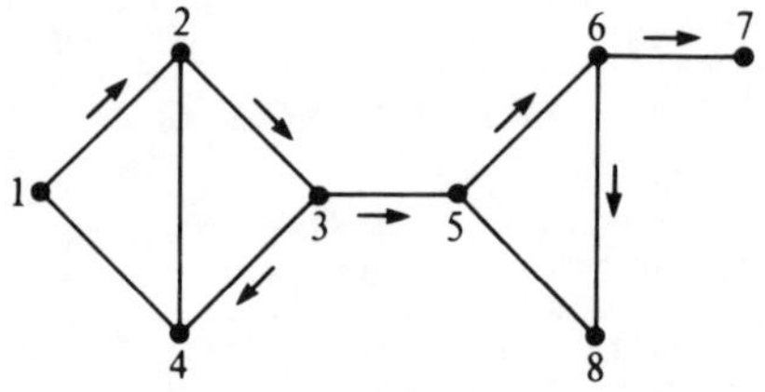

Notice that we have marked with an arrow those edges we used when going to each new vertex. These edges form a spanning tree, called a **DFS spanning tree**.

Breadth-first Search (BFS)

The basic idea of breadth-first search is to fan out to as many vertices as possible before penetrating deep into a tree. This means that we visit all vertices adjacent to the current vertex before going on to another one, as the following example shows.

Example 4 Consider the following tree:

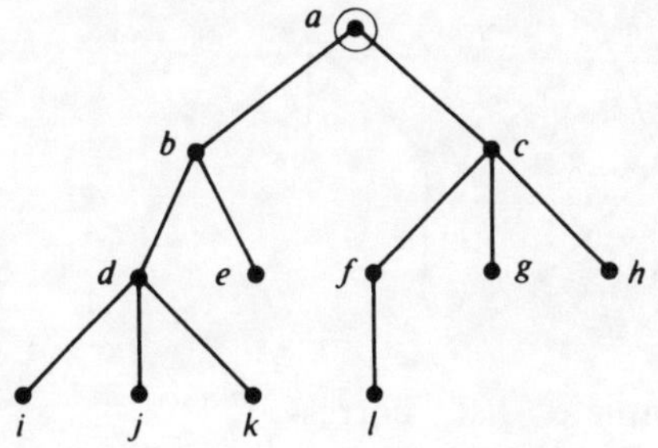

In order to perform a breadth-first search starting at the vertex a, we start by visiting the vertices b and c, which are adjacent to a. We then visit the vertices d and e adjacent to b, and the vertices f, g and h adjacent to c. If we label each new vertex as we come to it, we eventually obtain the following labelling of the vertices (the DFS labelling is shown for comparison):

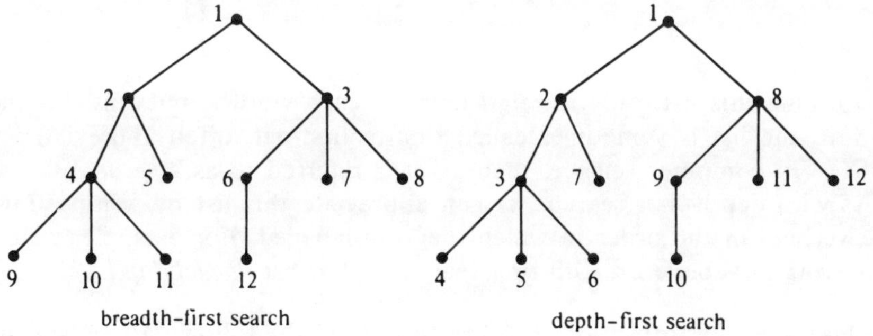

Note that if the vertices are laid out in horizontal 'levels', as in this example, then in a breadth-first search we must complete each level before proceeding to the next one.

This procedure can also be described by a list, obtained by the following rules:

1. if the first vertex in the word is adjacent to a new vertex, add this new vertex to the word;
2. if the first vertex in the word is not adjacent to a new vertex, delete the first vertex from the word.

Worked problem
Apply the rules to the following tree:

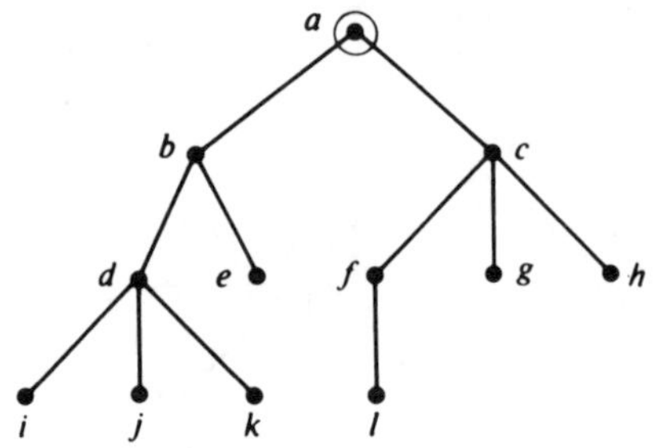

Solution The complete list corresponding to the above breadth-first search, together with the label assigned at each stage, is given on the right. Note that the word *ab* becomes *abc* (by rule 1), and the word *abc* becomes *bc* (by rule 2).

List	Label
a	$a = 1$
ab	$b = 2$
abc	$c = 3$
bc	
bcd	$d = 4$
bcde	$e = 5$
cde	
cdef	$f = 6$
cdefg	$g = 7$
cdefgh	$h = 8$
defgh	
defghi	$i = 9$
defghij	$j = 10$
defghijk	$k = 11$
efghijk	
fghijk	
fghijkl	$l = 12$
ghijkl	
hijkl	
ijkl	
jkl	
kl	
l	

In forming this list, only the *first* letter in each word is relevant. Because of this, the list is sometimes called **first-in first-out**, often abbreviated to FIFO. (In computer science, such a list is referred to as a *queue*.)

As with depth-first search, we can abbreviate this list by writing down the vertices in the order in which they are labelled. For example, we can represent the above breadth-first search by the list *abcdefghijkl*.

Problem 8.8 Perform a breadth-first search on the following tree, starting at the vertex *a*. Write down the corresponding list, the label of each vertex, and the abbreviated list.

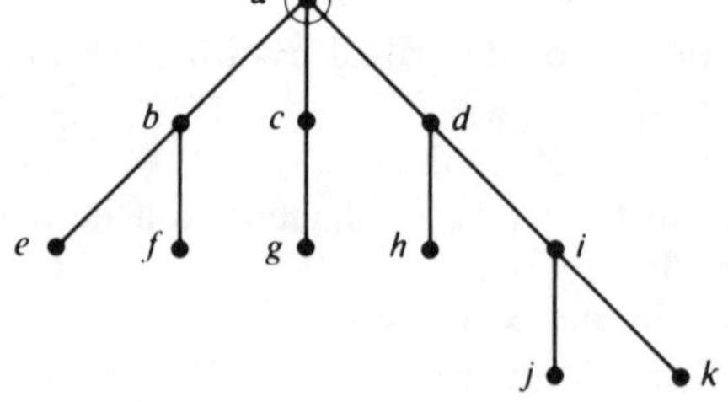

The above procedure can be extended to any graph, as the following example shows.

Example 5 Consider the following graph:

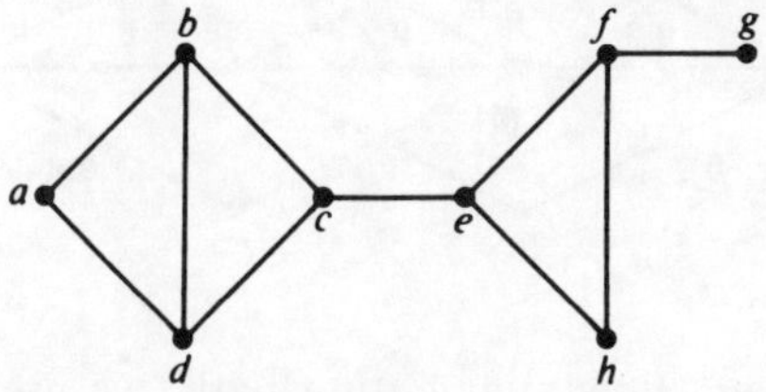

We can perform a breadth-first search on this graph by starting at a, visiting the vertices b and d adjacent to a, visiting the vertex c adjacent to b, and so on. This gives us the following labelling:

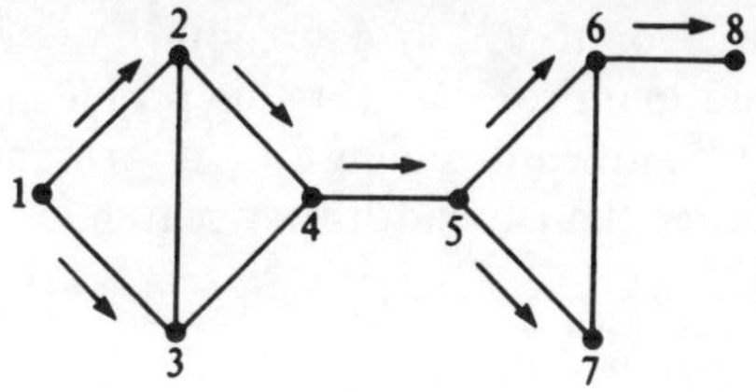

Notice that we have marked with an arrow those edges we used when going to each new vertex. These edges form a spanning tree, called a **BFS spanning tree**.

DFS versus BFS

For many problems involving graphs it is important that we choose the correct searching procedure, since the wrong decision can result in a great deal of unnecessary work and expense. However, no general rule can be given as to which procedure to choose, since each method has its advantages. We illustrate this by means of two algorithms introduced earlier.

The Maximum Flow Algorithm

In Chapter 3, we described the *maximum flow algorithm*—a systematic step-by-step method for building up a maximum flow in a given basic network by finding a succession of flow-augmenting paths. For example, in the following network, we can increase the flow by 3 along $SADT$, by 1 along

SBDT, by 2 along *SBET*, and by 1 along *SCEBFT*, giving a maximum flow of 7.

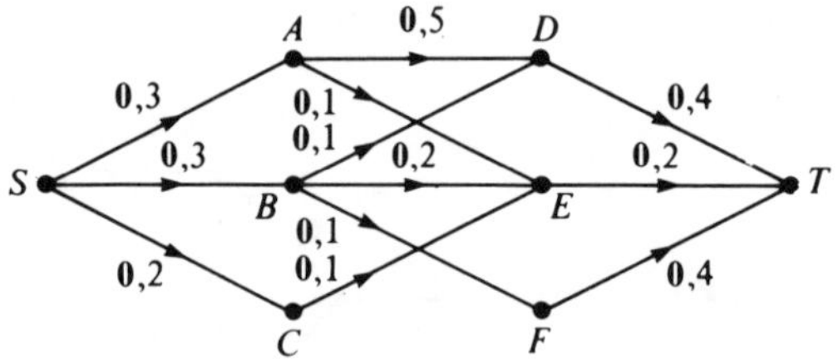

In order to find each flow-augmenting path, we use what amounts to a *depth-first search*. For example, we find the flow-augmenting path *SADT* by increasing the flow from *S* to *A*, calculating how much of this increase can be transmitted to *D*, and then calculating how much can be transmitted to *T*. If at any stage no increase is possible, then we simply backtrack and try another vertex. At no stage in finding this path are we concerned with the flows from *S* to *B* or to *C*, so a breadth-first search would be very wasteful of effort, and inappropriate for this problem. For large networks, we can often make significant savings in effort and expense by using depth-first search rather than breadth-first search.

The Shortest Path Algorithm

In Chapter 7, we discussed the *shortest path algorithm*—a method for finding the shortest path between two given vertices of a weighted graph or digraph. For example, in the following network, the shortest path from *S* to *T* is *SBCT*, with total length 10.

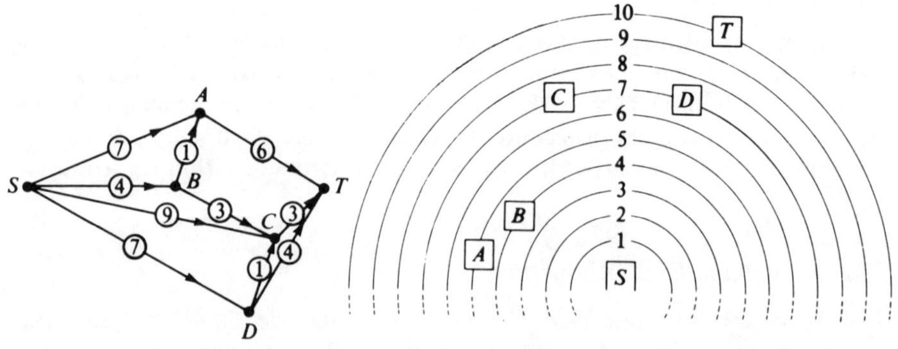

In order to find this shortest path, we construct a diagram of concentric circles indicating the distances from *S* to every other vertex. As we move outwards, we put each vertex on the appropriate circle—first B on circle 4, then A on circle 5, then C and D on circle 7, and finally T on circle 10. Since the distance from S to each vertex is dependent on the distances to the previous vertices, we first need to determine the distance from S to each of its neighbours, before going on to *their* neighbours, and so on. Thus the shortest path algorithm amounts to a *breadth-first search* on the graph or digraph. A depth-first search would involve taking a path directly to T, ignoring the other vertices closer to S on the way, and would be inappropriate for this problem.

8.4 COUNTING TREES

The subject of counting trees probably started with Cayley's attempts to count numbers of chemical molecules. In particular, in the 1870s Cayley endeavoured to count the number of alkanes $C_n H_{2n+2}$ with a given number of carbon atoms. This problem amounts to counting the number of non-isomorphic unlabelled trees in which the degree of every vertex is either 4 or 1.

When counting trees or other graphs, we need to be clear whether the problem we are concerned with is that of counting *labelled* graphs, or of counting *unlabelled* graphs. Both problems have been solved for trees, and the following table gives the numbers of trees with up to ten vertices.

	Number of Vertices									
	1	2	3	4	5	6	7	8	9	10
Unlabelled trees	1	1	1	2	3	6	11	23	47	106
Labelled trees	1	1	3	16	125	1296	16807	262144	4782969	10^8

In fact, there are exactly n^{n-2} labelled trees with n vertices. This result is known as *Cayley's theorem.*

Problem 8.9

(a) Use Cayley's theorem to calculate the number of labelled trees with four vertices. Draw all these trees.
(b) Draw all the non-isomorphic unlabelled trees with four vertices.

8.5 OFFSHORE NATURAL GAS PIPE-LINE SYSTEMS

Natural gas produced from a number of fields in the Gulf of Mexico is transported to the shore by underwater pipe-lines. To save costs, no unnecessary pipes are laid and so the pipe-line network is a tree structure. A typical gas pipe-line network off the coast of Louisiana has the following form, where vertices represent gas fields and edges represent the pipe-lines. The gas flows to an onshore delivery point represented by a small square.

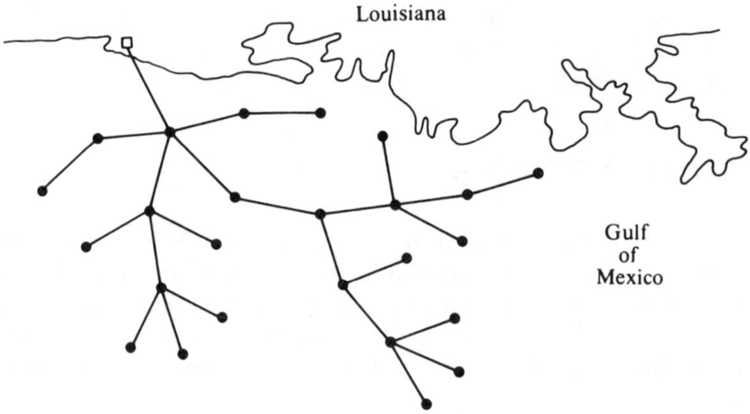

If the costs of the network depended solely on the lengths of the pipe-lines, the most economic network would be a minimum spanning tree connecting all the gas fields. However, things are not so simple because costs also depend on the pressure differences in the network and on the diameters of the pipes used.

A method for designing economical networks was developed in the late 1960s by a group of consultants working for the US government. (See the paper by B. Rothfarb, H. Frank and others, which is listed in the bibliography at the end of this book.)

The equation they used to describe the steady-state flow of gas in a pipe-line is:

$$Q = K[(P_1^2 - P_2^2)/L]^{0.5394}\ D^{2.6182}$$

where Q is the volume flowing in cubic feet per day, P_1 and P_2 are the input and output pressures, respectively, in pounds per square inch (absolute), D is the internal pipe diameter in inches, L is the length of pipe in miles, and K is a constant.

For a given tree pipe-line network, the pipe diameters to be used for each edge of the network are chosen to minimize the total cost. Smaller diameter pipes are cheaper, but they require a larger pressure-squared difference $P_1^2 - P_2^2$ in order to maintain the same flow. For larger pressure-squared differences, it is more expensive to install and operate the compressors which maintain the flow.

The consultants developed a procedure for finding an optimal allocation of pipe diameters for a given tree. They used this procedure to evaluate a number of different plausible trees in order to find a near-optimal tree.

In the allocation of pipe diameters, the problem is to balance the cost of the pipes (which increases with pipe diameter) against the cost of compression (which decreases as pipe diameter increases). For the network in question, 7 different pipe diameters were available, ranging from $10\frac{3}{4}$ inches to 30 inches. So for a tree with 13 edges, there are 7^{13}, or about 9.7×10^{10} possible allocations of pipe diameters. At a rate of one per second it would take more than 3000 years to evaluate them all! It was therefore necessary to eliminate most of the possible allocations from consideration.

A key point which enables us to eliminate a large proportion of diameter allocations is that somewhere in the network there must be a point of highest pressure, and this must occur at one of the vertices. Because the network is a tree, there is a unique path from that vertex to the onshore delivery point. This path is called the *critical path*. (The use of this term is NOT the same as in critical path analysis.) The pressure-squared differences along the critical path determine the delivery point pressure, and so it is these particular pressure-squared differences which determine the cost of compression (which is higher for larger pressures). It follows that for any edge not on the critical path, no savings will be made by using a larger diameter pipe to reduce the pressure-squared difference. All such allocations of diameters can therefore be eliminated from consideration. (If the diameter is reduced, however, the edge may then be on a critical path.)

For each edge, we can prepare a list of the pressure-squared differences (PSQ) and the costs of the pipe-line (PCOST) corresponding to the seven available pipe diameters. For convenience, we label the pipe diameters with a number from 1 to 7, where the label 1 indicates the smallest diameter and 7 the largest. For a particular edge b_1 the list might be as follows:

		Diameter size						
		1	2	3	4	5	6	7
b_1 list	PSQ	120	111	92	66	54	40	31
	PCOST	13	17	23	29	36	45	58

Notice that, as the diameter increases, the PSQ decreases, with correspondingly lower compression costs, but the pipe cost, PCOST, increases.

Rothfarb and his colleagues developed two techniques for combining lists which efficiently eliminate large numbers of possible allocations of diameters which cannot possibly be optimal. These techniques are the *parallel merge* and the *serial merge*.

The Parallel Merge

Suppose that we have three terminal branches leading into a common vertex as shown.

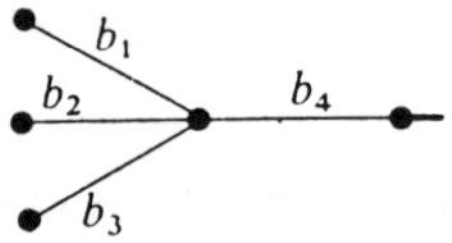

Let the three lists of values of PSQ and PCOST corresponding to all the possible allocations of diameters be as follows:

		Diameter size						
		1	2	3	4	5	6	7
b_1 list	PSQ	120	111	92	66	54	40	31
	PCOST	13	17	23	29	36	45	58
b_2 list	PSQ	150	139	118	87	75	70	67
	PCOST	6	9	14	21	30	40	56
b_3 list	PSQ	94	86	80	61	55	48	32
	PCOST	8	12	18	26	34	43	57

There is a total of $7^3 = 343$ possible allocations of diameters for the three pipe-lines. However, most of these can be eliminated and the three lists merged into a single list containing all the possibilities which we need to consider.

The crucial point is that *only one* of b_1, b_2 and b_3 can be on the critical path.

Suppose that b_2 has the smallest possible pipe diameter (size 1). We see from the list that it would have a PSQ value of 150, which is the largest of all the PSQ values. So b_2 may be on the critical path, but b_1 and b_3 will definitely not be on the critical path. There is therefore no point in giving b_1 and b_3 any pipe diameter other than the smallest; any other diameter would increase PCOST without reducing the maximum PSQ value. So if b_2 is allocated diameter size 1, then b_1 and b_3 must also have diameters of size 1. We have therefore eliminated 48 of the 49 possible allocations of diameters for which b_2 has diameter size 1.

The total cost of the three pipes b_1, b_2 and b_3 for this allocation of diameters is obtained by simply adding the three PCOST values, giving a total PCOST of $13 + 6 + 8 = 27$. We can record this possible allocation as follows:

PSQ	150
b_1 diameter size	1
b_2 diameter size	1
b_3 diameter size	1
PCOST	27

Note that we record the maximum PSQ value, since it is the maximum value which determines the compressor costs.

Next, consider the allocation of diameter size 2 to b_2. The corresponding PSQ of 139 is still larger than any in the b_1 or b_3 lists, so once again b_1 and b_3 cannot be on the critical path and so can both be given the smallest diameter. Again, there is only one allocation which need be considered. The PCOST total is $13 + 9 + 8 = 30$. We can now extend our merged list:

PSQ	150	139
b_1 diameter size	1	1
b_2 diameter size	1	2
b_3 diameter size	1	1
PCOST	27	30

Now consider the allocation of diameter 3 to b_2. The resulting PSQ of 118 is less than the PSQ of 120 in the b_1 list but larger than any PSQ in the b_3 list, so this time it is b_1 which may be on the critical path. If we allocate diameter 1 to b_1, then b_2 should have diameter 3 and b_3 diameter 1. The corresponding value of PCOST is $13 + 14 + 8 = 35$.

We have not yet covered all possibilities for the case when b_2 is allocated diameter 3. If b_1 is not given diameter 1, then b_2 once again has the largest PSQ value, and a possible allocation is b_1 with diameter 2, b_2 with diameter 3, and b_3 with diameter 1.

Continuing in this way, we obtain the final merged lists for b_1, b_2 and b_3. Each item is assigned an index number for easy reference.

	Index												
	1	2	3	4	5	6	7	8	9	10	11	12	13
PSQ	150	139	120	118	111	94	92	87	86	80	75	70	67
b_1	1	1	1	2	2	3	3	4	4	4	4	4	4
b_2	1	2	3	3	4	4	4	4	5	5	5	6	7
b_3	1	1	1	1	1	1	2	2	2	3	4	4	4
PCOST	27	30	35	39	46	52	56	62	71	77	85	95	111

From the original $7^3 = 343$ possible allocations, only the 13 combinations in the merged lists need to be considered in any optimal allocation of pipe diameters. We can regard the parallel merge list as belonging to an *equivalent edge* b^* which replaces b_1, b_2 and b_3.

Serial Merge

We now wish to merge b^* with b_4. The procedure for doing this is called the *serial merge*. Let us suppose that b_4 has the following list:

b_4 diameter	1	2	3	4	5	6	7
PSQ	133	124	104	78	65	51	42
PCOST	6	10	15	23	33	43	59

Each diameter of pipe b_4 may be combined with any of the items on the b^* list. We must examine systematically all these possible combinations, each of which corresponds to a particular allocation of pipe diameters to b_1, b_2, b_3 and b_4. To begin, we consider the b_4 list items in combination with the first item of the b^* list which has index number 1. Because the pipe b_4 and the equivalent pipe b^* are in series, we simply add the PSQ and PCOST values and thus obtain the following table:

b_4 diameter	1	2	3	4	5	6	7
PSQ	283	274	254	228	215	201	192
PCOST	33	37	42	50	60	70	86
b^* index	1	1	1	1	1	1	1

The entry with the highest PSQ value becomes the first candidate for inclusion on a new list representing the merging of b_4 and b^*.

candidate list:	PSQ	283
	PCOST	33

The table is now updated by removing this candidate item and promoting the index for the column from 1 to 2, as shown below.

b_4 diameter	1	2	3	4	5	6	7
PSQ:	272	274	254	228	215	201	192
PCOST:	36	37	42	50	60	70	86
b^* index	2	1	1	1	1	1	1

The maximum PSQ value now occurs in the second column, so this entry becomes another candidate for the merged lists.

candidate list:	PSQ	283	274
	PCOST	33	37

Updating the table, we obtain:

b_4 diameter	1	2	3	4	5	6	7
PSQ:	272	263	254	228	215	201	192
PCOST:	36	40	42	50	60	70	86
index	2	2	1	1	1	1	1

Again we put the item with the highest PSQ value on the candidate list, which now becomes

candidate list:				
	PSQ	283	274	272
	PCOST	33	37	36

We notice that the third candidate has both a lower PSQ and a lower PCOST than the second item, and so will always be preferred to it. We can therefore eliminate the second item from consideration. Since the candidate list is generated in decreasing order of PSQ, we can in general eliminate all earlier candidates with larger PCOST than a new addition.

Each of the seven items in the b_4 list can be combined with each of the thirteen items on the b^* list. There are therefore $9 \times 13 = 91$ possibilities. However, in this serial merge many of these possibilities will be eliminated. If the procedure as outlined above is continued, only 31 of the 91 possibilities will remain in the final merged lists. Rothfarb and his colleagues report that for larger examples an even greater proportion of the possibilities is rejected.

Allocation of Pipe-line Diameters

By successively repeating the parallel and series merge procedures, the whole tree can be reduced to a single equivalent edge. The total cost of the allocation of pipe diameters corresponding to an item on this final list is obtained by adding the PCOST value to the cost associated with the PSQ value. The smallest total cost gives the optimal allocation of pipe diameters.

For trees with up to 20 vertices, Rothfarb and his colleagues found that the final list never contained more than 1000 items. The maximum list size appears to grow approximately linearly with the number of vertices, whereas the number of possible diameter allocations increases exponentially with the number of vertices.

Comparison of Alternative Trees

The procedure just described for finding the optimal allocation of diameters takes about one minute on a UNIVAC 1105 computer. It is therefore feasible to compare a number of alternative trees. Clearly it is impossible to compare all trees, since there are so many of them. For example, the number of labelled trees with 20 vertices is 20^{18}, which is about 2.62×10^{23}!

It is instructive to notice that the problem is made vastly more complicated by the need to take into account pipe diameter and the pressure-squared differences. If the length of each pipe were the only consideration, all that would be required would be to determine a minimum spanning tree—this can be accomplished very quickly using the greedy algorithm or Prim's algorithm. For the practical problem, to obtain an optimal solution we need (in theory) to examine all possible trees, a procedure which, as we have already seen, is impossible in practice for a network of even modest size.

So the consultants used a heuristic algorithm (that is, one which generates a solution based on trial and error or intuition rather than on an exact procedure) to construct a number of trees which would be expected to be reasonable solutions but which were unlikely to be optimal ones. Small changes in the structure of these trees were then investigated in the search for better solutions.

From their experience, the consultants knew that trees which formed solutions close to optimal ones tended to have

low total pipe length;
a balanced flow structure in which flows in edges leading from the onshore delivery point are equal.

These two guidelines were incorporated in their heuristic algorithm. The next step is to examine small changes, called Δ-changes, in a tree structure.

The Δ-change Procedure

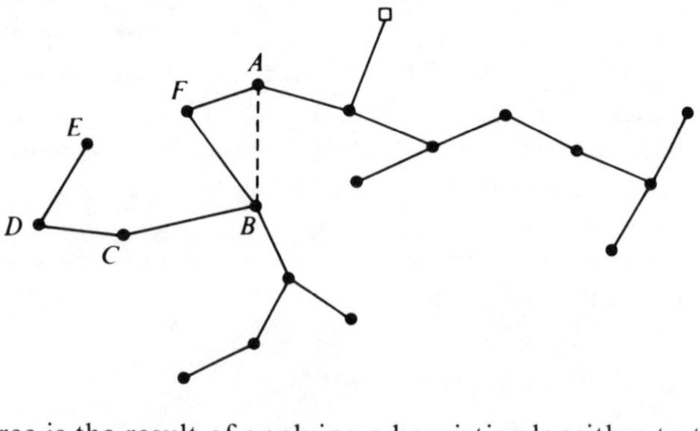

The above tree is the result of applying a heuristic algorithm to the problem of the gas fields in a section of the Gulf of Mexico. Assuming an optimal allocation of pipe diameters, the consultants calculated that the total cost over 20 years would be \$110 970 000. By replacing edge *FB* by edge *AB*, and then re-optimizing the allocation of pipe diameters, the cost is reduced to \$109 096 000, a saving of almost \$2 million. This is an example of a Δ-change. To see how Δ-changes are applied systematically, consider the following diagram:

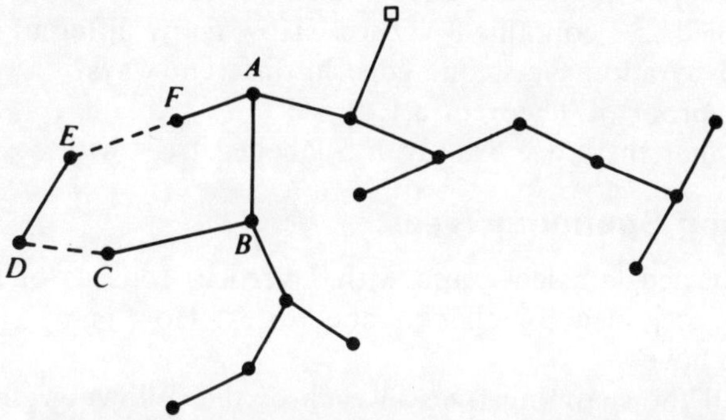

The new edge *AB* has been incorporated in place of *FB*. Now consider the vertex *F*. The nearest non-adjacent vertex is *E*. We introduce a new edge *FE* to form the cycle *ABCDEFA*. Now new trees can be produced by removing each edge of the cycle in turn. It turns out that the Δ-change resulting from removing edge *CD* gives the best result for this cycle and shows a further saving of nearly \$200 000.

By carrying out a series of ten Δ-changes in this way, the consultants finally obtained the following tree:

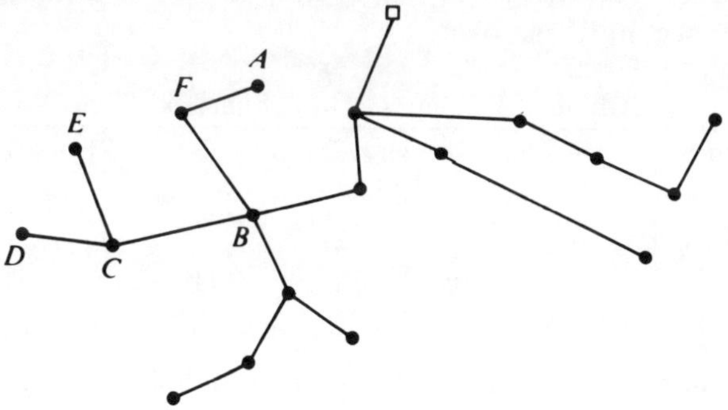

This shows a cost with optimal diameters of only \$101 340 000, a saving of \$9 630 000 over the cost of the original tree. This is the best result which can be obtained by a series of Δ-changes. It is called a *local optimum*. Tests have shown that for small trees the local optimum so obtained is very close to the global optimum.

EXERCISES 8

Mathematical Properties of Trees

8.1. By adding a new edge in all possible ways to each unlabelled tree with 6 vertices, find all the unlabelled trees with 7 vertices.

8.2. Draw the 23 unlabelled trees with 8 vertices.

8.3. A labelled tree contains 8 vertices. How many different cycles can be created by adding just one edge in different ways?

8.4. Give a proof of Theorem 8.1.

8.5. Verify that there are exactly 125 labelled trees with 5 vertices.

Constructing Spanning Trees

8.6. A connected labelled graph with 7 vertices consists of two cycles of equal size joined together at one vertex. How many spanning trees does it have?

8.7. Find all the spanning trees in each of the following graphs:

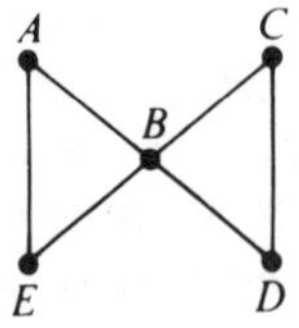

(a)

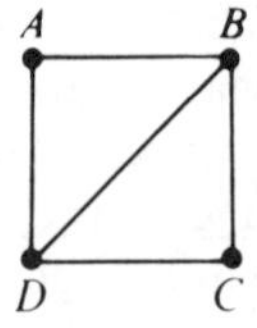

(b)

8.8. The following table gives the distances (in miles) between six places in Ireland. Use the greedy algorithm to find a minimum spanning tree connecting these places.

	Athlone	Dublin	Galway	Limerick	Sligo	Wexford
Athlone	—	78	56	73	71	114
Dublin	78	—	132	121	135	96
Galway	56	132	—	64	85	154
Limerick	73	121	64	—	144	116
Sligo	71	135	85	144	—	185
Wexford	114	96	154	116	185	—

8.9. The following table gives the distances (in miles) between five towns:

	A	*B*	*C*	*D*	*E*
A	—	9	7	5	7
B	9	—	9	9	8
C	7	9	—	7	6
D	5	9	7	—	6
E	7	8	6	6	—

Find a minimum spanning tree connecting these towns, using

(a) the greedy algorithm; (b) Prim's algorithm.

8.10. Find lower bounds for the solution of the travelling salesman problem for the cities given in Exercise 8.9, by removing

(a) town B; (b) town E.

Find the correct solution by inspection.

8.11. The following table gives the distances (in hundreds of miles) between six European cities:

	Berlin	London	Madrid	Moscow	Paris	Rome
Berlin	—	7	15	11	7	10
London	7	—	11	18	3	12
Madrid	15	11	—	27	8	13
Moscow	11	18	27	—	18	20
Paris	7	3	8	18	—	9
Rome	10	12	13	20	9	—

Find a minimum spanning tree connecting these cities, using

(a) the greedy algorithm; (b) Prim's algorithm.

8.12. Find lower bounds for the solution of the travelling salesman problem for the cities given in Exercise 8.11, obtained by removing

(a) the vertex London; (b) the vertex Moscow.

8.13. How would you attempt to find a good upper bound for the solution of the travelling salesman problem? Illustrate your method by applying it to the six European cities given in Exercise 8.11.

8.14. Describe an alternative algorithm for the minimum connector problem which involves the removal from the graph of edges of greatest weight.

8.15. (a) Prove that the greedy algorithm always produces a minimum spanning tree.
(*Hint*: Suppose that the algorithm produces a tree T, and that S is a spanning tree of smaller total weight than T. Let e be the edge of smallest weight lying in T but not in S, and consider the subgraph obtained by adding e to S.)

(b) How would you adapt your proof to show that a spanning tree of maximum weight can be constructed by successively choosing edges of maximum weight in such a way that no cycles are created?

Searching Trees

8.16. On the following tree, perform:

(a) a depth-first search; (b) a breadth-first search.

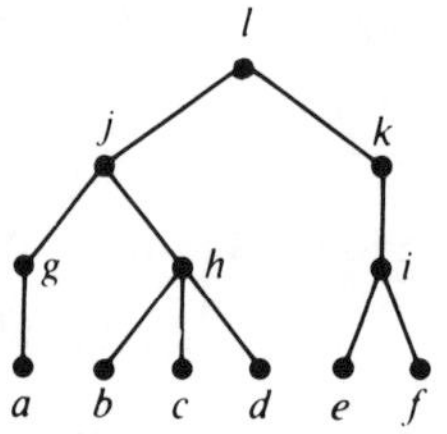

8.17. Find a tree which represents the expression

$$(ab + c) \times (d - e) \times f.$$

8.18. In the following graph, find:

(a) a DFS spanning tree; (b) a BFS spanning tree.

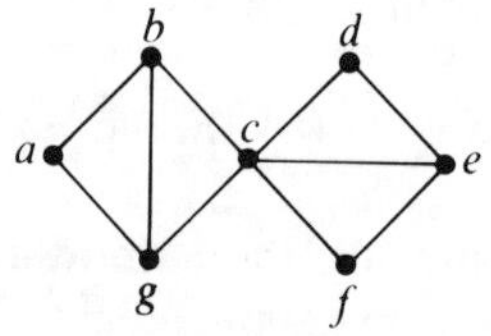

8.19. For the following tree, write down the order of the vertices visited

(a) in a depth-first search; (b) in a breadth-first search.

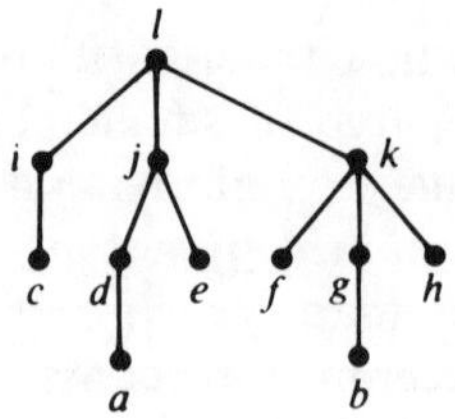

Counting Trees

8.20. Use Cayley's theorem to calculate the number of labelled trees with six vertices. How many different unlabelled trees are there with six vertices?

8.21. Show that the complete graph K_n has exactly n^{n-2} spanning trees.

Offshore Natural Gas Pipline Systems

8.22. Two terminal branches, b_1 and b_2, lead into a common branch b_3, as follows:

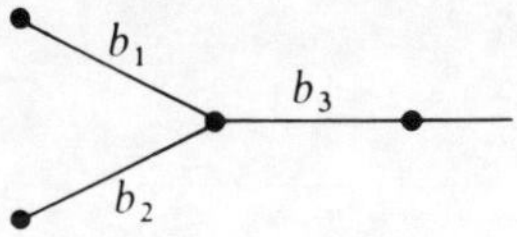

The corresponding PSQ and PCOST lists are:

b_1	PSQ	105	96	91	82	73
	PCOST	14	16	20	30	35
b_2	PSQ	94	89	85	78	70
	PCOST	10	12	15	21	27
b_3	PSQ	97	91	88	75	70
	PCOST	12	14	16	22	31

(a) By applying a *parallel merge*, obtain a merged list corresponding to a branch replacing b_1 and b_2.

(b) By then applying a *serial merge*, obtain a merged list corresponding to a branch replacing b_1, b_2 and b_3.

CHAPTER 9

PHYSICAL NETWORKS: MODELLING

After reading this chapter, you should be able to:

- understand what is meant by *through* and *across variables* and identify them in different physical systems;
- give examples of 2-terminal components and state the corresponding component equations;
- give examples of multi-terminal components and explain why a multi-terminal component can be regarded as a number of connected 2-terminal components;
- explain how physical systems can be represented by oriented graphs, and draw the oriented graph corresponding to a given system;
- state Kirchhoff's current and voltage laws, and apply them to simple electrical circuits.

Introduction

In this chapter and the next two, we discuss the use of graph theory to solve certain physical network problems. Most of the discussion is in terms of electrical networks, for which the method we give can be used to find all the voltages and currents in the network. Electric currents and voltages are examples of *through* and *across variables*, and they satisfy the well-known Kirchhoff's current and voltage laws. The method we describe is very general, and can be applied to any physical network for which the through and across variables satisfy analogues of Kirchhoff's laws. Examples include hydraulic networks, mechanical systems, pneumatic networks and thermal networks.

A physical network can be represented by a graph in which the edges represent the components of the network and the vertices correspond to the terminals at which the components are connected together. Associated with each edge are the *edge-variables*—the through and across variables (currents and voltages in the case of an electrical network) which satisfy Kirchhoff-type laws. Since we need to be able to determine the directions of the through and across variables, we assign a reference direction to each edge of the graph. We call a graph with reference directions an **oriented graph**.

If the physical behaviour of the components can be modelled by equations (called the *component equations*) then these equations, together with the equations resulting from applying Kirchhoff-type laws, can be solved to find the values of all the through and across variables for the network. For networks of small size, these equations can be solved by hand, but for larger networks it is more appropriate to use a computer. We show in these three chapters how the equations can be formulated as matrix equations; this procedure is appropriate for solving network problems on a computer. The required number of equations resulting from Kirchhoff's laws are obtained directly from the incidence matrix of the oriented graph. This means that the only information which needs to be fed into the computer is the incidence matrix together with the component equations. This is particularly convenient, because the incidence matrix can be written down directly from the oriented graph.

In Chapter 11 we describe a way of solving the matrix equations using Gaussian elimination. This method forms the basis of many computer programs.

In this chapter we explain what is meant by through and across variables, giving examples for different physical systems, and show how a physical system can be represented by an oriented graph. A physical system can be considered to be a number of individual components connected together. In the case of an electrical network, these components might be resistors, capacitors, transistors, transformers and so on. We first look at the representation of individual components, and the mathematical modelling of their behaviour, before going on to show how the whole connected network of components can be represented by an oriented graph.

9.1 COMPONENTS AND THROUGH AND ACROSS VARIABLES

Components

We assume that the physical systems we wish to analyse can be considered to consist of a number of interconnected components.

The point at which one component is connected to another is called a **terminal**. The following diagrams represent components with two, three and four terminals.

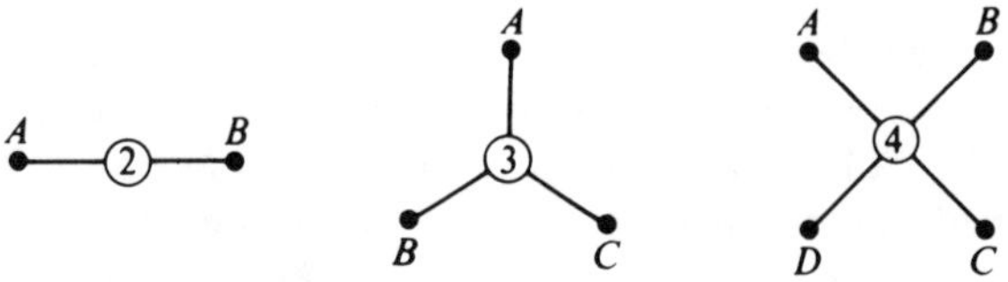

A **2-terminal component** might be a spring, a resistor, a length of wire or piping, or any other physical object which fits into a system via *two* connections. For example, a spring functions when both ends are appropriately connected so that it can be compressed or stretched. Similarly, a resistor functions when it is connected so that current may flow through it.

We shall use the following diagrams to denote electrical 2-terminal components:

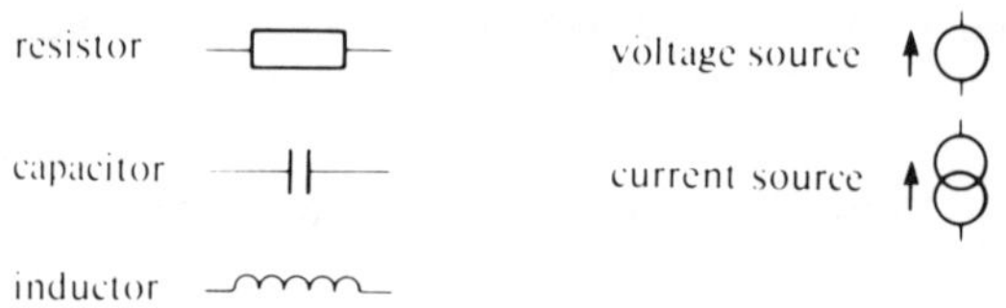

For example, diagram (a) below is the circuit diagram of an electrical circuit containing a voltage source connected to a resistor and a capacitor, and diagram (b) is the circuit diagram of a current source connected to a resistor and two inductors.

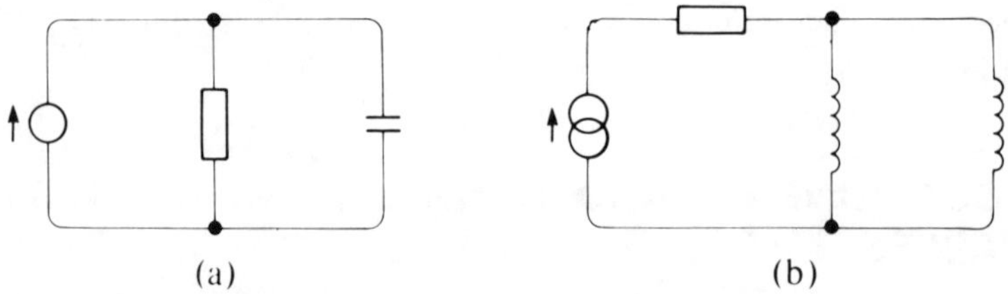

A **3-terminal component** might be a transistor. We discuss transistors in detail later in this chapter.

A **4-terminal component** might be a transformer.

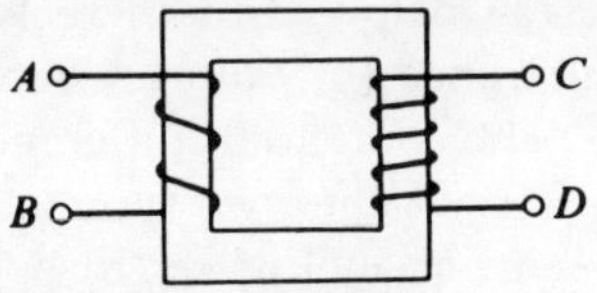

transformer

The much simplified diagram above shows a transformer in which two coils of wire, called the *primary* and *secondary*, are wound on a magnetic core so that they are in close proximity to each other but are not in electrical contact. The letters A and B denote the terminals of the primary winding, and C and D denote the terminals of the secondary winding. (Note that for some applications one of the input terminals, A or B, may be connected to one of the output terminals, C or D, so that the transformer operates as a 3-terminal component. See, for example, Exercise 9.2.)

We use a schematic diagram to show how the components of a particular system are connected together. For example, the following schematic diagram represents a system comprising five 2-terminal components, two 3-terminal components and one 4-terminal component:

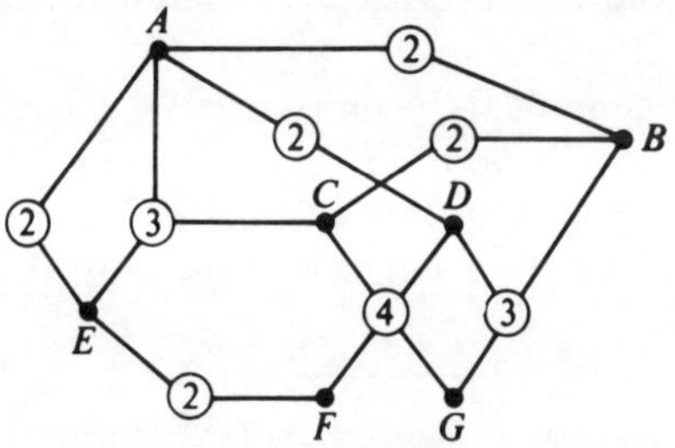

Through and Across Variables

In previous chapters we discussed flows in networks where the flow might, for example, be the flow of traffic in a road network. A flow in a network has the property that the amount of flow of a particular commodity into any vertex (other than a source or sink) is equal to the amount of flow out of it. A variable with this property is classified as a *through variable* because it represents an actual flow *through* part of a system.

Many types of network also involve a different kind of variable called an *across variable*. To make clear the distinction between these two types of variable, let us look at a specific example, an electrical network.

In an electrical network we are concerned with the amount of current which passes through a component in the network. This current is a flow, or a through variable, which is measured by a meter in series with the component as in Figure (a) below. But there is also an across variable which is the potential difference (or voltage) between (or across) any two points of the network. The potential difference can be measured by connecting a voltmeter across the two points under consideration, in parallel with the component, as in Figure (b) below.

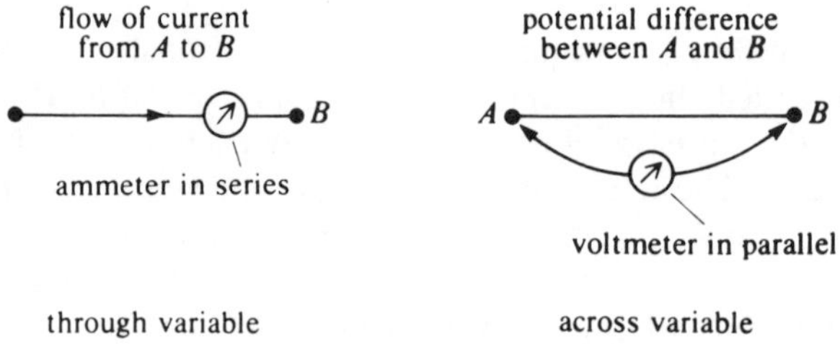

(a) (b)

An analogous example is that of a hydraulic network, which consists of a network of pipes carrying water or some other fluid. Here the through variable is the amount of water flowing through any part of the network and the across variable is the pressure difference between any two points of the network.

Consider now a mechanical system such as a car suspension. Here we have a mass (the car) supported by a spring and a shock absorber connected onto the wheel. As far as vertical motion is concerned, the wheel is more or less 'fixed' to the road. The following is a schematic diagram:

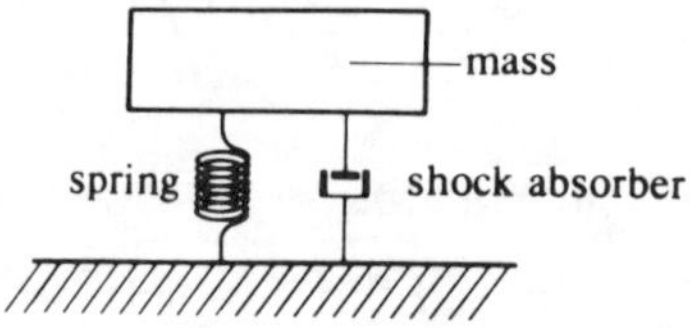

An engineer designing the suspension system is concerned both with the forces in the spring and shock absorber and with the velocity of the vertical motion of the car. Now the force in the spring could be measured by placing a load cell (a sophisticated version of bathroom scales) between the car and the spring, or the spring and the road; the force would then be carried by the load cell as well as by the spring (just as the bathroom scales bear your weight when you are standing on them). The force is thus measured by placing a meter in series with the component, and so is a through variable. The vertical velocity of the car can be measured, using some electronic timing mechanism, by measuring the change of the distance between a point on the car and the road in a given time: the measurement here is across the spring, from the car at one end to the ground at the other. Hence the velocity is an across variable.

In each of the above examples, there are two variables: the across variable, representing the difference of a quantity between one terminal of a component and another, and the through variable, representing the quantity flowing through the component. In the following table we list some examples of through and across variables for different types of physical system.

Some examples of through and across variables

type of system	example	through variable	across variable
electrical	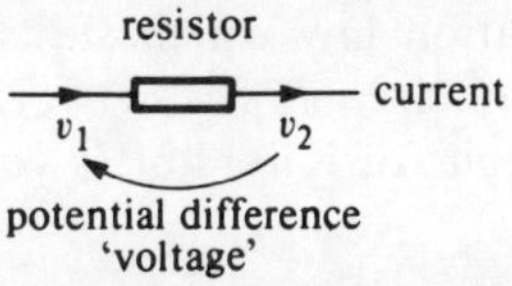	current	potential difference 'voltage' $v_1 - v_2$
mechanical	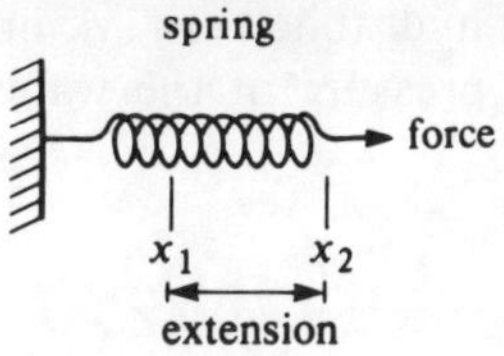	force	extension $x_2 - x_1$
hydraulic	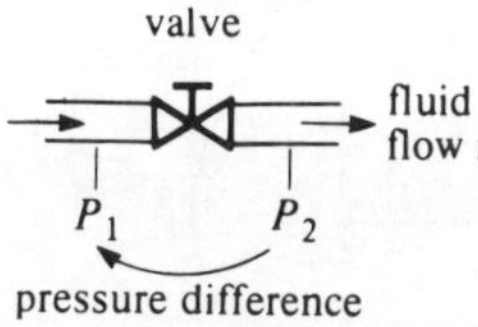	fluid flow rate	pressure difference $P_1 - P_2$
thermal	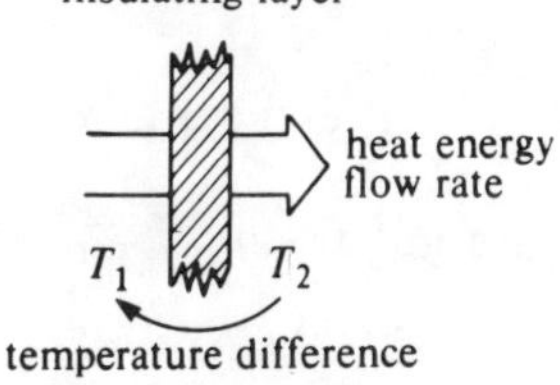	heat energy flow rate	temperature difference $T_1 - T_2$

Equations Relating Through and Across Variables

For electrical networks, the currents and voltages satisfy the following two laws:

Kirchhoff's current law *The current flow into any vertex of an electrical network is equal to the current flow out of it.*

Kirchhoff's voltage law *The algebraic sum of the potential differences across all the components around any circuit (cycle) in an electrical network is zero.*

Analogues of both these laws can be given for many types of network containing through and across variables.

For example, the analogue of Kirchhoff's current law for a network carrying a flow is simply the flow conservation law: what flows into a vertex of a network (other than a source or sink) must flow out of it. For the general physical network, the analogue of Kirchhoff's current law is called the *vertex law*.

The analogue of Kirchhoff's voltage law for a general physical network is the potential difference conservation law which states that the sum of the potential differences around any circuit of a network is zero. For the general physical network, the analogue of Kirchhoff's voltage law is called the *cycle law*.

Example 1 Consider the following hydraulic network in which the number next to each vertex represents the pressure of the water or other fluid at that vertex.

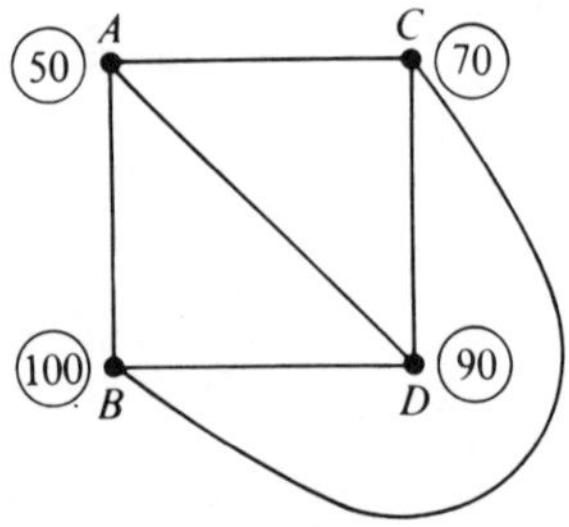

The pressure difference (or potential difference) across AC is equal to

(the pressure at C) – (the pressure at A) = $P_C - P_A = 70 - 50 = 20$.

Similarly,

the pressure difference across CD is $P_D - P_C = 90 - 70 = 20$;

the pressure difference across DA is $P_A - P_D = 50 - 90 = -40$.

Adding these numbers together, we see that the sum of the pressure differences around the circuit $ACDA$ is zero. In fact, the sum of the pressure differences around any circuit is zero—this is the analogue of Kirchhoff's voltage law for hydraulic networks.

In the next section we show how each component of a physical system—and hence the whole network—can be represented by an oriented graph. With each edge of the graph we can associate a through variable and an across variable. These variables are related by the component equations.

For example, if the component is an electrical resistor, the appropriate component equation is *Ohm's law*:

$$\frac{\text{voltage}}{\text{current}} = \text{a constant (the resistance).}$$

For a spring, the corresponding component equation is *Hooke's law*:

$$\frac{\text{tension}}{\text{extension}} = \text{a constant (stiffness).}$$

In this chapter and the next two, we consider only systems which obey Kirchhoff's laws or their analogues. We can use these laws to derive two further sets of equations, one set relating the through variables in the system (the vertex law equations) and the other relating the across variables (the cycle law equations).

Given these three sets of equations—the component equations, the vertex law equations, and the cycle law equations—we can set about solving them. That is, if we know the values of some of the edge variables of our graph (say the voltage at a particular point in an electrical network), then we can work out the values of all the other edge variables—the voltages and currents in this case. The aim of these three chapters is to describe how we tackle this type of problem.

Generally, we shall use electrical networks as examples to illustrate the solution techniques. You should, however, always bear in mind that the methods we describe are completely general, in that they can be used to work out the values of edge variables in any physical system that obeys Kirchhoff's laws or their analogues. When we have worked out values for the edge variables, we may then interpret them as currents and voltages, flows and pressures, forces and velocities, and so on, depending on the type of system in hand.

9.2 GRAPHICAL REPRESENTATION

The Graphical Representation of 2-terminal Components

In this section we show how 2-terminal components can be represented graphically. Later in this section we show that a component with more than two terminals (a *multi-terminal component*) can be represented graphically by treating it as a number of 2-terminal components.

A 2-terminal component can be represented by two vertices joined by a single edge. This is illustrated in the following diagram which shows a resistor and its representation as the edge of a graph:

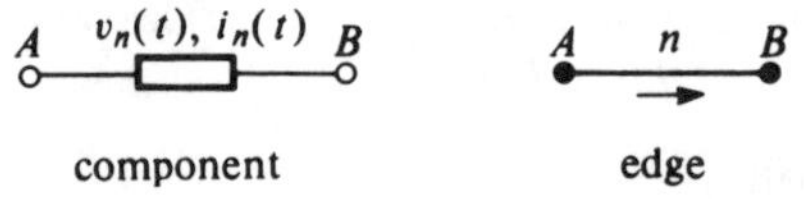

Associated with the edge of the graph are two *edge variables*—one through variable and one across variable. In the case of the resistor shown above, the across variable is the voltage $v_n(t)$ across the resistor, and the through variable is the current $i_n(t)$ flowing through the resistor. We use the subscript n when we wish to indicate that the resistor is the nth component of a particular system. To avoid cluttered diagrams, we label the corresponding edge simply by the letter n. (We have introduced the letter t to refer to time, since the currents and voltages are often time-dependent. However, we usually omit t and write just v_n and i_n.)

We also need a *reference direction* for the edge variables. We get this by assigning an arbitrary direction to each edge, and we indicate it by drawing an arrow alongside the edge. Currents which actually flow through the component corresponding to the edge in the direction of the arrow are regarded as positive, and currents in the opposite direction are regarded as negative. Similarly, voltage differences between the terminals corresponding to two vertices are always taken as the voltage at the tail of the arrow minus the voltage at the head. Thus if the voltage at the tail is greater than the voltage at the head, the voltage difference is positive: if the voltage at the tail is less, then the voltage difference across the edge is negative. It is important to note that the reference direction is purely arbitrary and does not necessarily represent the actual direction of the current flow or voltage difference.

Note that our convention of using the reference arrow to specify the direction of a voltage is the opposite to that normally used in electrical circuit theory: an arrow on a *circuit* diagram indicates a positive voltage if the voltage at the *head* is greater than the voltage at the *tail*. Here we use a different convention because we are using a single reference arrow to indicate the direction of both current and voltage. This usage is illustrated in the following example.

Example 2 Consider a system consisting of two 2-terminal components—a battery and a resistor. The following figure shows the circuit diagram and its graphical representation.

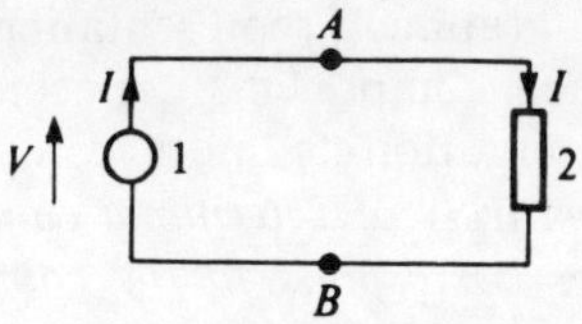

circuit diagram

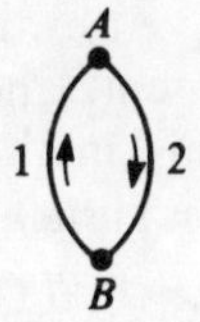

graphical representation

The circuit diagram shows the actual direction of flow of the current I and the actual direction of the voltage V across the battery. In the graph, the edge labelled 1 corresponds to the battery, and the edge labelled 2 corresponds to the resistor. Since both components are connected together in the circuit diagram, the corresponding edges are connected to the same vertices in the graph.

The edge variables associated with edge 1 are the current i_1 flowing through the battery, and the voltage v_1 across the battery. Similarly, the variables associated with edge 2 are the current i_2 flowing through the resistor, and the voltage v_2 across the resistor.

Consider edge 1. The reference direction is the same as the direction of flow of the current through the battery, so we have

$$i_1 = I.$$

In the circuit diagram, the arrow by the battery points towards terminal A, indicating that A is the positive terminal of the battery. Therefore the potential at terminal A is V volts higher than the potential at terminal B. In the graph representation, the arrow associated with edge 1 points from B towards A. The voltage v_1 is equal to the voltage at the tail of the arrow (B) minus the voltage at the head (A)—that is,

$$v_1 = -V.$$

Note that, although the reference arrow on the graph points in the same direction as the arrow on the circuit diagram, the expression for v_1 has a minus sign—this is because the sign conventions are different in the two cases.

Now consider edge 2. The reference direction is again the same as the direction of the flow of the current through the resistor, so we have

$$i_2 = I.$$

Terminal A of the resistor is connected to the positive battery terminal and so is at a higher potential than terminal B. Thus the voltage at the tail of the reference arrow on the graph is greater than the voltage at the head, so

$$v_2 = V.$$

We have expressed all four of the edge variables in terms of the two quantities given on the original circuit diagram.

We have now given examples of the graphical representation of a single 2-terminal component, and of two such components connected together. We can extend this method of representation to more complicated networks. In fact, *any system consisting entirely of 2-terminal components can be represented by a graph whose edges correspond to the components and whose vertices correspond to the terminals.*

Example 3 The following electrical network contains nine components—an independent voltage source, two resistors, two capacitors and four inductors. Replacing each of these nine components by an edge, and assigning an arbitrary direction to each of these edges, we obtain the corresponding graph:

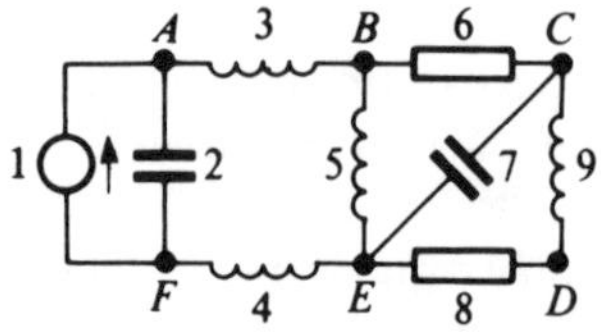

electrical network

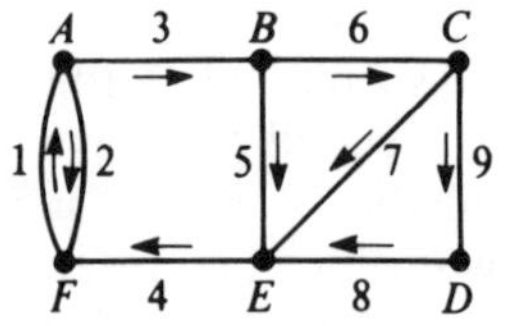

graphical representation

You may be wondering why we have not drawn the direction arrows *on* the edges (rather than beside them), referred to the edges as 'arcs', and called the diagram a 'digraph', rather than a graph. The reason is that for much of our discussion the directions are irrelevant: for example, we shall need to refer to cycles such as *BCEB* or *BCDEB*, and these can be considered as cycles only when we ignore the directions of the arrows. A diagram of this kind, which is something between a graph and a digraph, is called an **oriented graph**.

Oriented graphs can also be used to represent systems other than electrical systems. For example, the above oriented graph also correponds to the following hypothetical hydraulic power system, in which the terminals *A*, *B*, *C*, *D*, *E* and *F* are pipe junctions, and each segment of pipe contains a valve. The 'voltages' $v_1, \ldots, v_9$ are across variables representing the pressure drops across the valves, and the 'currents' $i_1, \ldots, i_9$ are through variables representing the flows through the corresponding valves.

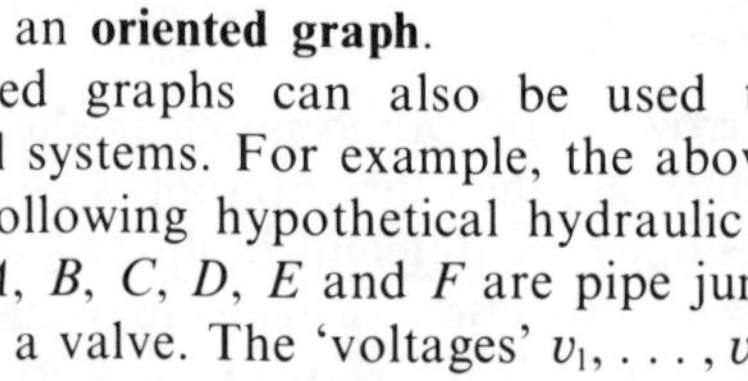

hydraulic power system

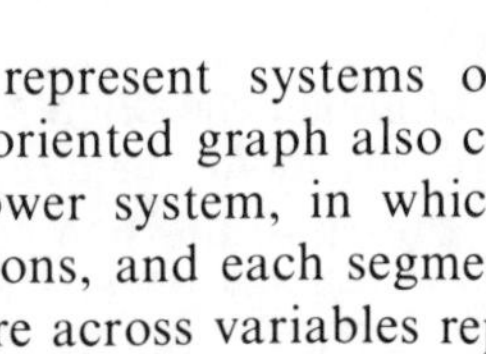

oriented graph

Problem 9.1 Label the components and the terminals of the following electrical network, and draw an oriented graph which represents it.

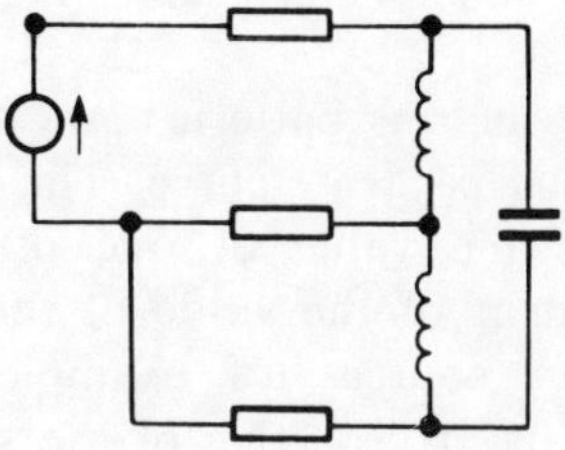

The Component Equations for 2-terminal Components

We now turn to the equations relating the two edge variables for each component—the **component equations**. Suppose that we measure the voltage across a 2-terminal component such as a resistor for different values of current flowing through it. We can then try to model the relationship between the voltage and current by an equation. Such an equation is called a *component equation* since it is characteristic of the component itself, as opposed to Kirchhoff's laws, which are a general property of any electrical network.

We start by looking at four types of electrical component—*resistors*, *capacitors*, *inductors* and *independent sources*.

A **resistor** is modelled by a 2-terminal component whose edge variables v and i satisfy a relationship of the form

$$v = Ri,$$

where R is a constant, usually called the **resistance**.

(This relationship is known as *Ohm's law*. It is sometimes written in the form $i = Gv$, where $G(= 1/R)$ is the *conductance* of the resistor.)

A **capacitor** is modelled by a 2-terminal component whose edge variables v and i satisfy a relationship of the form

$$i = C\frac{dv}{dt},$$

where C is a constant, usually called the **capacitance**.

An **inductor** is modelled by a 2-terminal component whose edge variables v and i satisfy a relationship of the form

$$v = L\frac{di}{dt},$$

where L is a constant, usually called the **inductance**.

Note that the component equations for a resistor, a capacitor and an inductor are all *linear* equations. Note also that R, C and L are *constants*—we are assuming that the behaviour of the components is *time-invariant*. The concepts of *linearity* and *time-invariance* are discussed further later in this chapter.

To model components such as batteries and electric generators, we use the concept of an **ideal independent source**. This is an idealized 2-terminal component which keeps the value of one of the edge variables at a prescribed level independent of the value of the other edge variable. An ideal independent *voltage* source, for example, maintains a prescribed *voltage* across its terminals, irrespective of the size of the current flowing between them. Under certain conditions, a battery can be modelled in terms of such a source. Similarly, an ideal independent *current* source maintains a prescribed *current* irrespective of the value of the voltage across its terminals.

The prescribed level of an independent source may not necessarily be constant; for example, for an independent voltage source, the prescribed voltage may be a sinusoidal voltage of the form $v_0 \sin \omega t$, where v_0 and ω are constants.

Some 2-terminal components of other types of system such as mechanical or hydraulic systems can be considered as analogues of these electrical components. The electrical analogues of two components of a mechanical system are considered in the following example and problem.

Example 4 Suppose that a force f accelerates a body of mass m. Then, according to Newton's Second Law of motion, the acceleration a is given by the equation

$$f = ma$$

or

$$f = m\frac{du}{dt},$$

where u is the velocity of the mass. In this case, f is the through variable and u is the across variable. The equation is of the same form as the equation

$$i = C\frac{dv}{dt}$$

for a capacitor. We may therefore consider the force f to be analogous to the current i (also a through variable), the velocity u to be analogous to the voltage v (also an across variable), and the mass m to be analogous to the capacitance C. The following table shows the analogous quantities for the two components.

	mechanical	electrical
component	mass	capacitor
through variable	f	i
across variable	u	v
component equation	$f = m\dfrac{du}{dt}$	$i = C\dfrac{dv}{dt}$

Problem 9.2 For a spring which obeys Hooke's law, the extension x of the spring produced by a force f is given by the equation

$$f = kx,$$

where k is a constant (the stiffness of the spring). If the rate of change of the extension of the spring (that is, the velocity $u = dx/dt$) is considered to be analogous to a voltage, which electrical component is analogous to the spring? Draw up a table of analogous quantities similar to the above table.

For the remainder of the book—with the exception of one example discussed in Chapter 11—all the examples we consider will be electrical. We can obtain corresponding results for other physical systems by considering the appropriate electrical analogues.

Multi-terminal Components

A **multi-terminal component** is a component which has more than two terminals, and which cannot therefore be represented by a single edge joining two vertices. Before showing how such a component can be represented graphically, we consider two important types of electrical multi-terminal component—*transformers* and *transistors*.

Transformers

A **transformer** is a device used extensively for changing the voltage at which electrical energy is supplied. A schematic diagram of a transformer consisting of two coils wound on an iron core is shown below.

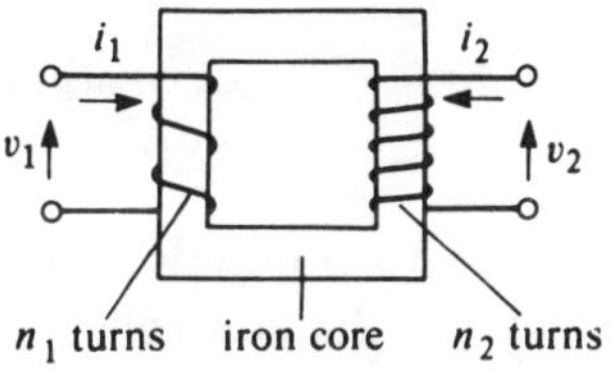

Under certain circumstances a transformer can be modelled by an *ideal transformer* which satisfies 'ideal' conditions such as having zero-resistance windings and no magnetic field losses.

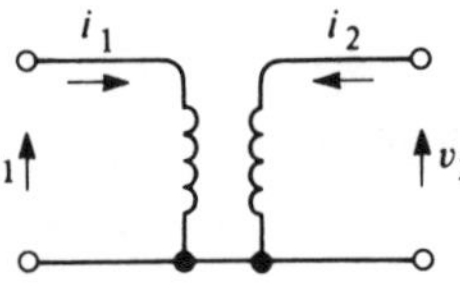

ideal transformer

A circuit diagram of an ideal transformer is shown above. Notice that one of the input terminals is connected to one of the output terminals; a transformer connected in this way can be regarded as a 3-terminal component. The equations relating the voltages and currents shown on the diagram, as derived from electromagnetic theory, are

$$\frac{v_1}{v_2} = \frac{n_1}{n_2} \quad \text{and} \quad \frac{i_1}{i_2} = -\frac{n_2}{n_1},$$

where n_1 and n_2 are the numbers of turns in the two coils. (Note that v_1, i_1, v_2 and i_2 are usually functions of time.)

Transistors

A **transistor** has three external connections called the *emitter*, the *base* and the *collector*. Under certain circumstances, its behaviour can be modelled by using the circuit shown in the following diagram.

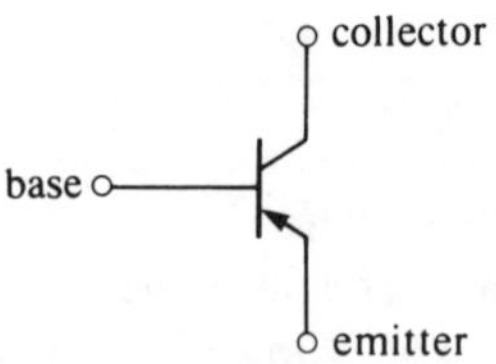

(a) pnp transistor

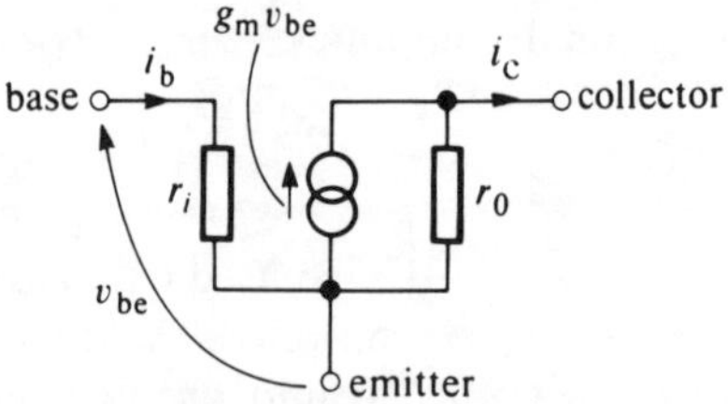

(b) controlled source representation. g_m is the transfer conductance or 'mutual conductance' of the transistor

Diagram (a) shows the symbol commonly used to represent a pnp type of transistor. Diagram (b) shows the controlled source representation. (A controlled source is an independent source whose output level is not constant but is controlled by the value of a particular voltage or current.) When a small voltage v_{be} is applied between the emitter and the base, under certain circumstances a greatly amplified voltage will appear between the emitter and the collector—in other words, the transistor acts as an amplifier. This is represented in diagram (b) by a current source which is controlled by the value of the input voltage v_{be}. The current flowing in the source is equal to $g_m v_{be}$, where g_m is a constant for particular operating conditions of the transistor. Notice that resistors have been added to the controlled source representation to model the input resistance r_i and the output resistance r_o of the transistor. We can use this representation to obtain relations between the voltages and currents. Since the input circuit consists of just a resistance r_i, the input current i_b is given by

$$i_b = \frac{v_{be}}{r_i}.$$

The output current i_c is equal to the source current $g_m v_{be}$ minus the current flowing through the resistor r_o, that is,

$$i_c = g_m v_{be} - \frac{v_{ce}}{r_o},$$

where v_{ce} is the output voltage between the collector and the emitter.

The Graphical Representation of Multi-terminal Components

We have seen how a 2-terminal component can be represented graphically by a single edge. We now show that a multi-terminal component can be regarded as a number of 2-terminal components, and can therefore be represented by a graph with a number of edges. To do this, we need to make use of Kirchhoff's current and voltage laws which we stated earlier, and repeat here for convenience.

Kirchhoff's current law *The current flow into any vertex of an electrical network is equal to the current flow out of it.*

Kirchhoff's voltage law *The algebraic sum of the potential differences across all the components around any circuit (cycle) in an electrical network is zero.*

These two laws can be formulated in graph-theoretical terms as follows.

The vertex law *The algebraic sum of the through variables associated with the edges at each vertex of an oriented graph is zero.*

Here ‘the algebraic sum’ means that the orientation of the edges must be taken into account, so if edges oriented *into* a vertex are counted as *positive*, then edges oriented *out* of a vertex are counted as *negative*; for example,

$$i_1 - i_2 + i_3 + i_4 - i_5 = 0$$

The cycle law *The algebraic sum of the across variables associated with the edges in any cycle of an oriented graph is zero.*

Here ‘the algebraic sum’ means that we choose a particular direction around a cycle (say clockwise) and count edges oriented in the *same* direction as *positive* and edges oriented in the *opposite* direction as *negative*; for example,

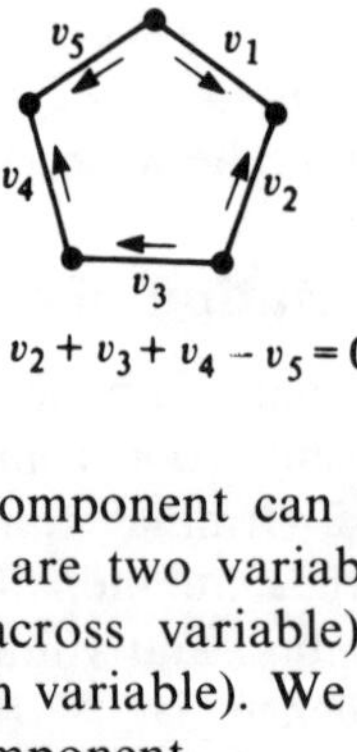

$$v_1 - v_2 + v_3 + v_4 - v_5 = 0$$

We know that a 2-terminal component can be represented by a single edge. Associated with the edge are two variables: the voltage across the terminals of the component (across variable) and the current flowing through the component (through variable). We now consider the variables associated with a 3-terminal component.

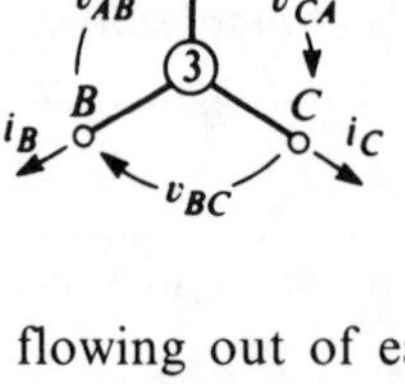

We can measure the current flowing out of each of the three terminals A, B and C; the currents are shown on the above diagram as i_A, i_B and i_C. Similarly, we can measure the voltage across each pair of terminals; the voltages are shown as v_{AB} for terminals A and B, v_{BC} for terminals B and C, and v_{CA} for terminals C and A.

At first sight it would seem that we need six variables to characterize a 3-terminal component—three voltages and three currents. However, these six variables are not all independent since we can obtain a relationship between them by applying Kirchhoff's two laws, as follows.

Kirchhoff's current law states that the total current flowing into a vertex is equal to the total current flowing out of the vertex. It can be deduced from Kirchhoff's law that the total current flowing into the region bounded by broken lines in the figure below is equal to the total current flowing out of that region. (We show in Chapter 10 that Kirchhoff's current law can be applied to *any* cutset in an oriented graph.)

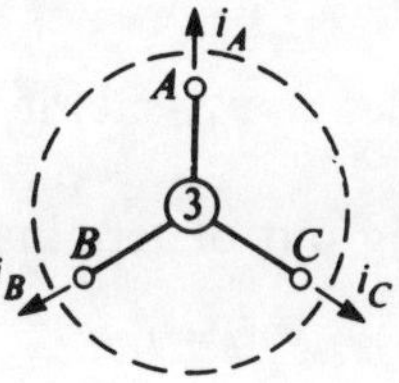

Applying Kirchhoff's current law to the currents flowing into the 3-terminal component, we obtain the equation

$$i_A + i_B + i_C = 0$$

or

$$i_A = -\, i_B - i_C.$$

A consequence of this last equation is that if we measure i_B and i_C then we do not need to measure i_A; we can calculate it from the equation. So only *two* of the three current variables are independent.

We can apply Kirchhoff's voltage law to the cycle $ABCA$. (We can assume that the component is connected to other components or to measuring instruments, so that there is such a cycle.) We thus obtain the equation

$$v_{AB} + v_{BC} + v_{CA} = 0$$

or

$$v_{BC} = -\, v_{AB} - v_{CA}.$$

Thus if we measure v_{AB} and v_{CA} we can calculate v_{BC} from this last equation. So only *two* of the three voltage variables are independent.

Of the six current and voltage variables, we have found that only *four* of them are independent—two current variables and two voltage variables. It follows that, to describe completely the behaviour of the 3-terminal component as measured at its terminals, we need to specify the values of only *two* of the current variables and *two* of the voltage variables.

For a 2-terminal component, the graphical representation is a single edge with two variables associated with it (a current and a voltage). For a 3-terminal component, we have *four* independent variables, so we need *two* edges for the graphical representation. One possibility is shown below:

3-terminal component graphical representation

We can express the variables i_A, v_{AB}, etc. in terms of these edge variables v_1, i_1 and v_2, i_2 as follows.

Clearly, the current flowing out of terminal B is equal to i_1, so

$$i_B = i_1.$$

(In the graphical representation, the arrow on edge 1 points towards the vertex B. This means that the current i_1 is directed towards vertex B and therefore flows *out* of the 3-terminal component via terminal B—it is therefore equal to the current i_B.)

Similarly,

$$i_C = i_2,$$

and so

$$i_A = -i_1 - i_2,$$

since $i_A = -i_B - i_C$.

The voltage across terminals A and B is clearly equal to v_1, so

$$v_{AB} = v_1.$$

Similarly,

$$v_{CA} = -v_2,$$

and so

$$v_{BC} = v_2 - v_1,$$

since $v_{BC} = -v_{AB} - v_{CA}$.

We showed earlier that a 2-terminal component is represented by *one* edge and has *one* component equation. We have just shown that a 3-terminal component is represented by *two* edges and, by analogy with the 2-terminal component, has *two* component equations. As a result, we can regard any 3-terminal component as being composed of two 2-terminal components, one corresponding to each edge of the graphical representation. To represent the 3-terminal component shown diagrammatically above, we chose the two edges *AB* and *AC*. Instead, we could have chosen the two edges *AB* and *BC*, or the two edges *BC* and *AC*. In the following problem we ask you to find relationships between the variables for the different choices of pairs of edges.

Problem 9.3 The following graphs can both be used to represent the same 3-terminal component:

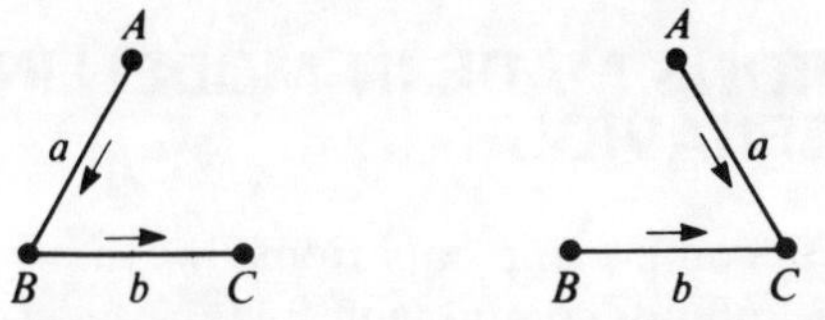

(a) (b)

In each case, find i_a, i_b, v_a and v_b in terms of the variables i_1, i_2, v_1 and v_2 defined on the previous page.

The method of representation just described can be extended to the graphical representation of *n*-terminal components for larger values of *n*. Just as a 3-terminal component can be represented by a tree with two edges, so an *n*-terminal component can be represented by a tree with $n - 1$ edges. Moreover, we have seen that each of the labelled trees with three vertices (shown below) is a suitable representation of the 3-terminal component.

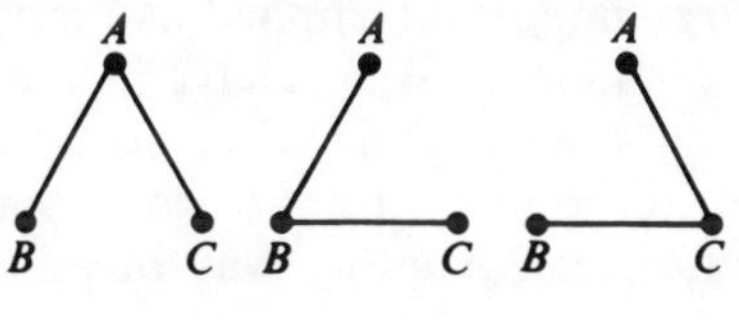

In the same way, any labelled tree with *n* vertices is a suitable representation of an *n*-terminal component. It follows that *any n-terminal component can be regarded as* $(n - 1)$ *2-terminal components*, one corresponding to each edge of the tree.

Problem 9.4

(a) Find a graphical representation of the following system.

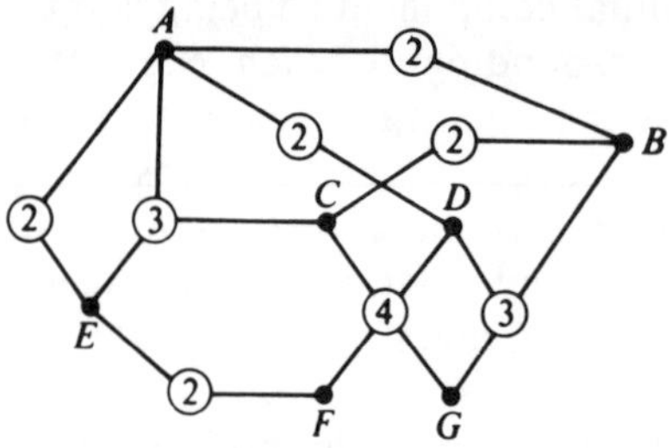

(b) How many component equations are associated with this system?

9.3 ASSUMPTIONS MADE IN MODELLING COMPONENT BEHAVIOUR

In our discussion of component equations, we have made a number of implicit assumptions about component behaviour. In particular, we assumed that the behaviour of a component does not change if it is moved from one system to another, and that the behaviour is time-invariant and linear. If these assumptions are not applicable to the components in hand, the mathematics will often turn out to be more complex and may require special manipulative techniques. We now comment briefly on the assumptions made in our treatment of physical networks.

System Independent Behaviour

One way of finding out how a component interacts with other components in a system is to remove it from the system and connect its terminals to a test system which can be used to take measurements. In this way we can construct a model of its behaviour, expressed as the component equation. There is, however, a very important implied assumption in doing this. We are assuming that the component will behave in the same way in our test system as in the original system. Otherwise, we would not be able to use the information obtained from the test to work out how the component behaves in the original system. Another way of putting this is to say that we must be careful that the model (or component equation) which we use is appropriate for the system in question. For example, one system may operate in a different type of environment, with a different ambient temperature, humidity or air pressure, than a second system, and this may significantly affect the component behaviour. Even if the environments of two systems are the same, we may not be able to use the same model of a component for both systems. For example, a model of a transistor which is suitable for describing its behaviour in a hi-fi amplifier may not be suitable for describing its behaviour in a computer.

Time-invariance

The assumption that the behaviour of a component is time-invariant means that we assume its behaviour is the same today as it was yesterday and will be tomorrow. We must be aware, however, that the behaviour of real components will change with time. If this change is insignificant in the application we are concerned with, then the assumption of time-invariance is reasonable.

Linearity

Mathematically, the assumption of linearity means that the equations describing the behaviour of a component are *linear* equations. A system made up of linear components is itself linear. This means that we can use the many mathematical techniques that have been developed to handle linear equations to work out how the system responds to changes at its inputs.

A linear system obeys the *principle of superposition*. This states that *if the response of a system to an input* x_1 *is* y_1*, and the response to an input* x_2 *is* y_2*, then the response of the system when both inputs are applied together will be* $y_1 + y_2$. Another way of looking at this is to consider the response of a system to a particular input. If the system is linear, and if, for example, the size of the input is doubled, then the resulting response will also be doubled.

The following diagrams illustrate this principle for a simple electrical system in which the behaviour of the components can, for our purposes, be modelled by linear equations.

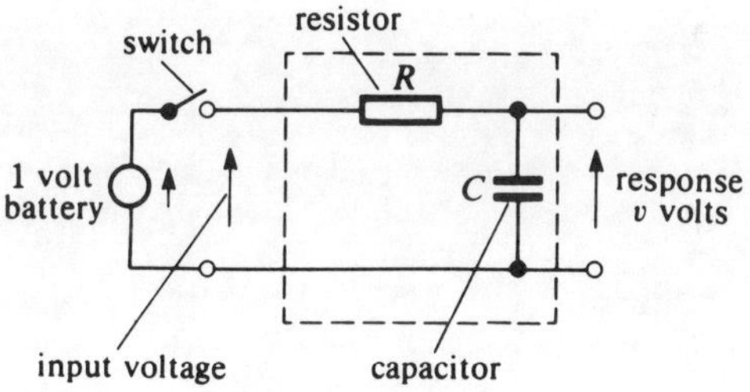

(a) first case

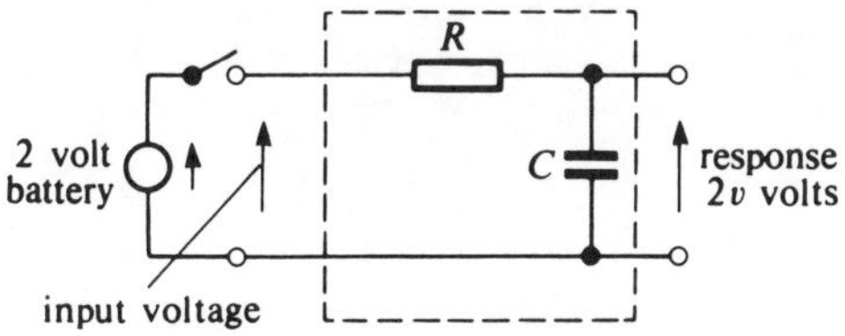

(b) second case

In each case, a capacitor and a resistor are connected via a switch to a battery.

In the first case, the voltage of the battery is 1 volt, and the system's response—the voltage v measured across the output terminals—on closing the switch is shown below. If we now replace the battery by one of 2 volts and repeat the experiment, then the output voltage at any instant will be double what it was in the first case.

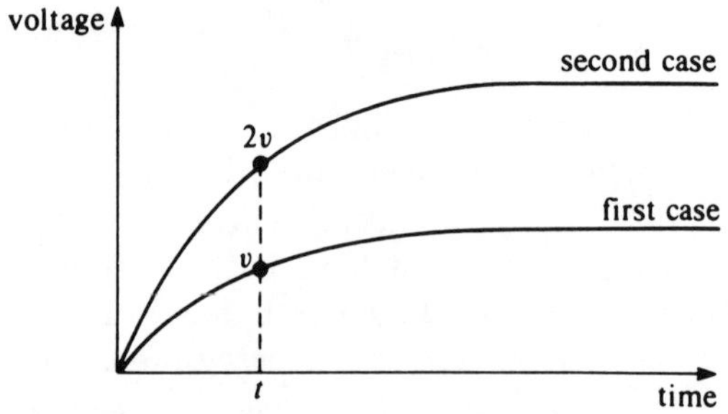

We conclude this chapter by listing some important types of component, together with their graphical representations and their corresponding component equations.

graphical representation 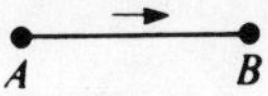

two-terminal component	through variable	across variable	component equation
resistor  resistance R	current i	voltage v	$v = Ri$
capacitor 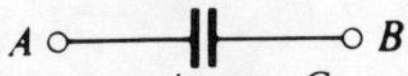 capacitance C	current i	voltage v	$i = C\dfrac{dv}{dt}$
inductor  inductance L	current i	voltage v	$v = L\dfrac{di}{dt}$
voltage source 	current i	voltage v	$v = V$
current source 	current i	voltage v	$i = I$
damper 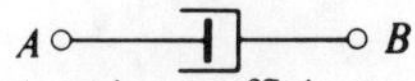 damping coefficient c	force f	velocity u	$f = cu$
mass 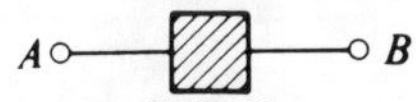 mass m	force f	velocity u	$f = m\dfrac{du}{dt}$
spring  stiffness k	force f	velocity u	$u = \dfrac{1}{k}\dfrac{df}{dt}$*
hydraulic valve  valve 'resistance' R	flow q	pressure p	$p = Rq$

* This relationship is the derivative of $f = kx$, known as Hooke's law, where x is the extension of the spring due to an applied force f.

Continued overleaf

<table>
<tr><td></td><td colspan="2">graphical representation</td><td>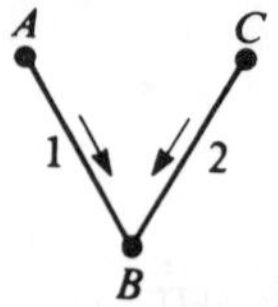
</td></tr>
<tr><td>three-terminal component</td><td>through variables</td><td>across variables</td><td>component equations</td></tr>
<tr><td>ideal transformer
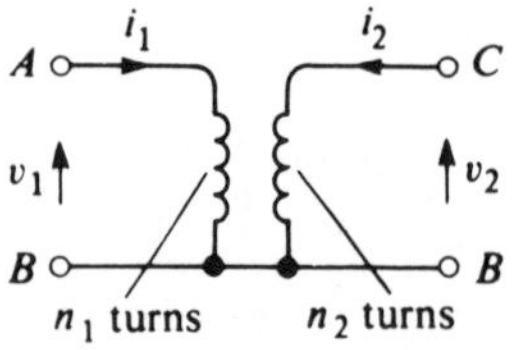
</td><td>currents
i_1 and i_2</td><td>voltages
v_1 and v_2</td><td>$v_2 = \frac{n_2}{n_1} v_1$
$i_2 = \frac{-n_1}{n_2} i_1$</td></tr>
<tr><td>gear train
(friction and inertia neglected)
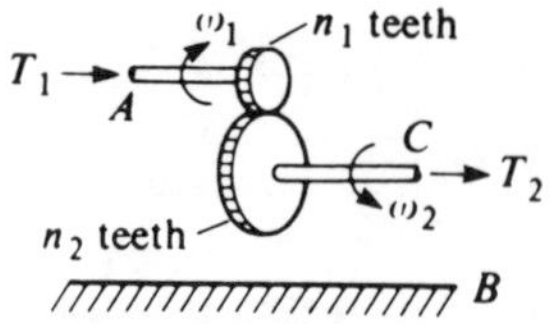

velocity reference</td><td>torques
T_1 and T_2</td><td>angular velocities
ω_1 and ω_2</td><td>$T_2 = \frac{n_2}{n_1} T_1$
$\omega_2 = \frac{-n_1}{n_2} \omega_1$</td></tr>
<tr><td>balance
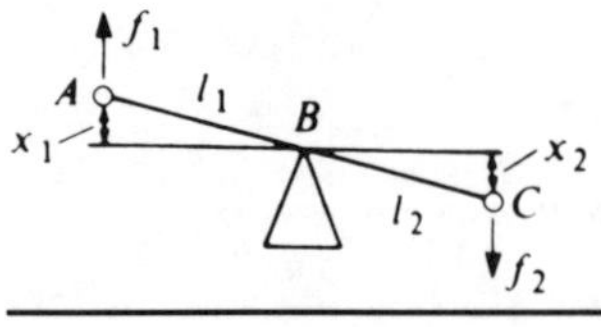
</td><td>forces
f_1 and f_2</td><td>displacements
x_1 and x_2</td><td>$f_2 = \frac{l_1}{l_2} f_1$
$x_2 = \frac{-l_2}{l_1} x_1$</td></tr>
</table>

EXERCISES 9

Components, and Through and Across Variables

9.1. Classify each of the following as a through variable or an across variable:

(a) the fluid flow rate for a hydraulic valve;
(b) the force acting on a mass;
(c) the derivative of the voltage for a capacitor.

9.2. Identify the through and across variables, and write down the component equation, for each of the following components:

(a) an electrical inductor;
(b) a spring;
(c) a transformer (considered as a 3-terminal component).

9.3. Consider the following system:

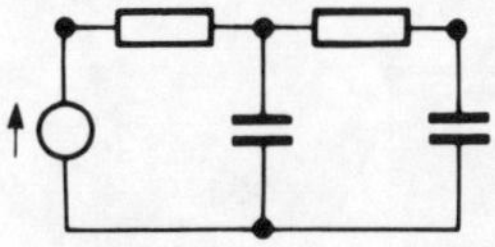

(a) Draw an oriented graph representation of the system.
(b) Apply Kirchhoff's voltage law to each cycle of the oriented graph.
(c) Apply Kirchhoff's current law to each vertex of the oriented graph.

9.4. Repeat Exercise 9.3 for the following system:

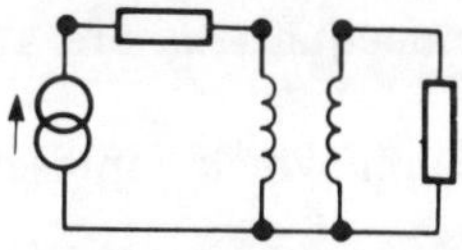

9.5. Repeat Exercise 9.3 for the following system (substituting cycle and vertex laws for the voltage and current laws):

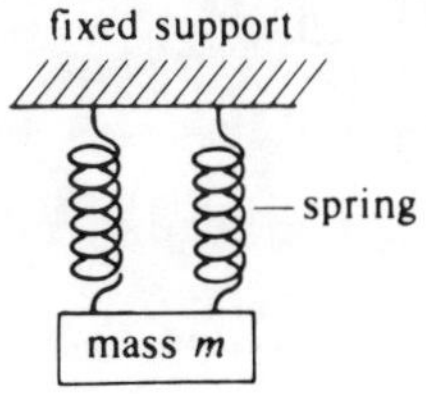

Graphical Representation

9.6. Decide whether each of the following statements is TRUE or FALSE.

(a) In an oriented graph, a 6-terminal component can always be represented by 6 vertices and 5 edges.

(b) In an oriented graph representation of an electrical network, the arrow next to an edge always points in the same direction as the actual direction of the current flow in the corresponding component.

(c) In an oriented graph, a 5-terminal component can be represented in exactly 5 different ways (ignoring edge directions).

9.7. A 4-terminal component may be represented as follows:

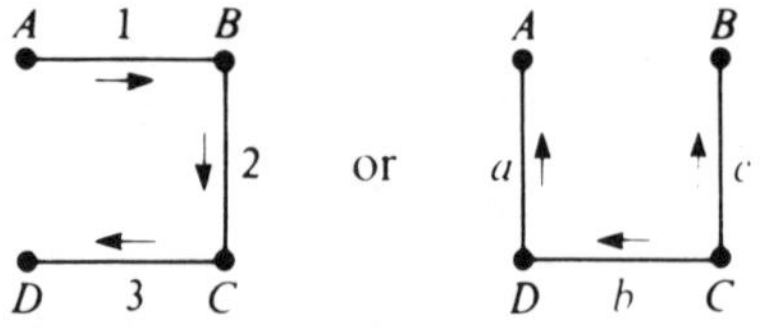

(a) Express each of the voltage variables, v_a, v_b and v_c in terms of v_1, v_2 and v_3.

(b) Express each of the current variables i_a, i_b and i_c in terms of i_1, i_2 and i_3.

9.8. Draw an oriented graph representation of a 5-terminal component. How many component equations are associated with such a component?

9.9. Draw an oriented graph which represents the following electrical network:

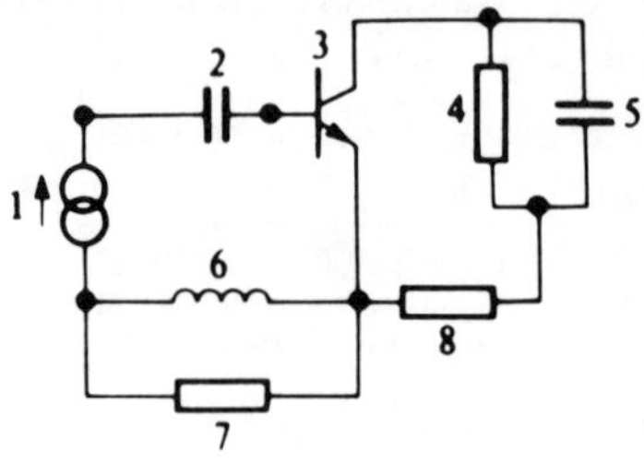

(Component 3 is a transistor)

9.10. Two springs of stiffness k_1 and k_2 are connected as shown below, and are extended by means of a force F.

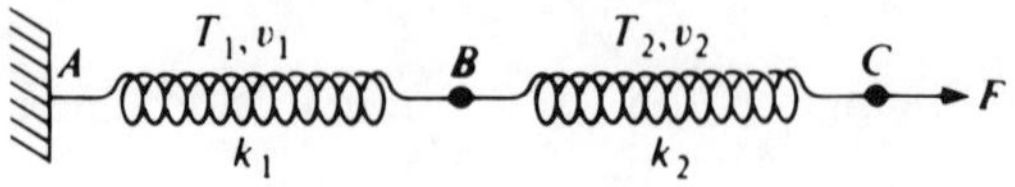

The tensions in the springs are T_1 and T_2, and the velocity of B relative to A, and of C relative to B, are v_1 and v_2 respectively.
It is desired to treat the two springs as a single component, with two terminals, A and C.
For this single component, write down, in terms of T_1, T_2, v_1 and v_2:

(a) the through variable;
(b) the across variable.

What is the component equation for this single component, if the velocity associated with it is v?

CHAPTER 10

ELECTRICAL NETWORKS: MATRIX EQUATIONS

After reading this chapter, you should be able to:

- write down the voltage and current law equations for a given network or oriented graph;
- find the fundamental cycles and the fundamental cycle matrix corresponding to a given spanning tree;
- find the fundamental cutsets and the fundamental cutset matrix corresponding to a given spanning tree;
- show how the fundamental cutset and cycle matrices can be obtained from the incidence matrix of an oriented graph.

Introduction

In the previous chapter we showed that an electrical network consisting of a number of interconnected components can be represented by an oriented graph. We also discussed the equations relating the voltages and currents associated with each component—the component equations. (Remember that *voltages* are *across* variables and *currents* are *through* variables.)

In this chapter we discuss two further sets of equations—the equations obtained by applying Kirchhoff's current law, and the equations obtained by applying Kirchhoff's voltage law to the oriented graph. In the next chapter we shall show that these three sets of equations:

the component equations,
Kirchhoff's current law equations,
Kirchhoff's voltage law equations,

are sufficient to enable us to solve electrical network problems (that is, to find the values of the through and across variables associated with all the components of a given network).

We can apply Kirchhoff's current law to every vertex of the oriented graph, and we can apply Kirchhoff's voltage law to every cycle of the graph. However, the equations which result contain a good deal of redundant information. What we should like is a systematic method for obtaining just enough equations to enable us to solve electrical network problems—or, to put it another way, to obtain sets of equations which contain all the information which Kirchhoff's laws can tell us about the network, without any redundancy.

Such a systematic method was given by Kirchhoff in 1847 in a paper entitled '*On the solution of the equations obtained from the investigation of the linear distribution of galvanic currents*'. His method used the concept of what is now called a *spanning tree* and involved finding a set of *fundamental cycles* and a set of *fundamental cutsets*. We use the same method, although the language in which we express it is somewhat different from that used by Kirchhoff. *In this chapter, we consider only connected graphs—this means that we can always find a spanning tree.*

Our aim is to describe a method of solving network problems which is appropriate for computer analysis, so we need to overcome the problem of obtaining the fundamental cycles and cutsets using a computer. We describe a method which enables us to obtain the fundamental cycles and cutsets directly from the incidence matrix of the oriented graph. The great advantage of this method is that the only information about the oriented graph which we have to feed into the computer is the incidence matrix itself, and we can obtain the incidence matrix by inspection from the graph.

10.1 KIRCHHOFF'S VOLTAGE LAW EQUATIONS

We can apply Kirchhoff's voltage law to any cycle of the oriented graph of an electrical network. But if we apply this law to *all* the cycles, we end up with a set of equations which contain a lot of redundant information. We demonstrate this for the following electrical network which we shall use as an illustrative example throughout this chapter.

Let us start by applying the voltage law to the cycles *BCEB*, *CDEC* and *BCDEB* of the oriented graph. (Recall that in going around each cycle we take a voltage to be positive if it is oriented in the same direction as the cycle, and negative if it is not.)

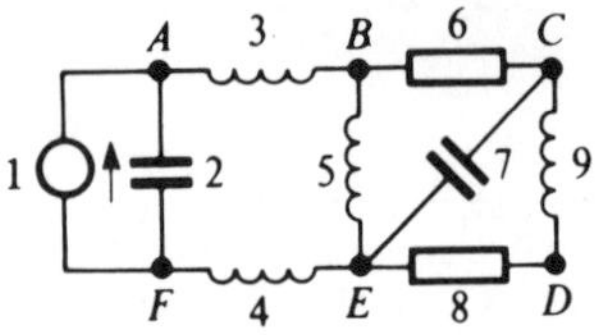

electrical network

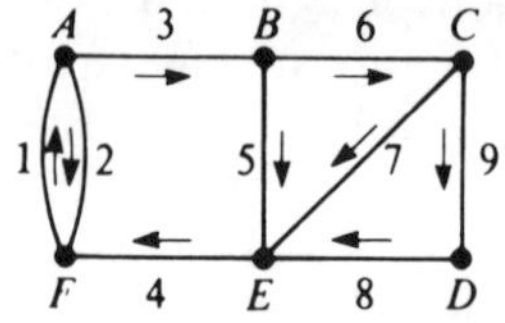

oriented graph

This gives us the following voltage equations:

BCEB: $v_6 + v_7 - v_5 = 0$
CDEC: $v_9 + v_8 - v_7 = 0$
BCDEB: $v_6 + v_9 + v_8 - v_5 = 0.$

The last of these equations is simply the sum of the other two, and so gives us no further information about the voltages.

Similarly, applying the voltage law to the cycles *ABEFA*, *BCEB*, *CDEC* and *ABCDEFA* (using the right-hand edge *AF* in each case), we get the following voltage equations:

ABEFA: $v_3 + v_5 + v_4 - v_2 = 0$
BCEB: $v_6 + v_7 - v_5 = 0$
CDEC: $v_9 + v_8 - v_7 = 0$
ABCDEFA: $v_3 + v_6 + v_9 + v_8 + v_4 - v_2 = 0$

Again, the last of these equations is the sum of the others, and so gives us no further information about the voltages.

In each of these cases, we can write each equation as the sum or difference of the other equations, and we express this by saying that the equations are *linearly dependent*. More generally, a number of equations are said to be **linearly dependent** if at least one of them 'depends' on the others, in the sense that it can be obtained by adding or subtracting multiples of the others. If none of the equations depends on the others in this way, then the equations are said to be **linearly independent**. For example, the voltage equations for the cycles *ABEFA*, *BCEB* and *CDEC* are linearly independent, since each contains a term contained in neither of the other two equations.

Problem 10.1 Consider the following oriented graph, which corresponds to the network in Problem 9.1.

(a) Write down the voltage equation for each of the seven cycles of the graph.
(b) What is the maximum number of these seven equations which are linearly independent?

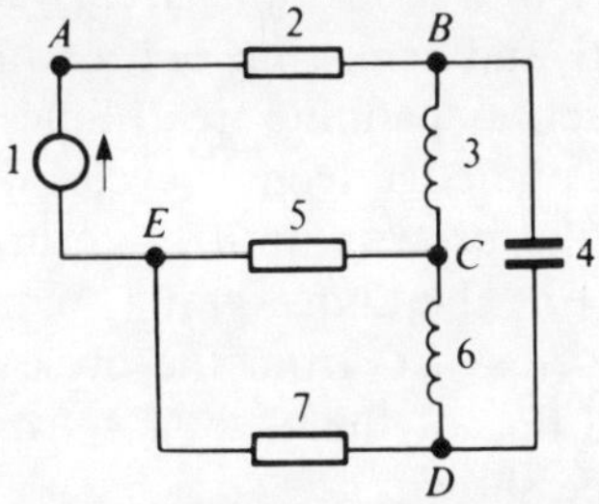

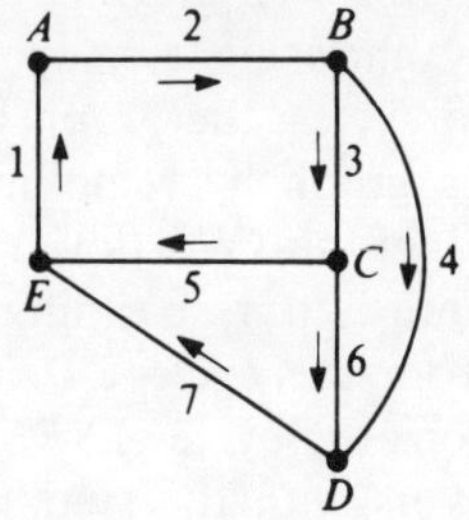

10.2 FUNDAMENTAL CYCLES

We now investigate the problem of finding a set containing the maximum number of Kirchhoff's voltage law equations which are linearly independent. What we need to do is to find a set of cycles in an oriented graph which give rise to such a set of linearly independent equations. We say that a number of cycles in a graph are **linearly independent** if the corresponding voltage equations are linearly independent. For instance, in our illustrative example, the cycles *ABEFA*, *BCEB* and *CDEC* are linearly independent, since they correspond to linearly independent equations. It follows that, in order to find a set of linearly independent equations for the network, it is enough to identify the corresponding linearly independent cycles. We can find such a set of cycles quite easily; to do so, we use Kirchhoff's idea of a *spanning tree.*

Definitions *Let G be a connected graph. A* ***spanning tree*** *in G is a subgraph of G which includes all the vertices of G and is also a tree. The edges of the tree are called* ***branches****. If we remove from G all the edges of a spanning tree, the remaining subgraph is called a* ***co-tree*** *and its edges are called* ***chords****.*

Remarks The number of spanning trees in a graph can be very large; for instance, the graph in our example (shown below) has exactly 50 different spanning trees.

Note that a co-tree is, in general, not a tree.

The following diagram shows the graph in our example, one of its spanning trees, and the associated co-tree.

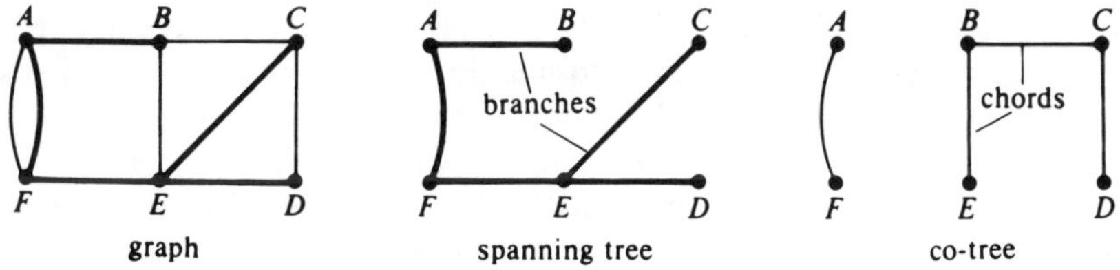

For any connected graph G, we can find a spanning tree systematically. We start by choosing any cycle in G and removing one of its edges. (If there are no cycles, the graph G is itself a spanning tree.) Since we cannot disconnect a graph by removing just one edge from a cycle, we still have a connected graph. We now repeat the process until there are no cycles left, and there is our spanning tree. For the above graph, removal of the edge CD from the cycle $CDEC$, the edge BC from the cycle $BCEB$, the edge BE from the cycle $ABEFA$, and one of the edges FA from the cycle AFA, leaves us with the spanning tree shown.

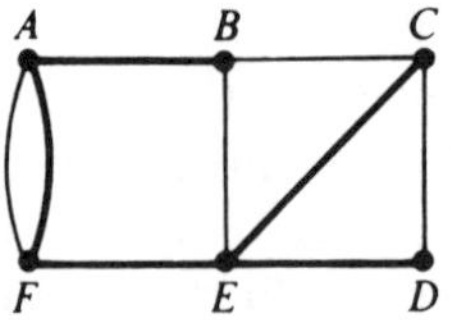

Problem 10.2

(a) Find two more spanning trees in the above graph, and list the branches and chords in each case.
(b) If a connected graph has n vertices and m edges, how many branches and chords are there associated with each spanning tree?
(c) What happens if we add a chord to a spanning tree? Illustrate your answer using the spanning tree given in our example.

The last part of Problem 10.2 shows us how to find a set of linearly independent cycles. We first choose a spanning tree; then, to find the first cycle in the set, we add one of the chords to the tree—this produces exactly one cycle. We repeat this procedure for each chord. If the graph has k chords of the spanning tree, then this procedure will produce k cycles and, since each cycle contains one chord which is not contained in any of the other cycles, the resulting set of cycles must be linearly independent.

If we were seeking a set of equations to solve by hand, we would choose the spanning tree with some care. In particular, we would choose the tree so that in general each edge for which the across variable was known would be a branch of the tree, and each edge for which the through variable was known would be a chord. However, this selection is not necessary in the mathematical sense; it serves only to make the resulting equations quicker to solve by hand. We will be describing a method appropriate for solving physical network problems on a computer where such considerations are unimportant, and so we will not need to follow a selection procedure: we will simply choose any convenient spanning tree.

Since the set of cycles we have found using our spanning tree are linearly independent, it follows that the corresponding voltage law equations must also be linearly independent. This leads us to make the following definitions.

Definitions *Let G be a connected graph, and let T be a spanning tree of G. The* ***set of fundamental cycles associated with T*** *consists of the cycles of G each of which is obtained by adding a single chord to T. The corresponding voltage law equations are called the* ***fundamental cycle equations****.*

For example, for the spanning tree shown below:

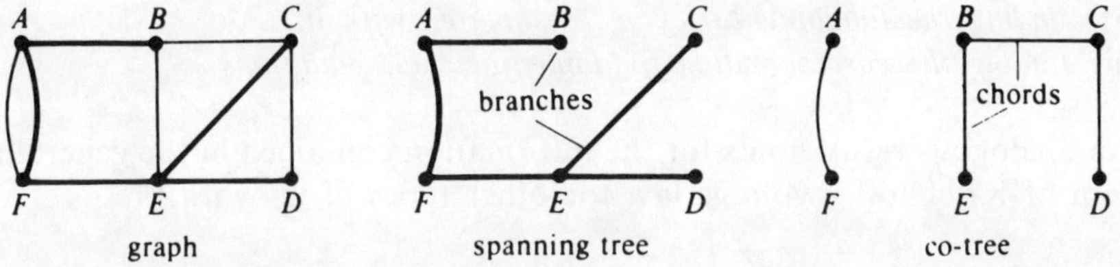

we have the following table:

Table 10.1

chord	fundamental cycle	diagram	fundamental cycle equation
BC	*ABCEFA*	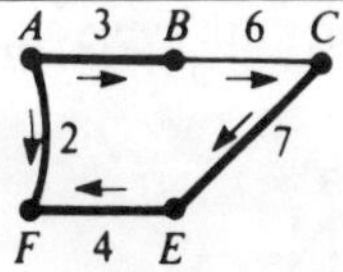	$v_3 + v_6 + v_7 + v_4 - v_2 = 0$
BE	*ABEFA*	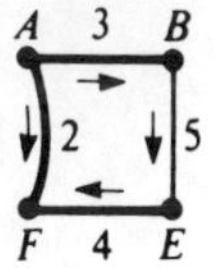	$v_3 + v_5 + v_4 - v_2 = 0$
CD	*CDEC*	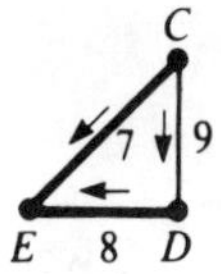	$v_9 + v_8 - v_7 = 0$
FA	*AFA*		$v_2 + v_1 = 0$

It follows from part (b) of the above problem that if G has n vertices and m edges, then any spanning tree gives rise to $m - n + 1$ chords, and so there are $m - n + 1$ cycles in a fundamental set. This number is the largest number of linearly independent cycles in G, and is called the **cycle rank** of G. It follows that there are $m - n + 1$ fundamental cycle equations, and that this is the largest possible number of linearly independent cycle equations. We therefore conclude that:

in any electrical network problem, we can obtain all the information contained in Kirchhoff's voltage law by:

(i) *choosing a spanning tree*;
(ii) *finding the fundamental cycles associated with it*;
(iii) *finding the corresponding fundamental cycle equations*.

(An analogous result holds for the information contained in the generalized form of Kirchhoff's voltage law for other types of network.)

Problem 10.3 For each of the spanning trees shown below, write down:

a list of the chords;
the corresponding fundamental cycles;
the corresponding fundamental cycle equations.

In each case, show how to obtain the voltage law equation for the cycle $ABCDEA$ from the fundamental cycle equations.

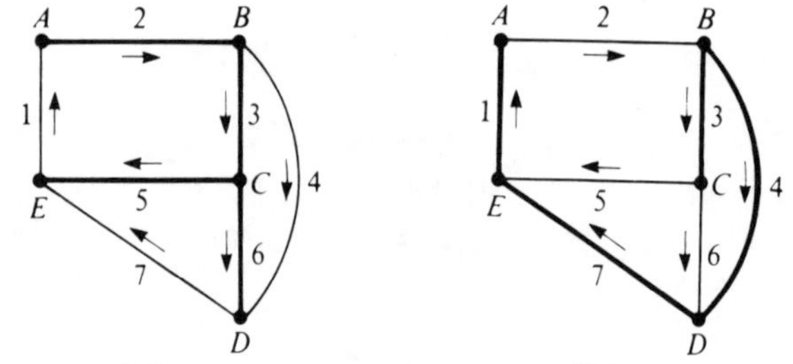

(a) (b)

Before leaving fundamental cycles, we show how the fundamental cycle equations can be represented in matrix form. The *rows* of the matrix correspond to the *chords* associated with a spanning tree, and the *columns* correspond to all the *edges* of the graph G. (It is usual to write the columns corresponding to the branches of the spanning tree first, followed by those for the chords.) If G has n vertices and m edges, we get an $(m - n + 1) \times m$

matrix. This matrix is called the **fundamental cycle matrix**, and is denoted by $\mathbf{C}_f$.

To construct this matrix, we add each chord to the spanning tree in turn, and look at the edges in the corresponding fundamental cycle. Tracing around this cycle in the direction of thc chord, we fill in the row corresponding to the given chord as follows: we write

1 in each column corresponding to an edge of the cycle oriented in the same direction as the chord;

-1 in each column corresponding to an edge traced in the opposite direction;

0 in each column corresponding to an edge which is not in the cycle.

For example, for the fundamental cycle *ABCEFA* illustrated below, the row corresponding to the chord *BC* has 1 in the columns corresponding to *BC*, *AB*, *CE* and *EF*, -1 in the column corresponding to *AF*, and 0 elsewhere.

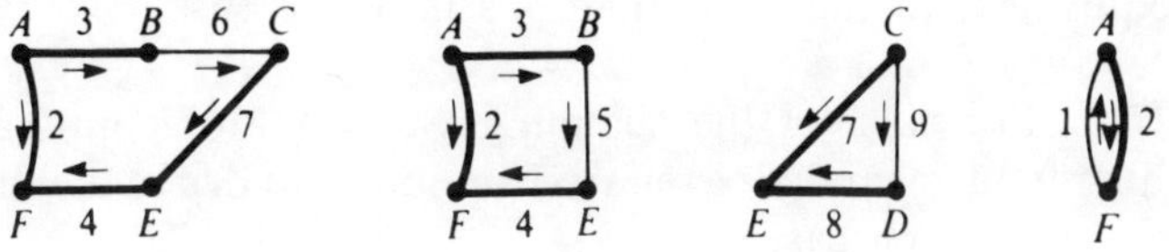

Repeating this procedure for all four fundamental cycles, we obtain the following matrix:

$$
\begin{matrix} \\ BC \\ BE \\ CD \\ FA \\ \\ \end{matrix}
\begin{matrix}
\begin{matrix} AB & AF & CE & DE & EF \end{matrix} & \begin{matrix} BC & BE & CD & FA \end{matrix} \\
\underbrace{\begin{matrix} 1 & -1 & 1 & 0 & 1 \\ 1 & -1 & 0 & 0 & 1 \\ 0 & 0 & -1 & 1 & 0 \\ 0 & 1 & 0 & 0 & 0 \end{matrix}}_{\text{Branches}} & \underbrace{\begin{matrix} 1 & 0 & 0 & 0 \\ 0 & 1 & 0 & 0 \\ 0 & 0 & 1 & 0 \\ 0 & 0 & 0 & 1 \end{matrix}}_{\text{Chords}}
\end{matrix}
\left.\begin{matrix} \\ \\ \\ \\ \end{matrix}\right\}\text{Chords}
$$

Having found the fundamental cycle matrix $\mathbf{C}_f$, we can write the fundamental cycle equations in matrix form as

$$\mathbf{C}_f\mathbf{v} = \mathbf{0},$$

where $\mathbf{v}$ is the column vector of edge voltages, written in the same order as the columns of the matrix, and $\mathbf{0}$ is the appropriate zero vector. For example, the above matrix gives

$$\begin{array}{ll} \text{edge} & \rightarrow \; 3 \quad 2 \quad 7 \quad 8 \quad 4 \quad 6 \quad 5 \quad 9 \quad 1 \\ \text{number} & \end{array}$$

$$\begin{bmatrix} 1 & -1 & 1 & 0 & 1 & 1 & 0 & 0 & 0 \\ 1 & -1 & 0 & 0 & 1 & 0 & 1 & 0 & 0 \\ 0 & 0 & -1 & 1 & 0 & 0 & 0 & 1 & 0 \\ 0 & 1 & 0 & 0 & 0 & 0 & 0 & 0 & 1 \end{bmatrix} \begin{bmatrix} v_3 \\ v_2 \\ v_7 \\ v_8 \\ v_4 \\ v_6 \\ v_5 \\ v_9 \\ v_1 \end{bmatrix} = \begin{bmatrix} 0 \\ 0 \\ 0 \\ 0 \end{bmatrix}$$

Multiplying the two matrices on the left, we obtain the equation

$$\begin{bmatrix} v_3 - v_2 + v_7 + v_4 + v_6 \\ v_3 - v_2 + v_4 + v_5 \\ -v_7 + v_8 + v_9 \\ v_2 + v_1 \end{bmatrix} = \begin{bmatrix} 0 \\ 0 \\ 0 \\ 0 \end{bmatrix},$$

which is equivalent to the four equations in Table 10.1.

Problem 10.4 For each of the spanning trees in Problem 10.3, find the corresponding fundamental cycle matrix, and use it to derive the fundamental cycle equations in each case.

10.3 KIRCHHOFF'S CURRENT LAW EQUATIONS

We now turn our attention to Kirchhoff's current law, which states that the algebraic sum of the currents at each vertex of the network is zero. As with the voltage law, we find that if we apply the current law to every vertex of a graph, then we end up with more equations than we need.

For example, consider our oriented graph:

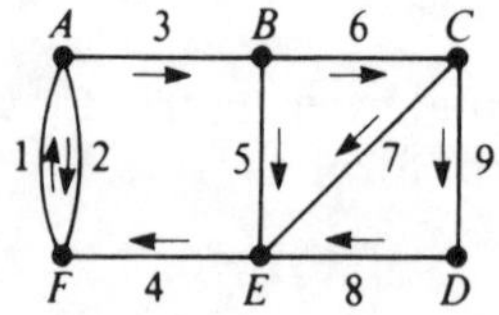

Applying the current law at each vertex, we obtain the following equations.

vertex A: $i_1 - i_2 - i_3 = 0$
vertex B: $i_3 - i_5 - i_6 = 0$
vertex C: $i_6 - i_7 - i_9 = 0$
vertex D: $i_9 - i_8 = 0$
vertex E: $i_5 + i_7 + i_8 - i_4 = 0$
vertex F: $i_2 + i_4 - i_1 = 0$

These six equations are not linearly independent: if we add any five of them, we get the sixth. However, any five of these equations are linearly independent. (We show later that for a graph with n vertices there are always exactly $n - 1$ linearly independent equations.) In fact, for the solution procedures which we use here, it is more appropriate to apply Kirchhoff's current law to the *cutsets* of the graph, rather than to the vertices. (We defined a *cutset* of a connected graph in Chapter 6.) To see what we mean by this, consider the cutset consisting of the edges BC, CE and DE of our usual graph:

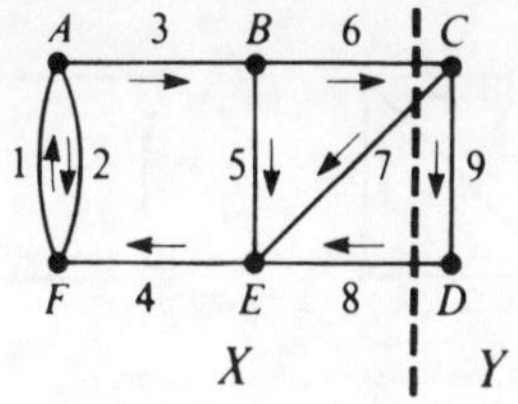

Removal of the edges of this cutset disconnects the graph into two parts, X and Y. If we add the equations for the vertices in just one of these parts (Y, say), we find that the currents along those edges lying entirely within that part cancel out, and we are left with an equation relating the currents in the edges of the cutset.

For example, if we add together the above equations for the vertices C and D, the current i_9 cancels out, leaving the equation

$$i_6 - i_7 - i_8 = 0$$

relating the currents in the cutset edges BC, CE and DE. Note that we get the same equation if we add the equations for the vertices A, B, E and F in the set X.

We can generalize this idea, and use the current law to find an equation relating the currents in any cutset. If the cutset separates the graph into two parts X and Y, then the equation has the following form:

(the sum of the currents from X to Y)
− (the sum of the currents from Y to X)
= 0

or simply,

the algebraic sum of the currents in any cutset is zero.

However, if we repeat this procedure for all the cutsets in the graph, we find that the resulting equations are not all linearly independent.

10.4 FUNDAMENTAL CUTSETS

As with the cycle equations, we wish to find the largest possible number of linearly independent cutset equations. To do this, we find a set of *fundamental cutsets*. As with fundamental cycles, these are constructed using a spanning tree.

We illustrate the method with the graph and spanning tree used previously:

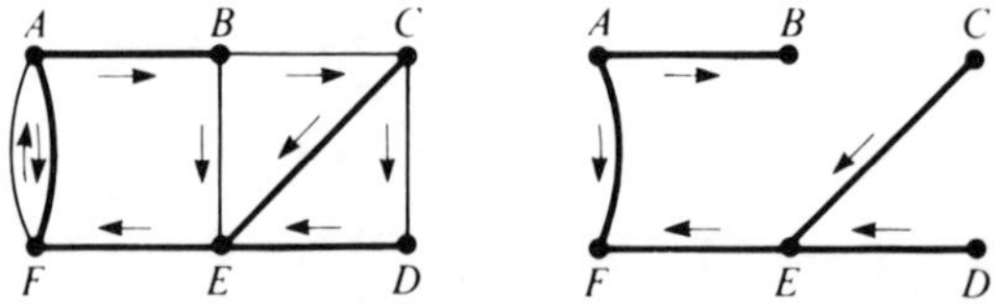

The removal of any branch separates the tree into two parts, X and Y; for example, the removal of the branch EF separates the tree into two parts, one containing the vertices C, D and E, and the other containing the vertices A, B and F. Since the branch EF is oriented from E to F, we let X be the part containing E, and Y be the part containing F. We now list all the edges of the oriented graph joining a vertex in X and a vertex in Y. (In our case, these are the edges EF, BE and BC.) One of these edges is the branch we removed to disconnect the tree (EF); the others will be chords (BE, BC). Together they form a cutset of the network graph. We repeat this procedure for each branch of the spanning tree. The resulting cutsets are called *fundamental cutsets*; since there are $n - 1$ branches in the spanning tree, there are $n - 1$ fundamental cutsets. This leads to the following general definition.

Definition *Let G be a connected graph, and let T be a spanning tree of G. The* ***set of fundamental cutsets associated with T*** *consists of the cutsets of G obtained by removing a branch of T, thus separating the tree into two parts, X and Y, and listing the edges of G joining a vertex in X and a vertex in Y. The corresponding current law equations are called the* ***fundamental cutset equations****.*

We illustrate this definition by our example, using the same spanning tree as before.

Table 10.2

branch	vertices in X	vertices in Y	fundamental cutset	diagram	fundamental cutset equation
AB	A, C, D, E, F	B	$\{AB, BC, BE\}$	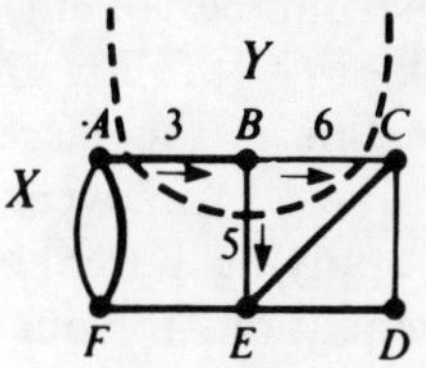	$i_3 - i_5 - i_6 = 0$
AF	A, B	C, D, E, F	$\{AF, BC, BE, FA\}$	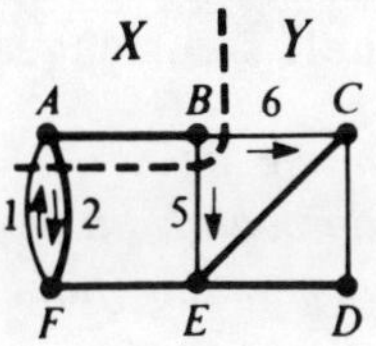	$i_2 + i_5 + i_6 - i_1 = 0$
CE	C	A, B, D, E, F	$\{BC, CD, CE\}$	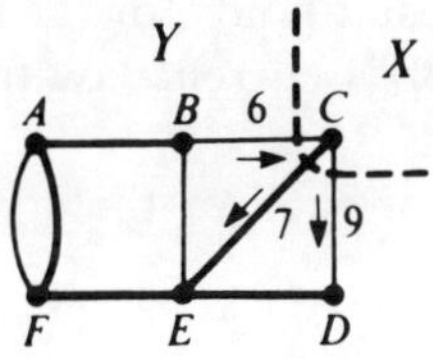	$i_7 + i_9 - i_6 = 0$
DE	D	A, B, C, E, F	$\{CD, DE\}$	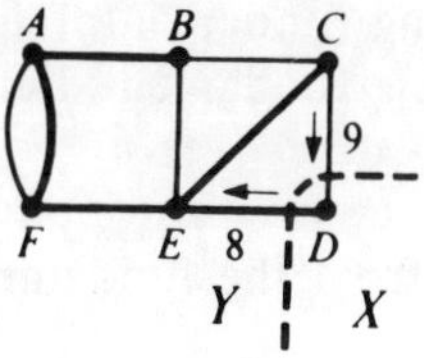	$i_8 - i_9 = 0$
EF	C, D, E	A, B, F	$\{BC, BE, EF\}$	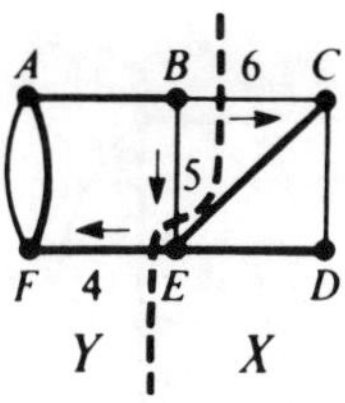	$i_4 - i_5 - i_6 = 0$

We say that a number of cutsets in a graph G are **linearly independent** if the corresponding current law equations are linearly independent. Thus *the cutsets in a set of fundamental cutsets are all linearly independent*, since each cutset contains one edge (the branch of the spanning tree) not contained in any of the other cutsets. If G has n vertices, then any spanning tree has $n - 1$ branches, and so there are $n - 1$ cutsets in a fundamental set. This number is the largest number of linearly independent cutsets in G, and is called the **cutset rank** of G. It follows that there are $n - 1$ fundamental cutset equations, and that this is the largest possible number of linearly independent cutset equations. Moreover, these equations, rather than the equations obtained by applying Kirchhoff's current law at the vertices of G, are the appropriate equations for our purposes. We therefore conclude that:

in any electrical network problem, we can obtain all the information contained in Kirchhoff's current law by:

(i) *choosing a spanning tree*;
(ii) *finding the fundamental cutsets associated with it*;
(iii) *finding the corresponding fundamental cutset equations.*

(An analogous result holds for the information contained in the generalized form of Kirchhoff's current law for other types of network.)

Problem 10.5 For each of the spanning trees shown below, write down:

a list of the branches;
the corresponding sets X and Y;
the corresponding fundamental cutsets;
the corresponding fundamental cutset equations.

In each case, show how to obtain the current law equation for the cutset $\{AB, BC, BD\}$ from the fundamental cutset equations.

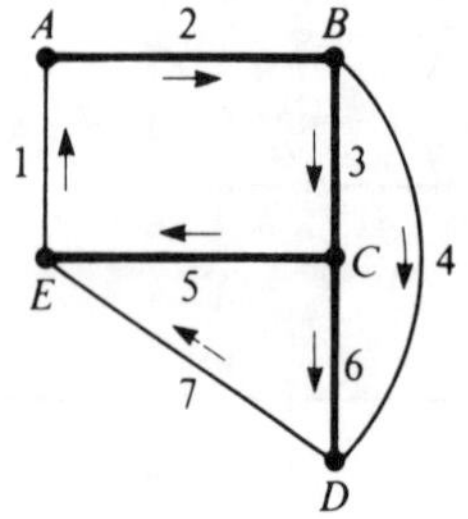

(a)

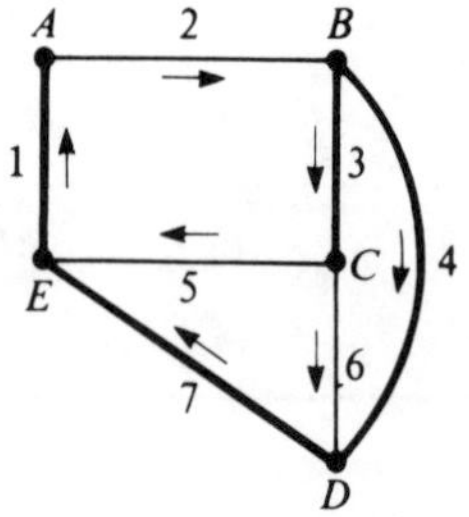

(b)

As with fundamental cycles, we can represent the fundamental cutset equations in matrix form. The rows of this matrix correspond to the branches of a spanning tree, and the columns correspond to all the edges of the graph G. (It is usual to write the columns corresponding to the branches first, followed by those for the chords.) If G has n vertices and m edges, we get an $(n-1) \times m$ matrix. This matrix is called the **fundamental cutset matrix**, and is denoted by $\mathbf{B_f}$.

To construct this matrix, we take each branch of the spanning tree in turn, and look at the edges in the corresponding fundamental cutset. We then fill in the row corresponding to the given branch as follows: we write

1 in each column corresponding to an edge of the cutset oriented from X to Y;
-1 in each column corresponding to an edge of the cutset oriented from Y to X;
0 in each column corresponding to an edge not in the cutset.

For example, for the fundamental cutset $\{AB, BC, BE\}$ illustrated below, the row corresponding to the branch AB has 1 in the column corresponding to AB, -1 in the columns corresponding to BC and BE, and 0 elsewhere.

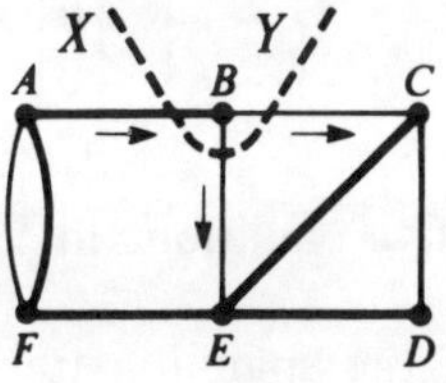

Repeating this procedure for all five fundamental cutsets (shown in Table 10.2), we obtain the following matrix:

$$
\begin{array}{c}
 \\ AB \\ AF \\ CE \\ DE \\ EF
\end{array}
\begin{array}{c}
\begin{array}{ccccccccc} AB & AF & CE & DE & EF & BC & BE & CD & FA \end{array} \\
\left[\begin{array}{ccccc|cccc}
1 & 0 & 0 & 0 & 0 & -1 & -1 & 0 & 0 \\
0 & 1 & 0 & 0 & 0 & 1 & 1 & 0 & -1 \\
0 & 0 & 1 & 0 & 0 & -1 & 0 & 1 & 0 \\
0 & 0 & 0 & 1 & 0 & 0 & 0 & -1 & 0 \\
0 & 0 & 0 & 0 & 1 & -1 & -1 & 0 & 0
\end{array}\right]
\end{array}
\Bigg\}\ \text{Branches}
$$

$$\underbrace{\qquad\qquad}_{\text{Branches}} \qquad \underbrace{\qquad\qquad}_{\text{Chords}}$$

Having found the fundamental cutset matrix $\mathbf{B_f}$, we can write the fundamental cutset equations in matrix form as

$$\mathbf{B_f i} = \mathbf{0},$$

where **i** is the column vector of edge currents, written in the same order as the columns of the matrix, and **0** is the appropriate zero vector. For example, the above matrix gives

$$\begin{matrix}\text{edge number} \rightarrow & 3 & 2 & 7 & 8 & 4 & 6 & 5 & 9 & 1\end{matrix}$$

$$\begin{bmatrix} 1 & 0 & 0 & 0 & 0 & -1 & -1 & 0 & 0 \\ 0 & 1 & 0 & 0 & 0 & 1 & 1 & 0 & -1 \\ 0 & 0 & 1 & 0 & 0 & -1 & 0 & 1 & 0 \\ 0 & 0 & 0 & 1 & 0 & 0 & 0 & -1 & 0 \\ 0 & 0 & 0 & 0 & 1 & -1 & -1 & 0 & 0 \end{bmatrix} \begin{bmatrix} i_3 \\ i_2 \\ i_7 \\ i_8 \\ i_4 \\ i_6 \\ i_5 \\ i_9 \\ i_1 \end{bmatrix} = \begin{bmatrix} 0 \\ 0 \\ 0 \\ 0 \\ 0 \end{bmatrix}$$

Multiplying the two matrices on the left, we obtain the equation

$$\begin{bmatrix} i_3 - i_6 - i_5 \\ i_2 + i_6 + i_5 - i_1 \\ i_7 - i_6 + i_9 \\ i_8 - i_9 \\ i_4 - i_6 - i_5 \end{bmatrix} = \begin{bmatrix} 0 \\ 0 \\ 0 \\ 0 \\ 0 \end{bmatrix}$$

which is equivalent to the five equations in Table 10.2.

Problem 10.6 For each of the spanning trees in Problem 10.5, find the corresponding fundamental cutset matrix, and use it to derive the fundamental cutset equations.

10.5 OBTAINING THE FUNDAMENTAL CYCLE AND CUTSET MATRICES

So far in this chapter we have shown that:

(i) all the information which can be obtained by applying Kirchhoff's current law to an electrical network is contained in the matrix equation

$$\mathbf{B_f i} = \mathbf{0},$$

where $\mathbf{B_f}$ is the fundamental cutset matrix;

(ii) all the information which can be obtained by applying Kirchhoff's voltage law to an electrical network is contained in the matrix equation

$$\mathbf{C_f v} = \mathbf{0},$$

where $\mathbf{C_f}$ is the fundamental cycle matrix.

We have shown how the fundamental cutset matrix $\mathbf{B_f}$ and the fundamental cycle matrix $\mathbf{C_f}$ with respect to a given spanning tree can be obtained by inspection of the oriented graph of an electrical network. Even for the small networks we have considered, this is a somewhat tedious process, so we would like to be able to get a computer to do this for us. The problem with this is that the method we have described is not very suitable for adaptation as a computer program. Fortunately, however, the matrices $\mathbf{B_f}$ and $\mathbf{C_f}$ can be obtained directly from the incidence matrix of the oriented graph by an elegant method which is suitable for translation into a computer program. We now describe this method, beginning with the definition of the *incidence matrix* of an oriented graph.

The Incidence Matrix of an Oriented Graph

We define the *incidence matrix* of an oriented graph by regarding the oriented graph as a digraph in which the direction of each arc is just the reference direction of the corresponding edge of the oriented graph. Adapting the definition of the incidence matrix of a digraph given in Chapter 2, we obtain the following definition.

Definition *Let G be an oriented graph with n vertices and m edges. The* ***incidence matrix*** $\mathbf{A}(G)$ *is the $n \times m$ matrix in which the entry in row i and column j is*

1 *if edge j is incident with, and oriented away from, vertex i;*
-1 *if edge j is incident with, and oriented towards, vertex i;*
0 *if edge j is not incident with vertex i.*

For our example, the oriented graph is

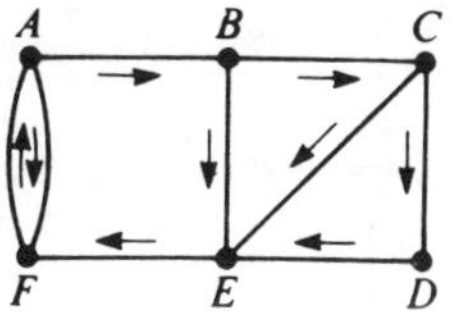

(repeated at the foot of the next page) and the incidence matrix is

$$\begin{array}{c} \\ A \\ B \\ C \\ D \\ E \\ F \end{array} \begin{array}{c} \begin{array}{ccccccccc} AB & AF & BC & BE & CD & CE & DE & EF & FA \end{array} \\ \left[\begin{array}{ccccccccc} 1 & 1 & 0 & 0 & 0 & 0 & 0 & 0 & -1 \\ -1 & 0 & 1 & 1 & 0 & 0 & 0 & 0 & 0 \\ 0 & 0 & -1 & 0 & 1 & 1 & 0 & 0 & 0 \\ 0 & 0 & 0 & 0 & -1 & 0 & 1 & 0 & 0 \\ 0 & 0 & 0 & -1 & 0 & -1 & -1 & 1 & 0 \\ 0 & -1 & 0 & 0 & 0 & 0 & 0 & -1 & 1 \end{array}\right] \end{array}$$

Each column of the incidence matrix contains one 1 and one -1, so that if the six rows are added together, the result is a row of zeros. Another way of expressing this fact is to say that each row is equal to minus the sum of all the other rows. This result holds for the incidence matrix of *any* oriented graph. We say that the rows of a matrix are **linearly dependent** if one of them can be obtained by adding or subtracting multiples of the others, and **linearly independent** otherwise. So *the rows of any incidence matrix are linearly dependent*. It follows that we can omit any single row of $\mathbf{A}(G)$ without loss of information. For example, in the above matrix, if we omit the row corresponding to the vertex C, then we get the following matrix. We shall denote it by $\mathbf{A}_0$:

$$\mathbf{A}_0 = \begin{array}{c} \\ A \\ B \\ D \\ E \\ F \end{array} \begin{array}{c} \begin{array}{ccccccccc} AB & AF & BC & BE & CD & CE & DE & EF & FA \end{array} \\ \left[\begin{array}{ccccccccc} 1 & 1 & 0 & 0 & 0 & 0 & 0 & 0 & -1 \\ -1 & 0 & 1 & 1 & 0 & 0 & 0 & 0 & 0 \\ 0 & 0 & 0 & 0 & -1 & 0 & 1 & 0 & 0 \\ 0 & 0 & 0 & -1 & 0 & -1 & -1 & 1 & 0 \\ 0 & -1 & 0 & 0 & 0 & 0 & 0 & -1 & 1 \end{array}\right] \end{array}$$

Such a matrix is called a **reduced incidence matrix**, and the vertex C is called the **reference vertex**.

If we are given the matrix $\mathbf{A}_0$, we can recover the original incidence matrix $\mathbf{A}(G)$ by inserting the missing row (calculated by taking minus the sum of the rows of $\mathbf{A}_0$). However, we cannot replace *two* missing rows, since there is no way of knowing which direction to assign to an edge joining the corresponding two vertices. It follows that if G has n vertices, then the **rank** of the incidence matrix $\mathbf{A}(G)$ (that is, the maximum number of linearly independent rows) is $n - 1$.

Suppose that we have chosen a spanning tree of G as follows:

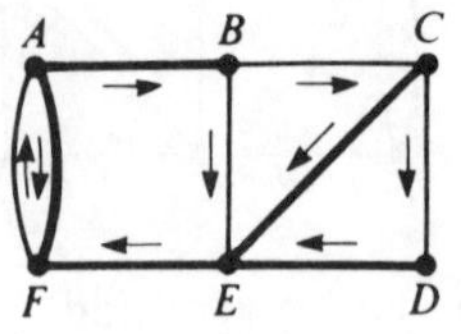

We can rearrange the columns of the reduced incidence matrix $\mathbf{A}_0$ so that those columns corresponding to branches of the tree come first, followed by those columns corresponding to the chords, as follows:

$$\mathbf{A}_0 = \begin{array}{c} \\ A \\ B \\ D \\ E \\ F \end{array} \begin{array}{c} \begin{array}{ccccc|cccc} AB & AF & CE & DE & EF & BC & BE & CD & FA \end{array} \\ \left[\begin{array}{ccccc|cccc} 1 & 1 & 0 & 0 & 0 & 0 & 0 & 0 & -1 \\ -1 & 0 & 0 & 0 & 0 & 1 & 1 & 0 & 0 \\ 0 & 0 & 0 & 1 & 0 & 0 & 0 & -1 & 0 \\ 0 & 0 & -1 & -1 & 1 & 0 & -1 & 0 & 0 \\ 0 & -1 & 0 & 0 & -1 & 0 & 0 & 0 & 1 \end{array}\right] \\ \underbrace{\hphantom{AB\ AF\ CE\ DE\ EF}}_{\text{Branches}} \quad \underbrace{\hphantom{BC\ BE\ CD\ FA}}_{\text{Chords}} \end{array}$$

We have partitioned $\mathbf{A}_0$ into two parts, one part corresponding to the branches, and one part corresponding to the chords. We can abbreviate this partitioned matrix by writing $\mathbf{A}_0 = [\mathbf{A}_t \mid \mathbf{A}_c]$, where $\mathbf{A}_t$ (the tree part) consists of the first five columns, and $\mathbf{A}_c$ (the co-tree part) consists of the last four columns:

$$\mathbf{A}_t = \begin{array}{c} \\ A \\ B \\ D \\ E \\ F \end{array} \begin{array}{c} \begin{array}{ccccc} AB & AF & CE & DE & EF \end{array} \\ \left[\begin{array}{ccccc} 1 & 1 & 0 & 0 & 0 \\ -1 & 0 & 0 & 0 & 0 \\ 0 & 0 & 0 & 1 & 0 \\ 0 & 0 & -1 & -1 & 1 \\ 0 & -1 & 0 & 0 & -1 \end{array}\right] \end{array} \quad \text{and} \quad \mathbf{A}_c = \begin{array}{c} \\ A \\ B \\ D \\ E \\ F \end{array} \begin{array}{c} \begin{array}{cccc} BC & BE & CD & FA \end{array} \\ \left[\begin{array}{cccc} 0 & 0 & 0 & -1 \\ 1 & 1 & 0 & 0 \\ 0 & 0 & -1 & 0 \\ 0 & -1 & 0 & 0 \\ 0 & 0 & 0 & 1 \end{array}\right] \end{array}$$

Problem 10.7 Consider the following oriented graph:

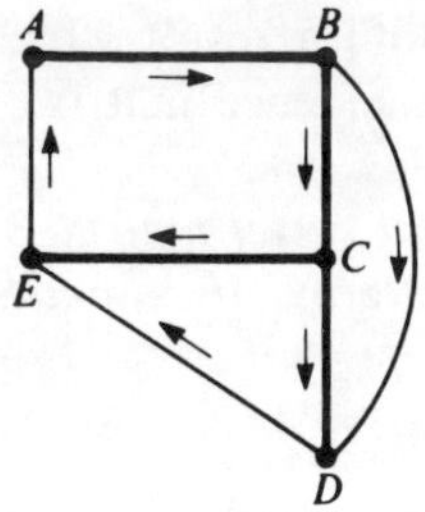

Write down:

(a) the incidence matrix of G;
(b) the reduced incidence matrix (using vertex D as the reference vertex);
(c) the matrices $\mathbf{A}_t$ and $\mathbf{A}_c$ (using the spanning tree shown).

Obtaining the Fundamental Cutset and Cycle Matrices from the Incidence Matrix

The reduced incidence matrix $\mathbf{A}_0$ completely specifies the corresponding oriented graph—that is, it tells us which edges join which vertices, and it tells us the reference direction of each edge. It is a convenient representation of the graph to use for computer analysis, since it can easily be written down by inspection of the oriented graph, and it is suitable for feeding into a computer. Because the reduced incidence matrix completely describes the oriented graph, it should theoretically be possible, once a spanning tree has been specified, for the computer to generate the fundamental cutset and cycle matrices without further reference to the graph itself. We now describe how this can be done.

First we must specify a spanning tree. Any spanning tree will serve our purpose, and an appropriate algorithm (such as the greedy algorithm or Prim's algorithm, given in Chapter 8) can be used to find a spanning tree using the information contained in the reduced incidence matrix. We can then arrange the reduced incidence matrix in the form $\mathbf{A}_0 = [\mathbf{A}_t \mid \mathbf{A}_c]$ which we described above.

Suppose that the oriented graph G has n vertices; then $\mathbf{A}_t$ has $n-1$ linearly independent rows (corresponding to the $n-1$ linearly independent rows of the reduced incidence matrix) and $n-1$ columns (corresponding to the $n-1$ edges in the spanning tree). So $\mathbf{A}_t$ is a square matrix whose rows are linearly independent. It follows from the theory of matrices that $\mathbf{A}_t$ has an inverse matrix; that is, there exists a matrix $\mathbf{A}_t^{-1}$ of order $(n-1) \times (n-1)$ such that

$$\mathbf{A}_t^{-1}\mathbf{A}_t = \mathbf{A}_t\mathbf{A}_t^{-1} = \mathbf{I}_{n-1},$$

where $\mathbf{I}_{n-1}$ is the identity matrix of order $(n-1) \times (n-1)$.

We now state (without proof) a result which enables us to obtain the matrices $\mathbf{B}_f$ and $\mathbf{C}_f$ from the incidence matrix.

Theorem 10.1 *The fundamental cutset matrix* $\mathbf{B}_f$ *and the fundamental cycle matrix* $\mathbf{C}_f$ *can be expressed in terms of* $\mathbf{A}_t$ *and* $\mathbf{A}_c$ *as follows*:

fundamental cutset matrix $\mathbf{B}_f = [\mathbf{I}_{n-1} \mid \mathbf{A}_t^{-1}\mathbf{A}_c]$,

fundamental cycle matrix $\mathbf{C}_f = [-(\mathbf{A}_t^{-1}\mathbf{A}_c)' \mid \mathbf{I}_{m-n+1}]$.

(Here $(\mathbf{A}_t^{-1}\mathbf{A}_c)'$ denotes the transpose of $\mathbf{A}_t^{-1}\mathbf{A}_c$.)

We shall explain in detail how these expressions are used to calculate $\mathbf{B}_f$ and $\mathbf{C}_f$, illustrating the procedure using our example. Unfortunately, the proofs of the validity of these expressions for $\mathbf{B}_f$ and $\mathbf{C}_f$ are too long and complicated to be included here.

We first turn to the problem of actually finding $\mathbf{A}_t^{-1}$. In general, it is not easy to find an inverse matrix. However, for this particular type of matrix $\mathbf{A}_t$, there is a simple method for finding its inverse matrix, $\mathbf{A}_t^{-1}$.

STEP 1 We first draw the spanning tree, marking in the orientations of the edges, and indicating the reference vertex, as follows:

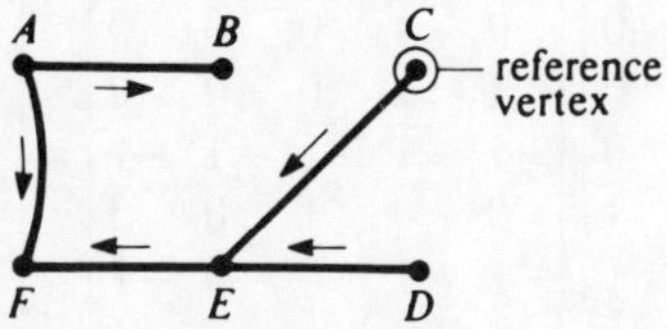

STEP 2 We next trace the path from each vertex of the tree to the reference vertex, noting carefully which branches are traversed in the same direction as the reference direction, and which branches are traversed in the opposite direction—for example, the path from vertex *A* to the reference vertex *C* consists of the branches

AF (forward), *EF* (backward), *CE* (backward).

STEP 3 Finally, we construct the matrix whose columns correspond to the vertices (other than the reference vertex) and whose rows correspond to the branches of the spanning tree. Each column represents the path from the corresponding vertex to the reference vertex; the element appearing in the column is

1 if the corresponding branch is traversed in the same direction as the reference direction;
−1 if the branch is traversed in the opposite direction;
0 if the branch is not included in the path.

For the above example, we obtain the following matrix:

$$\begin{array}{c} \\ AB \\ AF \\ CE \\ DE \\ EF \end{array} \begin{array}{c} \begin{array}{ccccc} A & B & D & E & F \end{array} \\ \begin{bmatrix} 0 & -1 & 0 & 0 & 0 \\ 1 & 1 & 0 & 0 & 0 \\ -1 & -1 & -1 & -1 & -1 \\ 0 & 0 & 1 & 0 & 0 \\ -1 & -1 & 0 & 0 & -1 \end{bmatrix} \end{array}$$

This is the required matrix $\mathbf{A}_t^{-1}$. We can check this by forming the matrix products $\mathbf{A}_t^{-1}\mathbf{A}_t$ and $\mathbf{A}_t\mathbf{A}_t^{-1}$:

$$\begin{bmatrix} 0 & -1 & 0 & 0 & 0 \\ 1 & 1 & 0 & 0 & 0 \\ -1 & -1 & -1 & -1 & -1 \\ 0 & 0 & 1 & 0 & 0 \\ -1 & -1 & 0 & 0 & -1 \end{bmatrix}\begin{bmatrix} 1 & 1 & 0 & 0 & 0 \\ -1 & 0 & 0 & 0 & 0 \\ 0 & 0 & 0 & 1 & 0 \\ 0 & 0 & -1 & -1 & 1 \\ 0 & -1 & 0 & 0 & -1 \end{bmatrix} = \begin{bmatrix} 1 & 0 & 0 & 0 & 0 \\ 0 & 1 & 0 & 0 & 0 \\ 0 & 0 & 1 & 0 & 0 \\ 0 & 0 & 0 & 1 & 0 \\ 0 & 0 & 0 & 0 & 1 \end{bmatrix} = \mathbf{I}_5$$

and

$$\begin{bmatrix} 1 & 1 & 0 & 0 & 0 \\ -1 & 0 & 0 & 0 & 0 \\ 0 & 0 & 0 & 1 & 0 \\ 0 & 0 & -1 & -1 & 1 \\ 0 & -1 & 0 & 0 & -1 \end{bmatrix}\begin{bmatrix} 0 & -1 & 0 & 0 & 0 \\ 1 & 1 & 0 & 0 & 0 \\ -1 & -1 & -1 & -1 & -1 \\ 0 & 0 & 1 & 0 & 0 \\ -1 & -1 & 0 & 0 & -1 \end{bmatrix} = \begin{bmatrix} 1 & 0 & 0 & 0 & 0 \\ 0 & 1 & 0 & 0 & 0 \\ 0 & 0 & 1 & 0 & 0 \\ 0 & 0 & 0 & 1 & 0 \\ 0 & 0 & 0 & 0 & 1 \end{bmatrix} = \mathbf{I}_5 .$$

(We do not really need to check *both* products, since *we know that the inverse* $\mathbf{A}_t^{-1}$ *exists*. If we call the square matrix which we have found $\mathbf{M}$, and show that $\mathbf{M}\mathbf{A}_t = \mathbf{I}$, then we have $\mathbf{M}\mathbf{A}_t\mathbf{A}_t^{-1} = \mathbf{I}\mathbf{A}_t^{-1}$. It follows that $\mathbf{M} = \mathbf{A}_t^{-1}$ since $\mathbf{A}_t\mathbf{A}_t^{-1} = \mathbf{I}$, by definition of the inverse matrix, and multiplying a matrix by $\mathbf{I}$ leaves the matrix unchanged.)

Problem 10.8 Use the above method to find the inverse $\mathbf{A}_t^{-1}$ of the matrix $\mathbf{A}_t$ in Problem 10.7, and check your answer by calculating $\mathbf{A}_t^{-1}\mathbf{A}_t$ and $\mathbf{A}_t\mathbf{A}_t^{-1}$.

Having constructed $\mathbf{A}_t^{-1}$, we can use it to calculate the fundamental cutset and cycle matrices using the expressions given in Theorem 10.1. We illustrate this procedure by applying it to our example.

First we form the matrix product $\mathbf{A}_t^{-1}\mathbf{A}_c$:

$$\mathbf{A}_t^{-1}\mathbf{A}_c = \begin{matrix} AB \\ AF \\ CE \\ DE \\ EF \end{matrix}\begin{bmatrix} 0 & -1 & 0 & 0 & 0 \\ 1 & 1 & 0 & 0 & 0 \\ -1 & -1 & -1 & -1 & -1 \\ 0 & 0 & 1 & 0 & 0 \\ -1 & -1 & 0 & 0 & -1 \end{bmatrix}\overset{\begin{matrix} BC & BE & CD & FA \end{matrix}}{\begin{bmatrix} 0 & 0 & 0 & -1 \\ 1 & 1 & 0 & 0 \\ 0 & 0 & -1 & 0 \\ 0 & -1 & 0 & 0 \\ 0 & 0 & 0 & 1 \end{bmatrix}}$$

$$= \begin{matrix} AB \\ AF \\ CE \\ DE \\ EF \end{matrix}\overset{\begin{matrix} BC & BE & CD & FA \end{matrix}}{\begin{bmatrix} -1 & -1 & 0 & 0 \\ 1 & 1 & 0 & -1 \\ -1 & 0 & 1 & 0 \\ 0 & 0 & -1 & 0 \\ -1 & -1 & 0 & 0 \end{bmatrix}}$$

We now insert the rows and columns of the identity matrix $\mathbf{I}_5$ at the left-hand side of the matrix:

$$[\mathbf{I}_5 \mid \mathbf{A}_t^{-1}\mathbf{A}_c] = \begin{array}{c|ccccc|cccc} & AB & AF & CE & DE & EF & BC & BE & CD & FA \\ \hline AB & 1 & 0 & 0 & 0 & 0 & -1 & -1 & 0 & 0 \\ AF & 0 & 1 & 0 & 0 & 0 & 1 & 1 & 0 & -1 \\ CE & 0 & 0 & 1 & 0 & 0 & -1 & 0 & 1 & 0 \\ DE & 0 & 0 & 0 & 1 & 0 & 0 & 0 & -1 & 0 \\ EF & 0 & 0 & 0 & 0 & 1 & -1 & -1 & 0 & 0 \end{array}$$

This is the fundamental cutset matrix $\mathbf{B}_f$, as can be seen by comparing it with the matrix we obtained on page 241 by inspection of the oriented graph.

To form the fundamental cycle matrix $\mathbf{C}_f$, we first take the transpose of the matrix $\mathbf{A}_t^{-1}\mathbf{A}_c$ above:

$$(\mathbf{A}_t^{-1}\mathbf{A}_c)' = \begin{array}{c|ccccc} & AB & AF & CE & DE & EF \\ \hline BC & -1 & 1 & -1 & 0 & -1 \\ BE & -1 & 1 & 0 & 0 & -1 \\ CD & 0 & 0 & 1 & -1 & 0 \\ FA & 0 & -1 & 0 & 0 & 0 \end{array}$$

We then multiply (each element of) this matrix by -1, and write the identity matrix $\mathbf{I}_4$ to the right of it:

$$[-(\mathbf{A}_t^{-1}\mathbf{A}_c)' \mid \mathbf{I}_4] = \begin{array}{c|ccccc|cccc} & AB & AF & CE & DE & EF & BC & BE & CD & FA \\ \hline BC & 1 & -1 & 1 & 0 & 1 & 1 & 0 & 0 & 0 \\ BE & 1 & -1 & 0 & 0 & 1 & 0 & 1 & 0 & 0 \\ CD & 0 & 0 & -1 & 1 & 0 & 0 & 0 & 1 & 0 \\ FA & 0 & 1 & 0 & 0 & 0 & 0 & 0 & 0 & 1 \end{array}$$

This is the fundamental cycle matrix $\mathbf{C}_f$, as can be seen by comparing it with the matrix we obtained on page 235 by inspection of the oriented graph.

Problem 10.9

(a) Use the matrices $\mathbf{A}_c$ and $\mathbf{A}_t^{-1}$ calculated in Problems 10.7 and 10.8 to find the corresponding fundamental cutset and cycle matrices $\mathbf{B}_f$ and $\mathbf{C}_f$, and check that your answers agree with those of Problems 10.6 and 10.4 respectively.

(b) Calculate the matrix product $\mathbf{C}_f\mathbf{B}_f'$.

10.6 TELLEGEN'S THEOREM

We conclude this chapter by showing how to prove a result of great theoretical importance, known as *Tellegen's theorem*, by making use of the incidence matrix. It may be stated in the following form.

Tellegen's Theorem *Suppose that there are two electrical networks which can be represented by the same oriented graph with n vertices and m edges. Then*

$$\sum_{k=1}^{m} v_k i_k = 0,$$

where v_k is the voltage associated with the kth edge for the first network, and i_k is the current associated with the kth edge for the second network.

Proof We start by applying Kirchhoff's current law to each of the n vertices of the oriented graph. It is not difficult to see that the resulting equations can be expressed in the following matrix form:

$$\mathbf{AJ} = \mathbf{0}_{n \times 1},$$

where

$\mathbf{A}$ is the $n \times m$ incidence matrix of the oriented graph;
$\mathbf{J}$ is the $m \times 1$ column vector of currents in the second network associated with the m edges of the graph;
$\mathbf{0}_{n \times 1}$ is the $n \times 1$ zero vector.

We now turn to Kirchhoff's voltage law. Since the voltages associated with the edges of the graph satisfy Kirchhoff's law, we can assign a unique potential to each of the n vertices such that the voltage v_k in the first network associated with edge k is given by

$$v_k = P_a - P_b,$$

where P_a and P_b are the potentials at the two vertices joined by the kth edge. The equation analogous to the above for all the vertices of the graph can be expressed in the following matrix form:

$$\mathbf{A'P} = \mathbf{V},$$

where

$\mathbf{A'}$ is the $m \times n$ transpose of the incidence matrix;
$\mathbf{P}$ is the $n \times 1$ column vector of potentials at the n vertices;
$\mathbf{V}$ is the $m \times 1$ column vector of voltages (or potential differences) associated with the m edges of the graph.

We can write the expression in Tellegen's theorem in matrix form thus:

$$\left[\sum_{k=1}^{m} v_k i_k\right] = \mathbf{V'J},$$

where

$\mathbf{V}'$ has order $1 \times m$, and $\mathbf{J}$ has order $m \times 1$, so $\mathbf{V}'\mathbf{J}$ has order 1×1.

(The brackets are inserted to indicate that the expression concerned is a 1×1 matrix.)

We know that $\mathbf{V} = \mathbf{A}'\mathbf{P}$, so we substitute this in the above matrix expression and thus obtain:

$$\left[\sum_{k=1}^{m} v_k i_k\right] = (\mathbf{A}'\,\mathbf{P})'\mathbf{J}$$

$$\begin{aligned}
&= (\mathbf{P}'\mathbf{A})\mathbf{J} && ((\mathbf{A}'\mathbf{P})' = \mathbf{P}'(\mathbf{A}')' = \mathbf{P}'\mathbf{A}) \\
&= \mathbf{P}'(\mathbf{A}\mathbf{J}) && \text{(property of matrix multiplication)} \\
&= \mathbf{P}'\mathbf{0}_{n \times 1} && (\text{since } \mathbf{A}\mathbf{J} = \mathbf{0}_{n \times 1}) \\
&= \mathbf{0}_{1 \times 1} && (\text{since } \mathbf{P}' \text{ has order } 1 \times n).
\end{aligned}$$

Hence

$$\sum_{k=1}^{m} v_k i_k = 0,$$

as required. □

Tellegen's theorem holds for any two networks which have the same oriented graph. Equally, we can apply it to just one network, in which case the result is that

$$\sum_{\text{all edges}} v_k\, i_k = 0,$$

where v_k and i_k are the voltage and current associated with the kth edge of this network. The product $v_k i_k$ is the *power* (or rate of change of energy associated with the kth edge, so the above equation tells us that energy is conserved in the network. (This is the *law of conservation of energy* for a network.)

Note that we have derived this well-known law of physics for an electrical network from Kirchhoff's current and voltage laws. In fact, given any two of these three laws (the conservation of energy law, Kirchhoff's current law and Kirchhoff's voltage law), it is possible to derive the third.

Although we have stated Tellegen's theorem for electrical networks, it also holds for the through and across variables of any physical systems which obey generalized forms of Kirchhoff's laws.

SUMMARY

On the next two pages, we summarize what we have learned about the two examples of this chapter. These results will be used in Chapter 11.

network

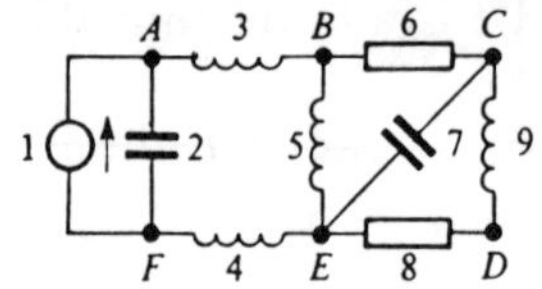

oriented graph:

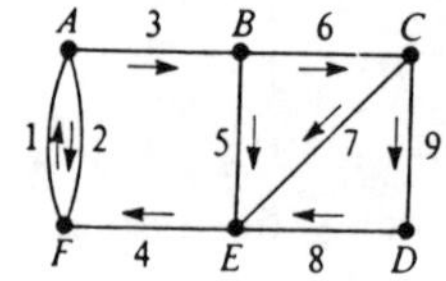

spanning tree

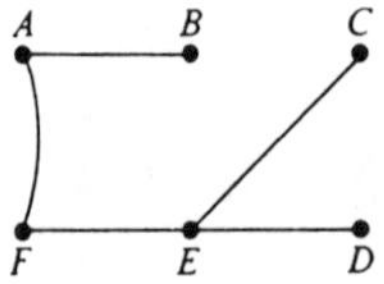

fundamental cycles

fundamental cycle matrix

$$\mathbf{C}_f = \begin{array}{c} \\ \\ \\ \\ \end{array} \begin{array}{c} \begin{array}{ccccc|cccc} AB & AF & CE & DE & EF & BC & BE & CD & FA \end{array} \\ \left[\begin{array}{ccccc|cccc} 1 & -1 & 1 & 0 & 1 & 1 & 0 & 0 & 0 \\ 1 & -1 & 0 & 0 & 1 & 0 & 1 & 0 & 0 \\ 0 & 0 & -1 & 1 & 0 & 0 & 0 & 1 & 0 \\ 0 & 1 & 0 & 0 & 0 & 0 & 0 & 0 & 1 \end{array}\right] \begin{array}{l} BC \\ BE \\ CD \\ FA \end{array}\Bigg\} \text{chords} \\ \underbrace{\hspace{4cm}}_{\text{branches}} \quad \underbrace{\hspace{3.5cm}}_{\text{chords}} \end{array}$$

fundamental cycle equations $\quad v_3 + v_6 + v_7 + v_4 - v_2 = 0,\ v_3 + v_5 + v_4 - v_2 = 0,\ v_9 + v_8 - v_7 = 0.\ v_2 + v_1 = 0$

fundamental cutsets $\quad \{AB, BC, BE\}, \{AF, BC, BE, FA\}, \{BC, CD, CE\}, \{CD, DE\}, \{BC, BE, EF\}$

fundamental cutset matrix

$$\mathbf{B}_f = \begin{array}{c} \begin{array}{ccccc|cccc} AB & AF & CE & DE & EF & BC & BE & CD & FA \end{array} \\ \left[\begin{array}{ccccc|cccc} 1 & 0 & 0 & 0 & 0 & -1 & -1 & 0 & 0 \\ 0 & 1 & 0 & 0 & 0 & 1 & 1 & 0 & -1 \\ 0 & 0 & 1 & 0 & 0 & -1 & 0 & 1 & 0 \\ 0 & 0 & 0 & 1 & 0 & 0 & 0 & -1 & 0 \\ 0 & 0 & 0 & 0 & 1 & -1 & -1 & 0 & 0 \end{array}\right] \begin{array}{l} AB \\ AF \\ CE \\ DE \\ EF \end{array}\Bigg\} \text{branches} \\ \underbrace{\hspace{4cm}}_{\text{branches}} \quad \underbrace{\hspace{3.5cm}}_{\text{chords}} \end{array}$$

fundamental cutset equations $\quad i_3 - i_5 - i_6 = 0,\ i_2 + i_6 + i_5 - i_1 = 0,\ i_7 - i_6 + i_9 = 0,\ i_8 - i_9 = 0.\ i_4 - i_6 - i_5 = 0$

network

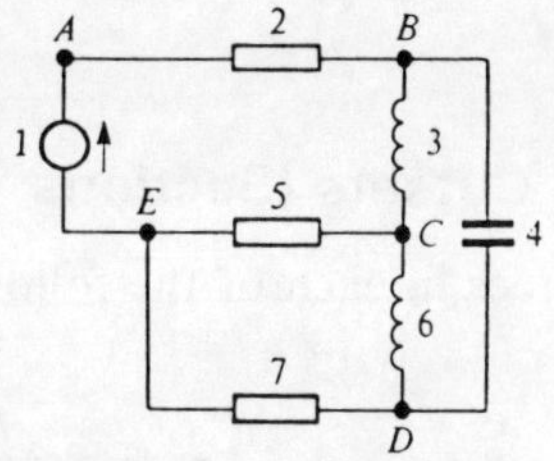

oriented graph:

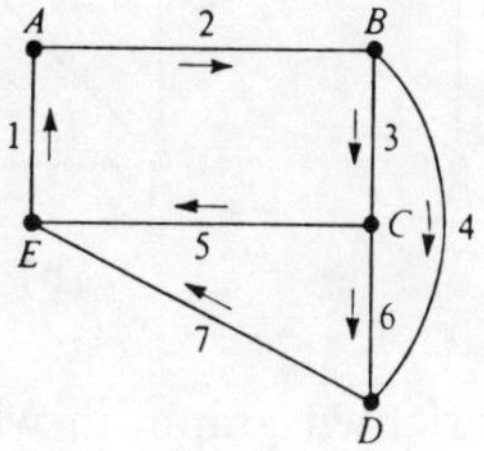

spanning tree

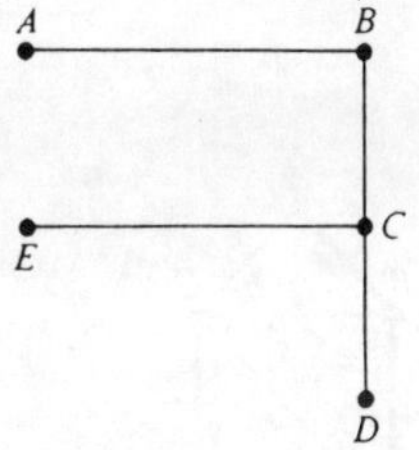

fundamental cycles

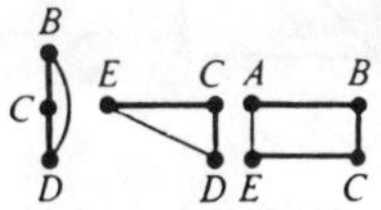

fundamental cycle matrix

$$\mathbf{C}_f = \begin{array}{c} \\ \\ \end{array}\begin{array}{cccc|ccc} AB & BC & CD & CE & BD & DE & EA \\ 0 & -1 & -1 & 0 & 1 & 0 & 0 \\ 0 & 0 & 1 & -1 & 0 & 1 & 0 \\ 1 & 1 & 0 & 1 & 0 & 0 & 1 \end{array} \begin{array}{c} \\ BD \\ DE \\ EA \end{array}\Bigg\}\text{ chords}$$

(columns AB–CE: branches; columns BD–EA: chords)

fundamental cycle equations $\quad v_4 - v_6 - v_3 = 0,\ \ v_6 + v_7 - v_5 = 0,\ \ v_2 + v_3 + v_5 + v_1 = 0$

fundamental cutsets $\quad \{AB, EA\}, \{BC, BD, EA\}, \{BD, CD, DE\}, \{CE, DE, EA\}$

fundamental cutset matrix

$$\mathbf{B}_f = \begin{array}{cccc|ccc} AB & BC & CD & CE & BD & DE & EA \\ 1 & 0 & 0 & 0 & 0 & 0 & -1 \\ 0 & 1 & 0 & 0 & 1 & 0 & -1 \\ 0 & 0 & 1 & 0 & 1 & -1 & 0 \\ 0 & 0 & 0 & 1 & 0 & 1 & -1 \end{array} \begin{array}{c} \\ AB \\ BC \\ CD \\ CE \end{array}\Bigg\}\text{ branches}$$

(columns AB–CE: branches; columns BD–EA: chords)

fundamental cutset equations $\quad i_2 - i_1 = 0,\ \ i_3 + i_4 - i_1 = 0,\ \ i_6 + i_4 - i_7 = 0,\ \ i_5 + i_7 - i_1 = 0$

EXERCISES 10

Fundamental Cycles and Cutsets (Sections 10.1–10.4)

10.1. Find all the spanning trees in each of the following oriented graphs:

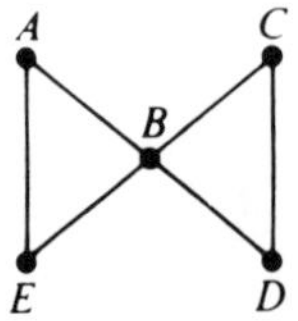

(a)

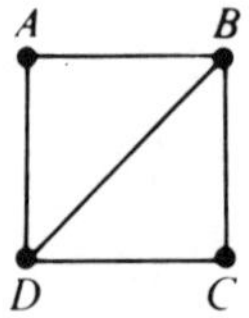

(b)

10.2. Consider the following oriented graph, in which a spanning tree is indicated by thick lines:

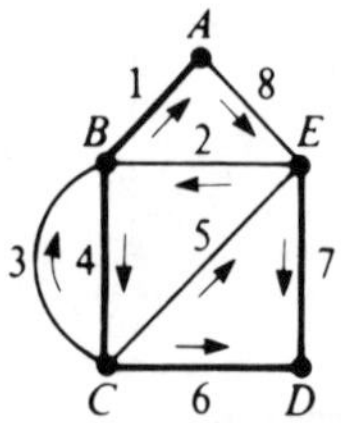

(a) List the fundamental cycles and write down the corresponding voltage law equations.
(b) Find the fundamental cycle matrix $\mathbf{C_f}$
(c) List the fundamental cutsets and write down the corresponding current law equations.
(d) Find the fundamental cutset matrix $\mathbf{B_f}$.

10.3. Repeat Exercise 10.2 for the following oriented graph, in which a spanning tree is indicated by thick lines:

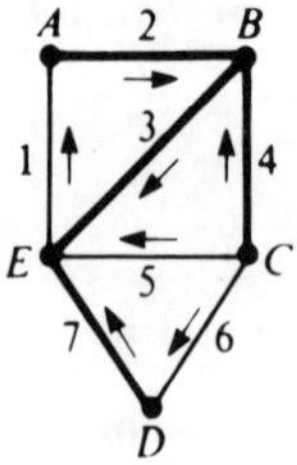

10.4. For the oriented graph and spanning tree shown below:

(a) find the fundamental cycles and fundamental cutsets;
(b) find the corresponding voltage and current equations;
(c) express these equations in matrix form.

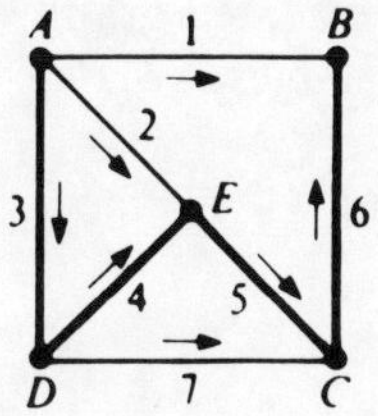

10.5. An oriented graph has 8 vertices and 12 edges. Write down:

(a) the number of edges in a spanning tree;
(b) the number of edges in each co-tree;
(c) the number of fundamental cycles;
(d) the number of fundamental cutsets.

10.6. Find the cycle and cutset ranks of each of the following:

(a) the complete graph K_n;
(b) the complete bipartite graph $K_{r,s}$.

10.7. Prove that, if G is a connected graph, then:

(a) every cutset of G has an edge in common with any spanning tree of G;
(b) every cycle of G has an edge in common with any co-tree of G.

Obtaining the Fundamental Cycle and Cutset Equations (Section 10.5)

10.8. Consider the oriented graph of Exercise 10.2, with vertex A as reference vertex.

(a) Write down the reduced incidence matrix, and partition it into a tree part $\mathbf{A_t}$ and a co-tree part $\mathbf{A_c}$.
(b) Use the method described in this chapter to find the inverse of the matrix $\mathbf{A_t}$.

(c) Use the matrices $\mathbf{A}_t^{-1}$ and $\mathbf{A}_c$ to find the fundamental cutset matrix $\mathbf{B}_f$ and the fundamental cycle matrix $\mathbf{C}_f$.

(d) Calculate the matrix product $\mathbf{C}_f\mathbf{B}_f'$.

10.9. Repeat Exercise 10.8 for the oriented graph in Exercise 10.4, with vertex *A* as reference vertex.

10.10. Prove that, if *G* is an oriented graph, then every fundamental cutset of *G* has either no edges or two edges in common with any fundamental cycle.

By expressing $\mathbf{B}_f$ and $\mathbf{C}_f$ in terms of $\mathbf{A}_t$ and $\mathbf{A}_c$, deduce that $\mathbf{C}_f\mathbf{B}_f' = 0$.

CHAPTER 11

ELECTRICAL NETWORKS: SOLVING THE EQUATIONS

After reading this chapter, you should be able to:

- find the component matrix for a given resistive network;
- combine the component matrix with the fundamental cycle and cutset matrices to form the matrix **H**;
- understand the use of Gaussian elimination in the solution of the matrix equation $\mathbf{Hx} = \mathbf{y}$;
- understand what is meant by *state equations*, and formulate the state equations from the component, cutset and cycle equations of a given network.

Introduction

In the previous two chapters we obtained a number of equations relating the voltages and currents in an electrical network. These equations are of three types:

1. **component equations**, relating the voltage and current in each component;
2. **fundamental cycle equations**, relating the voltages in each fundamental cycle of the network;
3. **fundamental cutset equations**, relating the currents in each fundamental cutset of the network.

These three sets of equations contain all the information we need to be able to work out all the voltages and currents in the network. If the oriented graph of the network has n vertices and m edges, then there are m voltages and m currents—a total of $2m$ unknowns.

These unknowns are related by

1. m component equations;
2. $m - n + 1$ fundamental cycle equations;
3. $n - 1$ fundamental cutset equations.

There are, therefore, $m + (m - n + 1) + (n - 1) = 2m$ equations in $2m$ unknowns.

In this chapter we show how these $2m$ equations can be organized systematically into a standard matrix form. We look first at networks that contain only resistors and independent sources, and show that, given the values of the independent sources in the network, we can manipulate the matrix equations to find the values of the current and voltage associated with each resistor. Remember that voltages, currents and resistors have analogues in other types of physical system; for example, in a hydraulic system we may wish to work out the pressure drops and fluid flow rates associated with valves that behave as hydraulic 'resistors'. We can use the method described in this chapter for any linear physical system that obeys Kirchhoff-type laws, simply by interpreting the currents and voltages as the through and across variables of the system under consideration.

The graphical approach we describe for the analysis of physical networks involves the construction and manipulation of large matrices—generally of order $2m \times 2m$ for a network whose oriented graph has m edges. So, for example, when using this method to work out all the voltages and currents in an electrical network containing only three components (say a battery and two resistors), we need to handle a 6×6 matrix. The solution procedure, therefore, is not particularly suited to hand calculation, but can be relatively easily programmed on a computer. We shall discuss a method called *Gaussian elimination*, which is a systematic procedure for solving a set of simultaneous equations in matrix form and which forms the basis of a number of computer solution algorithms.

In the final section, we extend our discussion to electrical networks that contain capacitors and inductors as well as resistors and independent sources. For such a network, the component equations involve not only v and i but also their derivatives with respect to time. To deal with this type of equation, we start by specifying certain of the voltages and currents as *state variables*. We then set up equations, called *state equations*, involving the state variables and their derivatives. We solve these equations for the state variables and show that the remaining network variables can be expressed in terms of these state variables.

11.1 FORMULATION OF THE MATRIX EQUATIONS

The Component Matrix

Our first task is to write down the component equations of our network in matrix form. In Chapter 9 we discussed four types of electrical component—resistors, capacitors, inductors and independent sources. For the moment, however, we restrict our attention to networks that contain only resistors and independent sources. Later, we discuss the modifications that must be made when capacitors and/or inductors are present.

Now the component equations corresponding to the resistors are all of the form

$$v = Ri,$$

and the component equations corresponding to the independent sources have the form

$$v = V \qquad \text{for a voltage source,}$$

or

$$i = I \qquad \text{for a current source.}$$

We begin by rewriting the component equations so that all the through and across variables appear on the left-hand side, and all the terms not involving these variables appear on the right, thus:

$$1v - Ri = 0 \qquad \text{for a resistor;}$$

$$1v + 0i = V \qquad \text{for a voltage source;}$$

$$0v + 1i = I \qquad \text{for a current source.}$$

Note that we have included the coefficients 0 and 1 where appropriate—this is because our next step is to rewrite the component equations in matrix form. We illustrate this procedure by two examples.

Example 1 Consider the following network and its oriented graph:

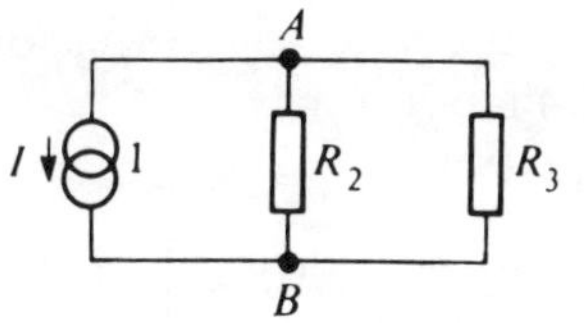

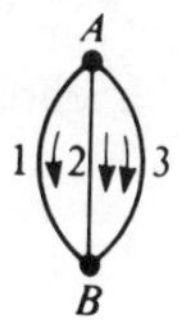

The component equations for the resistors are:

$$v_2 = R_2\, i_2$$

$$v_3 = R_3\, i_3,$$

which we can write as

$$1v_2 - R_2 i_2 = 0$$
$$1v_3 - R_3 i_3 = 0.$$

We can write these equations in matrix form as follows:

$$\left[\begin{array}{cc|cc} 1 & 0 & -R_2 & 0 \\ 0 & 1 & 0 & -R_3 \end{array}\right] \left[\begin{array}{c} v_2 \\ v_3 \\ \hline i_2 \\ i_3 \end{array}\right] = \begin{bmatrix} 0 \\ 0 \end{bmatrix} \tag{1}$$

There are 2 resistors, so on the left we have a 2×4 matrix of coefficients, multiplied by a 4×1 column vector of edge variables, and on the right we have the 2×1 zero column vector.

There is only one independent (current) source I, so we have only one further equation,

$$i_1 = I,$$

which we can write as

$$0v_1 + 1i_1 = I.$$

We can write this equation in matrix form as follows:

$$[0 \quad 1] \begin{bmatrix} v_1 \\ i_1 \end{bmatrix} = [I]. \tag{2}$$

We now combine the matrix equations (1) and (2). To do this, we incorporate all the edge variables into a single column vector $\mathbf{z}$ as follows:

$$\mathbf{z} = \left[\begin{array}{c} v_2 \\ v_3 \\ v_1 \\ \hline i_2 \\ i_3 \\ i_1 \end{array}\right] \begin{array}{l} \left.\vphantom{\begin{array}{c} v_2 \\ v_3 \\ v_1 \end{array}}\right\} \text{voltages} \\ \\ \left.\vphantom{\begin{array}{c} i_2 \\ i_3 \\ i_1 \end{array}}\right\} \text{currents} \end{array}$$

(We list the currents in the same order as the corresponding voltages; at this stage, the order chosen is unimportant.)

Thus we obtain the matrix equation

$$\left[\begin{array}{ccc|ccc} 1 & 0 & 0 & -R_2 & 0 & 0 \\ 0 & 1 & 0 & 0 & -R_3 & 0 \\ 0 & 0 & 0 & 0 & 0 & 1 \end{array}\right] \left[\begin{array}{c} v_2 \\ v_3 \\ v_1 \\ \hline i_2 \\ i_3 \\ i_1 \end{array}\right] = \begin{bmatrix} 0 \\ 0 \\ I \end{bmatrix}$$

We have shaded the partitions arising from equation (1).

Example 2 Consider the following network and its oriented graph:

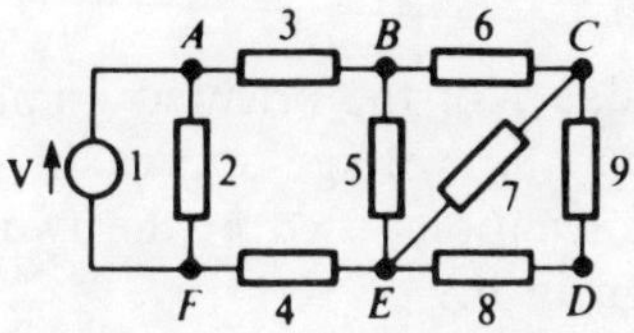

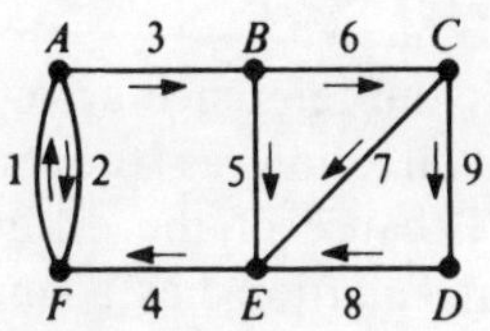

The component equations for the resistors are:

$$
\begin{array}{lcl}
v_2 = R_2\, i_2 & \text{or} & 1v_2 - R_2\, i_2 = 0\\
v_3 = R_3\, i_3 & \text{or} & 1v_3 - R_3\, i_3 = 0\\
\vdots & \vdots & \vdots\\
v_9 = R_9\, i_9 & \text{or} & 1v_9 - R_9\, i_9 = 0.
\end{array}
$$

We can write these equations in matrix form as follows:

$$
\left[\begin{array}{cccccccc|cccccccc}
1&0&0&0&0&0&0&0&-R_2&0&0&0&0&0&0&0\\
0&1&0&0&0&0&0&0&0&-R_3&0&0&0&0&0&0\\
0&0&1&0&0&0&0&0&0&0&-R_4&0&0&0&0&0\\
0&0&0&1&0&0&0&0&0&0&0&-R_5&0&0&0&0\\
0&0&0&0&1&0&0&0&0&0&0&0&-R_6&0&0&0\\
0&0&0&0&0&1&0&0&0&0&0&0&0&-R_7&0&0\\
0&0&0&0&0&0&1&0&0&0&0&0&0&0&-R_8&0\\
0&0&0&0&0&0&0&1&0&0&0&0&0&0&0&-R_9
\end{array}\right]
\left[\begin{array}{c}
v_2\\ v_3\\ v_4\\ v_5\\ v_6\\ v_7\\ v_8\\ v_9\\ \hline i_2\\ i_3\\ i_4\\ i_5\\ i_6\\ i_7\\ i_8\\ i_9
\end{array}\right]
=
\left[\begin{array}{c}
0\\0\\0\\0\\0\\0\\0\\0
\end{array}\right]
\qquad (1)
$$

There are 8 resistors, so on the left we have an 8×16 matrix of coefficients, multiplied by a 16×1 column vector of edge variables, and on the right we have the 8×1 zero column vector.

There is only one independent (voltage) source V, so we have only one further equation,

$$v_1 = -V,$$

which we can write as

$$1v_1 + 0i_1 = -V.$$

(Note that the reference arrow for edge 1 of the oriented graph points in the same direction as the arrow on the voltage source in the circuit diagram. Because of the different conventions used in the two cases, the component equation has a minus sign in it.)

We can write this equation in matrix form as follows:

$$[1 \quad 0]\begin{bmatrix} v_1 \\ i_1 \end{bmatrix} = [-V]. \qquad (2)$$

Combining the matrix equations (1) and (2) as before, we obtain the matrix equation

$$\left[\begin{array}{ccccccccc|ccccccccc}
1&0&0&0&0&0&0&0&0&-R_2&0&0&0&0&0&0&0&0\\
0&1&0&0&0&0&0&0&0&0&-R_3&0&0&0&0&0&0&0\\
0&0&1&0&0&0&0&0&0&0&0&-R_4&0&0&0&0&0&0\\
0&0&0&1&0&0&0&0&0&0&0&0&-R_5&0&0&0&0&0\\
0&0&0&0&1&0&0&0&0&0&0&0&0&-R_6&0&0&0&0\\
0&0&0&0&0&1&0&0&0&0&0&0&0&0&-R_7&0&0&0\\
0&0&0&0&0&0&1&0&0&0&0&0&0&0&0&-R_8&0&0\\
0&0&0&0&0&0&0&1&0&0&0&0&0&0&0&0&-R_9&0\\
0&0&0&0&0&0&0&0&1&0&0&0&0&0&0&0&0&0
\end{array}\right]
\left[\begin{array}{c} v_2\\ v_3\\ v_4\\ v_5\\ v_6\\ v_7\\ v_8\\ v_9\\ v_1\\ \hline i_2\\ i_3\\ i_4\\ i_5\\ i_6\\ i_7\\ i_8\\ i_9\\ i_1 \end{array}\right]
=
\begin{bmatrix} 0\\0\\0\\0\\0\\0\\0\\0\\-V \end{bmatrix}$$

The above examples show how we deal with resistors, independent current sources, and independent voltage sources. Note that, in each example, the matrix of coefficients on the left is very simple—most of the elements are zero, and each of the non-zero elements is either 1 or $-R_j$, for some resistance R_j.

More generally, for a given network of this type for which the oriented graph has m edges, we write down all the component equations and then rearrange them in matrix form

$$\mathbf{T}\begin{bmatrix}\mathbf{v}\\ \hdashline \mathbf{i}\end{bmatrix} = \mathbf{f},$$

where

$\mathbf{T}$ is the $m \times 2m$ matrix of coefficients which appear in the component equations;

$\begin{bmatrix}\mathbf{v}\\ \hdashline \mathbf{i}\end{bmatrix}$ is the $2m \times 1$ column vector of all the voltages, followed by all the currents, listed in the same order;

$\mathbf{f}$ is an $m \times 1$ column vector containing terms from the right-hand side of the component equations for independent sources (such as I, $-V$).

In fact, most of the elements of the matrix $\mathbf{T}$ are zero. Such a matrix is said to be *sparse*.

Problem 11.1 Consider the following network and its oriented graph:

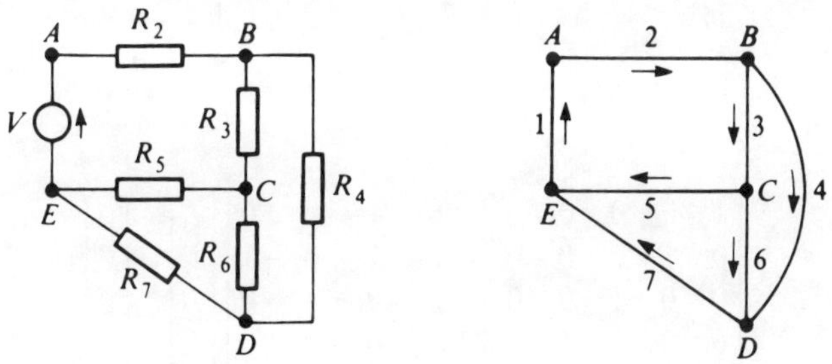

Assuming that R_2, R_3, . . . , R_7 and V are known, write down the component equations in matrix form.

We now have the component equations in matrix form. These equations describe the constraints on the voltages and currents in the network that are imposed by the components only. But we know that the voltages and currents in the network must also obey Kirchhoff's laws. Our next step, therefore, is to combine the information contained in the component equations with the information about how the components are connected together which is contained in the fundamental cycle and cutset equations.

The Fundamental Cycle and Cutset Matrices

We saw in Chapter 10 that the fundamental cycle and cutset equations can be expressed in matrix form—namely,

$$\mathbf{C_f}\,\mathbf{v} = \mathbf{0} \quad \text{and} \quad \mathbf{B_f}\,\mathbf{i} = \mathbf{0},$$

where $\mathbf{C_f}$ and $\mathbf{B_f}$ are the fundamental cycle and cutset matrices associated with a given spanning tree, and $\mathbf{v}$ and $\mathbf{i}$ are column vectors of edge voltages and currents written in the same order as the columns of the matrix.

For example, if we take the oriented graph of the network in Example 2, then the matrix equations are:

$$\underbrace{\left[\begin{array}{ccccc|cccc} 1 & -1 & 1 & 0 & 1 & 1 & 0 & 0 & 0 \\ 1 & -1 & 0 & 0 & 1 & 0 & 1 & 0 & 0 \\ 0 & 0 & -1 & 1 & 0 & 0 & 0 & 1 & 0 \\ 0 & 1 & 0 & 0 & 0 & 0 & 0 & 0 & 1 \end{array}\right]}_{\mathbf{C_f}\;(\mathbf{C}\ |\ \mathbf{I}_4)} \underbrace{\left[\begin{array}{c} v_3 \\ v_2 \\ v_7 \\ v_8 \\ v_4 \\ \hline v_6 \\ v_5 \\ v_9 \\ v_1 \end{array}\right]}_{\mathbf{v}} = \underbrace{\left[\begin{array}{c} 0 \\ 0 \\ 0 \\ 0 \end{array}\right]}_{\mathbf{0}}$$

and

$$\underbrace{\left[\begin{array}{ccccc|cccc} 1 & 0 & 0 & 0 & 0 & -1 & -1 & 0 & 0 \\ 0 & 1 & 0 & 0 & 0 & 1 & 1 & 0 & -1 \\ 0 & 0 & 1 & 0 & 0 & -1 & 0 & 1 & 0 \\ 0 & 0 & 0 & 1 & 0 & 0 & 0 & -1 & 0 \\ 0 & 0 & 0 & 0 & 1 & -1 & -1 & 0 & 0 \end{array}\right]}_{\mathbf{B_f}\;(\mathbf{I}_5\ |\ \mathbf{B})} \underbrace{\left[\begin{array}{c} i_3 \\ i_2 \\ i_7 \\ i_8 \\ i_4 \\ \hline i_6 \\ i_5 \\ i_9 \\ i_1 \end{array}\right]}_{\mathbf{i}} = \underbrace{\left[\begin{array}{c} 0 \\ 0 \\ 0 \\ 0 \\ 0 \end{array}\right]}_{\mathbf{0}}$$

The oriented graph for Example 2 is the same as the oriented graph for the example on page 252. We choose the same spanning tree, so these matrices are the matrices summarized on page 252.

In general, it is sometimes convenient to write these equations in abbreviated form as

$$[\mathbf{C} \mid \mathbf{I}_{m-n+1}]\ \mathbf{v} = \mathbf{0},$$

where $\mathbf{C}$ is an $(m - n + 1) \times (n - 1)$ matrix, and

$$[\mathbf{I}_{n-1} \mid \mathbf{B}]\ \mathbf{i} = \mathbf{0},$$

where $\mathbf{B}$ is an $(n - 1) \times (m - n + 1)$ matrix.

(Note that $\mathbf{C} = -(\mathbf{A}_t^{-1}\ \mathbf{A}_c)'$ and $\mathbf{B} = \mathbf{A}_t^{-1}\ \mathbf{A}_c$, in the notation of Chapter 10.)

Combining these Matrices

We now have three matrix equations, corresponding to the component equations, the fundamental cycles, and the fundamental cutsets. These matrix equations are:

Component equation: $\mathbf{T}\begin{bmatrix}\mathbf{v}\\ \hdashline \mathbf{i}\end{bmatrix} = \mathbf{f},$

Fundamental cycle equation: $\mathbf{C}_f\, \mathbf{v} = \mathbf{0},$

Fundamental cutset equation: $\mathbf{B}_f\, \mathbf{i} = \mathbf{0}.$

Our aim is to solve these equations—that is, to work out the voltages v and currents i that satisfy simultaneously the component equations, Kirchhoff's voltage law and Kirchhoff's current law. Because we want to solve these equations simultaneously, we start by combining them into just one matrix equation of the form

$$\mathbf{H}\begin{bmatrix}\mathbf{v}\\ \hdashline \mathbf{i}\end{bmatrix} = \begin{bmatrix}\mathbf{f}\\ \hdashline \mathbf{0}\\ \hdashline \mathbf{0}\end{bmatrix}.$$

This is possible, because we can write the left-hand side of each of the three equations as a matrix multiplying a vector of voltages and currents.

The component equation is already in this form:

$$\mathbf{T}\begin{bmatrix}\mathbf{v}\\ \hdashline \mathbf{i}\end{bmatrix} = \mathbf{f}.$$

The fundamental cycle equation can be written as

$$[\mathbf{C}_f \mid \mathbf{0}]\begin{bmatrix}\mathbf{v}\\ \hdashline \mathbf{i}\end{bmatrix} = \mathbf{0},$$

and, similarly, the fundamental cutset equation can be written as

$$[\mathbf{0} \mid \mathbf{B_f}]\begin{bmatrix}\mathbf{v}\\ \hdashline \mathbf{i}\end{bmatrix} = \mathbf{0}.$$

At this stage, a word of caution is necessary. The voltages which occur in the column vector **v** in the fundamental cycle equation appear *in a particular order* determined partly by the choice of spanning tree. The currents which occur in the column vector **i** in the fundamental cutset equation appear in the same order as the voltages in the fundamental cycle equation. It follows that if we want to combine these equations with the component equation, we must put the voltages and currents occuring in the column vector $\begin{bmatrix}\mathbf{v}\\ \mathbf{i}\end{bmatrix}$ in the component equation into the *same order* as the voltages and currents in the other two matrix equations.

Fortunately, this can be done very easily. All we need to do is to take the resistances R_k and the zero which occur in the diagonal of the right-hand partition of **T** in the component equation, and write them down in the required order. For example, in the resistor network of Example 2, the voltages appear in the fundamental cycle matrix equation in the order

$$v_3,\ v_2,\ v_7,\ v_8,\ v_4,\ v_6,\ v_5,\ v_9,\ v_1,$$

and similarly for the currents in the fundamental cutset matrix equation. The resistances and the zero on the diagonal in **T** must therefore appear in the corresponding order:

$$R_3,\ R_2,\ R_7,\ R_8,\ R_4,\ R_6,\ R_5,\ R_9,\ 0.$$

Once we have written the voltages and currents in each equation in the same order, we can combine the three matrix equations into one.

We have

$$[\mathbf{T}]\begin{bmatrix}\mathbf{v}\\ \hdashline \mathbf{i}\end{bmatrix} = \mathbf{f}$$

$$[\mathbf{C_f} \mid \mathbf{0}]\begin{bmatrix}\mathbf{v}\\ \hdashline \mathbf{i}\end{bmatrix} = \mathbf{0}$$

$$[\mathbf{0} \mid \mathbf{B_f}]\begin{bmatrix}\mathbf{v}\\ \hdashline \mathbf{i}\end{bmatrix} = \mathbf{0}$$

which can be written as one equation

$$\left[\begin{array}{c|c} \multicolumn{2}{c}{\mathbf{T}} \\ \hline \mathbf{C_f} & \mathbf{0} \\ \hline \mathbf{0} & \mathbf{B_f} \end{array}\right] \left[\begin{array}{c} \mathbf{v} \\ \hline \mathbf{i} \end{array}\right] = \left[\begin{array}{c} \mathbf{f} \\ \hline \mathbf{0} \\ \hline \mathbf{0} \end{array}\right]$$

or

$$\mathbf{Hx} = \mathbf{y},$$

where

- **H** is a $2m \times 2m$ matrix containing component, cycle and cutset information;
- **x** is a column vector which involves all the voltages and currents;
- **y** is a column vector consisting mainly of zeros, but also including terms such as I and $-V$ (source functions).

The resistor network in Example 2 has 9 components (8 resistors and one independent voltage source), so the matrix **H** corresponding to this network is an 18×18 matrix. Writing the matrix equation for the network out in full, we get the following:

$$\left[\begin{array}{ccccccccc|ccccccccc}
1 & 0 & 0 & 0 & 0 & 0 & 0 & 0 & 0 & -R_3 & 0 & 0 & 0 & 0 & 0 & 0 & 0 & 0 \\
0 & 1 & 0 & 0 & 0 & 0 & 0 & 0 & 0 & 0 & -R_2 & 0 & 0 & 0 & 0 & 0 & 0 & 0 \\
0 & 0 & 1 & 0 & 0 & 0 & 0 & 0 & 0 & 0 & 0 & -R_7 & 0 & 0 & 0 & 0 & 0 & 0 \\
0 & 0 & 0 & 1 & 0 & 0 & 0 & 0 & 0 & 0 & 0 & 0 & -R_8 & 0 & 0 & 0 & 0 & 0 \\
0 & 0 & 0 & 0 & 1 & 0 & 0 & 0 & 0 & 0 & 0 & 0 & 0 & -R_4 & 0 & 0 & 0 & 0 \\
0 & 0 & 0 & 0 & 0 & 1 & 0 & 0 & 0 & 0 & 0 & 0 & 0 & 0 & -R_6 & 0 & 0 & 0 \\
0 & 0 & 0 & 0 & 0 & 0 & 1 & 0 & 0 & 0 & 0 & 0 & 0 & 0 & 0 & -R_5 & 0 & 0 \\
0 & 0 & 0 & 0 & 0 & 0 & 0 & 1 & 0 & 0 & 0 & 0 & 0 & 0 & 0 & 0 & -R_9 & 0 \\
0 & 0 & 0 & 0 & 0 & 0 & 0 & 0 & 1 & 0 & 0 & 0 & 0 & 0 & 0 & 0 & 0 & 0 \\
\hline
1 & -1 & 1 & 0 & 1 & 1 & 0 & 0 & 0 & 0 & 0 & 0 & 0 & 0 & 0 & 0 & 0 & 0 \\
1 & -1 & 0 & 0 & 1 & 0 & 1 & 0 & 0 & 0 & 0 & 0 & 0 & 0 & 0 & 0 & 0 & 0 \\
0 & 0 & -1 & 1 & 0 & 0 & 0 & 1 & 0 & 0 & 0 & 0 & 0 & 0 & 0 & 0 & 0 & 0 \\
0 & 1 & 0 & 0 & 0 & 0 & 0 & 0 & 1 & 0 & 0 & 0 & 0 & 0 & 0 & 0 & 0 & 0 \\
\hline
0 & 0 & 0 & 0 & 0 & 0 & 0 & 0 & 0 & 1 & 0 & 0 & 0 & 0 & -1 & -1 & 0 & 0 \\
0 & 0 & 0 & 0 & 0 & 0 & 0 & 0 & 0 & 0 & 1 & 0 & 0 & 0 & 1 & 1 & 0 & -1 \\
0 & 0 & 0 & 0 & 0 & 0 & 0 & 0 & 0 & 0 & 0 & 1 & 0 & 0 & -1 & 0 & 1 & 0 \\
0 & 0 & 0 & 0 & 0 & 0 & 0 & 0 & 0 & 0 & 0 & 0 & 1 & 0 & 0 & 0 & -1 & 0 \\
0 & 0 & 0 & 0 & 0 & 0 & 0 & 0 & 0 & 0 & 0 & 0 & 0 & 1 & -1 & -1 & 0 & 0
\end{array}\right]
\left[\begin{array}{c} v_3 \\ v_2 \\ v_7 \\ v_8 \\ v_4 \\ v_6 \\ v_5 \\ v_9 \\ v_1 \\ \hline i_3 \\ i_2 \\ i_7 \\ i_8 \\ i_4 \\ i_6 \\ i_5 \\ i_9 \\ i_1 \end{array}\right]
=
\left[\begin{array}{c} 0 \\ 0 \\ 0 \\ 0 \\ 0 \\ 0 \\ 0 \\ 0 \\ -V \\ \hline 0 \\ 0 \\ 0 \\ 0 \\ \hline 0 \\ 0 \\ 0 \\ 0 \\ 0 \end{array}\right]$$

We have partitioned each of the matrices $\mathbf{C_f}$ and $\mathbf{B_f}$ into two, to reveal the identity matrices they contain.

Problem 11.2

(a) Consider the following network and its oriented graph:

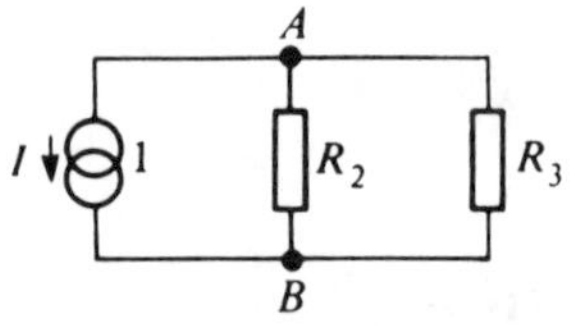

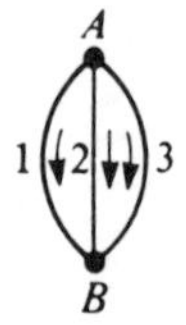

(This network was discussed earlier in Example 1.)
Taking the left-hand edge as the spanning tree, find the fundamental cycle and cutset matrices $\mathbf{C}_f$ and $\mathbf{B}_f$. Use these, together with the matrix on page 260, to find the corresponding matrix equation $\mathbf{Hx} = \mathbf{y}$.

(b) Use the fundamental cycle and cutset matrices given on page 253, to find the corresponding matrix equation for the network in Problem 11.1.

11.2 SOLVING THE MATRIX EQUATION

The matrix $\mathbf{H}$ and the column vector $\mathbf{y}$ contain all the information we need to be able to work out the $2m$ voltages and currents in the vector $\mathbf{x}$. One approach to solving the matrix equation is to rearrange the equation so that we express $\mathbf{x}$ in terms of $\mathbf{H}$ and $\mathbf{y}$. We have

$$\mathbf{Hx} = \mathbf{y}.$$

In formulating the component, cycle and cutset matrix equations, we have taken some care to ensure that the $2m$ equations in $2m$ unknowns that describe the network are linearly independent, so the matrix $\mathbf{H}$ has an inverse $\mathbf{H}^{-1}$. If we multiply each side of the above equation by $\mathbf{H}^{-1}$, we get

$$\mathbf{H}^{-1}\,\mathbf{Hx} = \mathbf{H}^{-1}\,\mathbf{y}.$$

Remember that we define the inverse matrix $\mathbf{H}^{-1}$ such that the product $\mathbf{H}^{-1}\,\mathbf{H}$ is equal to the identity matrix $\mathbf{I}_{2m}$. So we now have

$$\mathbf{I}_{2m}\,\mathbf{x} = \mathbf{H}^{-1}\,\mathbf{y}$$

or simply

$$\mathbf{x} = \mathbf{H}^{-1}\,\mathbf{y}.$$

From a theoretical point of view, the problem is now solved—in principle, we can go on to calculate $\mathbf{H}^{-1}$ for the network and hence work out each voltage and current in $\mathbf{x}$. However, even if we use a computer to perform the inversion and subsequent calculations, this is generally not the most efficient approach. One reason for this is that we are interested in $\mathbf{H}^{-1}$ only as a means to an end. What we are actually interested in are the voltages and currents in the network. So if we can work out the values of these variables without having to work out all the elements of $\mathbf{H}^{-1}$, then we can reduce the computational effort and hence the time required to analyse the network.

Instead of considering how to compute $\mathbf{H}^{-1}$, we shall briefly discuss a technique that forms the basis of many computer-based methods for solving a set of n linear equations in n unknowns. The technique is called *Gaussian elimination*.

Gaussian Elimination

Gaussian elimination is a systematic way of solving a linearly independent system of n simultaneous linear equations in n variables by eliminating the variables one at a time. This is done by performing certain allowable operations on the rows of the matrix of coefficients so as to reduce all the elements below the main diagonal (the diagonal from top left to bottom right) to zero. We can then successively solve the equations one at a time, starting with the last equation, until we have all the required variables.

The operations we allow on the rows of the matrix are:

(a) interchange two rows;
(b) multiply any row by a non-zero number;
(c) add any row to any other row.

Note that operations (b) and (c) mean that we can add (or subtract) any multiple of any row to (or from) any other.

If we also carry out these row operations on the column vector $\mathbf{y}$, then each of them corresponds to an operation on the corresponding equations which leaves the solution of the equations unchanged. For example, operation (a) corresponds to interchanging the equations, which clearly does not affect the solution.

Our aim is to describe the method of Gaussian elimination as simply and as briefly as possible. The example used is inevitably rather artificial. In practice, if solving this system of equations by hand, we would, of course, proceed differently. Moreover, when solving a system of equations by computer, we would design the program so that rounding errors are minimized.

Example Consider the set of equations

$$\begin{aligned} 2x_1 + 3x_2 - x_3 &= 5 \\ 4x_1 + 4x_2 - 3x_3 &= 3 \\ 2x_1 - 3x_2 + x_3 &= 1. \end{aligned}$$

These can be written in matrix form:

$$\begin{bmatrix} 2 & 3 & -1 \\ 4 & 4 & -3 \\ 2 & -3 & 1 \end{bmatrix} \begin{bmatrix} x_1 \\ x_2 \\ x_3 \end{bmatrix} = \begin{bmatrix} 5 \\ 3 \\ 1 \end{bmatrix}.$$

We want to work out the values of x_1, x_2 and x_3 without inverting the 3×3 matrix. We do this by performing the allowable row operations with the aim of changing the values of all the elements of the matrix below the main diagonal to zero. (It is possible to do this, and to obtain non-zero elements on the diagonal, because the equations are linearly independent.) The result is a matrix equation of the form

$$\begin{bmatrix} a_{11} & a_{12} & a_{13} \\ 0 & a_{22} & a_{23} \\ 0 & 0 & a_{33} \end{bmatrix} \begin{bmatrix} x_1 \\ x_2 \\ x_3 \end{bmatrix} = \begin{bmatrix} y_1 \\ y_2 \\ y_3 \end{bmatrix},$$

where $a_{11} \neq 0$, $a_{22} \neq 0$, $a_{33} \neq 0$.

This is equivalent to the three equations

$$a_{11} x_1 + a_{12} x_2 + a_{13} x_3 = y_1$$

$$a_{22} x_2 + a_{23} x_3 = y_2$$

$$a_{33} x_3 = y_3$$

From the last equation,

$$a_{33} x_3 = y_3$$

so

$$x_3 = \frac{y_3}{a_{33}} \qquad (a_{33} \neq 0).$$

From the second equation,

$$a_{22} x_2 + a_{23} x_3 = y_2$$

so

$$x_2 = \frac{y_2 - a_{23} x_3}{a_{22}} \qquad (a_{22} \neq 0)$$

Substituting $x_3 = y_3/a_{33}$, we get

$$x_2 = \frac{y_2}{a_{22}} - \frac{a_{23}\, y_3}{a_{22}\, a_{33}}.$$

Similarly, we can work out x_1 by substituting for x_2 and x_3 in the first equation.

This method of solution is called *back-substitution* and can be extended to systems of linearly independent linear equations of any order.

Let us consider how we use Gaussian elimination to reduce the 3×3 matrix of our example to triangular form. The matrix is:

$$\begin{bmatrix} 2 & 3 & -1 \\ 4 & 4 & -3 \\ 2 & -3 & 1 \end{bmatrix}.$$

We start with the first column. We require that all the elements in this column except the top left be zero. (This is to eliminate the term in x_1 from the second and third equations.) If we multiply the first row by -2, we get

$$-4 \quad -6 \quad 2;$$

adding this to the second row reduces that row to

$$0 \quad -2 \quad -1.$$

Turning our attention to the third row, we subtract the first row to give

$$0 \quad -6 \quad 2.$$

We must also remember to carry out the same operations on the vector $\begin{bmatrix} 5 \\ 3 \\ 1 \end{bmatrix}$ on the right-hand side of the matrix equation. Thus the first element 5 of this vector is unaffected, the second element 3 becomes

$$3 + (-2 \times 5) = -7,$$

and the third element 1 becomes

$$1 - 5 = -4.$$

The matrix equation can now be written as

$$\begin{bmatrix} 2 & 3 & -1 \\ 0 & -2 & -1 \\ 0 & -6 & 2 \end{bmatrix} \begin{bmatrix} x_1 \\ x_2 \\ x_3 \end{bmatrix} = \begin{bmatrix} 5 \\ -7 \\ -4 \end{bmatrix}.$$

Our operations to modify the first column using the first row are complete. We now consider the second column and the third row. Multiplying the second row by -3 and adding the result to the third row, we reduce the third row to

$$0 \quad 0 \quad 5.$$

The corresponding term -4 on the right-hand side becomes

$$-4 + (-3 \times -7) = 17.$$

Thus the final matrix equation is

$$\begin{bmatrix} 2 & 3 & -1 \\ 0 & -2 & -1 \\ 0 & 0 & 5 \end{bmatrix} \begin{bmatrix} x_1 \\ x_2 \\ x_3 \end{bmatrix} = \begin{bmatrix} 5 \\ -7 \\ 17 \end{bmatrix}.$$

From the third equation, we find that $5x_3 = 17$, or $x_3 = 3.4$. Substituting this value for x_3 into the middle equation and solving for x_2, we get $x_2 = 1.8$. Finally, knowing x_2 and x_3, we can solve the first equation for x_1, and obtain $x_1 = 1.5$.

Using Gaussian elimination to convert the matrix to triangular form, and then back-substitution, we have solved systematically the system of linear equations without having to invert the 3×3 matrix.

We now apply this technique to the solution of a network problem.

Worked problem Find the currents and voltages in each part of the following network:

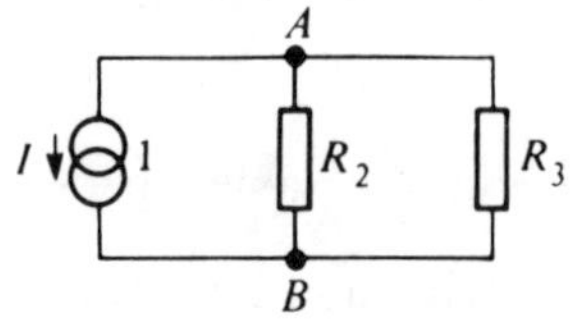

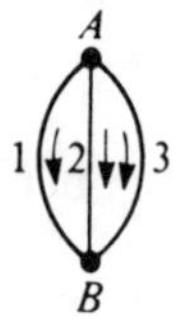

Solution The matrix equation, obtained in the solution of Problem 11.2, is

$$\left[\begin{array}{ccc|ccc} 0 & 0 & 0 & 1 & 0 & 0 \\ 0 & 1 & 0 & 0 & -R_2 & 0 \\ 0 & 0 & 1 & 0 & 0 & -R_3 \\ -1 & 1 & 0 & 0 & 0 & 0 \\ -1 & 0 & 1 & 0 & 0 & 0 \\ 0 & 0 & 0 & 1 & 1 & 1 \end{array}\right] \begin{bmatrix} v_1 \\ v_2 \\ v_3 \\ \hline i_1 \\ i_2 \\ i_3 \end{bmatrix} = \begin{bmatrix} I \\ 0 \\ 0 \\ 0 \\ 0 \\ 0 \end{bmatrix}.$$

Interchanging row 1 and row 4, and row 5 and row 6, we get

$$\left[\begin{array}{ccc|ccc} -1 & 1 & 0 & 0 & 0 & 0 \\ 0 & 1 & 0 & 0 & -R_2 & 0 \\ 0 & 0 & 1 & 0 & 0 & -R_3 \\ 0 & 0 & 0 & 1 & 0 & 0 \\ 0 & 0 & 0 & 1 & 1 & 1 \\ -1 & 0 & 1 & 0 & 0 & 0 \end{array}\right] \begin{bmatrix} v_1 \\ v_2 \\ v_3 \\ \hline i_1 \\ i_2 \\ i_3 \end{bmatrix} = \begin{bmatrix} 0 \\ 0 \\ 0 \\ I \\ 0 \\ 0 \end{bmatrix}$$

Subtracting rows 1 and 3 from row 6, and row 4 from row 5, we get

$$\left[\begin{array}{ccc|ccc} -1 & 1 & 0 & 0 & 0 & 0 \\ 0 & 1 & 0 & 0 & -R_2 & 0 \\ 0 & 0 & 1 & 0 & 0 & -R_3 \\ 0 & 0 & 0 & 1 & 0 & 0 \\ 0 & 0 & 0 & 0 & 1 & 1 \\ 0 & -1 & 0 & 0 & 0 & R_3 \end{array}\right] \left[\begin{array}{c} v_1 \\ v_2 \\ v_3 \\ \hline i_1 \\ i_2 \\ i_3 \end{array}\right] = \left[\begin{array}{c} 0 \\ 0 \\ 0 \\ I \\ -I \\ 0 \end{array}\right]$$

Finally, after adding row 2 and $R_2 \times$ (row 5) to row 6, we get the matrix equation in the required form:

$$\left[\begin{array}{ccc|ccc} -1 & 1 & 0 & 0 & 0 & 0 \\ 0 & 1 & 0 & 0 & -R_2 & 0 \\ 0 & 0 & 1 & 0 & 0 & -R_3 \\ 0 & 0 & 0 & 1 & 0 & 0 \\ 0 & 0 & 0 & 0 & 1 & 1 \\ 0 & 0 & 0 & 0 & 0 & (R_2 + R_3) \end{array}\right] \left[\begin{array}{c} v_1 \\ v_2 \\ v_3 \\ \hline i_1 \\ i_2 \\ i_3 \end{array}\right] = \left[\begin{array}{c} 0 \\ 0 \\ 0 \\ I \\ -I \\ -R_2 I \end{array}\right]$$

Successively solving these equations, beginning with the last one, we obtain

$$(R_2 + R_3)\, i_3 = -R_2 I, \quad \text{so} \quad i_3 = -R_2 I/(R_2 + R_3),$$

$$i_2 + i_3 = -I, \quad \text{so} \quad i_2 = -R_3 I/(R_2 + R_3),$$

$$i_1 = I, \quad \text{so} \quad i_1 = I,$$

$$v_3 - R_3 i_3 = 0, \quad \text{so} \quad v_3 = -R_2 R_3 I/(R_2 + R_3),$$

$$v_2 - R_2 i_2 = 0, \quad \text{so} \quad v_2 = -R_2 R_3 I/(R_2 + R_3),$$

$$-v_1 + v_2 = 0, \quad \text{so} \quad v_1 = -R_2 R_3 I/(R_2 + R_3).$$

Since i_2 and i_3 are negative, the current must flow from B to A in this part of the network.

We have now achieved our goal. Given a network containing resistors and independent sources, we can work out the edge variables—the voltage and current associated with each edge of the oriented graph of the network.

Remarks We have formulated the matrix equations so that they can be solved straightforwardly for cases in which we know the values of the source voltages or currents and wish to calculate the voltages and currents in all the other components. But suppose that we want to solve the converse problem and work out, say, the value of a source voltage which will result in a specified current flowing in one of the resistors of the network. To do this, we would have to rearrange the matrix equations. Essentially, we would have to treat the source as if it were a resistor and treat the resistor which is to have a specified current flowing in it as if it were a source. We mention this because such problems can arise in electrical network design, but we shall not discuss such a case in detail here.

All the examples we have discussed in this chapter have involved only 2-terminal components. However, the methods of solution we have given can be applied in a similar way to systems involving components with more than two terminals—for example, transformers or transistors. The cutset and cycle equation matrices will be of the same form, but the component equation matrix will be different.

11.3 STATE EQUATIONS

Up to now, we have confined our discussion to networks that contain only resistors and independent sources. For such networks, the component equations are particularly simple, being proportionality relations of the form

$$v = Ri.$$

We now turn our attention to networks involving capacitors and inductors. For such networks the component equations involve not only v and i, but also their derivatives. Our aim is to show how the matrix equations can be adapted to take the derivatives into account.

In Chapter 9 we saw that the component equations of a capacitor and an inductor are

$$i = C\frac{dv}{dt} \quad \text{and} \quad v = L\frac{di}{dt},$$

respectively, where C is the capacitance and L is the inductance of the component. Using a dot to denote a derivative with respect to time t, we may rewrite these equations in the form

$$\dot{v} = \frac{1}{C}i \quad \text{and} \quad \dot{i} = \frac{1}{L}v.$$

We shall use these equations to obtain the matrix equation in the form

$$\mathbf{Hx} = \mathbf{y} + \mathbf{K\dot{x}},$$

where

- **x** is a column vector which involves all the voltages and currents;
- $\dot{\mathbf{x}}$ is a column vector which involves their derivatives;
- **y** is the 'source function vector' we introduced earlier;
- **H** is a large matrix which involves the components and the fundamental cycles and cutsets;
- **K** is a matrix which isolates the derivatives of the edge variables.

We illustrate the ideas involved by considering the following simple network and its oriented graph:

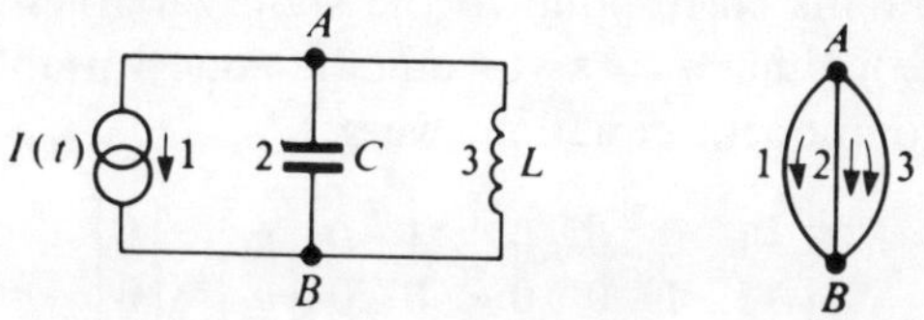

The component equations for this network are

$$i_1 = I(t),$$

$$\dot{v}_2 = \frac{1}{C} i_2,$$

$$\dot{i}_3 = \frac{1}{L} v_3.$$

We write these equations in matrix form as follows:

$$\begin{bmatrix} 0 & 0 & 1 & 0 & \vdots & 0 & 0 \\ 0 & 1/L & 0 & 0 & \vdots & 0 & 0 \\ 0 & 0 & 0 & 1/C & \vdots & 0 & 0 \end{bmatrix} \begin{bmatrix} v_1 \\ v_3 \\ i_1 \\ i_2 \\ \hline v_2 \\ i_3 \end{bmatrix} = \begin{bmatrix} I(t) \\ 0 \\ 0 \end{bmatrix} + \begin{bmatrix} 0 & 0 & 0 & 0 & \vdots & 0 & 0 \\ 0 & 0 & 0 & 0 & \vdots & 0 & 1 \\ 0 & 0 & 0 & 0 & \vdots & 1 & 0 \end{bmatrix} \begin{bmatrix} \dot{v}_1 \\ \dot{v}_3 \\ \dot{i}_1 \\ \dot{i}_2 \\ \hline \dot{v}_2 \\ \dot{i}_3 \end{bmatrix}$$

Notice that the variables whose derivatives occur in the component equations (v_2 and i_3) appear *last of all* in the column vectors **x** and $\dot{\mathbf{x}}$. These variables are called **state variables**. The values of the state variables represent the *state* of the system.

We now wish to combine this equation with the fundamental cycle and cutset equations for the network. These are, of course, the same as the fundamental cycle and cutset equations for the network of Example 1, and can be written as follows:

$$\left[\begin{array}{ccc|ccc} -1 & 1 & 0 & 0 & 0 & 0 \\ -1 & 0 & 1 & 0 & 0 & 0 \\ \hline 0 & 0 & 0 & 1 & 1 & 1 \end{array}\right] \left[\begin{array}{c} v_1 \\ v_2 \\ v_3 \\ \hline i_1 \\ i_2 \\ i_3 \end{array}\right] = \left[\begin{array}{c} 0 \\ 0 \\ \hline 0 \end{array}\right].$$

(This matrix is the lower half of the matrix in the solution of Problem 11.2(a).)

The shaded columns correspond to the state variables v_2 and i_3.

Rearranging the columns so as to put the edge variables into the same order as in the component equation, we get

$$\left[\begin{array}{cccc|cc} -1 & 0 & 0 & 0 & 1 & 0 \\ -1 & 1 & 0 & 0 & 0 & 0 \\ 0 & 0 & 1 & 1 & 0 & 1 \end{array}\right] \left[\begin{array}{c} v_1 \\ v_3 \\ i_1 \\ i_2 \\ \hline v_2 \\ i_3 \end{array}\right] = \left[\begin{array}{c} 0 \\ 0 \\ 0 \end{array}\right].$$

Combining this with the component matrix equation above, we obtain the required equation:

$$\underset{\mathbf{H}}{\left[\begin{array}{cccc|cc} 0 & 0 & 1 & 0 & 0 & 0 \\ 0 & 1/L & 0 & 0 & 0 & 0 \\ 0 & 0 & 0 & 1/C & 0 & 0 \\ -1 & 0 & 0 & 0 & 1 & 0 \\ -1 & 1 & 0 & 0 & 0 & 0 \\ 0 & 0 & 1 & 1 & 0 & 1 \end{array}\right]} \underset{\mathbf{x}}{\left[\begin{array}{c} v_1 \\ v_3 \\ i_1 \\ i_2 \\ \hline v_2 \\ i_3 \end{array}\right]} = \underset{\mathbf{y}}{\left[\begin{array}{c} I(t) \\ 0 \\ 0 \\ 0 \\ 0 \\ 0 \end{array}\right]} + \underset{\mathbf{K}}{\left[\begin{array}{cccc|cc} 0 & 0 & 0 & 0 & 0 & 0 \\ 0 & 0 & 0 & 0 & 0 & 1 \\ 0 & 0 & 0 & 0 & 1 & 0 \\ 0 & 0 & 0 & 0 & 0 & 0 \\ 0 & 0 & 0 & 0 & 0 & 0 \\ 0 & 0 & 0 & 0 & 0 & 0 \end{array}\right]} \underset{\dot{\mathbf{x}}}{\left[\begin{array}{c} \dot{v}_1 \\ \dot{v}_3 \\ \dot{i}_1 \\ \dot{i}_2 \\ \hline \dot{v}_2 \\ \dot{i}_3 \end{array}\right]}$$

Our next step is to manipulate these matrices so that we can write down two *state equations* for the network. The **state equations** are first-order differential equations that relate the derivatives of the state variables to the state variables and the independent sources. Although we shall not prove it here, it can be shown that two state equations are sufficient to describe completely the dynamic behaviour of this network. In other words, any voltage or current in the network may be expressed in terms of the state variables v_2 and i_3, their derivatives $\dot{v}_2$ and $\dot{i}_3$, and the independent current source $I(t)$.

To solve the matrix equations for v_2 and i_3, we use Gaussian elimination. But, because we are interested only in v_2 and i_3, we need concentrate only on the rows of the **H** matrix which involve these variables, that is, on rows 4 and 6. (Notice that row 4 has a 1 in column 5, corresponding to the 5th element in **x**, namely v_2, and that row 6 has a 1 in column 6 corresponding to i_3.)

We want the final matrix equation to be of the form

$$\left[\begin{array}{cccc:cc} & & & & & \\ & ? & & & ? & \\ & & & & & \\ 0 & 0 & 0 & 0 & 1 & 0 \\ \hdashline ? & ? & ? & ? & ? & ? \\ 0 & 0 & 0 & 0 & 0 & 1 \end{array}\right]\left[\begin{array}{c} v_1 \\ v_3 \\ i_1 \\ i_2 \\ \hdashline v_2 \\ i_3 \end{array}\right] = \left[\begin{array}{c} \\ ? \\ \\ \\ \hdashline \\ ? \end{array}\right] + \left[\begin{array}{cccc:cc} & & & & & \\ & ? & & & ? & \\ & & & & & \\ 0 & 0 & 0 & 0 & & \\ \hdashline ? & ? & ? & ? & & \\ 0 & 0 & 0 & 0 & ? & \end{array}\right]\left[\begin{array}{c} \dot{v}_1 \\ \dot{v}_3 \\ \dot{i}_1 \\ \dot{i}_2 \\ \hdashline \dot{v}_2 \\ \dot{i}_3 \end{array}\right].$$

Those regions denoted by ? are to be determined.

We wish first to eliminate the term -1 in the first column of row 4 of **H**. We do this in two stages. First we subtract row 5 from row 4, giving the new row 4 of **H**:

$$0 \quad -1 \quad 0 \quad 0 \quad 1 \quad 0.$$

A new term, -1, has now appeared in the second column, but we can eliminate that by multiplying row 2 by L, and then adding it to the new row 4, giving

$$0 \quad 0 \quad 0 \quad 0 \quad 1 \quad 0.$$

Next, we perform the same row operations on the right-hand side, and the matrix equation becomes:

$$\left[\begin{array}{cccc:cc} 0 & 0 & 1 & 0 & 0 & 0 \\ 0 & 1/L & 0 & 0 & 0 & 0 \\ 0 & 0 & 0 & 1/C & 0 & 0 \\ 0 & 0 & 0 & 0 & 1 & 0 \\ \hdashline -1 & 1 & 0 & 0 & 0 & 0 \\ 0 & 0 & 1 & 1 & 0 & 1 \end{array}\right]\left[\begin{array}{c} v_1 \\ v_3 \\ i_1 \\ i_2 \\ \hdashline v_2 \\ i_3 \end{array}\right] = \left[\begin{array}{c} I(t) \\ 0 \\ 0 \\ 0 \\ \hdashline 0 \\ 0 \end{array}\right] + \left[\begin{array}{cccc:cc} 0 & 0 & 0 & 0 & 0 & 0 \\ 0 & 0 & 0 & 0 & 0 & 1 \\ 0 & 0 & 0 & 0 & 1 & 0 \\ 0 & 0 & 0 & 0 & 0 & L \\ \hdashline 0 & 0 & 0 & 0 & 0 & 0 \\ 0 & 0 & 0 & 0 & 0 & 0 \end{array}\right]\left[\begin{array}{c} \dot{v}_1 \\ \dot{v}_3 \\ \dot{i}_1 \\ \dot{i}_2 \\ \hdashline \dot{v}_2 \\ \dot{i}_3 \end{array}\right].$$

The first four numbers in row 4 are now zero, as required, and we turn our attention to row 6.

Adding $-1 \times$ (row 1) and $-C \times$ (row 3) to row 6, we get

$$\left[\begin{array}{cccc|cc} 0 & 0 & 1 & 0 & 0 & 0 \\ 0 & 1/L & 0 & 0 & 0 & 0 \\ 0 & 0 & 0 & 1/C & 0 & 0 \\ 0 & 0 & 0 & 0 & 1 & 0 \\ \hline -1 & 1 & 0 & 0 & 0 & 0 \\ 0 & 0 & 0 & 0 & 0 & 1 \end{array}\right]\left[\begin{array}{c} v_1 \\ v_3 \\ i_1 \\ i_2 \\ \hline v_2 \\ i_3 \end{array}\right] = \left[\begin{array}{c} I(t) \\ 0 \\ 0 \\ 0 \\ \hline 0 \\ -I(t) \end{array}\right] + \left[\begin{array}{cccc|cc} 0 & 0 & 0 & 0 & 0 & 0 \\ 0 & 0 & 0 & 0 & 0 & 1 \\ 0 & 0 & 0 & 0 & 1 & 0 \\ 0 & 0 & 0 & 0 & 0 & L \\ \hline 0 & 0 & 0 & 0 & 0 & 0 \\ 0 & 0 & 0 & 0 & -C & 0 \end{array}\right]\left[\begin{array}{c} \dot{v}_1 \\ \dot{v}_3 \\ \dot{i}_1 \\ \dot{i}_2 \\ \hline \dot{v}_2 \\ \dot{i}_3 \end{array}\right]. \quad (*)$$

This completes the elimination procedure. From the rows 4 and 6 of the matrix equation we can write down the two equations relating the two state variables of the network:

$$v_2 = L\dot{i}_3$$

$$i_3 = -I(t) - C\dot{v}_2$$

or, in the more usual form,

$$\dot{i}_3 = \frac{1}{L} v_2$$

$$\dot{v}_2 = \frac{1}{C} i_3 - \frac{1}{C} I(t).$$

The general matrix form of the state equations is

$$\dot{\mathbf{x}} = \mathbf{A}\mathbf{x} + \mathbf{u},$$

where

A is a $k \times k$ matrix,
x is a column vector of the k state variables of a system,
$\dot{\mathbf{x}}$ is a column vector of their derivatives,
u is a column vector involving the independent sources.

Thus the state equations in our example can be written as

$$\begin{bmatrix} \dot{v}_2 \\ \dot{i}_3 \end{bmatrix} = \begin{bmatrix} 0 & -1/C \\ 1/L & 0 \end{bmatrix}\begin{bmatrix} v_2 \\ i_3 \end{bmatrix} + \begin{bmatrix} -(1/C)I(t) \\ 0 \end{bmatrix}.$$

We now solve the state equations. The exact form of the solution depends on how the current $I(t)$ varies with time, on the initial conditions of the problem (that is, the values of v_2 and i_3 at time $t = 0$), and on the values of L and C.

In this case, the solution has the form

$$v_2(t) = A \cos \omega t + B \sin \omega t + f(t),$$

$$i_3(t) = A_1 \cos \omega t + B_1 \sin \omega t + g(t),$$

where A, A_1, B and B_1 are constants, $\omega = (LC)^{-1/2}$, and f and g are functions which depend on $I(t)$ and its derivative.

In fact,

$$A_1 = -BC\omega,$$

$$B_1 = AC\omega,$$

$$g(t) = -Cf'(t) - I(t),$$

and f is a function satisfying the differential equation

$$f''(t) + \frac{1}{LC} f(t) + \frac{1}{C} I'(t) = 0.$$

Finding the other edge variables i_1, v_3, i_2 and v_1 is now straightforward, since we can obtain them in terms of the state variables v_2 and i_3 and their derivatives. To do this, we look back at the matrix equation (*). The first four equations are

$$i_1 = I(t),$$

$$\frac{1}{L} v_3 = \dot{i}_3,$$

$$\frac{1}{C} i_2 = \dot{v}_2,$$

$$-v_1 + v_2 = 0,$$

and i_1, v_3, i_2 and v_1 can easily be calculated from these equations.

Problem 11.3

(a) Express the following component equations in matrix form:

$$v_1 = k, \qquad i_2 = C\frac{dv_2}{dt},$$

$$v_3 = L\frac{di_3}{dt}, \qquad v_4 = R_4 i_4.$$

(b) Write down the **H**-matrix equation for a system which has the component equations given in part (a), and the following fundamental cycle and cutset equations:

$$\begin{bmatrix} 1 & 1 & 1 & 0 \\ 0 & -1 & 0 & 1 \end{bmatrix} \begin{bmatrix} v_1 \\ v_3 \\ v_2 \\ v_4 \end{bmatrix} = \begin{bmatrix} 0 \\ 0 \end{bmatrix}$$

and

$$\begin{bmatrix} 1 & 1 & -1 & 0 \\ 0 & 0 & -1 & 1 \end{bmatrix} \begin{bmatrix} i_1 \\ i_3 \\ i_2 \\ i_4 \end{bmatrix} = \begin{bmatrix} 0 \\ 0 \end{bmatrix}.$$

(c) Find the state equations.

The procedure carried out above for our simple example can be extended to any network made up of capacitors, inductors, resistors and independent sources, or their analogues in other types of physical system. The number of state variables, and hence state equations, needed to describe the dynamic behaviour of a system completely is related to the number of energy storage components in the system. In our simple example, we had two such components: a capacitor, which stores energy in the form of an electric field, and an inductor, which stores energy in the form of a magnetic field. We were able to write down the general form of the solution to the state equations, but for more complex networks the state equations would be solved numerically using a computer. There are various computational techniques available to tackle the problem, but a discussion of their relative advantages and drawbacks is beyond the scope of this book.

Numerical Computing Methods

A detailed discussion of the numerical solution of the network equations is outside the scope of this book. However, we make a few remarks about computing techniques.

In the matrix equation $\mathbf{Hx} = \mathbf{y}$, the matrix $\mathbf{H}$ contains a large number of zero elements. Such a matrix is called a *sparse* matrix. It is obviously inefficient for a computer to store and manipulate all these zero elements, so special numerical methods for dealing with sparse matrices are often used to analyse network problems. Their use is particularly important for large networks, where the saving in computer storage and computing time may be very large.

Care must be taken in numerical solution procedures to ensure that rounding errors do not significantly affect the final result. As an example of the sort of problem which may arise, suppose that at some stage we obtain the following equations:

$$3x + 4y = 7$$

$$3.1x + 3.9y = 7.$$

The solution is $x = 1$ and $y = 1$. Suppose, however, that because of numerical rounding errors the equations calculated by the computer are:

$$3x + 4y = 7$$

$$3.1x + 4y = 7.002.$$

The solution of these equations is $x = 0.002$ and $y = 1.7485$. So small errors in the numbers in the equations have resulted in large changes in the solution. Such simultaneous equations which are sensitive to rounding errors are said to be *ill-conditioned.*

EXERCISES 11

Formulation of the Matrix Equations

11.1. The following diagram shows an electrical network and an associated oriented graph:

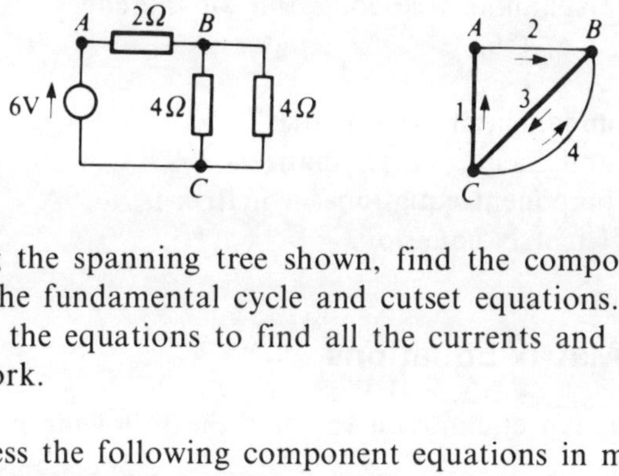

(a) Using the spanning tree shown, find the component equations and the fundamental cycle and cutset equations.
(b) Solve the equations to find all the currents and voltages in the network.

11.2. (a) Express the following component equations in matrix form:

$$v_1 = R_1 i_1, \quad v_2 = 3, \quad v_3 = R_3 i_3, \quad v_4 = R_4 i_4.$$

(b) Write down the **H**-matrix equation for a system which has the component equations given in part (a), and the following fundamental cycle and cutset equations:

$$\begin{bmatrix} 1 & 1 & 1 & 0 \\ 0 & -1 & 0 & 1 \end{bmatrix} \begin{bmatrix} v_1 \\ v_3 \\ v_2 \\ v_4 \end{bmatrix} = \begin{bmatrix} 0 \\ 0 \end{bmatrix}$$

and

$$\begin{bmatrix} 1 & 0 & -1 & 0 \\ 0 & 1 & -1 & 1 \end{bmatrix} \begin{bmatrix} i_1 \\ i_3 \\ i_2 \\ i_4 \end{bmatrix} = \begin{bmatrix} 0 \\ 0 \end{bmatrix}.$$

11.3. The following diagram shows a resistive network and an associated oriented graph:

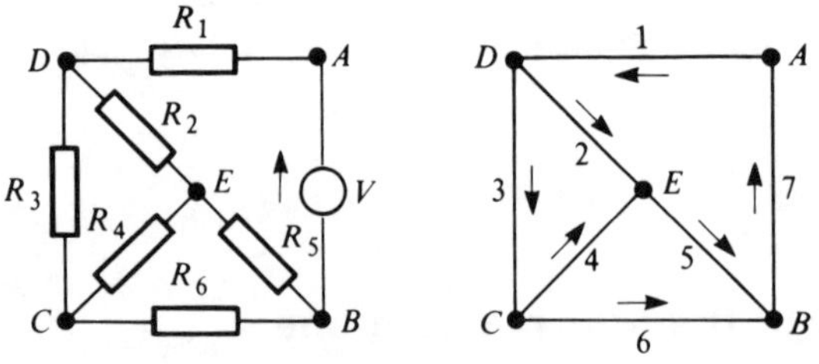

network

oriented graph

Using the spanning tree consisting of the edges
AD, *DC*, *CE*, *CB*,
find:
(a) the fundamental cycle matrix;
(b) the fundamental cutset matrix;
(c) the component equations in matrix form;
(d) the **H**-matrix equation.

Solving the Matrix Equations

11.4. Use Gaussian elimination to solve the following matrix equation:

$$\begin{bmatrix} 1 & 0 & 0 & 1 \\ 1 & 2 & 0 & 0 \\ 1 & 0 & 1 & 1 \\ 0 & 0 & 3 & -1 \end{bmatrix} \begin{bmatrix} x_1 \\ x_2 \\ x_3 \\ x_4 \end{bmatrix} = \begin{bmatrix} 3 \\ 3 \\ 0 \\ -13 \end{bmatrix}$$

State Equations

11.5. The following diagram shows an electrical network and an associated oriented graph:

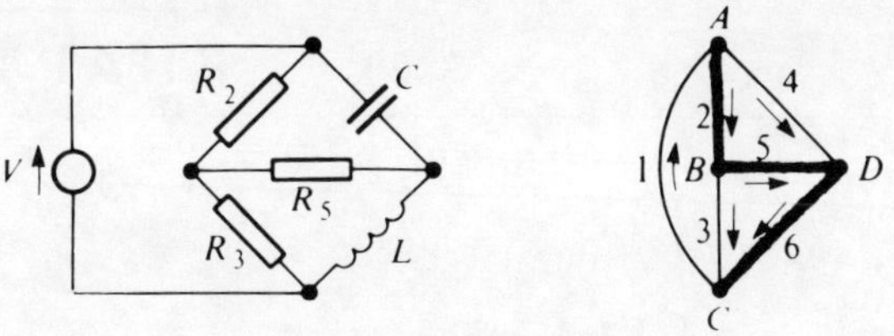

Using the spanning tree shown, find:
(a) the fundamental cycle matrix;
(b) the fundamental cutset matrix;
(c) the component equations in matrix form;
(d) the **H**-matrix equation;
(e) one of the state equations.

11.6. The following diagram shows an electrical network and an associated oriented graph.

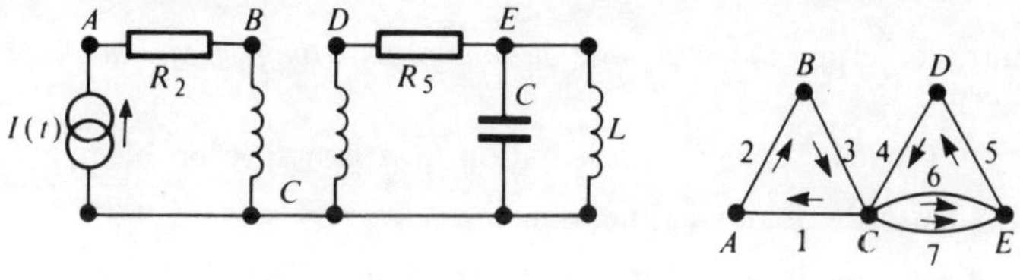

network oriented graph

Using the spanning tree which consists of the edges 2, 3, 4 and 5, find the **H**-matrix equation, and hence find the state equations.

CHAPTER 12

MATCHING PROBLEMS

After reading this chapter, you should be able to:

- explain the terms *the marriage problem*, *matching*, *maximum matching* and *transversal*;
- draw the bipartite graph representation of a marriage problem;
- state and use the marriage theorem;
- state and use the modified marriage theorem;
- identify *alternating paths* with respect to a matching in a bipartite graph;
- explain briefly the procedure used in the algorithm for finding a maximum matching in a bipartite graph;
- use the algorithm to find a maximum matching in a bipartite graph.

Introduction

The common feature of the network problems which we consider in this and the next two chapters is that they can all be modelled by bipartite graphs. Some of these problems (such as the marriage problem discussed in this chapter) originated as problems in pure mathematics, whereas others (for example, the transportation problem discussed in Chapter 14) arose from practical problems in industry and commerce.

The three basic optimization problems of these three chapters—the marriage problem, the assignment problem and the transportation problem—are closely related to the flow problems considered in earlier chapters, and can in fact be solved by using the maximum flow algorithm given in

Chapter 3 or the maximum-flow minimum-cost algorithm given in Chapter 7. However, the algorithms given here are more efficient, since they make use of the fact that all the graphs involved are bipartite graphs.

In this chapter we discuss the marriage problem and some related problems involving matchings in bipartite graphs, and give an algorithm for finding a maximum matching. We also explain how such problems can be expressed in the language of transversal theory.

12.1 THE MARRIAGE PROBLEM

We begin by presenting three examples of the type of problem which we consider in this chapter.

Example 1 Four tasks are to be assigned to four workers. Worker 1 is qualified to do only tasks *A* and *B*; worker 2 is qualified to do tasks *A*, *B*, *C* and *D*; worker 3 is qualified to do only tasks *A* and *B*; and worker 4 is qualified to do only task *B*. Can all four workers be assigned to different tasks for which they are qualified?

Example 2 Four men each know some of four women. Man 1 knows women *A* and *B*; man 2 knows women *A*, *B*, *C* and *D*; man 3 knows women *A* and *B*; and man 4 knows woman *B*. Is it possible for all the men to marry women they know?

Example 3 A company has four lorries and four depots. Lorry 1 can carry loads up to $1\frac{1}{2}$ tons; lorry 2 can carry loads up to 2 tons; and lorries 3 and 4 can each carry loads up to 1 ton. Depot 1 needs a lorry to carry loads up to $1\frac{1}{2}$ tons; depot 2 needs a lorry to carry loads up to 1 ton; depot 3 needs a lorry to carry loads up to $1\frac{1}{2}$ tons; and depot 4 needs a lorry to carry up to 2 tons. Can each depot be allocated a suitable lorry?

In fact, all three examples can be expressed as exactly the same mathematical problem. This becomes obvious if we represent each of the examples by a graph. For Example 1, the following table shows the tasks for which each worker is qualified. A graphical representation of this information is given on the right:

workers	tasks
1	*A*, *B*
2	*A*, *B*, *C*, *D*
3	*A*, *B*
4	*B*

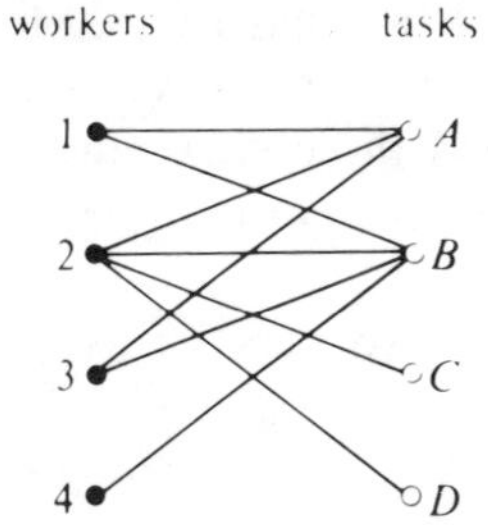

In the above graph, the black vertices represent workers and the white vertices represent tasks. A black vertex is joined by an edge to a white vertex if the corresponding worker is qualified to do the corresponding task. Since only vertices of different colours are joined, the graph is a bipartite graph.

If we represent Examples 2 and 3 by graphs in a similar way, we find that the graphs are identical to the one obtained above. For Example 2, we simply substitute 'men' for 'workers', and 'women' for 'tasks'. For Example 3, we substitute 'depots' for 'workers', and 'lorries' for 'tasks'.

As stated, each problem is an existence problem; the actual assignment is a construction problem. Since the three problems are essentially the same, we restrict our discussion to the solution of Example 1.

Assigning a worker to a task corresponds to choosing an edge of the bipartite graph. We can easily assign three of the workers to different tasks; for example, we can assign worker 1 to task *A*, worker 2 to task *D*, and worker 3 to task *B*.

This arrangement is shown in the following graph, where thick edges represent assignments of workers to tasks.

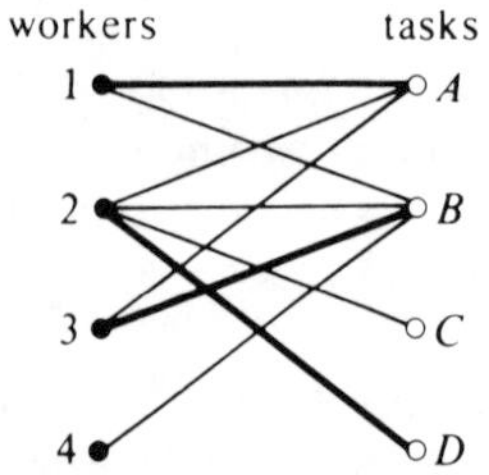

The pairing of some or all of the elements of one set with elements of a second set is called a **matching**. The matching represented by the thick lines of the above graph comprises the following pairings:

worker 1 with task *A*
worker 2 with task *D*
worker 3 with task *B*.

A matching in a bipartite graph corresponds to a set of edges no two of which meet at the same vertex. The matching represented by the above graph contains three such edges, as shown by the thick lines.

We now see that our original question:

can all four workers be assigned to different tasks for which they are qualified?

is equivalent to the following question:

can we find a matching containing four edges in the bipartite graph?

In fact, it is not possible to find such a matching in this graph. We say that the matching with three edges is a **maximum matching**—that is, it is a matching with the largest possible number of edges.

Problem 12.1 Suppose that four applicants apply for four jobs and that each applicant is qualified for some of these jobs, as shown in the following table:

applicants	jobs
1	*A, D*
2	*A, B*
3	*A, B, C*
4	*B*

(a) Draw the bipartite graph which represents this situation.
(b) By finding a maximum matching in the bipartite graph, find the maximum number of applicants that can be assigned to different jobs for which they are qualified.

In the above problem, it is possible for all four applicants to be assigned to jobs for which they are qualified. This is in contrast to our previous example, where at most three of the four workers can be assigned to different tasks for which they are qualified. For the general problem in which we are trying to allocate a set of workers to various tasks (or jobs), we should like to know under what conditions it is possible for all the workers to be assigned to tasks for which they are qualified.

This problem was formulated in 1935 as a problem in group theory. It was later restated in terms of marriages, as in Example 2 at the beginning of this chapter. For this reason, it is often called the **marriage problem**. We shall follow this tradition of discussing matching problems in bipartite graphs in terms of marriages, rather than in terms of more practical problems such as the assignment of workers to tasks or applicants to jobs. This has the advantage that it helps us to keep clear the distinction between matching problems and the related, but more general, assignment problems which we discuss in Chapter 13.

We can state the marriage problem as follows:

THE MARRIAGE PROBLEM
Given a set of men, each of whom knows some women from a given set of women, under what conditions is it possible for all the men to marry

women they know? (It is assumed that each man can marry only one woman, and each woman can marry only one man.)

One condition which must obviously be satisfied is that the total number of women must be equal to or greater than the total number of men. We can go further than this, and state that any two men must collectively know at least two women. More generally, for each subset of m men, the m men must collectively know at least m women, for all possible values of m. This is a *necessary* condition for it to be possible for all the men to marry women they know. What is not so obvious is that it is also a *sufficient* condition. This result is known as the *marriage theorem.*

The marriage theorem *A necessary and sufficient condition for there to be a solution to the marriage problem (that is, for every man to be able to marry a woman he knows) is that for every subset of m men, the m men collectively know at least m women, for all values of m in the range* $1 \leqslant m \leqslant n$*, where n is the total number of men.*

The marriage theorem was proved (in a different formulation) by Phillip Hall in 1935, and it is often referred to as *Hall's theorem.*

We shall prove this theorem later in this chapter. First, however, let us see how it applies to some specific cases.

We can apply the marriage theorem to Example 2 at the beginning of this chapter. For convenience, we change the notation: we denote the men by A_1, A_2, A_3 and A_4, and the women by B_1, B_2, B_3 and B_4. The following table shows which women are known to each of the four men:

men	women known
A_1	B_1, B_2
A_2	B_1, B_2, B_3, B_4
A_3	B_1, B_2
A_4	B_2

The marriage theorem states that for all the men to be able to marry women they know, the men in every subset of $\{A_1, A_2, A_3, A_4\}$ of size m must collectively know at least m women, for each possible value of m. (This condition is known as the *marriage condition.*) So, to apply the condition systematically, we must list all possible subsets of men. The number of subsets of a set of n elements is 2^n. There are four men, so the total number of subsets is 2^4. However, we need look at only $2^4 - 1 = 15$ subsets, since we obviously do not need to consider the empty set.

The four men collectively know four women, so the marriage condition is satisfied for the subset $\{A_1, A_2, A_3, A_4\}$. The following table shows whether the marriage condition is satisfied for each subset of three men.

subset of men	women known	number of men (m)	number of women known	is marriage condition satisfied?
$\{A_1, A_2, A_3\}$	B_1, B_2, B_3, B_4	3	4	yes
$\{A_1, A_2, A_4\}$	B_1, B_2, B_3, B_4	3	4	yes
$\{A_1, A_3, A_4\}$	B_1, B_2	3	2	no
$\{A_2, A_3, A_4\}$	B_1, B_2, B_3, B_4	3	4	yes

For the third of these subsets, the three men concerned collectively know only two women, so the marriage condition is not satisfied, and therefore all four men cannot marry women they know. For this example, we do not need to examine any more subsets, but if the marriage condition had been satisfied for all four subsets of three men, then we should have had to continue with the subsets containing two men, and if necessary, the subsets containing only one man.

Problem 12.2 Suppose that each of the three men, A_1, A_2 and A_3, know some of the four women B_1, B_2, B_3 and B_4, as shown in the following table:

men	women known
A_1	B_1, B_3, B_4
A_2	B_4
A_3	B_2, B_3

(a) Draw up a table (similar to the one for the example in the text) showing for each possible subset of men, the number of men in that subset, and the number of women the men collectively know. For each subset, indicate whether the marriage condition is satisfied. Use the table to decide whether it is possible for all three men to marry women they know.
(b) If (in the general case) there are n men, how many subsets of men do you need to consider when applying the marriage theorem? Check that your answer for the case $n = 3$ agrees with the number of entries in your table.
(c) Draw the bipartite graph representation of this problem. Find a maximum matching in this graph corresponding to the maximum number of men who can marry women they know.

The marriage problem is closely related to a branch of mathematics called *transversal theory*. We now digress briefly to explain the idea of a transversal.

Suppose that E is a non-empty set, and let $S_1, S_2, \ldots, S_m$ be non-empty subsets of E. (These subsets do not necessarily have to be different.) This collection of subsets is called a *family* $\mathcal{F}$ of subsets, and we write $\mathcal{F} = (S_1, S_2, \ldots, S_m)$. Note that a family is not a set, since the same element may appear more than once. For example, $\{a, b, c\}$ is a set, and (a, b, a, a, c, b) is a family. Parentheses (rather than curly brackets) are used to denote a family.

We can now define a *transversal*.

Definition *A* ***transversal*** *(or* ***system of distinct representatives****) of $\mathcal{F}$ is a set of m distinct elements of E, one drawn from each set S_i.*

For example, let E be the set of women in the last problem, that is, $E = \{B_1, B_2, B_3, B_4\}$. We take the subset S_1 to be the set of women known to the man A_1, the subset S_2 to be the set of women known to A_2, and the subset S_3 to be the set of women known to A_3. Thus

$$S_1 = \{B_1, B_3, B_4\}, \quad S_2 = \{B_4\}, \quad S_3 = \{B_2, B_3\}.$$

Then the family $\mathcal{F} = (S_1, S_2, S_3)$ has a transversal $\{B_1, B_4, B_3\}$.

We can represent this transversal of $\mathcal{F}$ by certain edges of a bipartite graph as follows:

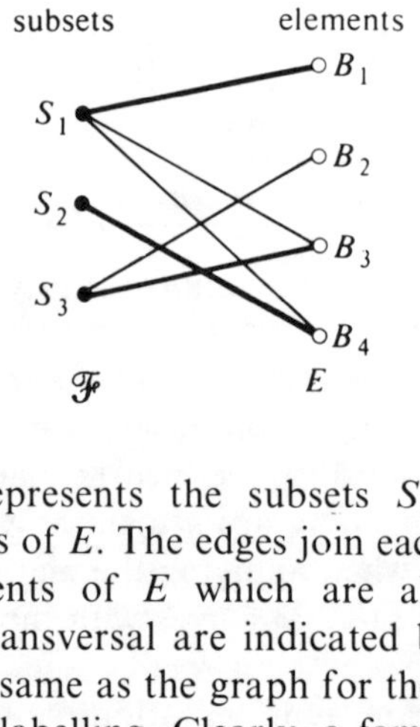

One set of vertices represents the subsets S_1, S_2, S_3, and the other set represents the elements of E. The edges join each vertex S_i to the vertices representing those elements of E which are also elements of S_i. The edges representing this transversal are indicated by thick lines.

The above graph is the same as the graph for the solution to the previous problem, apart from the labelling. Clearly, a family $\mathcal{F}$ has a transversal if and only if there exists a matching in the bipartite graph such that the number of edges in the matching is equal to the number of subsets in $\mathcal{F}$ (in this case, 3). The marriage problem is equivalent to the following problem in transversal theory:

> under what conditions does the family $\mathcal{F} = (S_1, S_2, \ldots, S_m)$ have a transversal?

Problem 12.3 Let E be the set $\{B_1, B_2, B_3, B_4\}$ and $\mathcal{F}$ be the family (S_1, S_2, S_3, S_4), where

$$S_1 = \{B_2\}, \quad S_2 = \{B_2, B_3\}, \quad S_3 = \{B_1, B_2, B_4\}, \quad S_4 = \{B_3\}.$$

(a) Draw the bipartite graph representation of this situation.
(b) Let $\{A_1, A_2, A_3, A_4\}$ be a set of men. Suppose that the elements of the set E above represent women, and that S_i is the set of women known to man A_i (for $i = 1, 2, 3, 4$). Write down the subsets of men (if any) for which the marriage condition is not satisfied.
(c) By trial and error, find a maximum matching in the bipartite graph, and hence find the maximum number of men who can marry women they know.
(d) Does the family of $\mathcal{F}$ sets have a transversal?

So far in this chapter we have been mainly concerned with the question posed by the marriage problem:

> under what condition is it possible for all men to marry women they know?

We have seen that this question is answered by the marriage theorem. When, as in the previous problem, it is not possible for all the men to marry, it is natural to ask a second question, called the *modified marriage problem*:

> what is the maximum number of men who can marry women they know?

In a more practical version of this type of problem (such as assigning workers to tasks), we should probably be most interested in actually constructing a maximum matching. We express this problem as a third question:

> how can we determine which man should marry which woman so that the maximum possible number of marriages can take place?

In the next section we shall be mainly concerned with the second question. Our answer to the third question takes the form of an algorithm, which we give in Section 12.4.

12.2 THE MODIFIED MARRIAGE PROBLEM

In the solution to Problem 12.3, we found by trial and error that of the four men A_1, A_2, A_3 and A_4, only three can marry women they know. We now show how to prove that the maximum number of men who can marry is three, without having to construct a possible scheme of marriages.

For convenience, we set out below a table and a bipartite graph showing which men know which women.

men	women known
A_1	B_2
A_2	B_2, B_3
A_3	B_1, B_2, B_4
A_4	B_3

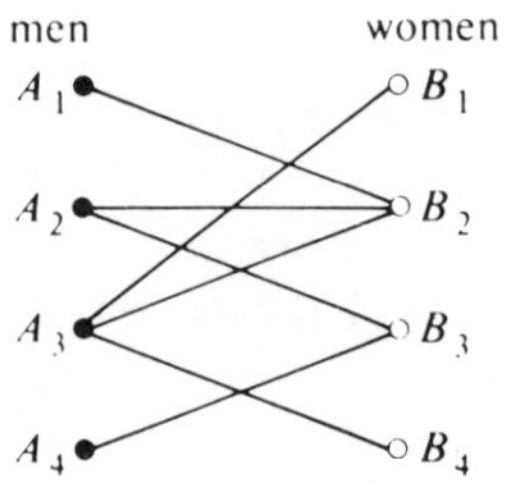

We found in Problem 12.3 that the marriage condition is satisfied for all subsets of men except for the subset containing the three men A_1, A_2 and A_4, who collectively know only two women, B_2 and B_3. By the marriage theorem, it is not possible for all four of the men to marry.

Since the marriage condition breaks down because the three men A_1, A_2 and A_4 know only two women, it seems likely that if we remove one of these men, then the marriage condition will be satisfied for all possible subsets of the remaining three men. This is indeed the case. For example, if we remove A_1, then the subset $\{A_1, A_2, A_4\}$ becomes the subset $\{A_2, A_4\}$ and A_2 and A_4 collectively know two women.

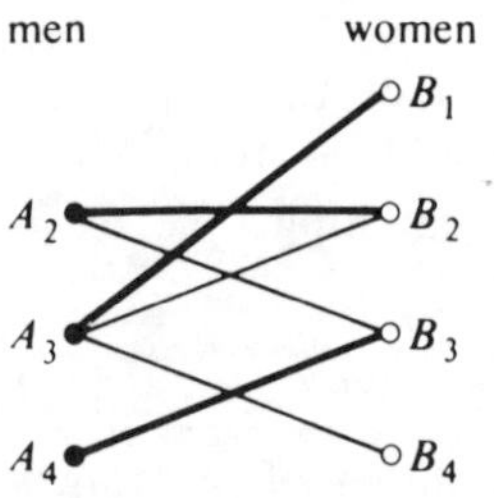

All the other subsets which contained A_1 and which previously satisfied the marriage condition will still satisfy the marriage condition. Although this is fairly obvious if we remove A_1, it is not so obvious if we remove A_2, and it is certainly not obvious that we can apply similar reasoning to a more complicated example.

Let us try another approach. Faced with a situation where the marriage condition was not satisfied, we converted the situation to one where the marriage condition was satisfied, by removing one of the men. However, we can also satisfy the marriage condition for all subsets of men by adding an extra woman, B_5, who is known by all the men. The men in the subset $\{A_1, A_2, A_4\}$ now know three women, B_2, B_3 and B_5, and all the other subsets of men now know one more woman than before. Therefore, by the marriage theorem, we can marry all four men to women they know. If we remove B_5, we are left with three men who are married and one (the one who was married to B_5) unmarried. We have thus shown that it is possible for three men to marry.

We now turn our attention to the general problem of finding the maximum number of men who can marry when the marriage condition is not satisfied.

If there are n men, then the marriage condition states that they can all marry women they know if

(the number of women known by each subset of m men) $\geq m$, for all values of m in the range $1 \leq m \leq n$.

If the marriage condition is not satisfied, we can write instead

(the number of women known by each subset of m men) $\geq m - d$, for all values of m in the range $1 \leq m \leq n$, and some positive integer d.

This is called the *modified marriage condition*. If the modified marriage condition is satisfied for a given value of d, then $n - d$ men can marry women they know. We prove this result by an argument similar to the one we used above for the example of Problem 12.3. All we have to do to prove the result is to add a group of d women who are known by all the men. With the enlarged group of women, the marriage condition is satisfied, so all the men can marry. Removal of the extra d women leaves $n - d$ men who are married, and d men who are not married. In other words, in the original problem, $n - d$ men can marry.

If we let $r = n - d$, we can write the modified marriage condition as follows:

(the number of women known by each subset of m men) $\geq m - (n - r)$, for all values of m in the range $1 \leq m \leq n$.

Note that if r is equal to n, then this condition reduces to the original marriage condition.

The maximum value of r for which this modified marriage condition is satisfied is the largest number of men who can marry women they know. We can rearrange the inequality in this marriage condition so that only r appears on the right-hand side:

(the number of women known by each subset of m men) + $(n - m) \geq r$, for all values of m in the range $1 \leq m \leq n$.

It is clear that the *maximum* value of r for which this condition is satisfied is equal to the *minimum* value of the expression on the left-hand side over all subsets of m men, over all values of m. We summarize this result in the following theorem.

The modified marriage theorem *If a group of n men each know some of a group of women, the maximum number of men who can marry women they know is equal to the minimum value of the expression*

$$(\text{number of women known by a subset of } m \text{ men}) + (n - m)$$

for any subset of m men, and for all values of m in the range $1 \leqslant m \leqslant n$.

Remark This theorem is a minimax theorem, like the max-flow min-cut theorem in Chapter 3, and Menger's theorem in Chapter 6.

(In stating this theorem, we assume that the original marriage condition is not satisfied. Otherwise the minimum value of the expression may be greater than n.)

Worked problem Suppose that five men A_1, A_2, A_3, A_4 and A_5 know four women B_1, B_2, B_3 and B_4, as shown in the following table and bipartite graph. What is the maximum number of men who can marry women they know?

men	women known
A_1	B_1, B_3
A_2	B_2, B_3
A_3	B_1, B_2
A_4	B_1, B_2, B_3, B_4
A_5	B_1

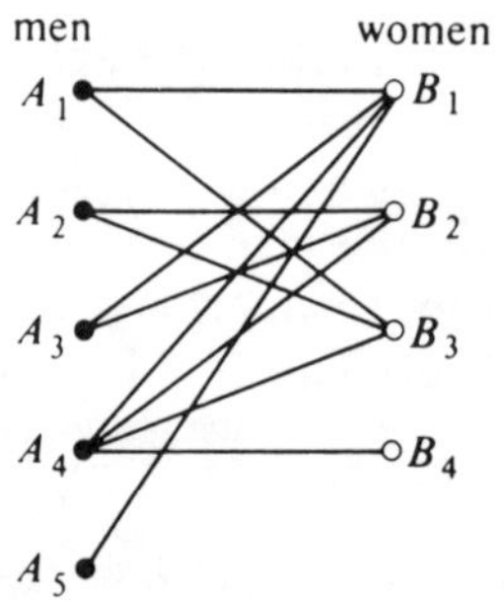

Solution To answer this question, we need to find the minimum value of the expression in the modified marriage theorem. To do this, we need consider only those subsets of men for which the original marriage condition is not satisfied. By inspection of the table or the graph, we see that there are only two such subsets—namely, $\{A_1, A_2, A_3, A_5\}$ and $\{A_1, A_2, A_3, A_4, A_5\}$. We draw up the following table:

subset of men	women known	number of men (m)	number of women known (p)	$p + (n - m)$
$\{A_1, A_2, A_3, A_5\}$	B_1, B_2, B_3	4	3	4
$\{A_1, A_2, A_3, A_4, A_5\}$	B_1, B_2, B_3, B_4	5	4	4

For both subsets, the value of the expression $p + (n - m)$ is 4. Hence exactly four men can marry women they know.

In the above worked problem, we found that at most four men can marry. The next question which arises is: which four of the five men can marry women they know? Or, to put it another way, which one man should we remove so that the remaining four can marry? To answer this question, we look at the two subsets for which the marriage condition is not satisfied. We must remove a man who is in both of these subsets. Clearly, there are four possibilities: A_1, A_2, A_3 or A_5. We remove each of A_1, A_2, A_3 and A_5 in turn, and find a marriage scheme for the remaining four men in each case. All such marriage schemes involving the maximum number of men (four) are shown below:

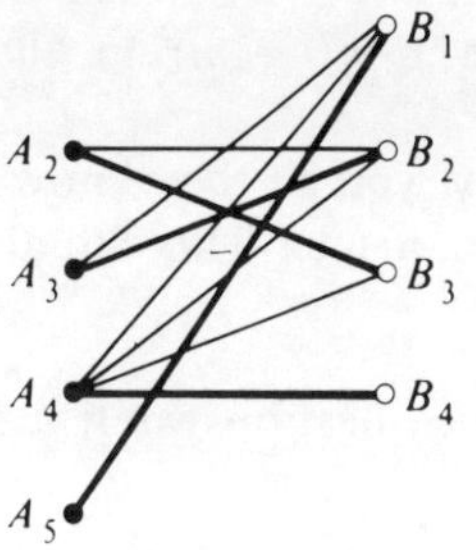

A_1 removed

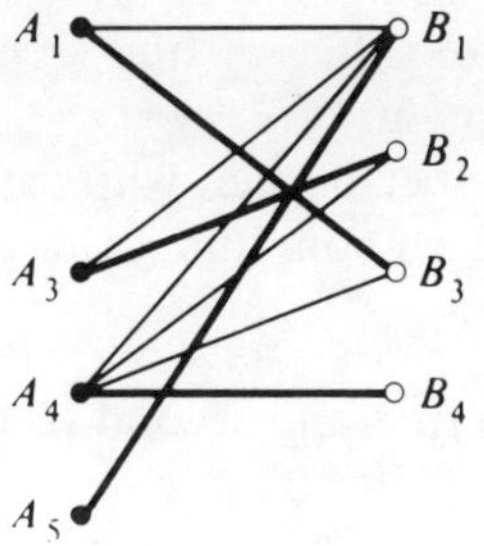

A_2 removed

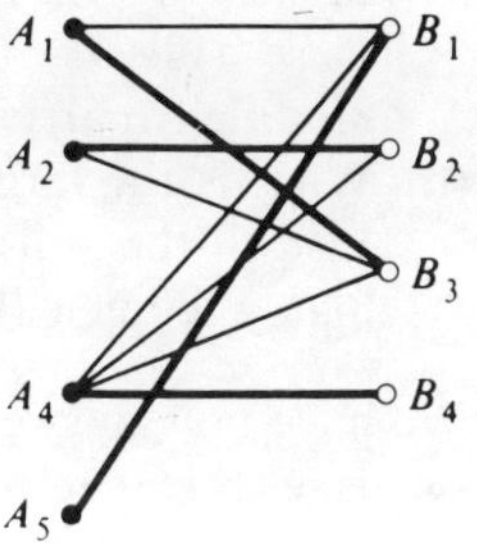

A_3 removed

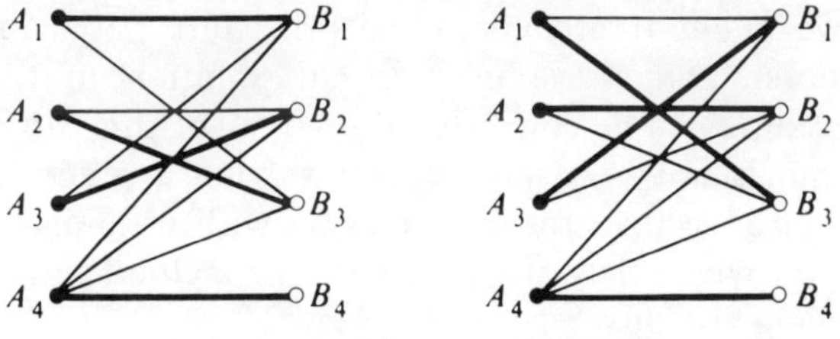

A_5 removed (two possibilities)

If A_4 is removed, then it is obvious that the remaining four men cannot all marry women they know, since none of them knows B_4

Problem 12.4 Suppose that five men A_1, A_2, A_3, A_4 and A_5 know women B_1, B_2, B_3 and B_4 as follows:

men	women known
A_1	B_1, B_3
A_2	B_2, B_3
A_3	B_2
A_4	B_3
A_5	B_1, B_2, B_4

(a) Show that the marriage condition is not satisfied.
(b) Draw a bipartite graph to represent this problem. (You may find the graph representation helpful in answering the rest of this problem.)
(c) List all the subsets of men for which the marriage condition is not satisfied. (One way of doing this is to construct a table as in the solution to Problem 12.2. Alternatively, you may find the required subsets by inspection, using the graph representation.) For each of these subsets, find the value of the expression $p + (n - m)$ in the modified marriage theorem.
(d) What is the largest number of men who can marry women they know? List all the subsets of men containing this number of men who can all marry women they know.

We have now answered the first and second of the three questions we listed on page 291. The remaining question is:

> how can we determine which man should marry which woman so that the maximum possible number of marriages can take place?

To answer this question, we might try to adapt the procedure based on the modified marriage theorem which we used in the solution to the previous problem. This procedure can be converted into an algorithm fairly easily, but such an algorithm is not very suitable for solving a practical form of the marriage problem in which, for example, we wish to assign a number of workers to tasks for which they are qualified. The method would be unsuitable for the following reasons:

- Although the procedure tells us which subsets of men can all marry (or which workers can be assigned to tasks), it does not tell us which man should marry which woman. For small groups of men it is a fairly simple matter to construct a suitable marriage scheme, but for larger numbers we need a systematic method.
- The procedure involves looking at all possible non-empty subsets of the set of n men. The number of such subsets is $2^n - 1$, which increases very rapidly as n increases. The time taken to perform an

algorithm based on this procedure is therefore roughly proportional to 2^n—that is, to an exponential function of n. Since the time taken increases so rapidly with n, it would not be practicable to use the procedure for problems involving large values of n, even when using a computer. (We shall consider the time taken to perform algorithms in Chapter 20.)

- The procedure is wasteful, since it effectively carries out a search of all possibilities and finds *all* the subsets of men which contain the largest number who can marry (or the largest number of workers who can all be assigned to different tasks). In a practical situation, we normally require only *one* solution.

What we need is an efficient algorithm which constructs just one maximum matching. We shall describe such an algorithm later in this chapter. Before doing this, we give two different proofs of the marriage theorem, and discuss the connection between matching problems and maximum flow problems.

12.3 TWO PROOFS OF THE MARRIAGE THEOREM

For convenience, we repeat the statement of the marriage theorem:

The marriage theorem *A necessary and sufficient condition for there to be a solution to the marriage problem (that is, for every man to be able to marry a woman he knows) is that for every subset of m men, the m men collectively know at least m women, for all values of m in the range* $1 \leqslant m \leqslant n$, *where n is the total number of men.*

Since it is obvious that the condition given in the statement of the marriage problem is necessary, we have only to prove that it is also sufficient. We prove this first by induction and then by a constructive proof. (By a *constructive proof*, we mean a proof which actually shows how to *construct* a solution to a problem, as contrasted with a proof (such as a proof by induction) which establishes that a solution *exists*, but does not provide a procedure for finding one.)

Proof by induction To prove by induction that the condition is sufficient, we carry out two steps:

STEP 1 *Verify that the theorem is true when* $n = 1$.

In this case, there is one man who knows at least one woman, so obviously the marriage problem has a solution.

STEP 2 *Show that if the theorem is true for every number of men less than n, then it must also be true for n men.*

We begin by assuming that the statement is true for every number of men less than n. We consider n men. There are two cases which together cover all possibilities.

Case (i) *We suppose that for all $m < n$ every set of m men collectively know at least $m + 1$ women.*

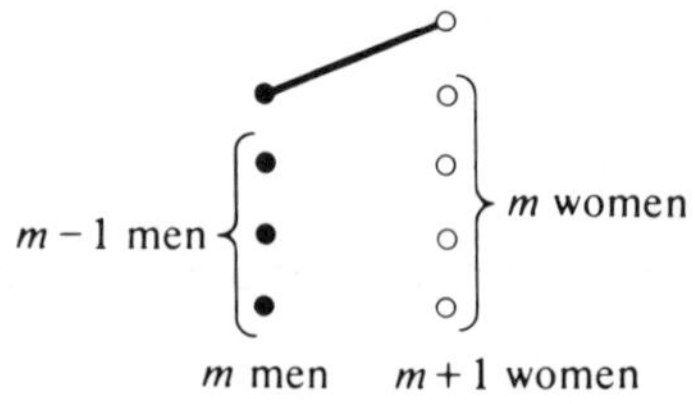

In this case, the marriage condition is satisfied with 'at least one woman to spare' for every set of m men. We can therefore take any man, and marry him to any woman he knows, and the marriage condition will still be satisfied for the remaining $n - 1$ men. We are assuming that the theorem is true for every number of men less than n, so it must be true for these $n - 1$ men. We can therefore marry off the $n - 1$ men appropriately. We have now married off all n men, so this completes the proof of Step 2 in this case.

Case (ii) *We suppose that $m < n$ and that there is at least one set of m men who collectively know exactly m women.*

From our assumption that the theorem is true for every number of men less than n, it follows immediately that we can marry these m men to the m women, leaving $n - m$ men and (since we are also assuming that the marriage condition is satisfied) at least $n - m$ women.

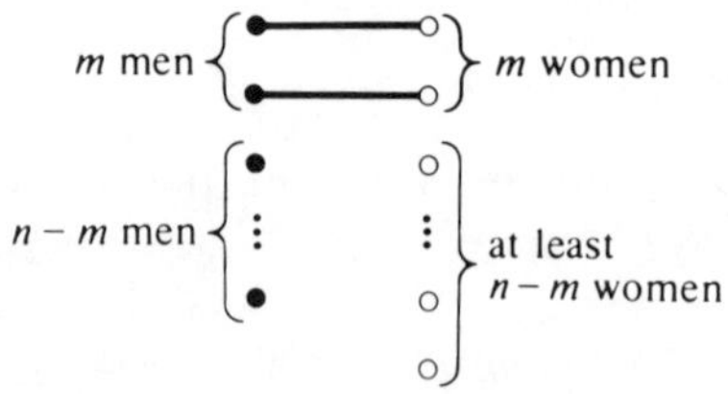

Now any collection of r men from these $n - m$ men must collectively know at least r women; otherwise these r men, together with the set of m

men, would collectively know less than $r + m$ women, which is contrary to the marriage condition. It follows that, since we are assuming that the marriage theorem is true for fewer than n men, we can marry the remaining $n - m$ men to women they know. This completes the proof of Step 2.

We have thus proved that if the statement is true for all numbers of men less than n, then it is true for n men. Thus, since we know from Step 1 that the statement is true for $n = 1$, it must be true for all positive integers n. This completes the proof of the theorem. □

Constructive proof To prove that the marriage condition is a sufficient condition for there to be a solution to the marriage problem (we already know that it is necessary) we show that, starting with m men paired with m women (where $m < n$), it is possible to increase this to a pairing of $m + 1$ men. There is no difficulty in finding an initial pairing, since we can if necessary start with $m = 1$—that is, by pairing any man with a woman he knows. We can then successively increase the number of men paired until all n men are paired with women they know.

We begin, then, with m men paired with m women whom they know. If there is a man left who knows a woman not already paired, then an $(m + 1)$th pairing is immediately possible. Otherwise we proceed as follows.

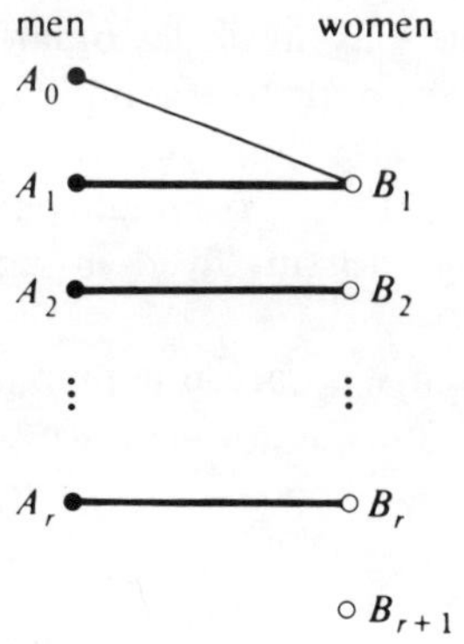

We choose any unpaired man, A_0. Since the marriage condition is satisfied, there must be at least one woman whom he knows—we choose one of these women, B_1. Suppose that B_1 is paired with a man, A_1. (This pairing is indicated on the above diagram by a thick line, and the fact that A_0 knows B_1 is indicated by the thin line joining the corresponding vertices.) By the marriage condition, A_0 and A_1 must collectively know at least two women—namely, B_1 and at least one other woman, B_2. If B_2 is not already paired, we stop.

If B_2 is paired with A_2, then the men A_0, A_1 and A_2 must collectively know a third woman, B_3. If B_3 is paired with A_3, we continue until an unpaired woman B_{r+1} is reached. We can represent a possible result of this procedure by the following diagram:

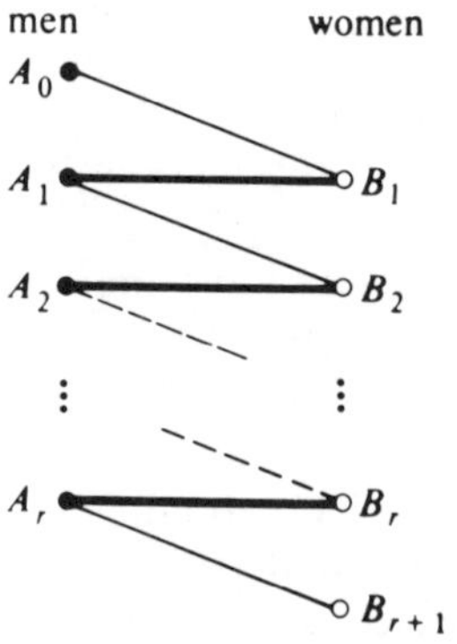

Note that each woman B_i involved in this process must be known by at least one man A_j, where $j < i$. This fact enables us to rearrange the pairings as follows. First we pair woman B_{r+1} with a man A_i who knows her (where $i < r + 1$). This frees woman B_i, who can be paired with some man A_j (where $j < i$). We continue in this way until a woman is free who can be paired with A_0. This process corresponds to tracing back a path which starts from B_{r+1} and proceeds by alternate thin and thick lines back to A_0.

We have now replaced the r original pairings with $r + 1$ pairings. Together with the pairings not involved in this process, we now have a total of $m + 1$ pairings.

If there are still any men who are not paired (that is, if $m + 1 < n$), we continue with this process until all m men are paired with women they can marry. □

The constructive process of the second proof provides a method for finding a matching of the n men to n of the women. The algorithm which we shall use for finding a maximum matching is based on the procedure of the above proof.

The Relation Between Matching Problems and Maximum Flow Problems

The problem of finding a maximum matching in a bipartite graph is closely related to the problem of finding a maximum flow in a network. In fact, a problem of finding a maximum matching can always be formulated as a maximum flow problem. To see how this is done, consider the following diagrams:

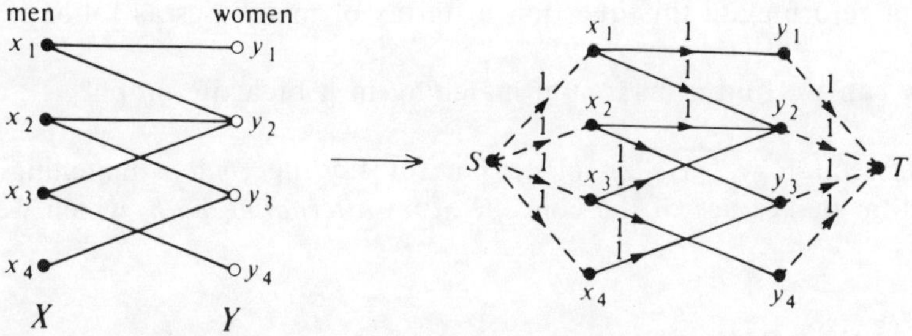

The bipartite graph on the left is converted to a basic network by:

(i) adding a source S and a sink T;
(ii) replacing each edge $x_i y_j$ by an arc of capacity 1 directed from x_i to y_j;
(iii) adding an arc Sx_i with capacity 1 for each of the vertices x_i of X, and adding an arc $y_j T$ of capacity 1 for each of the vertices y_j of Y.

Suppose that we use the algorithm for finding a maximum flow to find a maximum flow in this network, as described in Chapter 3. Then the flow along each arc will have the value 0 or 1. Furthermore, no two of the arcs shown as solid lines which then have a *flow* of 1 will meet at the same vertex, since the flow into any vertex of X or out of any vertex of Y cannot be greater than 1. It follows that, when the procedure of the algorithm has been carried out, the solid arcs with a flow of 1 in the network correspond to the edges of a maximum matching in the original bipartite graph.

In other words, we can solve a maximum matching problem on a bipartite graph by converting the bipartite graph to a basic network and using the labelling procedure to find a maximum flow. In fact, the algorithm we give for finding a maximum matching is a special case of the maximum-flow algorithm of Chapter 3.

We can also deduce the marriage theorem from the max-flow min-cut theorem of Chapter 3 by using similar reasoning.

12.4 AN ALGORITHM FOR FINDING A MAXIMUM MATCHING

We showed earlier that the marriage problem can be expressed in terms of a matching in a bipartite graph. We posed the following question (our third question), which is related to the marriage problem:

how can we determine which man should marry which woman so that the maximum possible number of marriages can take place?

We can reformulate this question in terms of matchings as follows:

how can we find a maximum matching in a bipartite graph?

In this section we give an algorithm for finding such a matching. This algorithm makes use of the concept of an *alternating path*, which we now discuss.

Alternating Paths

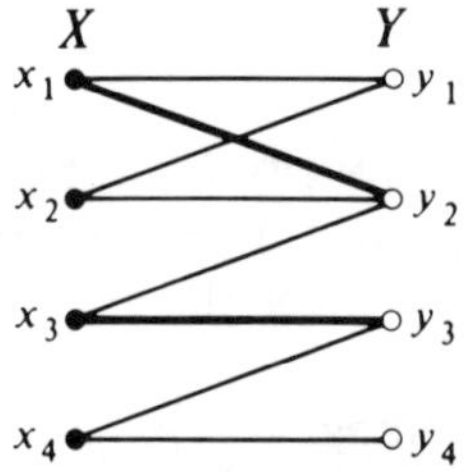

bipartite graph

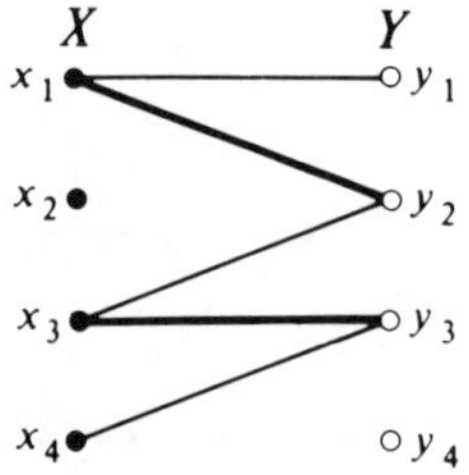

an alternating path

In the bipartite graph shown above, a matching M consists of the two edges $x_1 y_2$ and $x_3 y_3$ (indicated by thick lines). The diagram on the right illustrates an example of an alternating path with respect to M. As you can see, only alternate edges of this path are edges of the matching.

Definition *Let G be a bipartite graph in which the vertices are divided into two sets, X and Y. An* **alternating path** *with respect to a matching M in G is a path which satisfies the following conditions*:

(i) *the path joins a vertex x from one set of vertices, X, to a vertex y from the other set, Y*;
(ii) *alternate edges of the path are in M, and the other edges are not in M*;
(iii) *the initial and final vertices x and y are not incident with any edge of M.*

Conditions (ii) and (iii) of the definition mean that the 1st, 3rd, 5th, . . ., edges are not in M, and the 2nd, 4th, 6th, . . ., edges are in M. Condition (i) tells us that any alternating path is of odd length.

There are three other alternating paths in the above bipartite graph—the paths $x_2y_2x_1y_1$, x_2y_1 and x_4y_4.

Problem 12.5 Consider the following bipartite graphs:

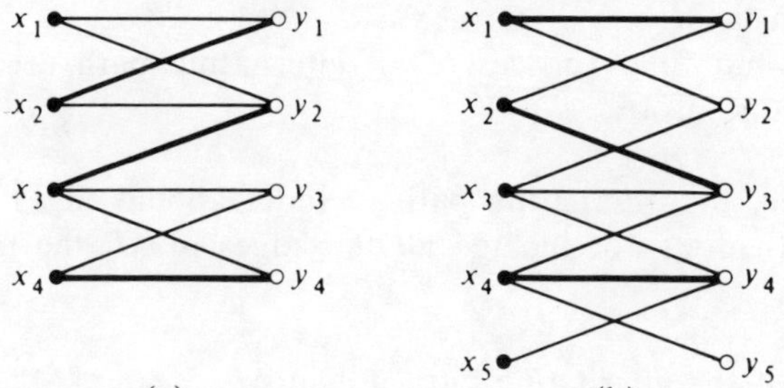

(a) (b)

(a) Which of the following paths are alternating paths with respect to the matching shown in graph (a)?

(i) $x_3y_2x_2y_1$;
(ii) $y_3x_3y_2x_2y_1x_1$;
(iii) $x_3y_4x_4y_3$.

(b) Find all the alternating paths with respect to the matching shown for graph (b).

To see why the concept of an alternating path is useful for finding a maximum matching, consider the following diagrams.

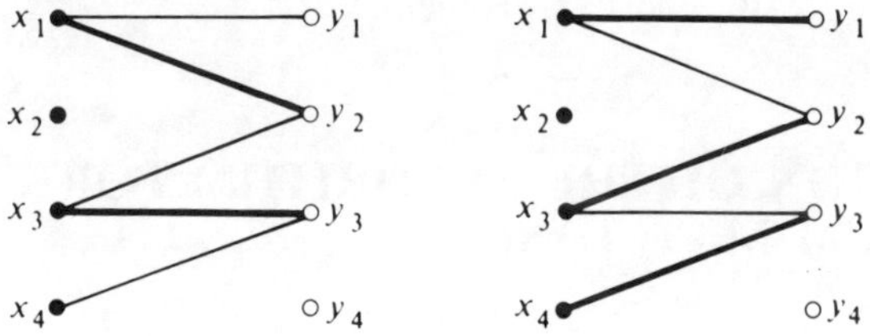

alternating path new matching

The diagram on the left shows our earlier example of an alternating path with respect to a matching M. In the diagram on the right, we have replaced the thick edges of the path by thin edges, and *vice versa.* In this way we have constructed a new matching M' consisting of three edges—one more edge than in the original matching M. It follows immediately from the definition of an alternating path that if we can find an alternating path with respect to a matching M, then we can always construct a matching M' which has one more edge than M. The properties of an alternating path which ensure that this is the case are:

(i) for an alternating path with respect to a matching M, the number of edges of the path not in M is always one more than the number of edges of the path in M;
(ii) the initial and final vertices of an alternating path are not incident with any edge in M.

In our example, the alternating path includes all the edges of M. If the alternating path does not include all the edges of M, the new matching M' consists of:

(i) the edges of the alternating path not in M

and

(ii) the edges of M which are not edges of the path.

This method of using an alternating path to form a new matching is the basis of the following algorithm for finding a maximum matching.

We first give a formal statement of the algorithm and then work through an example. To find a maximum matching, we start with a matching M, look for an alternating path, and, if one is found, construct a new matching with one more edge than before. We repeat the procedure until no more alternating paths can be found; the current matching is then a maximum matching.

We suggest that you read briefly through the statement of the algorithm before going on to the worked problem.

DESCRIPTION OF THE ALGORITHM FOR FINDING A MAXIMUM MATCHING

The algorithm is in two parts:

Part A: The labelling procedure for finding an alternating path.
Part B: The matching improvement procedure for finding a new matching with one more edge.

START with any matching M in a bipartite graph whose sets of vertices are $X = \{x_1, x_2, \ldots\}$ and $Y = \{y_1, y_2, \ldots\}$.

(There is no difficulty in finding the initial matching M. If necessary, it can consist of just one edge, or of no edges.)

Part A: The Labelling Procedure

STEP 1 Label with (∗) each vertex in X which is not incident with any edge in the current matching M. If no vertex in X can be labelled, STOP.

STEP 2 Follow instructions (a) and (b) below alternately until no further labelling is possible:

(a) Choose a newly-labelled vertex in X, say x_i, and label with (x_i) all the unlabelled vertices in Y joined to x_i by an edge NOT IN M.
Repeat this for all newly-labelled vertices in X, before going on to instruction (b).
(b) Choose a newly-labelled vertex in Y, say y_i, and label with (y_i) all unlabelled vertices in X joined to y_i by an edge IN M.
Repeat this for all newly-labelled vertices in Y, before returning to instruction (a).

When no further labelling is possible, go on to Step 3.

STEP 3 Look for a labelled vertex in Y which is not incident with an edge in M. If there is no such vertex, STOP. The current matching has the maximum possible number of edges.
If there is such a vertex, we say that *breakthrough has occurred*: go on to Part B, Step 4.

Part B: Matching Improvement Procedure

STEP 4 Find an alternating path as follows. Start at the breakthrough vertex, and go to the vertex indicated by its label. From this vertex, go to the vertex indicated by *its* label, and so on, until a vertex labelled (∗) is reached. This path is an alternating path P.

STEP 5 Form a new matching from:

(i) the edges of the current matching M which are not in the path P;
(ii) the edges of the path P which are not in the current matching M.

Remove all labels and go back to Step 1 to see if the new matching can be improved further.

Worked problem using the algorithm for finding a maximum matching

Find a maximum matching in the following bipartite graph with vertices x_1, x_2, x_3, x_4 in X and y_1, y_2, y_3, y_4 in Y.

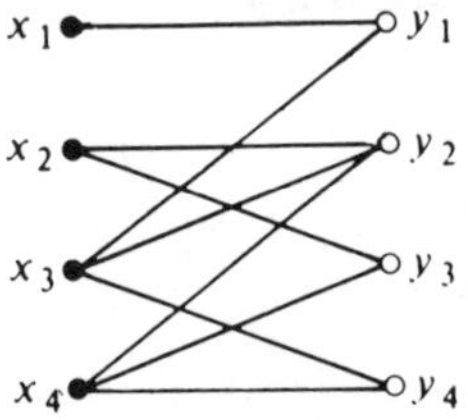

Solution The algorithm works by starting with any matching, and then looking systematically for a new matching with one more edge than the previous one.

Let us start with the initial matching x_1y_1, x_2y_2, x_4y_4 shown below.

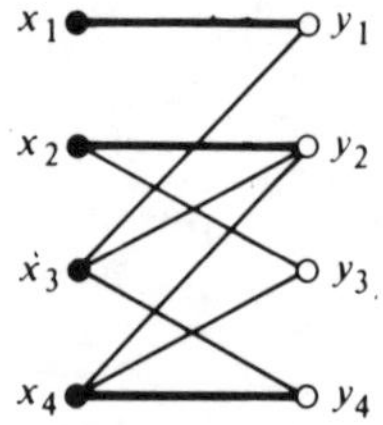

bipartite graph with initial matching M

This matching contains three edges, shown by the thick lines in the diagram. We now use the algorithm to find an alternating path, and hence to construct a new matching with four edges.

Part A of the algorithm is the labelling procedure. The first thing we do (Step 1) is to label with (*) all the vertices in X which are not incident with any edge in the initial matching M. The only such vertex is x_3, which is labelled in the following diagram with (*).

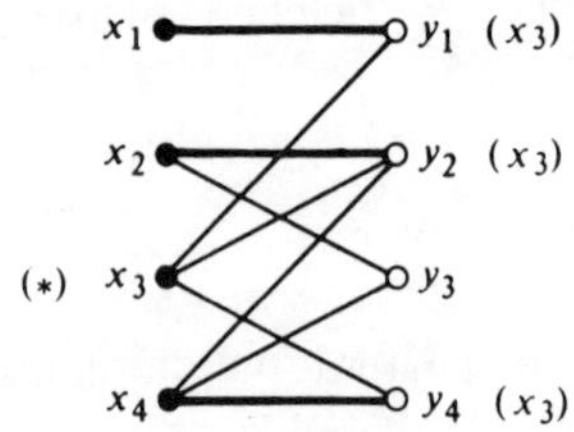

labelling the graph: steps 1 and 2(a)

Now we continue with Step 2, part (a). The only newly-labelled vertex in X is x_3, so we have to label the vertices in Y which are joined to x_3 by an edge not in M (a thin line) with (x_3). There are three such vertices (y_1, y_2 and y_4) and these are labelled (x_3) in the above diagram.

Step 2, part (b) now requires us to label vertices in X which are joined to a newly-labelled vertex in Y by an edge in M (a thick line). We choose, say, newly-labelled vertex y_1, and label the vertex x_1 with (y_1). Continuing in this fashion, we label vertex x_2 with (y_2) and vertex x_4 with (y_4). This completes Step 2 part (b), so we go back to Step 2, part (a).

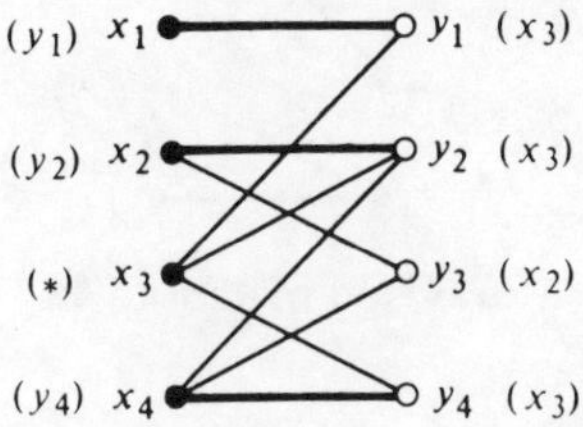

labelling the graph: steps 2(b) and 2(a)

To follow Step 2, part (a), we note that there is only one unlabelled vertex in Y, namely y_3. So we can choose x_2 as a newly-labelled vertex in X, and we label vertex y_3 with (x_2).

All the vertices are now labelled as shown in the figure above, so no more labelling is possible and we go on to Step 3.

Step 3 asks us to look for a labelled vertex in Y which is not incident with an edge in M. From the figure we see that vertex y_3 is such a vertex, and so breakthrough has occurred.

We now go on to Part B, Step 4.

In Step 4, we find an alternating path. Now we can see the point of the labelling procedure, for all we have to do is to backtrack through the labels to find the required alternating path. Thus vertex y_3 is labelled (x_2), so we go to vertex x_2 which is labelled (y_2). Then we go to vertex y_2 which is labelled (x_3), and finally to x_3 which is labelled $(*)$. Thus the alternating path P is $y_3x_2y_2x_3$, as shown by the dashed lines in the following figure.

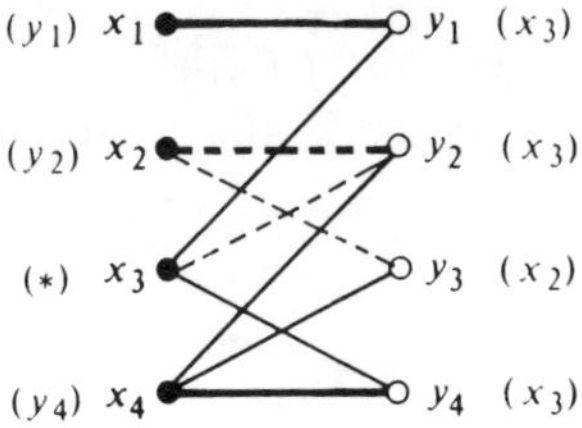

an alternating path found

We go on to Step 5 and form a new matching from the edges in M not in P, together with the edges of P not in M. Thus in effect we change the thick lines of the alternating path to thin ones, and *vice versa*, so in the new matching M' we replace the edge x_2y_2 in M by the two edges x_2y_3 and x_3y_2. The new matching M' is shown in the following diagram.

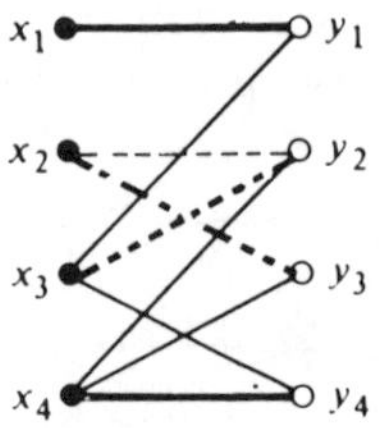

revised matching M'

Formally, the algorithm now requires us to go back to Step 1 to see if the matching can be improved further. However, it is clear that, since no vertex in X is not incident with M', no further improvement can be made. So M' is an optimum matching with four edges.

For convenience, we summarize the algorithm as follows.

Summary of the Algorithm for Finding a Maximum Matching

START with any matching.

Part A: Follow the labelling procedure to look for an alternating path.
Part B: Construct an alternating path by backtracking through the labels obtained in Part A. Form a new matching from the edges in the old matching that were *not in* the alternating path, plus the edges in that path *not in* the old matching.

REPEAT Parts A and B until there are no more alternating paths, then STOP.

Problem 12.6 Use the algorithm to find a maximum matching in the following graph, starting with the initial matching shown.

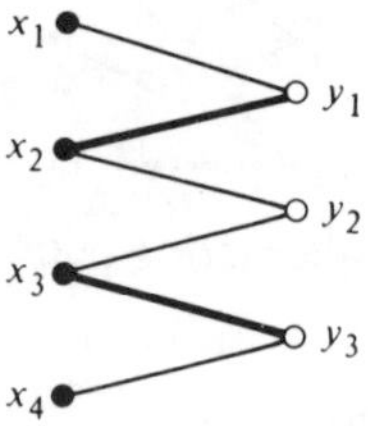

The algorithm, in common with others presented in this book, is written in a form suitable for implementing as a computer program. Its advantages may not be apparent from the simple examples we have given, where the solutions can be quickly found by trial and error, but its usefulness is that it is a precise and systematic method which can be used to solve very complicated examples with the aid of a computer.

EXERCISES 12

The Marriage Problem

12.1. Five people have each won a holiday to one of five resorts—London, Rome, Edinburgh, Paris and San Francisco. Of the winners, one will go only to Britain, one will not go to France or Scotland, two will not go to Britain, and the fifth will only go to San Francisco.

(a) Draw the bipartite graph representation of this situation.
(b) By finding a maximum matching in this graph, determine whether it is possible for each winner to go to a different resort.

12.2. A computer agency advertises for a systems analyst (S), a programmer (P) and a computer manager (M). There are four applicants—one is a qualified systems analyst (A_1), two are both qualified systems analysts and programmers (A_2 and A_3), and the other is a qualified programmer and computer manager (A_4).

(a) Draw up a table showing, for each possible subset of jobs, the number of jobs in that subset and the number of suitable applicants for that subset of jobs.
(b) For each subset, indicate whether the marriage condition is satisfied, and decide whether it is possible for all three jobs to be filled.
(c) Draw the bipartite graph representation of this problem and find, by trial and error, a maximum matching in this graph corresponding to the maximum number of jobs that can be filled.

12.3. Does there exist a maximum matching containing 4 edges in the following bipartite graph? If so, find such a matching; if not, use Hall's Theorem to prove that such a matching does not exist.

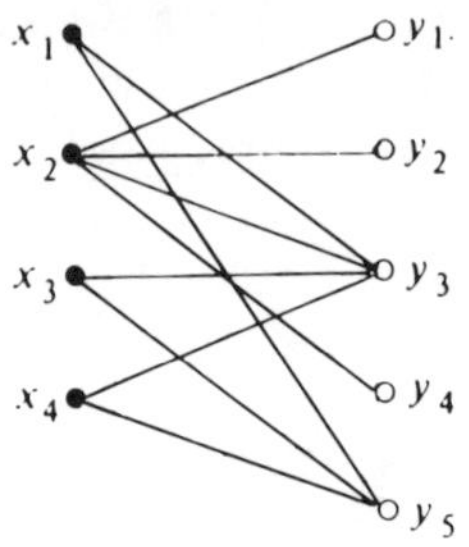

12.4. Which of the following families of subsets of $S = \{T, H, E, O, R, Y\}$ have transversals? In each case, either give an example or find all the collections of subsets for which the marriage condition fails.

(a) $(\{T\}, \{H, T\}, \{E, O\}, \{E, R\}, \{H, E\})$;
(b) $(\{H\}, \{H, T\}, \{E, T\}, \{E, H\})$;
(c) $(\{T, H\}, \{H, O, R\}, \{T, Y\}, \{H, R\})$;
(d) $(\{H, O\}, \{H, O\}, \{H, T\}, \{H\})$.

12.5. Suppose that there are r readers and r books, and that each reader wishes to read one of k books ($r > k$), and each book has k readers who wish to read it. Show that it is possible to distribute the books so that every reader has a book that he or she wishes to read.

12.6. Deduce the marriage theorem from the max-flow min-cut theorem.

The Modified Marriage Problem

12.7. Five tutors T_1, T_2, T_3, T_4, T_5, apply to teach five courses; just one tutor is required for each course. To find out how many of the tutors can be appointed to courses which they are competent to teach, the marriage theorem is applied to each subset of tutors. It is found that the marriage condition is satisfied for all except the following subsets:

tutors T_1, T_2 and T_3 are between them competent to teach on only 2 courses;
tutors T_1 and T_3 are between them competent to teach on only 1 course.

(a) How many of the five tutors can simultaneously be appointed to courses which they are competent to teach? Explain your answer.
(b) For which of the following subsets can all the tutors in the subset be appointed to courses which they are competent to teach? Explain your answer.

(i) $\{T_1, T_2, T_3, T_4\}$ (ii) $\{T_2, T_3, T_4\}$
(iii) $\{T_1, T_2, T_4, T_5\}$ (iv) $\{T_1, T_3, T_5\}$
(v) $\{T_1, T_2, T_3, T_5\}$ (vi) $\{T_1, T_3, T_4, T_5\}$

12.8. A university has 5 vacancies, V_1, V_2, V_3, V_4, V_5, and there are 5 applicants, A_1, A_2, A_3, A_4, A_5, whose suitability for the different jobs is shown in the following table:

job	suitable applicants
V_1	A_1, A_2, A_3
V_2	A_3, A_4
V_3	A_4
V_4	A_4
V_5	A_1, A_2, A_5

(a) Draw a bipartite graph to represent this problem.
(b) List all the subsets of vacancies for which the marriage condition is not satisfied.
(c) For each of these subsets, find the value of the expression in the modified marriage theorem, and hence find the largest number of vacancies which can be filled.
(d) Find, by trial and error, a matching in the bipartite graph with this number of edges.

12.9. Five applicants apply for five jobs. The jobs for which each applicant is qualified are shown in the following bipartite graph:

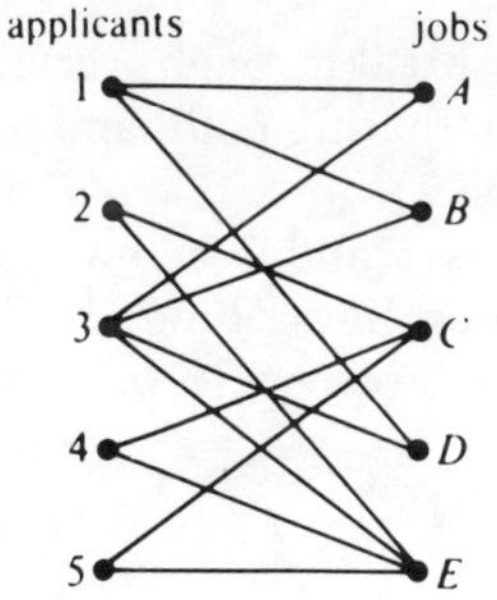

(a) List all the subsets of applicants for which the marriage condition is not satisfied.
(b) Use the modified marriage theorem to determine the largest number of applicants who can be assigned to jobs for which they are qualified.
(c) Find by inspection an arrangement which allocates the largest possible number of applicants to jobs.

The Matching Algorithm

12.10. Find, by inspection, an alternating path in the following bipartite graph, and hence obtain a new matching with one more edge.

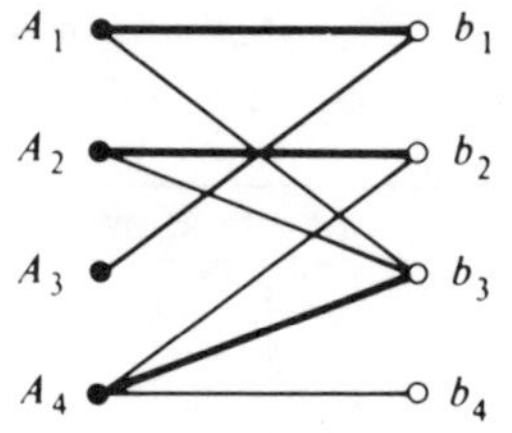

12.11. Carry out the labelling procedure for the graph in Exercise 12.10.

12.12. Consider the following labelled bipartite graph:

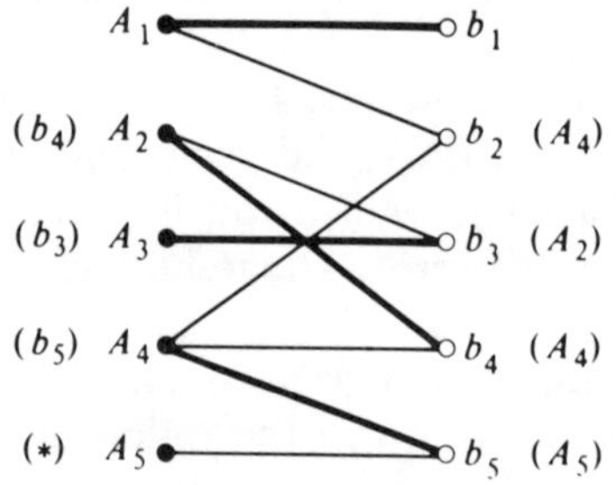

Starting from the breakthrough vertex, trace back through the labels to find an alternating path, and hence obtain a new matching with one more edge.

12.13. The bipartite graphs 1 and 2 below each have a partial matching denoted by the thick lines. Find, by inspection, a total of three alternating paths with respect to these matchings in the two graphs.

Graph 1

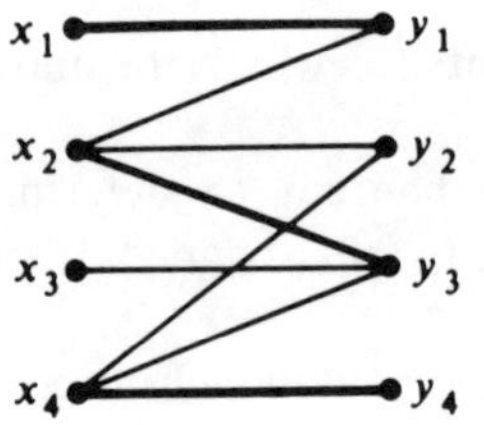

Graph 2

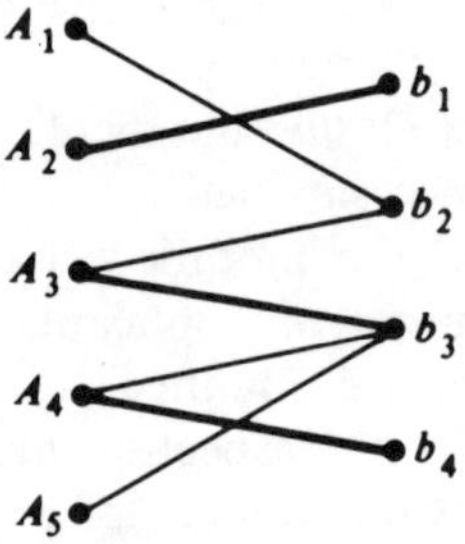

12.14. The thick lines in the following bipartite graph denote an initial matching:

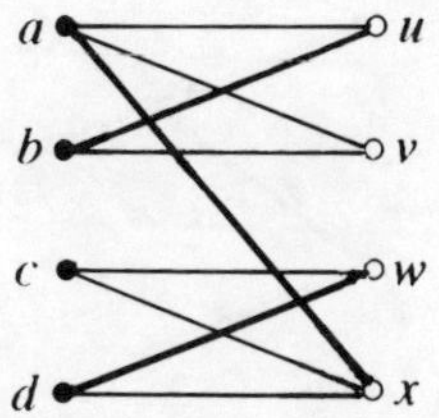

(a) Write down all the alternating paths with respect to this matching.
(b) For each alternating path, find the resulting maximum matching in the graph.

12.15. Use the matching algorithm to find a maximum matching in the bipartite graph of Exercise 12.14, starting from the initial matching shown.

12.16. Use the matching algorithm to find a maximum matching in the following bipartite graph, starting from the given initial matching.

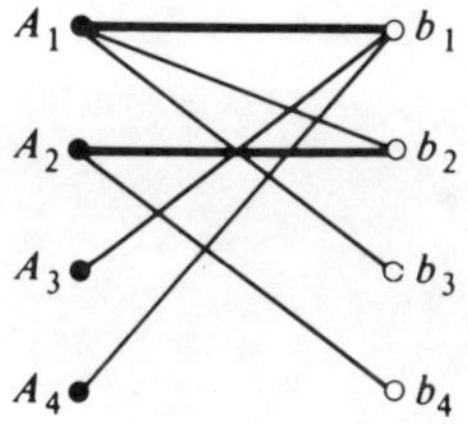

12.17. Use the matching algorithm to find a maximum matching in the following bipartite graph, starting from the given initial matching.

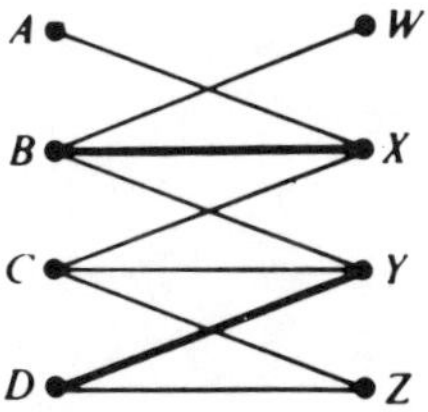

12.18. Use the matching algorithm to find a maximum matching in the following bipartite graph, starting from the given initial matching.

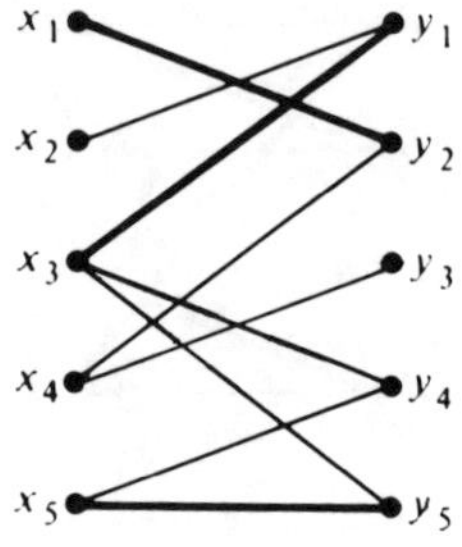

CHAPTER 13

THE ASSIGNMENT PROBLEM

After reading this chapter, you should be able to:

- explain what is meant by *the assignment problem*, *the bottle-neck assignment problem* and *the cabinet reshuffle problem*;
- explain briefly the procedure of the assignment problem algorithm;
- explain how the assignment problem algorithm can be modified to solve the bottle-neck assignment problem and the cabinet reshuffle problem;
- use the appropriate algorithm to solve simple examples of the assignment problem, the bottle-neck assignment problem and the cabinet reshuffle problem.

Introduction

In the previous chapter we were concerned with finding a matching with the maximum possible number of edges in a bipartite graph. Often, there are several such maximum matchings. In the matching problem, we were concerned with finding just one maximum matching, and we had no reason to prefer one maximum matching to any other. However, if the bipartite graph represents a practical problem, some of the matchings may represent more desirable solutions to the problem than others. We can express this fact by assigning costs to the edges of the bipartite graph—we then look for a maximum matching with the lowest total cost. (In Chapter 14 we shall have weights on both edges and vertices. To distinguish between the two cases, we refer to the weights on edges as *costs*.) This type of problem is called an **assignment problem**. Let us look at an example.

Example Three men apply for three jobs. The suitability of each candidate for each job is represented by a cost—the lower the cost, the more suitable the candidate is for the job. We can represent this situation by the following bipartite graph:

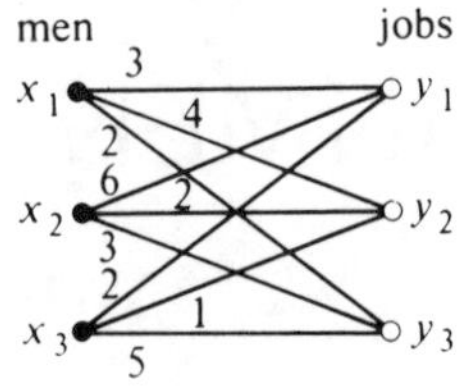

bipartite graph

		jobs		
		y_1	y_2	y_3
	x_1	3	4	2
men	x_2	6	2	3
	x_3	2	1	5

cost matrix

We have written the costs associated with the assignment of each man to each job next to the corresponding edges. Thus, for example, the cost associated with assigning the man x_1 to job y_2 is 4. For larger graphs it is more convenient to display these costs in the form of a **cost matrix**. The cost matrix for this example is shown underneath the bipartite graph.

The problem is to find a maximum matching for which the sum of the costs on the edges of the matching is a minimum. One way of proceeding would be to find all possible maximum matchings and then to choose one with minimum total cost. However, this is not an efficient procedure: for a graph with a large number of vertices, it would take a long time, even using a computer. What is needed is an efficient algorithm. Such an algorithm—called the *Hungarian algorithm*—is given below. (The algorithm is called 'Hungarian' in recognition of the Hungarian mathematician E. Egerváry who introduced the idea in 1931.)

For this example, the number of men is equal to the number of jobs, and the graph is a complete bipartite graph. Other assignment problems may involve incomplete graphs or different numbers of vertices in the two sets. The algorithm we give can accommodate such problems, although it may not be possible to assign all the men to jobs.

13.1 THE HUNGARIAN ALGORITHM FOR THE ASSIGNMENT PROBLEM

We first give a statement of the algorithm, and then illustrate its use by working through an example. We suggest that you read through the description given here before studying the worked problem.

The algorithm starts with no initial matching in the bipartite graph containing the sets of vertices $X = \{x_1, x_2, \ldots\}$ and $Y = \{y_1, y_2, \ldots\}$. It selects a set of edges of the graph in such a way that a matching involving these edges is at minimum cost—that is, it is not possible to find a matching of the same size involving other edges of the graph and with a lower total cost. The graph containing the original vertices, together with this set of edges, is an example of a **partial graph**. If the matching involving the largest possible number of edges in the partial graph is not a maximum matching of the original graph, then further edges are selected and a larger matching is found. This process is repeated until a maximum matching at minimum cost is found.

The algorithm is in four parts. Parts B and C are essentially the same as in the algorithm for finding a maximum matching given in Chapter 12.

DESCRIPTION OF THE HUNGARIAN ALGORITHM FOR THE ASSIGNMENT PROBLEM

START with no initial matching in a bipartite graph whose sets of vertices are $X = \{x_1, x_2, \ldots\}$ and $Y = \{y_1, y_2, \ldots\}$.

Part A: Construction of the Initial Partial Graph

STEP 1 Assign weights to the vertices as follows.

To each vertex of X, assign a weight equal to the lowest cost on any edge incident to that vertex.

Decrease the cost on each edge by the weight on the vertex to which it is incident.

(You can think of an original edge-cost as being the total cost of a trip from one vertex to the other. We are dividing this cost into three parts: the cost of leaving the vertex in X; the cost of the journey (the new edge-cost); and the cost of arriving at the vertex in Y.)

Repeat this procedure for each vertex in Y.

(When this procedure has been carried out, for any edge, the sum of the weights on the end-vertices plus the new cost is equal to the original cost.)

STEP 2 Form a partial graph consisting of the vertices of the original bipartite graph, together with only those edges whose current cost is zero.

Part B: The Labelling Procedure

STEP 3 Label with (∗) each vertex in X which is not incident with any edge of the current matching M.

(Note that the first time we meet Step 3, the current matching has no edges.)

If no such vertex exists, STOP.

Otherwise, go on to Step 4.

STEP 4 Choose a newly-labelled vertex in X, say x_i, and label with (x_i) all the unlabelled vertices in Y which are joined to x_i by an edge NOT IN M.

Repeat for all newly-labelled vertices in X, and then go on to Step 5.

STEP 5 Choose a newly-labelled vertex in Y, say y_i, and label with (y_i) all unlabelled vertices in X which are joined to y_i by an edge IN M. Repeat for all newly-labelled vertices in Y.

Repeat Steps 4 and 5 until

EITHER a demand vertex whose demand is not satisfied is labelled (this is called *breakthrough*), in which case go on to Part C, Step 6,

OR no more labelling is possible, in which case go on to Part D, Step 8.

Part C: Matching Improvement

STEP 6 Find an alternating path as follows. Start at the breakthrough vertex, and go to the vertex indicated by its label. From this vertex, go to the vertex indicated by *its* label, and so on, until a vertex labelled (∗) is reached. This path is an alternating path P.

STEP 7 Form a new matching from:

(i) the edges of the current matching M which are not in the path P;
(ii) the edges of the path P which are not in the current matching M.

Remove all labels and go back to Step 3 to see if the new matching can be improved further.

Part D: Modification of the Partial Graph

STEP 8 Find the lowest current cost on any edge of the original bipartite graph which

(i) starts at a LABELLED vertex in X;
(ii) ends at an UNLABELLED vertex in Y;
(iii) has a non-zero current cost.

Call this lowest cost δ.

STEP 9 (a) Increase the weight on each labelled vertex in X by δ.
(b) Decrease the weight on each labelled vertex in Y by δ.
(c) For each edge which joins a labelled vertex in X to an unlabelled vertex in Y, and whose current cost is not zero, decrease the current cost by δ.
(d) For each edge which joins an unlabelled vertex in X to a labelled vertex in Y, and whose current cost is not zero, increase the current cost by δ.

Part (c) above should have produced at least one more edge with current cost zero. Incorporate all such edges with zero cost in the partial graph, delete all labels, and return to Part B, Step 3.

(Note that in Step 9 no negative costs are produced in the cost matrix. Also, the total cost of a 'trip' from a vertex in X to a vertex in Y via any edge not in the partial graph—that is, the sum of the two vertex weights and the edge cost—is kept constant throughout Step 9.)

This completes the description of the Hungarian algorithm. It is instructive at this point to compare this algorithm and the algorithm for finding a maximum matching given in the previous chapter with two of the algorithms which we described earlier—the algorithm for finding a maximum flow (described in Chapter 3) and the max-flow min-cost flow algorithm (described in Chapter 7). As we pointed out in Chapter 12, the problem of finding a maximum matching in a bipartite graph is closely related to the problem of finding a maximum flow in a network. The process of finding an alternating path is essentially the same as that of finding a flow-augmenting path in the algorithm for finding a maximum flow. Similarly, the assignment problem is analogous to the problem of finding a maximum flow at minimum cost. The algorithms for these two problems find a flow or a matching at minimum cost, and then increase the flow or the matching in stages, using a procedure which ensures that the cost is a minimum at each stage. Such a procedure is called a **least-cost flow-augmenting cycle**.

Worked problem using the Hungarian algorithm for the assignment problem
Four applicants x_1, x_2, x_3 and x_4 apply for four jobs y_1, y_2, y_3 and y_4. The suitability of each applicant for each job is represented by a cost as given in the cost matrix below, such that the lower the cost, the more suitable the applicant.

	y_1	y_2	y_3	y_4
x_1	6	12	15	15
x_2	4	8	9	11
x_3	10	5	7	8
x_4	12	10	6	9

cost matrix

Find the optimum (that is, the least total cost) assignment of applicants to jobs.

Solution Since each applicant can do each job, the problem is represented by the complete bipartite graph $K_{4,4}$ shown below, where the cost associated with each edge is given in the cost matrix.

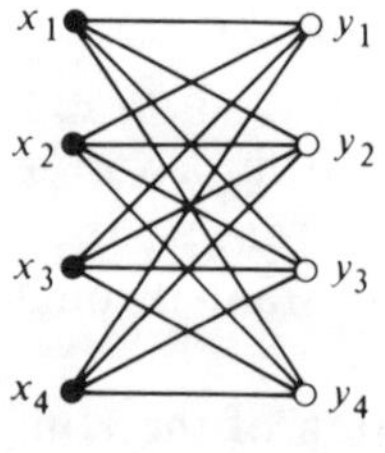

bipartite graph ($K_{4,4}$)

Part A of the algorithm requires us to construct an initial partial graph, starting with no initial matching. This initial partial graph is constructed so that a matching formed by using its edges is a lowest-cost matching, but not necessarily a complete matching.

In Step 1 we assign weights to all the vertices. We start with x_1, and find the lowest cost of any incident edge. Looking along the first row of the cost matrix, we see that the lowest cost is 6, associated with the edge x_1y_1. So we assign a weight of 6 to vertex x_1, and reduce the weight on each incident edge by 6. We repeat this procedure for each vertex in X. Similarly, by looking along the columns, we repeat this procedure for each vertex in Y, to obtain the revised cost matrix shown below.

weights →		0	0	0	3
↓		y_1	y_2	y_3	y_4
6	x_1	0	6	9	6
4	x_2	0	4	5	4
5	x_3	5	0	2	0
6	x_4	6	4	0	0

first revised cost matrix

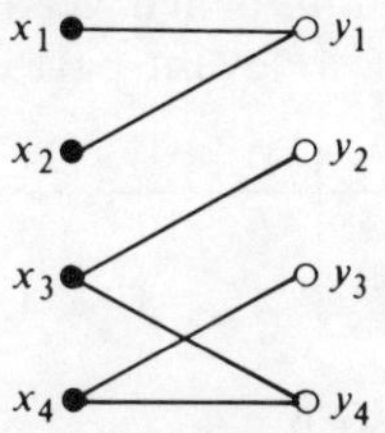

initial partial graph

From the revised cost matrix we construct the initial partial graph according to Step 2. The partial graph, shown above, consists of all the vertices, together with the edges of the revised cost matrix which have zero current cost.

It is useful for understanding the above procedure to think of an original cost as the cost of a trip from one vertex to another. In forming the revised costs, we are expressing the cost of the trip as the sum of three parts: the cost of leaving the vertex in X; the cost of the journey (the new edge cost) and the cost of arriving at the vertex in Y. During the algorithm, the edge costs and the vertex costs are adjusted, but (for edges not already in the current partial graph) the total cost of each trip remains the same.

We now continue with Parts B and C of the algorithm, which are essentially the same as in the algorithm for finding a maximum matching given in the previous chapter. We thus obtain a maximum matching, one possibility being shown by the thick lines in the following figure.

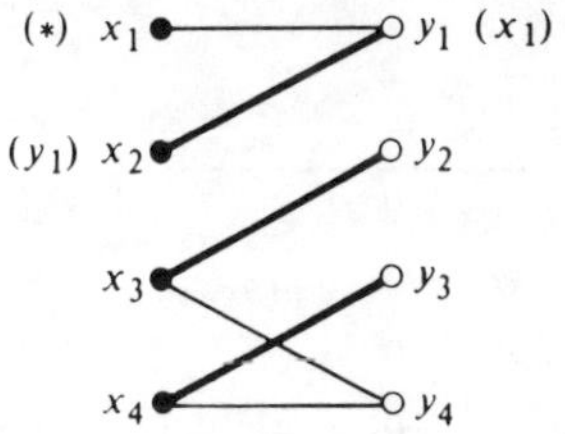

a maximum matching

(The other possible maximum matching contains edge x_1y_1 instead of edge x_2y_1.)

After this matching has been obtained (Step 7), the algorithm requires us to return to Part B and to label the graph, resulting in the labels shown on the diagram above. To proceed, we need to incorporate further edges into the partial graph in Part D. So we now go on to Part D. In Step 8 we find the lowest non-zero current cost associated with any edge of the original bipartite graph which joins a labelled vertex in X to an unlabelled vertex in Y. The labelled vertices in X are x_1 and x_2, so we need to look at the top two rows of the cost matrix. The unlabelled vertices in Y are all those except for y_1. We thus write down the relevant part of the cost matrix as follows:

	y_2	y_3	y_4
x_1	6	9	6
x_2	4	5	4

The lowest cost is clearly 4, so $\delta = 4$.

We now need to revise the current cost matrix according to the four instructions in Step 9. So we

(a) increase the weight on each labelled vertex in X by δ;
(b) decrease the weight on each labelled vertex in Y by δ;
(c) decrease by δ the cost on each edge joining a labelled vertex in X to an unlabelled vertex in Y;
(d) increase by δ the cost on each edge joining an unlabelled vertex in X to a labelled vertex in Y.

Note that, using our previous analogy of a trip, in Step 9 we maintain the cost of each trip from one vertex to another, but distribute the costs differently between vertices and edges. At each stage of revising the cost matrix, this process enables us to identify the next-cheapest edges which can be used in a matching. (Note that in general the cost of the trip is maintained for new edges, that is, for edges not yet in the partial graph, but not necessarily for edges already incorporated in the current partial graph.) Also, no negative costs are introduced on edges, but vertices in Y may have negative weights.

The second revised cost matrix is shown below:

		−4	0	0	3
		y_1	y_2	y_3	y_4
10	x_1	0	2	5	2
8	x_2	0	0	1	0
5	x_3	9	0	2	0
6	x_4	10	4	0	0

second revised cost matrix

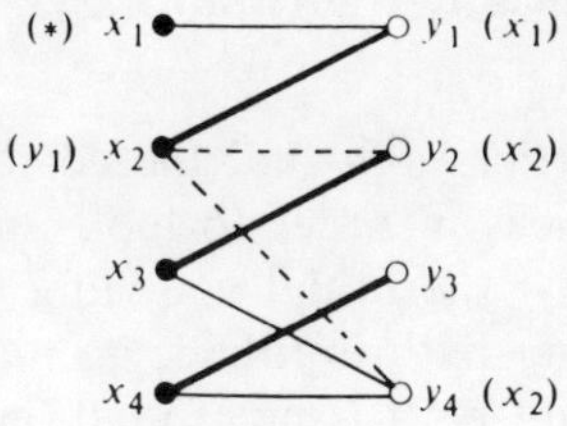

partial graph,
breakthrough at y_4

The final part of Step 9 is to form the revised partial graph (shown above) by incorporating all the additional edges which now have zero current cost. At least one such edge should have been produced by part (c) of Step 9, and in this particular example there are two new edges, x_2y_2 and x_2y_4.

We now return to Part B, beginning with Step 3, and carry out the labelling procedure again. The resulting labels are shown on the above partial graph. Breakthrough is achieved with vertex y_4, and so in Step 6 we find the alternating path $y_4x_2y_1x_1$. The revised matching formed in Step 7 by replacing the thick edge x_2y_1 in the alternating path by the two edges x_1y_1 and x_2y_4 is shown below.

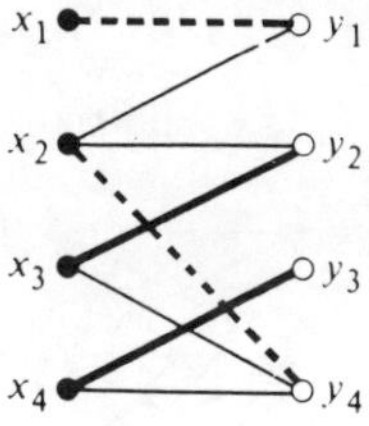

optimum assignment: x_1y_1, x_2y_4, x_3y_2, x_4y_3

This is the optimum assignment, and all that remains is to find the cost. We can do this in two ways. We could go back to the original cost matrix and add up the costs of all the edges in the matching, giving $6 + 11 + 5 + 6 = 28$. Or we can go to the last revised cost matrix and add the weights of the vertices incident with the edges of the matching. This works because the edges of the matching all have zero weight. Since all the vertices are incident to an edge of the matching, we simply add the weights of all the vertices to give

$$\text{total cost} = 10 + 8 + 5 + 6 - 4 + 3 = 28.$$

We can summarize the algorithm as follows.

Summary of the Hungarian Algorithm

START with no matching.

PART A Weight the vertices and find the corresponding partial graph.

PART B Label the vertices in order to look for an alternating path.
(First time only: the initial matching has no edges.)
If an alternating path is found, go on to Part C.
If an alternating path is not found, go on to Part D.
STOP when all the vertices on one side of the graph are incident with an edge in the matching.

PART C Modify the matching.
Return to Part B.

PART D Modify the weights and find a new partial graph.
Return to Part B.

Problem 13.1 Use the Hungarian algorithm to find the optimum assignment and its total cost for the case represented by the following cost matrix and bipartite graph.

	y_1	y_2	y_3	y_4
x_1	8	4	6	7
x_2	10	12	8	14
x_3	9	6	11	15
x_4	12	8	14	8

cost matrix

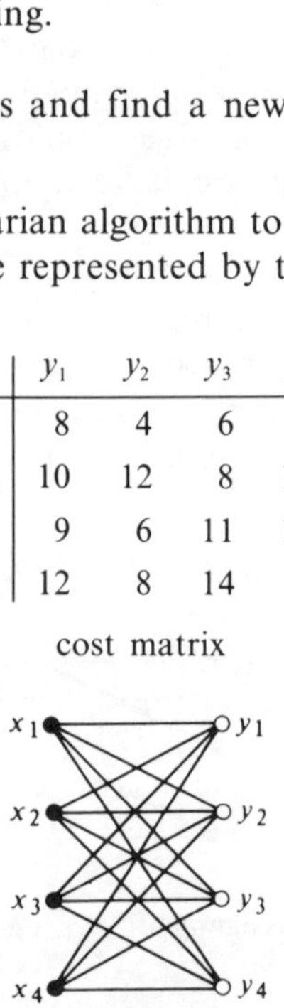

bipartite graph

Remark Some people prefer not to draw the partial graphs when working through the algorithm and instead record everything on the cost matrices. This is very simple to do; the edges of the partial graph correspond to zero entries in the cost matrix, and those zero entries in the matrix corresponding to a matching in the partial graph are indicated by underlining them or printing them in bold. The labels required during the labelling procedure can also be indicated on the matrix by placing them by the vertex labels. This method is illustrated in the solution to the above problem; the partial graphs are also included to show the correspondence between the two methods of recording the steps of the algorithm.

13.2 VARIATIONS ON THE ASSIGNMENT PROBLEM

Many problems which occur in practice are variations of the basic assignment problem, and can be solved by modifying the Hungarian algorithm. We now give two examples of such problems and show how the basic algorithm can be adapted to solve them.

The Bottle-neck Assignment Problem

Suppose that a product is to be made on a serial production line which involves a number of activities, each of which must be completed once to produce one item of the product. A number of people are to be assigned to these activities. If we know how long it takes each person to complete each activity, how can we assign the people to the activities so that the time taken to produce one item of the product is as short as possible?

('The time taken to complete one item of the product' is ambiguous here. What is meant is the time-interval between two items rolling off the production line. This is not the total sum of the times for the various activities, because the workers are working on the activities at the same time.)

It is clear that the production line can operate only as fast as the rate of working of the slowest person on the line—the bottle-neck. What we need to do, therefore, is to find an assignment of people to activities such that the longest time taken by any person to do his assigned activity is a minimum. In the following example, we show how to solve such a problem by modifying the algorithm for the assignment problem so that, instead of finding a maximum matching at minimum total cost, we find a maximum matching for which the largest individual cost (in this case the time taken by the slowest worker) is a minimum.

Worked problem Suppose that three people are to be assigned to three activities on a production line, and that the times taken for the people to do the various jobs are given by the following cost matrix:

		activities		
		y_1	y_2	y_3
	x_1	3	2	2
people	x_2	3	4	3
	x_3	2	2	1

How can we assign the three people to the three activities in such a way that the production line can operate as quickly as possible?

Solution We denote the time taken to complete one item of the product by p, so that, in a particular assignment of people to activities, p is the longest time taken by any one person to do his or her assigned activity. At each stage of the modified algorithm for this problem, we need to keep account only of the parameter p. This makes the procedure considerably simpler than that of the algorithm for the basic assignment problem, where at each stage we had to keep account of *total cost*, which we did by assigning weights to all the vertices of the graph.

Note that when we have a partial assignment, the current value of p will be the longest time taken by any person to do the activities assigned to him or her so far.

The lowest cost in the above cost matrix is 1, corresponding to worker x_3 doing activity y_3. We set p initially to this value, because it cannot be lower, and subtract 1 from all the entries in the cost matrix.

This gives a new cost matrix from which we can construct the corresponding partial graph, as we did in the assignment problem, using only those entries in the cost matrix which are zero. The new cost matrix and partial graph are:

	y_1	y_2	y_3	
x_1	2	1	1	
x_2	2	3	2	$p = 1$
x_3	1	1	0	

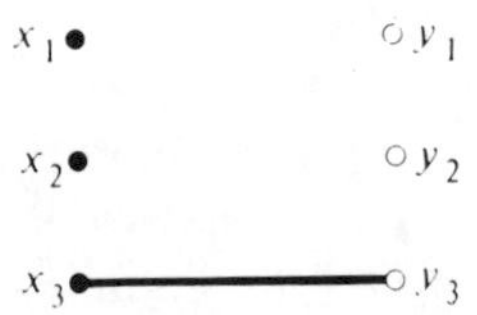

A complete assignment has not been found, so we must enlarge the partial graph. The minimum of all the non-zero costs is now 1, so we reduce the costs of all non-zero entries in the matrix by 1 and increase the value of p by the same amount. This gives a new cost matrix and partial graph:

	y_1	y_2	y_3	
x_1	1	0	0	
x_2	1	2	1	$p = 2$
x_3	0	0	0	

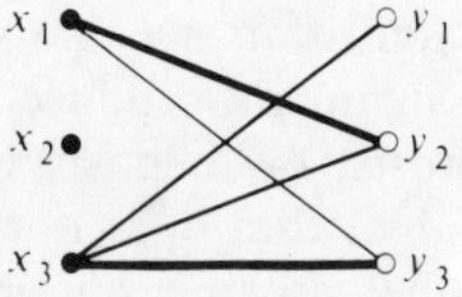

Again, a complete assignment has not been found, so we continue updating the cost matrix, finding the minimum non-zero entry (again 1) and reducing each non-zero entry by this amount and increasing p by the same amount. Our new cost matrix and partial graph are:

	y_1	y_2	y_3	
x_1	0	0	0	
x_2	0	1	0	$p = 3$
x_3	0	0	0	

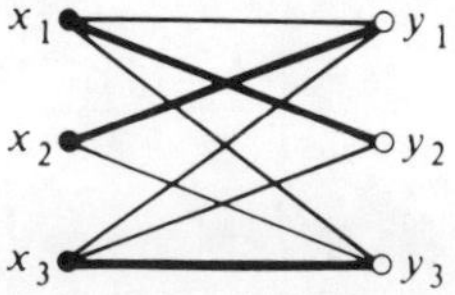

The resulting partial graph shows that a complete assignment has now been made, and so the minimum time for the completion of one item is 3 (the final value of p). Note that in this case there are other possible optimum matchings of the final partial graph corresponding to alternative optimum assignments.

The Cabinet Reshuffle Problem

Suppose that a Prime Minister wishes to reshuffle the Cabinet which consists (apart from himself) of k ministers holding k posts. He knows the preferences of each minister for each post, and he would like to reshuffle the Cabinet so that each minister is assigned to a post which he prefers to his original one. Such an ideal reshuffle may not be possible in a particular case, depending on the preferences of the ministers for the posts. What we need to solve this problem is a procedure which either finds an ideal reshuffle, or, if this is not possible, finds a new assignment in which the maximum number of ministers are assigned to posts which they prefer.

We assume that a minimum-cost solution is required, where the costs in this case are numbers representing the preferences of the ministers for the posts, a lower cost indicating a higher preference.

The assignment algorithm can be used to solve this type of problem, but a single application of the algorithm may not produce an ideal reshuffle, since the minimum-cost assignment produced may assign some of the ministers to their original posts, or to posts which are not preferable. If the first application of the algorithm does not produce an ideal reshuffle, we must use it to provide other possible minimum-cost assignments. If an ideal reshuffle is still not obtained, we must increase the values of some of the costs representing the preferences until either an ideal reshuffle is obtained, or we demonstrate that one is not possible, in which case we accept a minimum-cost assignment in which the maximum number of ministers have posts they prefer.

We illustrate the procedure in the following worked problem. You may like to attempt the problem yourself before reading the solution.

Worked problem The preferences for posts of the ministers are shown in the following cost matrix, where a lower cost indicates a higher preference:

		posts			
		p_1	p_2	p_3	p_4
ministers	m_1	3	2	1	4
	m_2	2	4	1	3
	m_3	1	3	4	2
	m_4	3	4	1	2

In the original assignment of ministers to posts, minister m_1 held post p_1, minister m_2 held post p_2, and so on.

Use the assignment algorithm to attempt to reshuffle the Cabinet so that each minister has a post he prefers to his original post. Look for a new assignment which satisfies this requirement at minimum total cost.

Solution We apply the assignment algorithm to the problem. The first step is to assign weights to the vertices of the graph as shown in the following diagram:

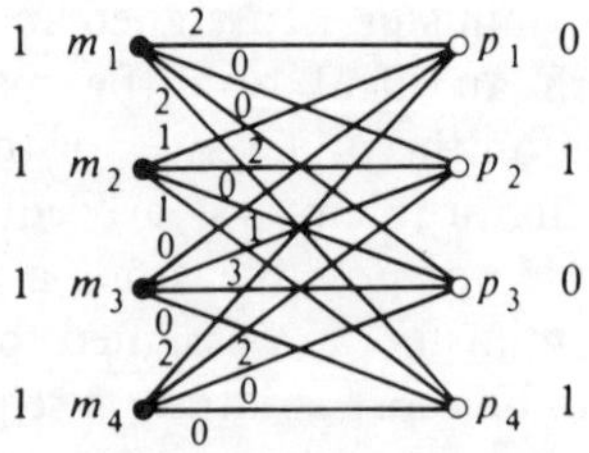

The cost matrix and the resulting partial graph then become:

		0	1	0	1
		p_1	p_2	p_3	p_4
1	m_1	2	0	0	2
1	m_2	1	2	0	1
1	m_3	0	1	3	0
1	m_4	2	2	0	0

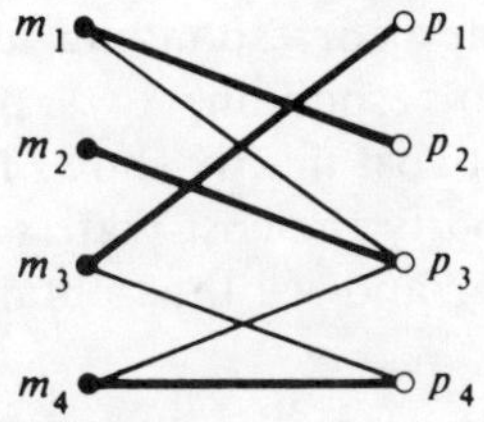

Only one assignment involving four edges is possible, and this is indicated by the thick lines on the partial graph. The cost of this assignment is equal to the sum of the weights of the vertices—that is, $1 + 1 + 1 + 1 + 1 + 1 = 6$. However, this assignment is not an ideal reshuffle, since minister m_4 is assigned to his original post.

We must now look for an assignment with higher total cost which satisfies the requirement that every minister has a post he prefers to his original post. To do this, we increase the cost representing the preference of minister m_4 for post p_4 by 1, and increase the costs of all posts for which his preference is lower than for p_4 by 1. (A lower preference corresponds to a higher cost.)

We now repeat the algorithm. The new cost matrix after weights have been assigned to the vertices is as follows:

		0	1	0	1
		p_1	p_2	p_3	p_4
1	m_1	2	0	0	2
1	m_2	1	2	0	1
1	m_3	0	1	3	0
1	m_4	3	3	0	1

We can now draw the corresponding partial graph, find a maximum matching, and label the vertices in accordance with the algorithm, as shown below. (Note that several other 3-edge matchings are possible.)

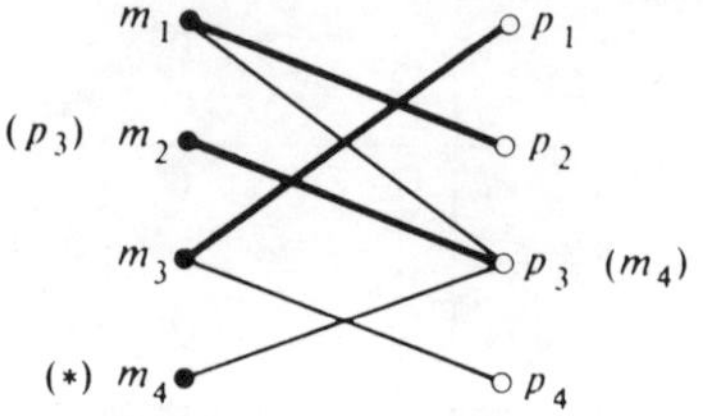

Here no breakthrough is possible, so we must again modify the cost matrix. We must first find the lowest non-zero current cost of any edge which joins a labelled vertex corresponding to a minister (here m_2 or m_4) to an unlabelled vertex corresponding to a post (here any vertex except p_3). A glance at the current cost matrix shows that the lowest of these costs is 1. We may therefore modify the cost matrix and partial graph according to Step 9 of the algorithm, and we thus obtain:

		0	1	−1	1
		p_1	p_2	p_3	p_4
1	m_1	2	0	0	2
2	m_2	0	1	0	0
1	m_3	0	1	4	0
2	m_4	2	2	0	0

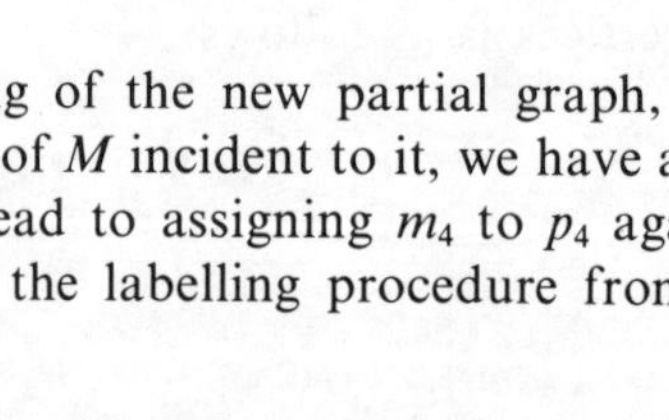

In the initial labelling of the new partial graph, we label p_4 with (m_4), and, as p_4 has no edges of M incident to it, we have a case of breakthrough. However, this would lead to assigning m_4 to p_4 again, which we want to avoid, so we continue the labelling procedure from p_3 to look for other possible assignments:

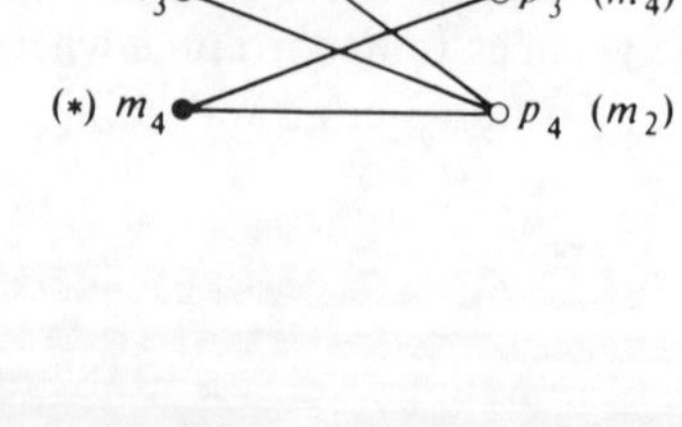

The labelling of p_4 with m_2 provides an alternative breakthrough. Backtracking, we find an alternating path $p_4m_2p_3m_4$. This produces the following improved matching:

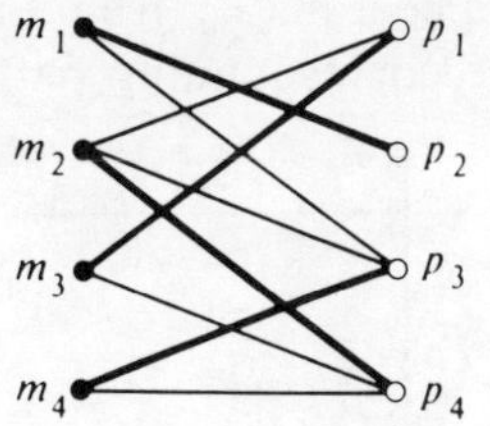

This is an ideal reshuffle, since every minister now has a post he prefers to his original one. The total cost of this assignment is equal to the sum of the weights of all the vertices—that is, $1 + 2 + 1 + 2 + 1 - 1 + 1 = 7$.

Another minimum-cost solution with a total cost of 7 is also possible. This is shown in the following graph. This solution is obtained if a different choice of initial matching is made in the second application of the assignment algorithm.

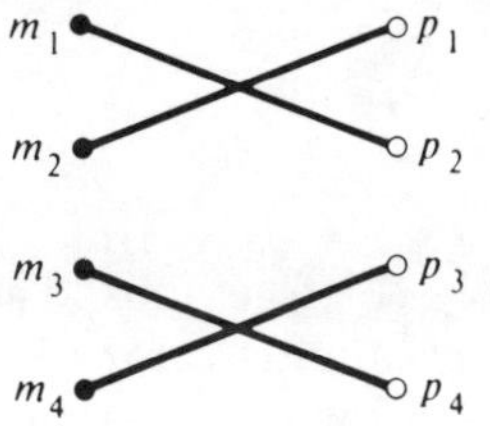

In the above problem we obtained a solution in which every minister is assigned to a post he prefers to his original post. In general, such an ideal reshuffle may not be possible. As the algorithm progresses, we can tell whether or not an ideal reshuffle is possible by examining the cost of the assignments produced. For example, in the problem above, the total cost of the original assignment is $3 + 4 + 4 + 2 = 13$. If, in a new assignment, every minister has a post he prefers to his original one, then the total cost of this assignment cannot be greater than

(the total cost of the original assignment) – (the number of ministers)

—that is, $13 - 4 = 9$. Thus if we had arrived at a minimum-cost assignment with total cost greater than 9, it would have meant that there was no ideal reshuffle possible and we would have had to accept a minimum-cost solution as close as possible to an ideal reshuffle.

EXERCISES 13

The Assignment Problem

13.1. Construct the revised cost matrix from the original cost matrix given below, and draw the corresponding partial graph.

	b_1	b_2	b_3	b_4
A_1	4	6	8	3
A_2	2	5	1	4
A_3	6	5	3	2

13.2. For the following partial graph and current cost matrix, find the lowest current cost δ and construct the revised cost matrix.

		3	1	0	2	4
		b_1	b_2	b_3	b_4	b_5
5	A_1	0	0	6	4	3
6	A_2	0	0	4	6	2
2	A_3	5	4	0	0	0

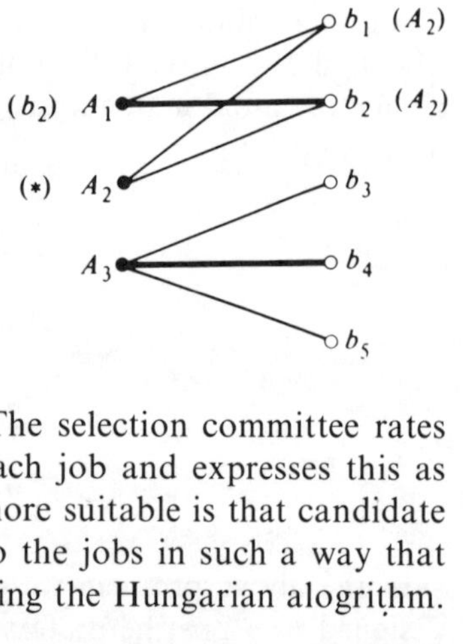

13.3. Four candidates apply for four jobs. The selection committee rates the suitability of each candidate for each job and expresses this as a cost, where the lowest the cost, the more suitable is that candidate for that job. Allocate the candidates to the jobs in such a way that the total cost is as small as possible, using the Hungarian alogrithm. The costs are as follows:

		jobs			
		b_1	b_2	b_3	b_4
applicants	A_1	3	1	5	10
	A_2	7	4	2	3
	A_3	8	8	5	2
	A_4	8	8	5	2

13.4. The minutes of a business meeting must be typed, photocopied, collated and dispatched to all the participants. These tasks, denoted by w, x, y and z respectively, are to be carried out by Alice, Bill, Carol and Don, each of whom can do any of the tasks. However, they are not equally good at performing each task. The time taken by each person to do each of the four tasks is given in the following table:

tasks

persons		w	x	y	z
	A	7	2	3	5
	B	8	3	5	2
	C	5	5	6	3
	D	3	4	2	5

Use the Hungarian algorithm to allocate each person to a task in such a way that the total time taken is a minimum.

13.5. Show how the assignment problem can be treated as a maximum-flow minimum-cost problem, and hence solved by the maximum-flow minimum-cost algorithm.

The Bottle-neck Assignment Problem

13.6. Four people are to be assigned to four activities on a serial production line. The times taken for the people to do the various jobs are as follows:

activities

people		b_1	b_2	b_3	b_4
	A_1	3	1	5	10
	A_2	7	4	2	3
	A_3	8	8	5	2
	A_4	8	8	5	2

Use an appropriate algorithm to assign the four people to the four activities in such a way that the production line can operate as quickly as possible.

13.7. If the same four people as in Exercise 13.4 process the minutes of several meetings, then the time taken between the dispatch of one set of minutes and the dispatch of the next set is effectively the longest time taken by any one of them to do the appointed task.

Use an appropriate algorithm (and the table in Exercise 13.4) to assign the people to tasks so as to minimize this longest time.

The Cabinet Reshuffle Problem

13.8. A company has four senior managers whose preferences for four posts are shown in the following table, in which a lower cost indicates a higher preference.

		posts			
		p_1	p_2	p_3	p_4
managers	m_1	4	3	2	1
	m_2	4	3	1	2
	m_3	2	4	3	1
	m_4	3	4	1	2

Originally, manager m_1 held post p_1, manager m_2 held post p_2, and so on.

Use an appropriate algorithm to attempt to reshuffle the posts so that each manager has a post which he prefers to his original one. Find a new allocation which satisfies this requirement at minimum total cost.

CHAPTER 14

THE TRANSPORTATION PROBLEM

After reading this chapter, you should be able to:

- explain what is meant by *the transportation problem* and *the transhipment problem*;
- explain briefly the procedures used in the Hungarian algorithm and in the cost-reducing cycle algorithm for the transportation problem;
- find minimum-cost solutions to simple examples of the transportation and transhipment problems by applying the Hungarian algorithm.

Introduction

The well-known transportation problem, sometimes called the *Hitchcock problem* after F. L. Hitchcock, who studied the problem in 1941, can be stated as follows:

THE TRANSPORTATION PROBLEM

A manufacturer has a number of factories, each of which can supply a particular product to a number of warehouses. Each factory has a fixed output, and there is a transportation cost involved in sending the product from each factory to each warehouse. To which warehouses should the products of each factory be sent so that the requirements of all the warehouses are satisfied at minimum total cost?

An example of the transportation problem involving three factories and four warehouses is represented by the following bipartite graph. The costs of supplying each warehouse from each factory are given in the cost matrix below the graph.

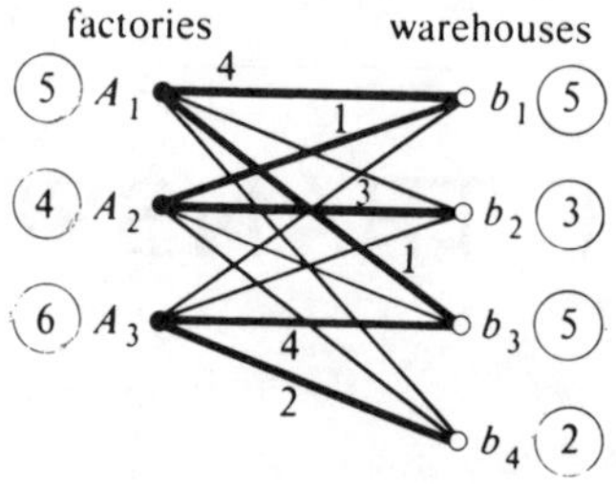

demand

			⑤ b_1	③ b_2	⑤ b_3	② b_4
	⑤	A_1	5	4	7	6
supply	④	A_2	2	5	3	2
	⑥	A_3	6	3	4	4

In the bipartite graph, the factories A_1, A_2 and A_3 are represented by the black vertices, which we call *supply vertices*. The warehouses b_1, b_2, b_3 and b_4 are represented by the white vertices, which we call *demand vertices*.

The circled numbers next to the supply and demand vertices on the graph, and also on the cost matrix, are the numbers of units of the product supplied by the factories or demanded at the warehouses. In this example, the total supply and the total demand are both equal to 15 units. Note that for a solution to be possible, the total supply must be at least as large as the total demand.

A flow pattern which satisfies the supply and demand constraints is indicated by the thick edges of the bipartite graph. The number beside each of these edges is the number of units of the product to be sent along the route represented by that edge. The other edges represent unused routes. We can calculate the total transportation cost for this flow pattern by multiplying the amount of flow along each edge by the cost associated with that edge. Thus the total cost is

$$(4 \times 5) + (1 \times 7) + (1 \times 2) + (3 \times 5) + (4 \times 4) + (2 \times 4) = 68.$$

Although this flow pattern satisfies the flow constraints, it is not a minimum-cost solution. We shall find a minimum-cost solution by using an algorithm in Problem 14.2.

The transportation problem is closely related to the assignment problem which we discussed in the previous chapter. If, in the assignment problem, we interpret the assignment of a worker to a job as being a 'flow' of one unit between the corresponding vertices of the bipartite graph representation, then we see that the problem is a special case of the transportation problem in which the supplies and demands at the vertices are all 1. The transportation problem is more complicated, because we need to specify the amount of flow along each edge included in a solution, and also because a single supply vertex (factory) may send goods to more than one demand vertex (warehouse).

The Hungarian algorithm for the transportation problem which we give in this chapter is based on the algorithm for the assignment problem.

14.1 THE HUNGARIAN ALGORITHM FOR THE TRANSPORTATION PROBLEM

Before setting down the detailed procedure of the algorithm, we describe in general terms how the method works.

The first step is to construct a partial graph, using the same procedure as in the algorithm for the assignment problem. We then assign the maximum possible flow to the edges of the partial graph, subject to the supply and demand constraints at each vertex. The way in which the partial graph is constructed ensures that this is a minimum-cost flow—in other words, we cannot obtain the same value of total flow at lower cost by using edges not in the partial graph. If the flow thus obtained does not satisfy the demands at all the demand vertices, then we enlarge the partial graph, again using the same procedure as in the algorithm for the assignment problem. We can then increase the value of the total flow by using the new edges. We continue in this way until we eventually obtain a minimum-cost solution in which all the demands are satisfied. It is important to note that, as with the assignment problem, there may be other minimum-cost solutions.

The Hungarian algorithm thus builds up the flow in stages, ensuring that at each stage the total flow is at minimum cost—thus it uses a least-cost flow-augmenting cycle.

In the Hungarian algorithm for the transportation problem, we are not seeking a matching, as we were in the algorithm for the assignment problem. (This is because there may be flows from one supply vertex to a number of demand vertices.) So, instead of looking for alternating paths, we look for paths in which there is a flow on even edges—that is, on the 2nd, 4th, 6th, . . . edges. There may or may not be a flow on odd edges of the path. Such a path is analogous to a flow-augmenting path used in the algorithm for finding a maximum flow discussed in Chapter 3.

We suggest that you read quickly through the following description of the algorithm before studying the Worked Problem.

DESCRIPTION OF THE HUNGARIAN ALGORITHM FOR THE TRANSPORTATION PROBLEM

START with no flow in a bipartite graph whose vertices are divided into two sets: the supply vertices and the demand vertices.

Part A: Construction of the Initial Partial Graph

STEP 1 Assign weights to the supply and demand vertices as follows.
To each supply vertex, assign a weight equal to the lowest cost on any edge incident to that vertex.
Decrease the cost on each edge by the weight on the supply vertex to which it is incident.
Repeat this procedure for each demand vertex.

STEP 2 Form a partial graph consisting of the vertices of the original bipartite graph, together with only those edges whose current cost is zero.

Part B: The Labelling Procedure

STEP 3 Label with (*) each supply vertex whose supply has not all been allocated to edges.
If no such vertex exists, STOP.
Otherwise, go on to Step 4.

STEP 4 Choose a newly-labelled supply vertex A_i, and label with (A_i) all unlabelled demand vertices which are joined by an edge of the partial graph.
Repeat for all newly-labelled supply vertices, and then go on to Step 5.

STEP 5 Choose a newly-labelled demand vertex b_j, and label with (b_j) all unlabelled supply vertices which are joined to b_j by an edge which has been allocated a flow.
Repeat for all newly-labelled demand vertices.

Repeat Steps 4 and 5 until

EITHER a demand vertex whose demand is not satisfied is labelled (this is called *breakthrough*), in which case go on to Part C, Step 6,

OR no more labelling is possible, in which case go on to Part D, Step 8.

Part C: The Flow-augmenting Procedure

STEP 6 Start with the breakthrough vertex, and backtrack through the labels until a supply vertex which is labelled (∗) is reached. A flow-augmenting path has been found.

(The flow-augmenting path plays a similar role in this algorithm to that of the alternating path in the algorithm for the assignment problem.)

Calculate the maximum flow F which can be sent along this path as follows.

First, find the minimum of all flows assigned to even edges of the path (that is, the 2nd, 4th, . . . , edges). This is the maximum possible backward flow.

Next, find the minimum value of:

- the available supply at the start vertex of the path;
- the required demand at the end vertex;
- the maximum backward flow.

This is the required value of F.

(Note that for a flow-augmenting path with just one edge, there is no backward flow, so we just take the minimum of the supply and the demand.)

STEP 7 Form a new flow pattern as follows.

(i) Increase the flows on the odd edges of the path by this value of flow, F.
Decrease the flows on the remaining edges by F.

(ii) Decrease the available supply at the start vertex by F.
Decrease the demand at the end vertex by F.

Remove all the labels and return to Part B, Step 3.

Part D: Modification of the Partial Graph

STEP 8 Find the lowest current cost on any edge of the original bipartite graph which

(i) starts at a LABELLED supply vertex;
(ii) ends at an UNLABELLED demand vertex;
(iii) has a non-zero current cost.

Call this lowest cost δ.

STEP 9 (a) Increase the weight on each labelled supply vertex by δ.
(b) Decrease the weight on each labelled demand vertex by δ.
(c) For each edge which joins a labelled supply vertex to an unlabelled demand vertex, and whose current cost is not zero, decrease the current cost by δ.

(d) For each edge which joins an unlabelled supply vertex to a labelled demand vertex, and whose current cost is not zero, increase the current cost by δ.

Incorporate all edges which now have a current cost of zero in the partial graph, delete all labels, and return to Part B, Step 3.

Worked problem using the Hungarian algorithm for transportation problem Two factories A_1 and A_2 can supply a certain product to three warehouses b_1, b_2 and b_3 with transportation costs as shown in the cost matrix below. The number of units of the product which can be supplied by each factory, and the number of units required by each warehouse are shown by circled numbers. Which factories should supply which warehouses for minimum total transportation cost?

demand

			④ b_1	⑥ b_2	⑤ b_3
supply	⑥	A_1	3	5	7
	⑨	A_2	6	4	3

cost matrix

Solution Notice that the total supply is equal to the total demand, so a solution is possible. The bipartite graph for the problem is:

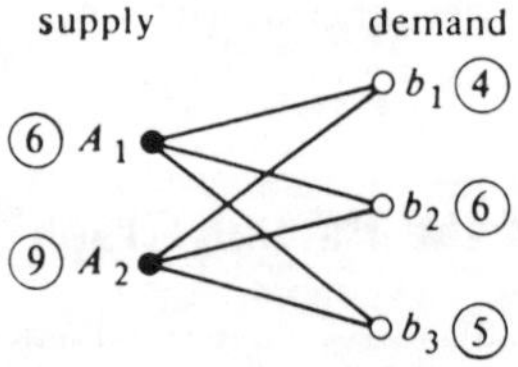

bipartite graph

Part A: Construction of the Initial Partial Graph

These first two steps of the algorithm are the same as the first two steps of the assignment algorithm. We revise the cost matrix by assigning weights to each vertex and reducing the edge costs. Note that in this process, no edge cost becomes negative, and the sum of each new edge cost plus the weights on the two incident vertices is equal to the original edge cost.

STEP 1 The costs on the edges incident to A_1 are 3, 5 and 7. The lowest of these is 3, so we assign a weight of 3 to A_1.

Similarly, we assign a weight of 3 to A_2. So we reduce all edge costs by 3.

			④	⑥	⑤
			b_1	b_2	b_3
3	⑥	A_1	0	2	4
3	⑨	A_2	3	1	0

When we reduce the edge costs, there are zeros in the columns corresponding to b_1 and b_3, so the weights for these vertices are zero. The costs in the second column are 2 and 1 at this stage, so the weight for b_2 is 1. We then have to further reduce the edge costs in this column by 1, giving the first revised cost matrix shown below.

weights →			0	1	0
↓			④	⑥	⑤
			b_1	b_2	b_3
3	⑥	A_1	0	1	4
3	⑨	A_2	3	0	0

first revised cost matrix

STEP 2 The first partial graph, containing only those edges for which the current cost is zero, is shown below.

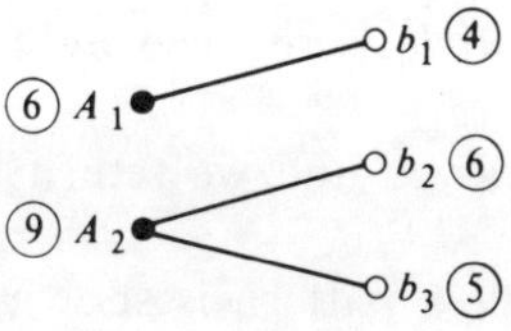

first partial graph

We go on to Part B, Step 3.

Part B: The Labelling Procedure (*first time*)

This part is slightly different from the labelling procedure for the assignment algorithm, because we are now dealing with flows rather than with matchings, and so look for a flow-augmenting path instead of an alternating path.

STEP 3 We label with (∗) all the supply vertices whose supply has not been all allocated; so to start with we label both A_1 and A_2.

STEP 4 We choose a newly-labelled supply vertex and label all demand vertices joined to it by an edge in the partial graph. Choose A_1, and label vertex b_1 with A_1, as shown in the following partial graph.

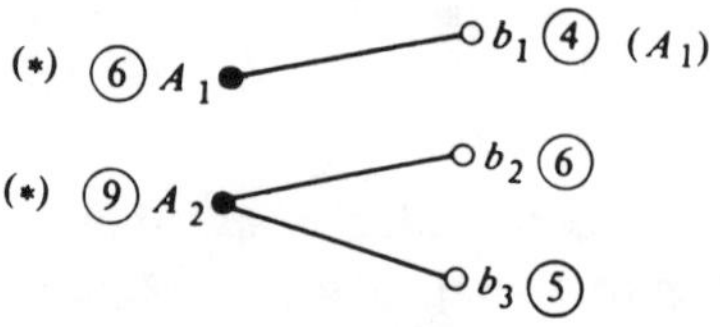

first labelling: breakthrough at b_1

Now b_1 is a demand vertex whose demand is not satisfied, so breakthrough has been achieved, and we go on to Part C, Step 6.

Part C: The Flow-augmenting Procedure (*first time*)

STEP 6 The flow-augmenting path is simply the edge A_1b_1.
The maximum flow F which can be sent along this path is 4 units, since this is the minimum of the supply at A_1 and the demand at b_1.

STEP 7 (i) We send a flow of 4 units along A_1b_1.
(ii) The available supply at A_1 is decreased by 4 units to 2 units. The demand at b_1 is reduced by 4 units to zero.

All labels are removed and we return to Part B, Step 3.

The current flow pattern is shown below.

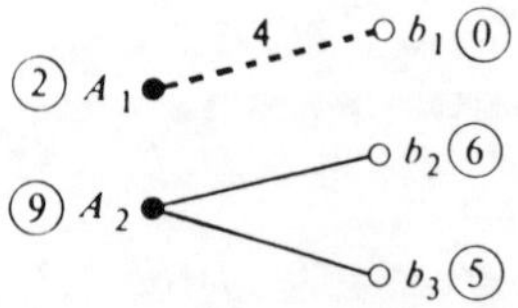

first modified flow pattern

Part B: The Labelling Procedure (*second time*)

STEP 3 Label A_1 and A_2 with (*).
STEP 4 Label b_1 with (A_1).

Label b_2 and b_3 with (A_2).
This labelling is shown on the following figure.

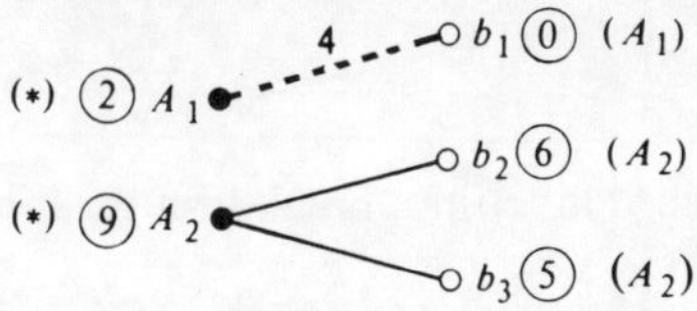

second labelling: breakthrough at b_2

Both b_2 and b_3 are labelled demand vertices whose demand is not satisfied—we choose one of them, say b_2, as a case of breakthrough, and go on to Part C, Step 6.

Part C: The Flow-augmenting Procedure (*second time*)

STEP 6 The flow-augmenting path is simply A_2b_2.
There is a supply of 9 at A_2, and a demand of 6 at b_2, and no backward flows, so the maximum flow which can be sent along this path is 6 units.

STEP 7 (i) We send 6 units along A_2b_2.
(ii) Reduce the supply at A_2 to 3 units and reduce the demand at b_2 to zero units.

The current flow pattern is shown below.

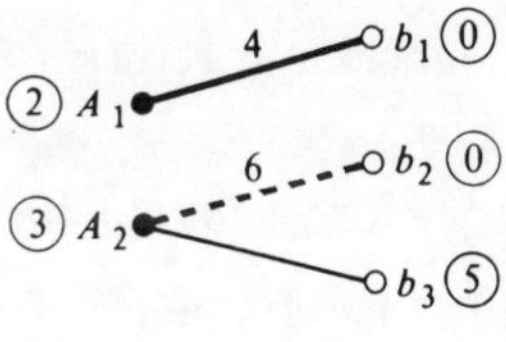

second modified flow pattern

We remove all the labels and return to Part B, Step 3.

Part B: The Labelling Procedure (*third time*)

STEP 3 Label A_1 and A_2 with $(*)$.
STEP 4 Label b_1 with A_1.
Label b_2 and b_2 with A_2.
This labelling is shown on the following figure.

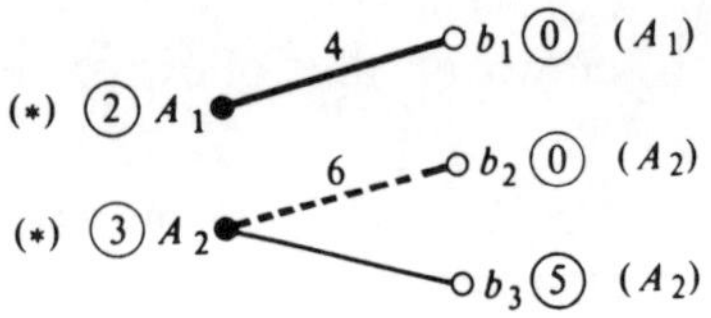

third labelling: breakthrough at b_3

Breakthrough is achieved with vertex b_3, so we go on to Part C, Step 6.

Part C: The Flow-augmenting Procedure (*third time*)

STEP 6 The flow-augmenting path is simply A_2b_3.
The supply of 3 units at A_2 limits the possible flows, so the maximum flow is 3.

STEP 7 (i) We send a flow of 3 units along A_2b_3.
(ii) We reduce the supply at A_2 and demand at b_3 by 3 units.
The resulting flow pattern is shown below.

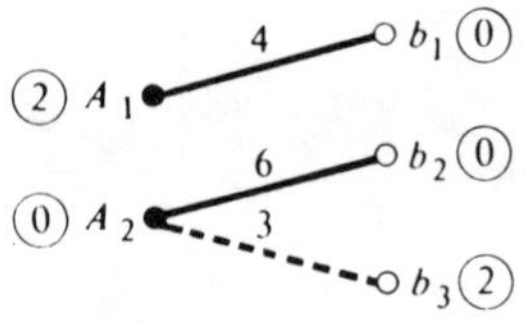

third modified flow pattern

We remove all the labels and return to Part B, Step 3.

Part B: The Labelling Procedure (*fourth time*)

This time we can label only A_1 with a (*), since the remaining supply at A_2 is now zero. We then label b_1 with A_1, and no more labelling is possible.

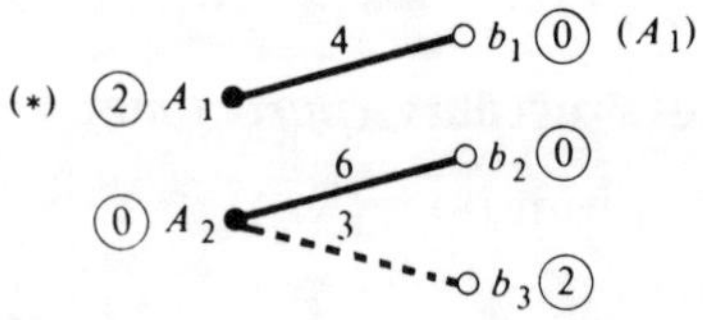

final labelling: no breakthrough

Breakthrough has not been achieved, so we go on to Part D, Step 8 to revise the partial graph.

Part D: Modification of the Partial Graph (*first time*)

At this stage, we need to incorporate more edges into the partial graph to satisfy the remaining demand of 2 units at b_3. We do this by revising the cost matrix.

STEP 8 We need to find the lowest non-zero current cost δ on an edge which starts at a labelled supply vertex and ends at an unlabelled demand vertex.

weights →			0	1	0
↓			④	⑤	⑥
			b_1	b_2	b_3
3	⑥	A_1	0	1	4
3	⑨	A_2	3	0	0

first revised cost matrix

Looking at the current cost matrix shown above, we see that there are two edges satisfying the above condition, with costs of 1 and 4. The lowest of these is 1, so $\delta = 1$.

STEP 9 (a) We increase the weight on labelled supply vertices by δ; that is, we increase the weight on A_1 to 4.

(b) We decrease the weight on labelled demand vertices by δ; that is, we decrease the weight on b_1 to -1.

(c) We decrease the cost by δ on each edge joining a labelled supply vertex to an unlabelled demand vertex, provided this cost is not already zero. So we decrease the costs of A_1b_2 and A_1b_3 (as shown in the first revised cost matrix) to 0 and 3 (as shown in the second revised cost matrix below).

(d) We increase by δ the cost on each edge joining an unlabelled supply vertex to a labelled demand vertex, provided this cost is not already zero. So we increase the cost on A_2b_1 from 3 to 4.

			−1	1	0
			⓪	⓪	②
			b_1	b_2	b_3
4	②	A_1	0	0	3
3	⓪	A_2	4	0	0

second revised cost matrix

The second revised cost matrix produced by Step 9 is shown above. There is an additional edge with zero cost, A_1b_2, so we incorporate this to produce the revised partial graph shown below.

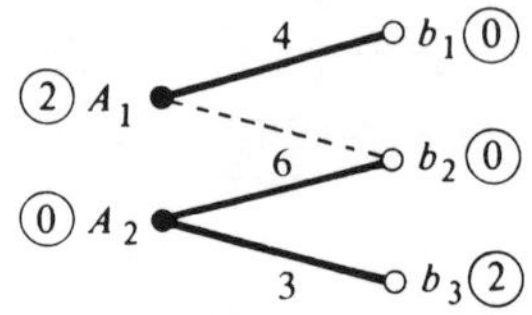

new partial graph

We now remove all the labels and return to Part B, Step 3.

Part B: The Labelling Procedure (*fifth time*)

STEP 3 Label A_1 with (∗).
STEP 4 Label b_1 and b_2 with (A_1).
STEP 5 Label A_2 with (b_2).
STEP 4 Label b_3 with (A_2).

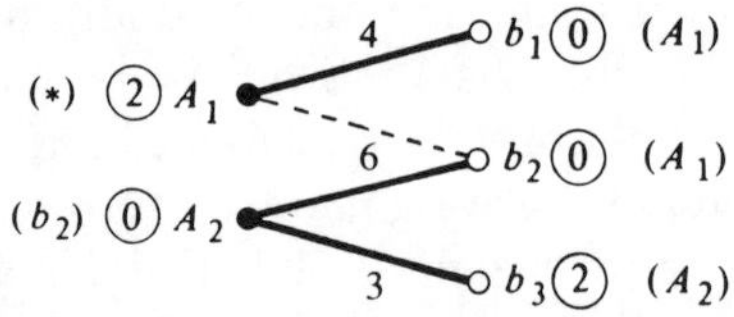

new partial graph with first labelling:
breakthrough at b_3

We have achieved breakthrough at b_3, so we go on to Part C, Step 6.

Part C: The Flow-augmenting Procedure (*fourth time*)

STEP 6 Backtracking through the labels, we find the flow-augmenting path $b_3A_2b_2A_1$.
We note that there is a supply of 2 at A_1, a demand of 2 at b_3, and a flow of 6 along edge A_2b_2. The maximum possible backward flow is therefore 6, and the maximum flow which can be sent along the flow-augmenting path is 2.

STEP 7 (i) We increase the flow along odd edges—that is, along A_1b_2 and

A_2b_3, by 2, and decrease the flow along even edges—that is, along A_2b_2, by 2.

(ii) Decrease the supply at A_1 by 2 to zero, and decrease the demand at b_3 by 2 to zero.

We remove all the labels and return to Part B, Step 3.

The revised flow pattern is shown below.

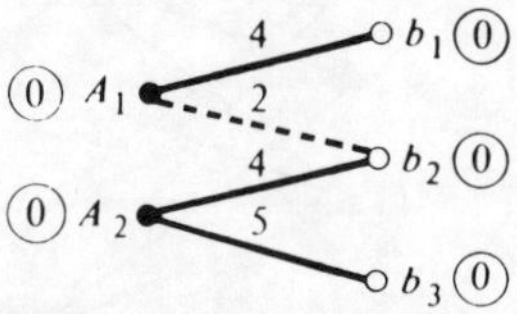

Part B: The Labelling Procedure (*sixth time*)

STEP 3 No vertex has any remaining supply, so the algorithm is completed.

We now have our minimum-cost solution. The total cost is found by multiplying the flow along each edge by the original edge cost, that is,

$$(4 \times 3) + (2 \times 5) + (0 \times 7) + (0 \times 6) + (4 \times 4) + (5 \times 3) = 53$$

We can summarize the Hungarian algorithm for the transportation problem as follows.

Summary of the Hungarian Algorithm

START with no flow.

PART A: Weight the vertices and find the initial partial graph.

PART B: Label the vertices in order to look for a flow-augmenting path.
STOP if no labelling is possible.
If a flow-augmenting path is found, go on to Part C.
If a flow-augmenting path is not found, go on to Part D.

PART C: Augment the flow.
Return to Part B.

PART D: Modify the weights and find a new partial graph.
Return to Part B.

Problem 14.1 Construct a flow chart for the Hungarian algorithm for the transportation problem. In each box of the flow chart, give the step numbers and a brief description of the process represented by that box.

Problem 14.2 Use the Hungarian algorithm to find a minimum-cost solution to the transportation problem described in the text at the beginning of this chapter, and calculate the total transportation cost. The bipartite graph representation and the cost matrix for this problem are given below.

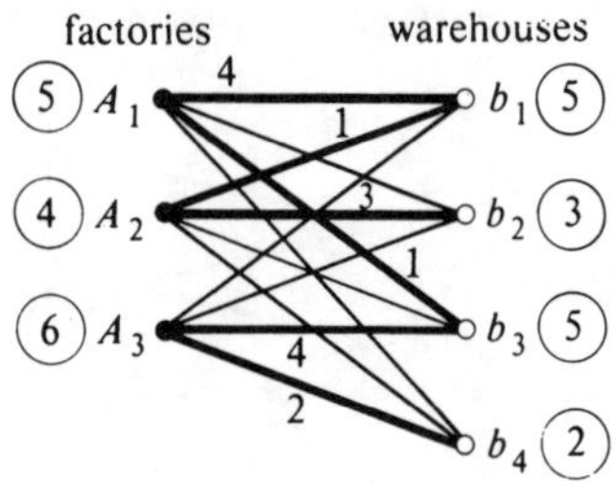

		demand			
		⑤	③	⑤	②
		b_1	b_2	b_3	b_4
	⑤ A_1	5	4	7	6
supply	④ A_2	2	5	3	2
	⑥ A_3	6	3	4	4

As with the assignment algorithm of the previous chapter, it is possible to do all the working on the cost matrix without drawing the various graphs. To illustrate this method, we now use it to solve the problem of the previous worked example.

Worked problem using matrix only Solve the previous example (on page 340), without drawing the partial graphs.

Solution The steps are the same as before, except that we write the various flows as circled numbers adjacent to the costs of the edges, and put the labels near the vertices on the matrix.

Part A and Part B (*first time*)

				(A_1)		
				0	1	0
				④	⑥	⑤
				b_1	b_2	b_3
(*)	3	⑥	A_1	0	1	4
(*)	3	⑨	A_2	3	0	0

Breakthrough is achieved at b_1.

Part C (*first time*) **and Part B** (*second time*)

The flow-augmenting path is simply A_1b_1, so we send the maximum possible flow of 4 units along this edge, as indicated by a circled 4 in the cost matrix. We then carry out the labelling to obtain the following:

				(A_1)	(A_2)	(A_2)
				0	1	0
				⓪	⑥	⑤
				b_1	b_2	b_3
(*)	3	②	A_1	0 ④	1	4
(*)	3	⑨	A_2	3	0	0

Breakthrough is achieved at b_2, so we proceed to Part C, Step 6.

Part C (*second time*) **and Part B** (*third time*)

The flow-augmenting path is A_2b_2, so we send a flow of 6 units along this path. We then repeat the labelling procedure to obtain the following:

				(A_1)	(A_2)	(A_2)
				0	1	0
				⓪	⓪	⑤
				b_1	b_2	b_3
(*)	3	②	A_1	0 ④	1	4
(*)	3	③	A_2	3	0 ⑥	0

Breakthrough is achieved at b_3, so we proceed to Part C, Step 6.

Part C (*third time*) **and Part B** (*fourth time*)

The flow-augmenting path is A_2b_3, so we send 3 units along this edge, and then relabel to obtain the following:

				(A_1)		
				0	1	0
				⓪	⓪	②
				b_1	b_2	b_3
(*)	3	②	A_1	0 ④	1	4
	3	⓪	A_2	3	0 ⑥	0 ③

Breakthrough has not been achieved, and no more labelling is possible, so we go to Part D, Step 8 to modify the partial graph.

Part D (*first time*) and Part B (*fifth time*)

By inspection of the above matrix, $\delta = 1$, so we obtain the revised cost matrix below. We now have an additional edge A_1b_2 in the partial graph, and we repeat the labelling procedure.

				(A_1)	(A_1)	(A_2)
				-1	1	0
				⓪	⓪	②
				b_1	b_2	b_3
$(*)$	4	②	A_1	0 ④	0	3
(b_2)	3	⓪	A_2	4	0 ⑥	0 ③

Breakthrough is achieved at b_3, so we go on to Part C, Step 6.

Part C (*fourth time*) and Part B (*sixth time*)

The flow-augmenting path is $b_3A_2b_2A_1$, and we send a flow of 2 along it. We then repeat the labelling procedure to obtain the following:

			-1	1	0
			⓪	⓪	⓪
			b_1	b_2	b_3
4	⓪	A_1	0 ④	0 ②	3
3	⓪	A_2	4	0 ④	0 ⑤

No vertex has any supply remaining, so we STOP.

Problem 14.3 Use the matrix-only form of the Hungarian algorithm to find a minimum-cost solution to the transportation problem described in the text at the beginning of this chapter, and calculate the total transportation cost. The cost matrix for this problem is given below.

			demand			
			⑤	③	⑤	②
			b_1	b_2	b_3	b_4
	⑤	A_1	5	4	7	6
supply	④	A_2	2	5	3	2
	⑥	A_3	6	3	4	4

The Use of a Dummy Vertex

In each of the examples of the transportation problems which we have considered, the total supply is equal to the total demand. In other cases, the total supply may be greater than the total demand—for example, in the previous problem it may be a simple matter to increase the supply of each factory by 2 units. To enable us to use the Hungarian algorithm to solve such problems, we introduce a *dummy vertex*. This represents an artificial warehouse (or other demand vertex) whose demand is chosen to be equal to the excess supply. Thus, for the example of Problem 14.3, if each factory can supply 2 additional units and the warehouse demands are the same, we represent the problem by the following bipartite graph and cost matrix:

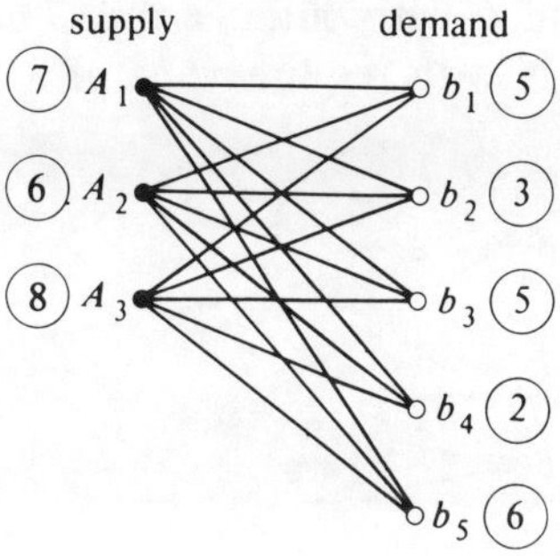

demand

		⑤	③	⑤	②	⑥
		b_1	b_2	b_3	b_4	b_5
	⑦ A_1	5	4	7	6	0
supply	⑥ A_2	2	5	3	2	0
	⑧ A_3	6	3	4	4	0

Vertex b_5 is the dummy vertex with an artificial demand of 6, equal to the excess supply. In the cost matrix, we must set the transportation costs from each supply vertex to this dummy vertex to be equal—we have chosen here to set them all equal to 0. We can then apply the algorithm in the usual way. The flow pattern produced will send 6 units to the dummy vertex. To obtain a flow pattern which is a solution to the original problem, we simply remove all flows to the dummy vertex.

14.2 AN ALTERNATIVE ALGORITHM FOR THE TRANSPORTATION PROBLEM

We have already noted that the Hungarian algorithm uses a least-cost flow-augmenting cycle. This means that it first constructs a flow pattern which is at minimum cost, but which normally does not satisfy all the demand, and then it increases the flow in stages until all the demand is met. At each stage, the flow is at minimum cost.

An alternative strategy is to start with a flow which satisfies all the demand, but which is not necessarily at minimum cost. The cost is then successively reduced by diverting some of the flow along different routes. This method is an example of a cost-reducing cycle. We illustrate this method in the following example.

Worked problem using the alternative algorithm for the transportation problem Three factories A, B and C with outputs of 4, 7 and 6 units per day supply two customers M and N with requirements of 8 and 9 units per day.

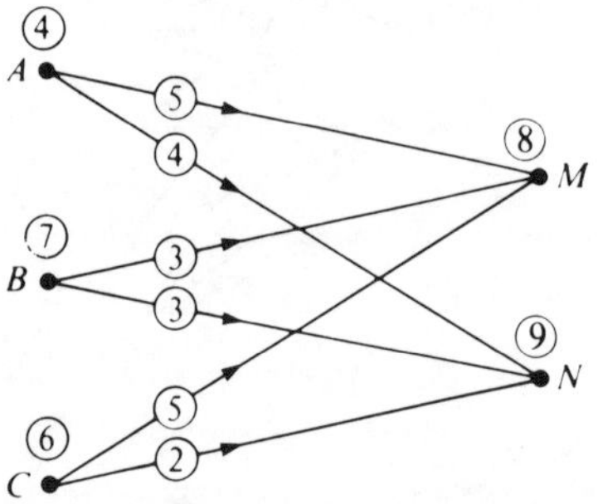

These supplies and demands are represented on the above figure by the circled numbers near the vertices. Each supply route has its transportation cost indicated on the graph by the circled number on the corresponding edge. Find a flow pattern of goods along the edges of the network which is at minimum total transportation cost.

Solution We proceed as follows.

STEP 1: Finding a Feasible Solution

We begin by finding a *feasible solution*, that is, one which satisfies the customers' requirements and the outputs from the factories, but which is probably not at minimum cost. The total supply from the factories is equal to the total demand, so no dummy vertices are needed.

We note that the cheapest route is *CN*, so it is reasonable to start by sending the maximum possible flow of 6 units along *CN*. A feasible solution may then be found by following a path through the bipartite graph, choosing the cheapest edge at each stage. We thus obtain the following flow pattern:

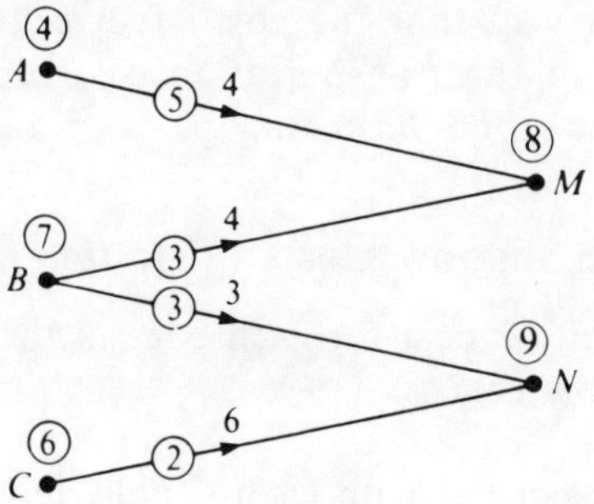

The total cost of this feasible solution, obtained by multiplying the flow on each edge by the cost and adding, is

$$(4 \times 5) + (4 \times 3) + (3 \times 3) + (6 \times 2) = 53.$$

We now need to work out whether we can improve on this feasible solution by diverting some of the flow along unused routes.

For example, we can send 1 unit of goods along the unused route *AN*. To do this, we also need to reduce the flow along the path *AMBN* by 1 unit, since vertex *A* can supply a total of only 4 units. This results in the following flow pattern:

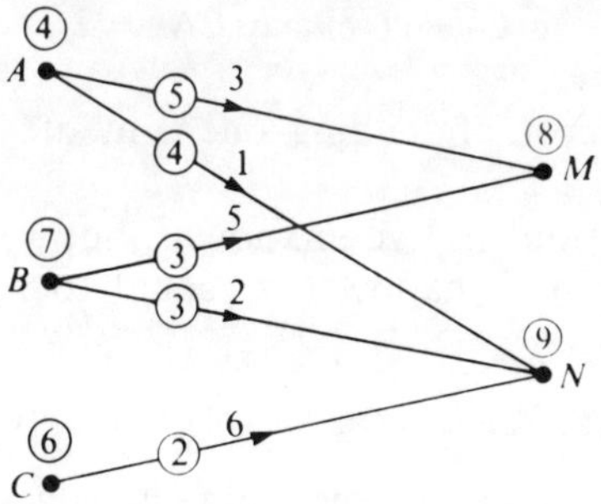

The total cost of transportation is now

$$(3 \times 5) + (1 \times 4) + (5 \times 3) + (2 \times 3) + (6 \times 2) = 52.$$

This is one less than before. We could proceed by calculating the change in cost for each unused route in the same way, but for a problem involving a large number of supply and demand vertices this would be very cumbersome. What we need is a simple and straightforward method for calculating the change in cost of opening up each route in turn. Such a method is employed in the remaining steps of our algorithm, and makes use of the concept of *shadow costs*. Shadow costs are costs which we associate with the vertices in such a way that the cost of diverting one unit along an unused edge is equal to the transportation cost for that edge minus the shadow costs on the two end vertices.

STEP 2: Calculating Shadow Costs (*first time*)

We split the transportation cost for each edge in the feasible solution into two parts called shadow costs:

(i) a dispatch cost associated with each supply vertex, assumed to be the same for each route out of that vertex;
(ii) a reception cost associated with each demand vertex, assumed to be the same for all routes into that vertex.

Let $c(A)$, $c(B)$ and $c(C)$ denote the dispatch costs, and $c(M)$ and $c(N)$ denote the receipt costs; then for the four routes in our original feasible solution we have:

$$c(A) + c(M) = \Delta(AM) = 5,$$

$$c(B) + c(M) = \Delta(BM) = 3,$$

$$c(B) + c(N) = \Delta(BN) = 3,$$

$$c(C) + c(N) = \Delta(CN) = 2,$$

where $\Delta(AM)$ is the cost of sending one unit of goods along AM, and so on.

There are four equations in five unknowns, but we can obtain a solution by arbitrarily putting one shadow cost equal to zero—for instance, put $c(A) = 0$. This gives

$$c(M) = 5, \quad c(B) = -2, \quad c(N) = 5, \quad c(C) = -3.$$

This calculation depends on the fact that the edges in the network form a tree, and in fact we can always arrange for this to be the case, since no cycles are necessary. If initially a network is composed of several isolated trees, then extra unused routes can be introduced which have no effect on the flow or cost but link up the parts of the network into a single tree.

STEP 3: Assessing Unused Routes (*first time*)

The shadow costs can be used to calculate the change in transportation cost brought about by diverting one unit of flow along an unused route, so we can assess whether or not that route should be used instead of one of the existing ones.

The change in transportation cost brought about by diverting one unit of flow along the unused route AN is

$$\Delta(AN) - \Delta(AM) + \Delta(BM) - \Delta(BN) = 4 - 5 + 3 - 3 = -1$$

since if we divert one unit along AN, we must reduce the flow along the route $AMBN$ to compensate. So we must reduce the flow along AM, increase it along BM, and reduce it along BN.

Now, substituting the shadow costs where possible, we obtain

$$\Delta(AN) - \Delta(AM) + \Delta(BM) - \Delta(BN)$$

$$= \Delta(AN) - (c(A) + c(M)) + (c(B) + c(M)) - (c(B) + c(N))$$

$$= \Delta(AN) - c(A) - c(N) = 4 - 0 - 5 = -1$$

For the other unused route, CM, we have

$$\Delta(CM) - c(C) - c(M) = 5 - (-3) - 5 = 3.$$

In general, for any unused route XY, the value of the expression

$$\Delta(XY) - c(X) - c(Y)$$

tells us the change in transportation cost brought about by diverting one unit of flow along the unused route XY. If the value of this expression is negative for any route, then it is advantageous to divert as much flow as possible along that route. (The diverted flow will be taken from another route, so when one of the edges on this route has lost all its flow, no more can be diverted.) So we wish to use the unused route AN, but not the route CM.

STEP 4: Exploiting the New Route

We wish to divert as much flow as possible along AN. The effect of sending a flow of δ along AN is:

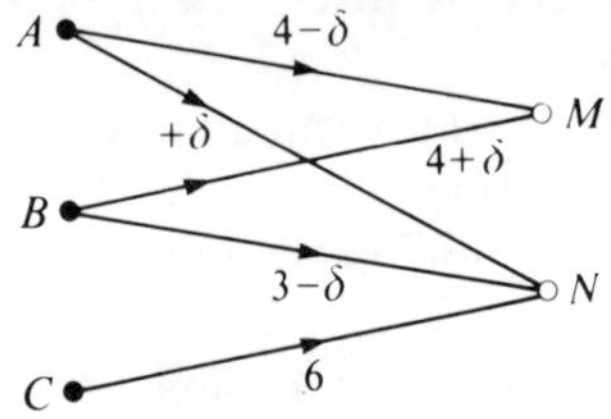

Two of the routes have a decrease in flow, *AM* and *BN*. It is *BN* which goes out of use first, when $\delta = 3$. The new feasible solution is then:

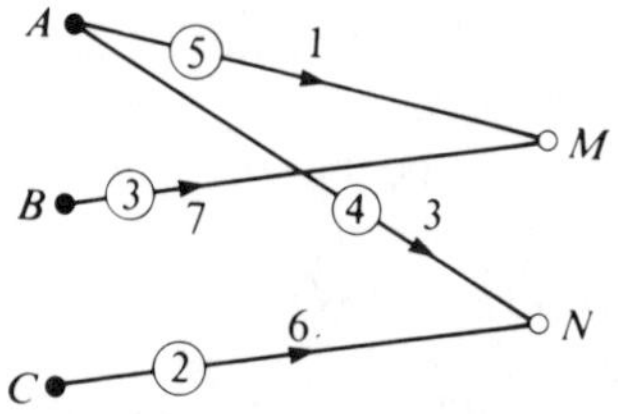

The total transportation cost is now

$$(1 \times 5) + (3 \times 4) + (7 \times 3) + (6 \times 2) = 50.$$

To find out whether this is the optimum solution, we must return to Step 2 to check whether any unused route can be brought in to reduce the cost further.

Step 2: Calculating Shadow Costs (*second time*)

We must now recalculate the shadow costs for the new feasible solution given in the above diagram. Setting $c(A) = 0$, we obtain

$$c(B) = -2, \quad c(C) = -2, \quad c(M) = 5, \quad c(N) = 4.$$

Step 3: Assessing Unused Routes (*second time*)

The two unused routes are *BN* and *CM*.

For *BN*, we have $\Delta(BN) - c(B) - c(N) = 3 - (-2) - 4 = +1$,
For *CM*, we have $\Delta(CM) - c(C) - c(M) = 5 - (-2) - 5 = +2$,

so there is no cost advantage in bringing either of these routes into use. We have found the optimum solution, and the algorithm stops.

A flow chart for this cost-reducing cycle algorithm is given overleaf.

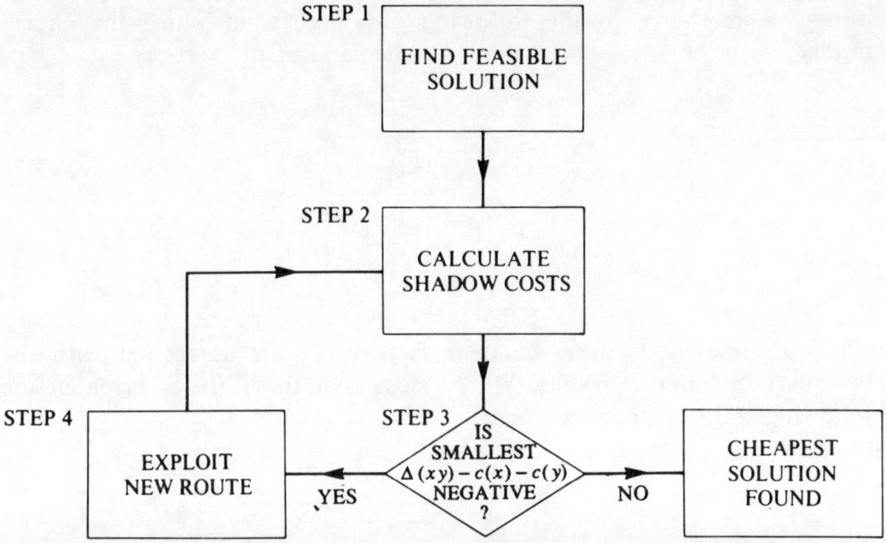

Is there any reason for choosing one algorithm for the transportation problem rather than the other? The second algorithm is probably easier to work by hand for small problems, but it seems from experience that the Hungarian algorithm is quicker for larger examples, and is probably more efficient for use with a computer for large-scale problems.

14.3 THE TRANSHIPMENT PROBLEM

In the transportation problem discussed above, we assumed that the products from each factory are sent directly to one or more warehouses. However, in a practical situation, it may be cheaper to send the products from a factory first to an intermediate point (one of the other factories or warehouses) and then on to the destination warehouse.

The sending of goods via intermediate points is called *transhipment*, and a problem in which transhipment is allowed is called a *transportation problem with transhipment*, or, more simply, a *transhipment problem*. The graphical representation of a transhipment problem is not a bipartite graph, since we can send goods from factory to factory, or from warehouse to warehouse. However, we can easily transform the graphical representation into a bipartite graph—to do this, we represent each factory or warehouse by *two* vertices, one representing the factory or warehouse as a supply point, and one representing it as a demand point. To illustrate how this works, we consider a specific example.

Suppose we have a transportation problem for which the transportation costs are as shown in the following cost matrix, in which the circled numbers are the amounts of supply or demand at the vertices.

			demand		
			⑤	④	⑥
			b_1	b_2	b_3
	②	A_1	15	11	15
supply	⑥	A_2	17	14	11
	⑦	A_3	18	18	15

If transhipment is allowed, then there are costs associated with the various transhipment routes. We suppose that these are as given in the following transhipment cost matrix:

					demand			
			⓪	⓪	⓪	⑤	④	⑥
			a_1	a_2	a_3	b_1	b_2	b_3
	②	A_1	0	6	5	15	11	15
	⑥	A_2	5	0	4	17	14	11
	⑦	A_3	4	4	0	18	18	15
supply	⓪	B_1	14	19	20	0	5	4
	⓪	B_2	10	12	17	5	0	2
	⓪	B_3	17	13	14	4	2	0

In this matrix, A_1 represents the first factory considered as a supply vertex, and a_1 represents the *same* factory considered as a demand vertex. Similarly, B_1 represents the first warehouse considered as a supply vertex, and b_1 represents that warehouse considered as a demand vertex. Since we must consider each factory and warehouse as both demand and supply vertices, we have a total of six supply vertices and six demand vertices. Note that the part of the cost matrix for the original vertices A_1, A_2, A_3 and b_1, b_2, b_3 is the same as the original cost matrix.

As before, the circled numbers indicate the amounts demanded or available at the vertices considered as demand or supply vertices. Factories considered as demand vertices have zero demand, and warehouses considered as supply vertices have zero supply. The fact that the cost matrix is not symmetric means that outgoing shipments and returning shipments do not cost the same. The entries on the diagonal of the matrix are all zero, since they are the costs of transporting goods between supply and demand vertices which represent the same factory or warehouse.

By splitting the vertex representing each factory or warehouse into two vertices—a supply vertex and a demand vertex—we have cast the transportation problem with transhipment into the same form as a transportation problem without transhipment. We do this so that we can apply an algorithm for the transportation problem to the problem with transhipment. However, before we can apply the algorithm, a further modification is required. The reason for this is the following. When we apply the transportation algorithm, we can ship goods from a demand vertex to a supply vertex only if there is already a flow from that supply vertex to that demand vertex. This means that, for the transportation problem with transhipment, we cannot ship goods from a particular factory considered as a demand vertex to the same factory considered as a supply vertex. To overcome this problem, we add to all the demand and supply vertices an amount of demand or supply at least as great as the total demand or supply. This ensures that any vertex considered as a demand vertex can accommodate any transhipment which could occur in the solution to the problem. When the procedure of the algorithm has been completed, we simply remove these artificial supplies and demands (and the corresponding artificial flows) to obtain a solution to the original problem.

Worked problem Use the Hungarian algorithm for the transportation problem to obtain a minimum-cost solution to the above transhipment problem.

Solution The total supply and the total demand are both equal to 15. We must add an amount greater than this to all the supply and demand vertices—we choose 20 as this amount to be added.

When we apply the algorithm, we form a partial graph, using the cost matrix given in the above text. Since all the entries on the diagonal are zero, and all other entries are non-zero, this partial graph consists only of the edges joining A_1 to a_1, A_2 to a_2, and so on, as follows:

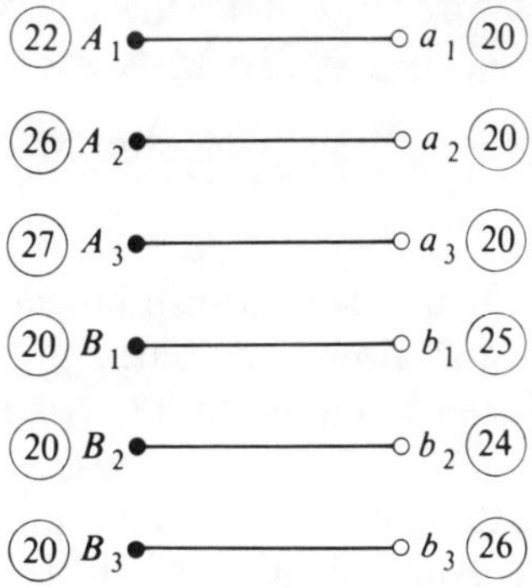

The labelling procedure and the procedure for modifying the flow pattern result in a flow of 20 being sent along each edge—these are the artificial flows necessary to ensure that any transhipment can be accommodated. The resulting flow pattern is shown below. The circled numbers next to the vertices represent unallocated supplies or unsatisfied demands—these are just the original supplies and demands.

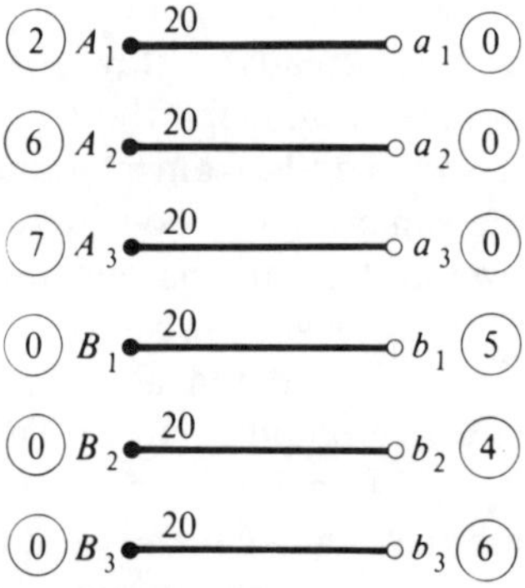

We continue to apply the algorithm in the same way as for a transportation problem without transhipment: the procedure is exactly the same as that used in the first worked problem. We thus obtain the following flow pattern:

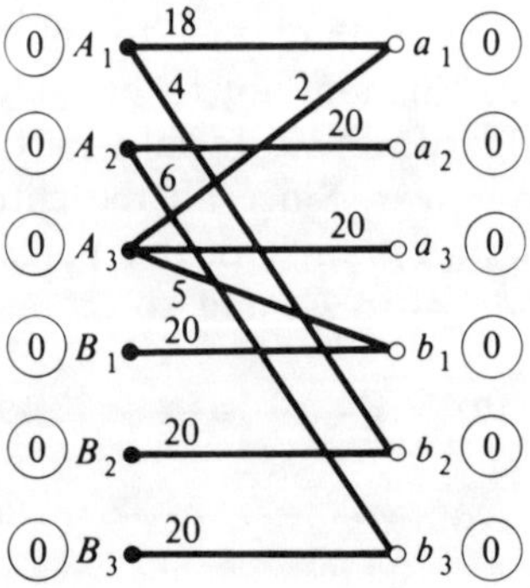

The flow path $A_3a_1A_1b_2$ involves transhipment via the factory A_1. To accommodate the flow of 2 along this path, the artificial flow along the edge A_1a_1 has been reduced from 20 to 18. No other flow path involves transhipment.

To obtain the flow pattern which is a solution to the original problem, we must subtract all the artificial flows of 20 from the flow pattern given above. This gives the following solution, in which we have indicated the direction of flow along each edge by an arrow. The original supplies and demands are shown next to the vertices.

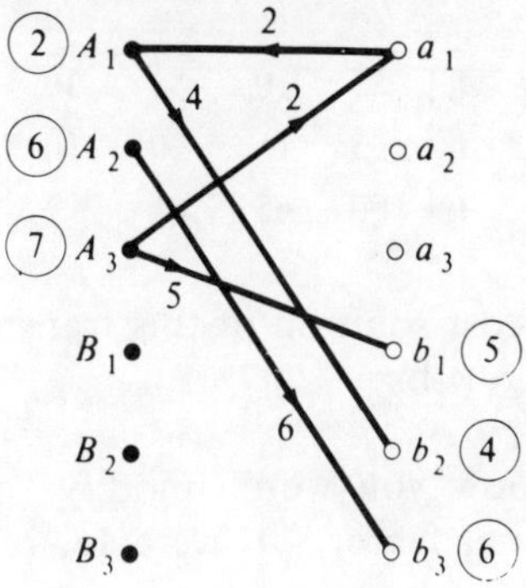

EXERCISES 14

The Hungarian Algorithm for the Transportation Problem

14.1. Given the following partial graph of a transportation problem, carry out the labelling procedure.

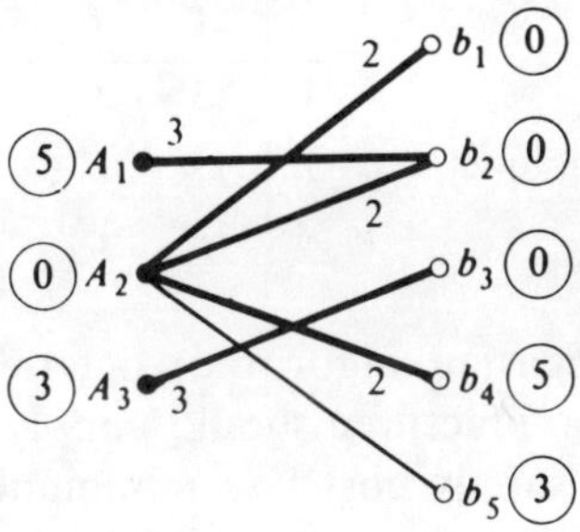

14.2. (a) Explain briefly what is meant by an alternating path with respect to a matching in a bipartite graph, and give an example of a graph containing such a path with 5 edges.

(b) Explain briefly why the concept of an alternating path cannot be used in an algorithm for the transportation problem.

14.3. Three factories, A_1, A_2 and A_3, can supply 10, 15 and 20 units of a product, respectively. The demands for these products at four warehouses, b_1, b_2, b_3 and b_4, are 5,12,13 and 15 units, respectively. The costs of transporting one unit of product from each factory to each warehouse are given in the following cost matrix:

	b_1	b_2	b_3	b_4
A_1	25	10	2	30
A_2	5	15	20	10
A_3	100	65	1	2

Find a minimum-cost solution of this transportation problem, using the Hungarian algorithm.

14.4. Describe briefly how you would modify your solution to Exercise 14.3 if the supply at factory A_2 were 18, rather than 15.

14.5. A printer with three printing works, A, B and C, requires 10 batches of paper for A, 10 batches for B, and 13 batches for C. Three paper manufacturers, X, Y and Z, can supply 8 batches, 13 batches and 12 batches, respectively. The cost (per batch) of supplying paper from each manufacturer to each printing works is shown in the following cost matrix:

		printing works		
		A	B	C
	X	13	9	11
manufacturer	Y	13	17	14
	Z	12	19	15

Using the Hungarian algorithm, determine how many batches of paper each manufacturer should supply to each printing works so that the total supply cost is a minimum. What is the minimum cost?

(Your answer must show clearly how each step of the algorithm is implemented at each stage.)

14.6. Suppose that the manufacturer X in Exercise 14.5 can supply 20 batches of paper instead of the original 8 batches. Explain briefly how you would tackle the problem with this new value for the available supply. Draw a bipartite graph and cost matrix for this new situation, and find the first revised cost matrix and partial graph. (Do not attempt to find a full solution.)

14.7. A transportation problem is modelled by the complete bipartite graph $K_{4,4}$. The cost matrix is as follows, with the supplies and demands indicated by circled numbers next to the appropriate vertex labels.

			demand ⑤ y_1	④ y_2	⑧ y_3	④ y_4
supply	⑦	x_1	5	7	8	10
	④	x_2	7	9	3	5
	⑤	x_3	7	5	4	8
	⑤	x_4	10	12	5	8

Use the Hungarian algorithm to find a minimum-cost allocation of items from the supply vertices to the demand vertices. What is the value of this minimum cost?

14.8. The following cost matrix is for a transportation problem with four supply points and four demand points. The supplies available and the amounts demanded are shown as circled numbers next to the appropriate vertex labels.

			demand ⑤ y_1	④ y_2	⑧ y_3	④ y_4
supply	⑦	x_1	4	4	8	9
	④	x_2	8	5	10	6
	⑤	x_3	9	6	7	5
	⑤	x_4	3	14	7	11

Use the Hungarian algorithm to find a minimum-cost solution, and determine the value of the minimum cost.

14.9. Show how the transportation problem can be treated as a maximum-flow minimum-cost problem, and hence solved by using the max-flow min-cost algorithm given in Chapter 7.

The Alternative (Cost-reducing cycle) Algorithm for the Transportation Problem

14.10. What are the main differences between the Hungarian algorithm for the transportation problem and the alternative algorithm using a cost-reducing cycle?

14.11. Use the alternative algorithm to solve Exercise 14.3.

14.12. Use the alternative algorithm to solve Exercise 14.5.

The Transhipment Problem

14.13. A transhipment problem has the following cost matrix:

			demand			
			⓪	⓪	③	②
			a_1	a_2	b_1	b_2
supply	②	A_1	0	1	3	6
	③	A_2	1	0	4	5
	⓪	B_1	6	4	0	1
	⓪	B_2	5	3	1	0

Use the Hungarian algorithm to find a minimum-cost solution, and determine the value of this minimum cost.

CHAPTER 15

CRITICAL PATH ANALYSIS

After reading this chapter, you should be able to:

- explain the terms *activity network*, *critical path* and *float*;
- distinguish between the two types of activity network in common use;
- construct an activity network from a table of precedence relations;
- find a critical path in an activity network, and calculate the float of each activity.

Introduction

In this chapter and the next three we discuss a number of problems concerned with the optimum use of resources in industrial projects. This chapter and the next are concerned with the use of network analysis in scheduling the activities involved in large-scale projects such as those occurring in industry. In Chapter 17 we discuss packing problems, which are related to scheduling problems, and in Chapter 18 we consider the problem of finding the optimum locations on a network for supply points for a product or a service. The methods of solution to these problems are given in the form of algorithms.

These network problems are, in general, considerably more complex than those we have studied in earlier chapters. A characteristic of many of the scheduling and location problems which we discuss is that they are very difficult, in the sense that there are no known efficient algorithms for finding optimum solutions.

In the present chapter we are concerned with the use of network analysis to provide a method for planning and scheduling industrial projects. Any project which can be broken down into separate activities which are interrelated in time can be represented by a network called an **activity network**. An example (to be discussed more fully later) is shown in the following figure which is an activity network for the construction of a small building such as a garage. Here the vertices represent separate activities involved in the project, and the numbers next to the arcs represent the completion times of the activities.

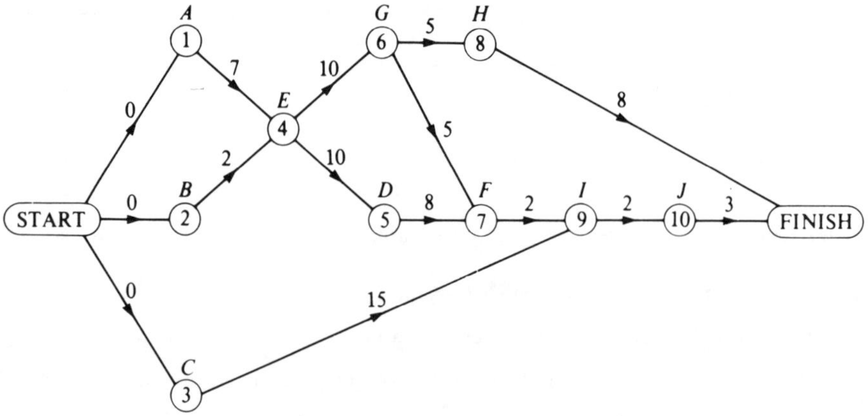

activity network for building construction

We shall show later how the minimum completion time for the whole project can be determined by finding a longest path, called a **critical path**, through the corresponding activity network.

Many companies use network analysis as an aid to the planning and control of large-scale projects. The first forms of network analysis were developed and used in the USA in the late 1950s. Since that time, many different systems of network analysis have been developed which use the same basic techniques.

An essential part of these methods is the use of an activity network to represent a project. Once an activity network has been constructed, we can find a critical path. We can then calculate the earliest and latest times by which each activity must begin if the project is not to be delayed.

Two types of activity network are at present in common use. We shall mostly use the type of activity network in which *the vertices represent activities.* Later in this chapter we shall describe the other type in which *each vertex represents an event* (a stage in the process corresponding to the start or finish of one or more activities) and give a brief comparison of the two types.

15.1 ACTIVITY NETWORKS WHICH USE VERTICES TO REPRESENT ACTIVITIES

We begin our discussion by considering a project to construct a garage. The separate activities involved, and their estimated durations, are given in the following table:

	Activity	Duration (in days)
A	prepare foundations	7
B	make and position door-frame	2
C	lay drains, floor base and screed	15
D	install services and fitting	8
E	erect walls	10
F	plaster ceiling	2
G	erect roof	5
H	install doors and windows, and paint outside	8
I	fit gutters and pipes	2
J	paint inside	3

Clearly, some of these activities cannot be started until other activities have been completed. For example, activity *G* (erect roof) cannot begin until activity *E* (erect walls) has been completed.

The following table of **precedence relations** shows which activities must precede which:

	Precedence relations
D	must follow *E*
E	must follow *A* and *B*
F	must follow *D* and *G*
G	must follow *E*
H	must follow *G*
I	must follow *C* and *F*
J	must follow *I*

In each of our examples, we simply *state* a list of precedence relations. We do not discuss the problem of actually obtaining a minimal list (that is, a list containing no redundant information) of precedence relations for a given project.

Throughout this chapter, we assume that there is no restriction on the number of workers available.

We can represent this process of construction by the following activity network:

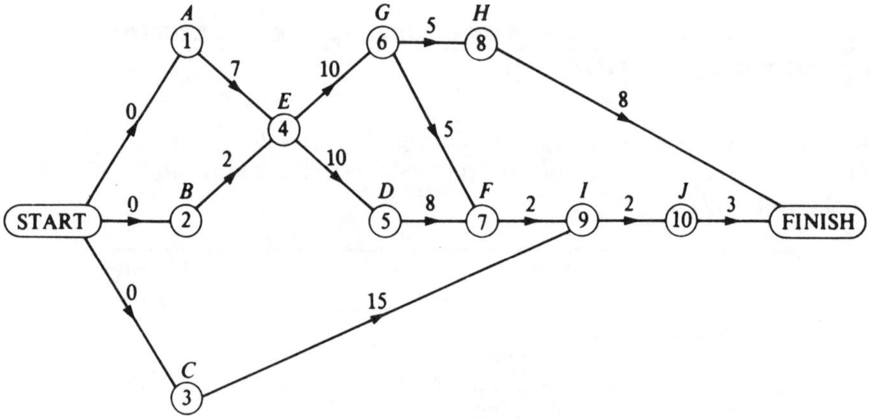

activity network for building construction

In this type of activity network, each activity is represented by a numbered vertex. An arc joining a vertex X to a vertex Y indicates that the activity represented by the vertex X must be completed before the activity represented by vertex Y can be started. The activity network thus represents the precedence relations in diagrammatic form.

The number associated with an arc is the duration of the activity from which that arc is incident. For example, the arc joining vertex A to vertex E indicates that activity E must follow activity A, and the number 7 next to the arc is the duration (in days) of activity A.

Notice also that the activity network has a START vertex and a FINISH vertex—these do not represent any activities, but are included to complete the network. Since the START vertex does not represent an activity, the duration time associated with each arc leading from that vertex is zero.

An activity network can be used to find the minimum time necessary for the completion of the project it represents. Before showing how this is done, we present a systematic method for constructing an activity network from a table of precedence relations.

Algorithm for Constructing an Activity Network

The following algorithm is a systematic procedure for numbering the activities and then drawing the activity network. One of the advantages of using this method rather than drawing the activity network by trial and error is that it produces a clear diagram in which the vertices are placed in a logical order.

The first part of the algorithm is a procedure developed by D. R. Fulkerson for numbering the activities and the corresponding vertices. The activities are numbered in such a way that each activity is assigned a number greater than the number assigned to any activity which must precede it.

The second part of the algorithm is a procedure for drawing the activity network.

DESCRIPTION OF THE ALGORITHM FOR CONSTRUCTING AN ACTIVITY NETWORK

Part A: Procedure for Numbering the Vertices

START Represent each activity by a vertex. For each such vertex, create a *shadow vertex*, so that for each activity there correspond two vertices—the original vertex and the shadow vertex.

Construct a bipartite graph in which one set of vertices consists of the original vertices, and the other set consists of the shadow vertices.

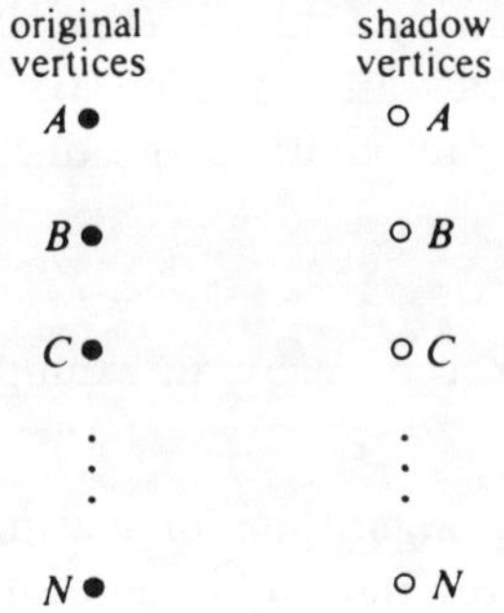

If an activity X must follow an activity Y, draw an edge joining the original vertex representing X to the shadow vertex representing Y.

STEP 1 Number consecutively all the original vertices (chosen in any order) which have no edges incident to them.

STEP 2 Delete all numbered vertices, their corresponding shadow vertices, and all edges incident to these vertices.

Repeat Steps 1 and 2, recording the numbering at each stage, until all the vertices have been numbered.

Part B: Procedure for Drawing the Activity Network

START with an artificial START vertex.

STEP 3 Draw an arc from the START vertex to each vertex which was numbered in the first iteration of the numbering procedure, and assign a weight of zero to each arc.

STEP 4 Consider the vertices which were numbered in the next iteration of the numbering procedure. To each such vertex X, draw an arc from each previously-numbered vertex Y if there is an edge joining the original vertex X to the shadow vertex Y in the original bipartite graph constructed in Part A. Assign a weight to each arc YX equal to the duration of the activity Y.

Repeat until all vertices have been included in the activity network.

STEP 5 From each terminal vertex Z (that is, each vertex whose out-degree is zero), draw an arc to an artificial FINISH vertex. Assign a weight to each arc equal to the duration of the corresponding activity Z.

We illustrate the above procedure by an example.

Worked problem using the algorithm for constructing an activity network
Find an activity network for the building-construction example.

Solution To start Part A, we draw a bipartite graph with a vertex in each set corresponding to each activity—the vertex representing the activity in one set, and a corresponding shadow vertex in the other set. We draw an edge joining an original vertex to one of the shadow vertices if the activity corresponding to the original vertex must follow the activity represented by the shadow vertex. We thus obtain the following bipartite graph representing the precedence relations.

	Precedence relations
D	must follow *E*
E	must follow *A* and *B*
F	must follow *D* and *G*
G	must follow *E*
H	must follow *G*
I	must follow *C* and *F*
J	must follow *I*

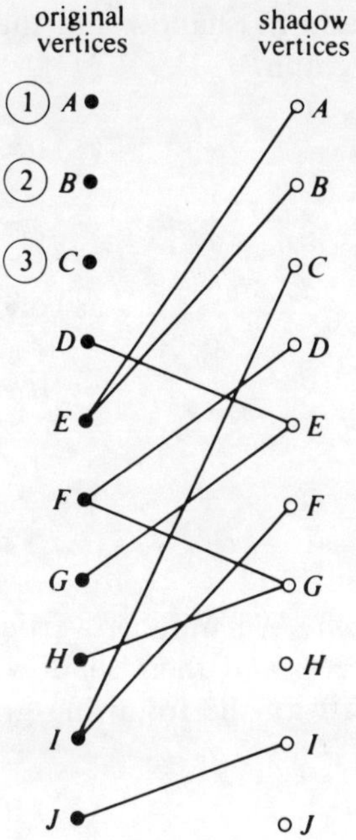

1st iteration We number all the original vertices which have no edges incident to them. Thus we assign the number 1 to vertex *A*, 2 to vertex *B*, and 3 to vertex *C*, as shown on the above diagram. Next we delete all numbered vertices, and the corresponding shadow vertices and all edges incident to them, and we are left with the following graph:

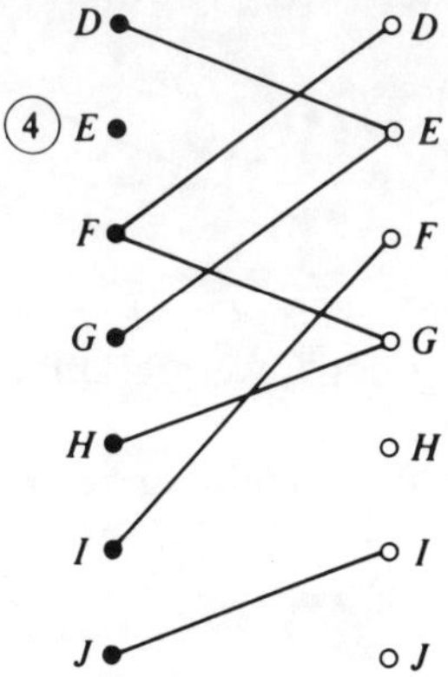

2nd iteration The only original vertex which has no edge incident to it is vertex *E*, so we number this vertex 4, as shown above. Removing the vertex E, and its shadow and the edges incident to it, we are left with the following graph:

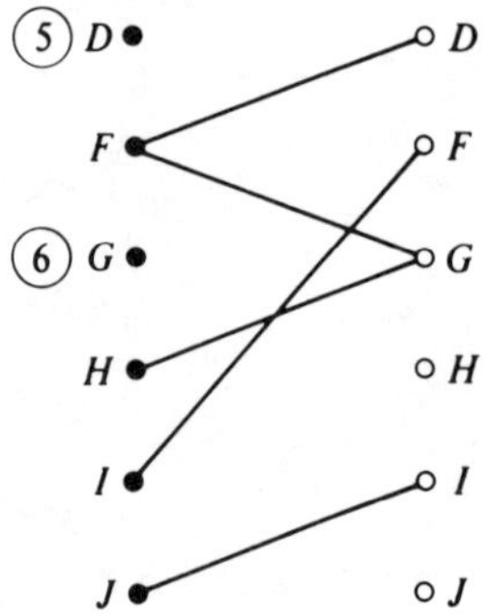

3rd iteration We number vertices D and G as shown above. Removing these vertices, and their shadow vertices and the edges incident to them, we are left with the following graph:

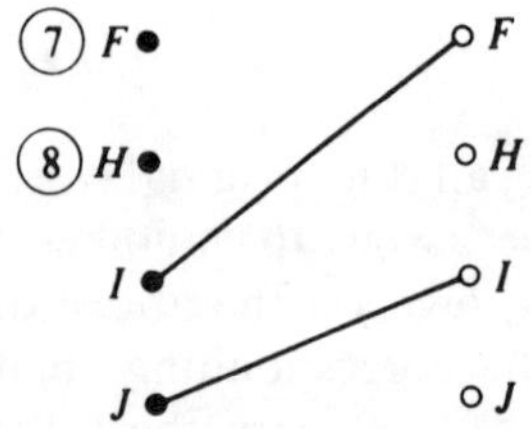

4th iteration After numbering vertices F and H and removing the appropriate vertices and edge, we are left with the following graph:

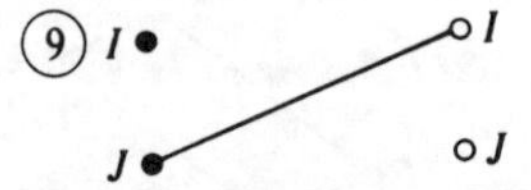

5th iteration We number vertex I and, after removing this vertex, its shadow, and the remaining edge, we are left with just vertex J.

6th iteration We number the remaining vertex J:

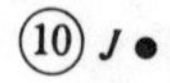

This completes Part A of the algorithm.

We can now proceed with Part B of the algorithm. This results in the activity network which we gave at the beginning of this chapter:

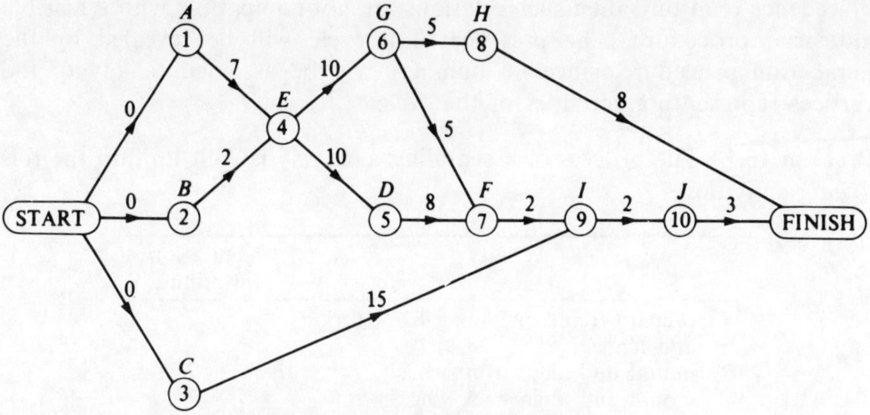

activity network for building construction

Notice that we have arranged the vertices in vertical columns corresponding to the iterations of the numbering procedure in which they were numbered: vertices *A,B* and *C* were numbered in the first iteration, *E* was numbered in the second iteration, and so on.

Inconsistencies in the Precedence Relations

Note that, as well as ordering the activities, the procedure in Part A enables us to detect any inconsistencies in the precedence relations. For example, suppose that we are given the precedence relations:

B must follow *A*
A must follow *C*
C must follow *B*

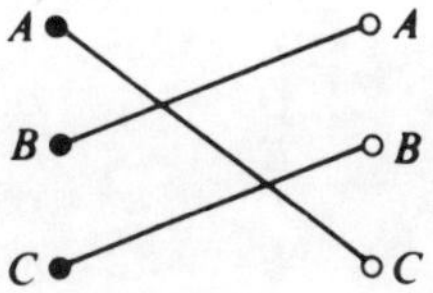

To start the algorithm, we construct the above graph, but we cannot proceed with Step 1, because there are no isolated vertices.

We can represent these relations by the following digraph.

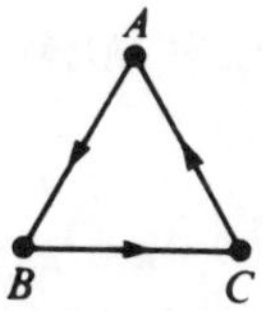

The activities form a cycle, so it is impossible to schedule them, since none of the activities can be started. If a cycle arises from a list of precedence relations, then such relations are not compatible with a feasible industrial procedure. The presence of a cycle will be revealed by the numbering procedure, since no numbers will be assigned to any of the vertices representing activities of the cycle.

Problem 15.1 The process of assembling a bicycle is split up into the following activities:

	Activity	Duration (in minutes)
A	prepare frame, including front fork and fenders	7
B	mount and align front wheel	7
C	mount and align back wheel	7
D	attach the derailer to the frame	2
E	install the gear cluster	3
F	attach the chain-wheel to the crank	2
G	attach the crank and chain-wheel to the frame	2
H	mount the right pedal and toe-clip	8
I	mount the left pedal and toe-clip	8
J	make final attachments; fix and adjust handle-bars, seat, brakes, etc.	18

The precedence relations of these activities are shown in the following table:

Activity	Preceding activities
J	*A*, *B*, *C*, *D*, *E*
C	*D*, *E*
E	*D*
F	*D*
H	*E*, *F*, *G*
I	*E*, *F*, *G*
G	*F*

Use the algorithm to construct an activity network for this process.

15.2 ACTIVITY NETWORKS WHICH USE VERTICES TO REPRESENT EVENTS

An alternative type of activity network uses arcs to represent activities and vertices to represent events. An **event** is a stage in the process corresponding to the start or finish of one or more activities. For example, if activities *B* and *C* must follow activity *A*, this can be represented as follows:

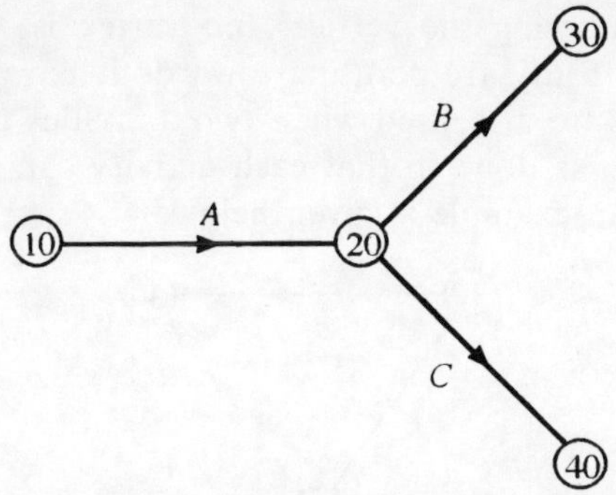

Vertices 10 and 20 represent the start and finish of activity *A*. The event 20 is also the start event for activities *B* and *C*. (Activities are commonly referred to by quoting their start and finish event numbers; so, for example, activity *C* is referred to as activity 2040.)

With this type of activity network, it is sometimes necessary to introduce *dummy activities*. For example, if activity *C* must follow activity *A*, and activity *D* must follow activities *A* and *B*, we would draw the activity network as follows:

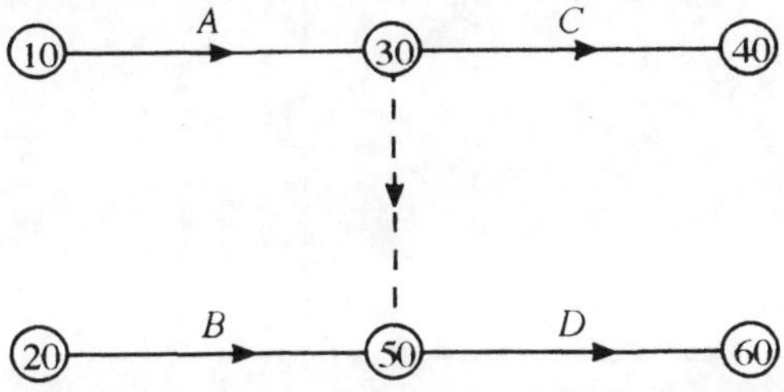

The dummy activity 3050, represented by the broken line, is introduced to show that activity *D* cannot start until activity *A* has been completed. Dummy activities have no time duration. Note that this activity network is not the same as

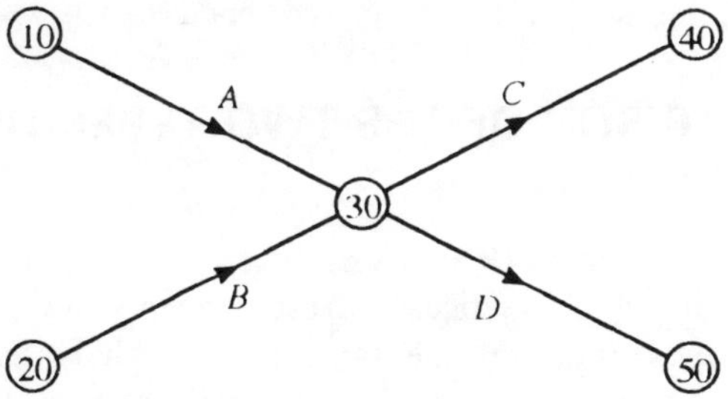

which represents the case when both *C* and *D* must follow both activities *A* and *B*.

Note that in numbering the vertices, no vertex is numbered until the vertices at the tail of any arc pointing towards it have been numbered.

Dummy activities are also used when two activities have the same start and end events. This is done so that each activity can be referred to by a different number. An example is given below:

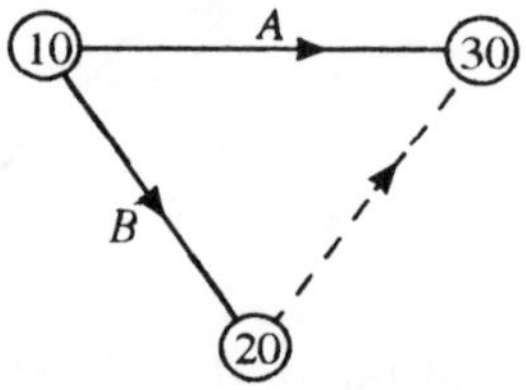

As an illustration we give below an activity network with vertices representing events for the building construction example which we introduced earlier in this chapter.

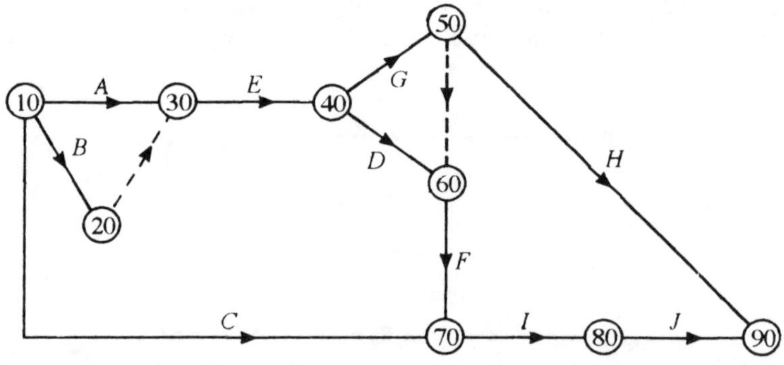

You may be wondering why the vertices of the activity network are numbered with such large numbers. The reason for this is that during the planning stage or later it may be necessary to include additional activities. The style of numbering adopted allows any additional vertices to be conveniently numbered in the correct sequence.

15.3 COMPARISON OF THE TWO TYPES OF ACTIVITY NETWORK

The first methods of network analysis to be developed used the type of activity network in which vertices represent events. Two methods of this type were developed at about the same time in the late 1950s and were the first to be used to plan large-scale industrial projects. These were PERT (Program Evaluation and Review Technique) and CPM (Critical Path Method). PERT was used by the US Navy to plan the Polaris missile project, where the main objective was to complete the project in the shortest possible time. This was a complex project involving activities whose times could not be accurately predicted. CPM was developed by the Du Pont de Nemours Company; here the primary concern was to minimize the total cost of a project. The most widely accepted distinction between PERT and CPM concerns the performance times for the activities. CPM is used for cases where these times can be predicted reasonably accurately. The important distinctive feature of PERT is that the technique permits probabilistic estimates of activity times, and so goes some way towards accommodating research and developmental projects in which times for activities cannot be predicted with confidence. It can also be used for projects which may suffer disruption through strike action or late delivery of materials, etc.

The type of activity network in which vertices represent activities was developed by Bernard Roy of Metra International, a management-consultancy firm. This type of activity network is easier to understand and easier to construct than the other type. In particular, it does not require the use of dummy activities. A disadvantage for some applications is that events are not included in the network, which makes it more difficult to plan a number of concurrent projects.

In this book we have chosen to adopt the type of activity network in which vertices represent activities, because its construction is more straightforward than for the other type. The network analysis for both types of activity network have the same essential features and involve finding a *critical path* in the network. Our aim is to explain the basic principles and techniques upon which practical methods of network analysis such as PERT and CPM are based. However, both PERT and CPM involve complexities which cannot be included here.

The algorithms we give are in a form suitable for the type of activity network in which vertices represent activities. However, basically the same algorithms can be used, with only minor modifications, for the other type of activity network.

15.4 CRITICAL PATHS

When planning a project, we usually need to know the minimum time necessary to complete the project. We can calculate this minimum time by using the activity network of the project. To do this, we find a longest path, called a **critical path**, from the START vertex to the FINISH vertex. By a **longest path** we mean a path for which the sum of the times associated with the arcs of the path (that is, the length of the path) is the largest possible. The *minimum completion time* is equal to the *length of a critical path*. Any delay in completing an activity on a critical path delays the completion of the whole project by the same amount. Note that an activity network may contain more than one critical path.

Problem 15.2 What is the effect on the completion time of a project of a delay in the completion of an activity which is not on a critical path?

In the solution to the above problem, we saw that an activity not on a critical path may be delayed by a certain amount without delaying the whole project. The maximum time by which an activity may be delayed without delaying the project is called the **float** of that activity.

If a project is to be completed in the shortest possible time, then particular attention must be paid to activities on any critical path. For other activities, there is some leeway in their starting time or durations—the amount of leeway for such an activity is its float.

In Chapter 7 we discussed the problem of finding the longest path in a network. This problem is obviously closely related to the problem of finding a critical path in an activity network, and we could use the longest path algorithm given in Chapter 7 to find a critical path in an activity network. However, because of the special structure of an *activity* network—namely, the way the vertices are numbered—we can improve on this method. In other words, the algorithm we now give for finding a critical path is more efficient than the longest path algorithm.

The Critical Path Algorithm

This algorithm is similar to the longest path algorithm given in Chapter 7, except that we make use of the numbering of the vertices in the activity diagram. We saw earlier that this numbering is possible only because an activity network contains no cycles. The algorithm has two parts:

Part A: the **forward scan**, in which vertices are labelled consecutively;
Part B: the **backward scan**, in which a critical path is found.

We give a formal statement of the algorithm, and then illustrate its use with an example. We adopt the following notation.

Conventions We assume that the network involves n activities (that is, n vertices). We regard the START vertex as the 0th vertex and the FINISH vertex as the $(n+1)$th vertex.

The duration of the activity represented by the arc ij is denoted by $c_{i,j}$. The algorithm assigns weights p_j and e_j to each vertex j, for $j = 0, 1, \ldots, n+1$. When the algorithm has been completed, e_j is the length of the longest path from the START vertex to the vertex j, and p_j is the number of the preceding vertex on this longest path.

DESCRIPTION OF THE CRITICAL PATH ALGORITHM

Part A: Forward Scan

STEP 1 Label the START vertex with $p_0 = 0$ and $e_0 = 0$.

STEP 2 Carry out the following procedure for each vertex j, starting with $j = 1$ and continuing with $j = 2, j = 3$, and so on, until all the vertices (including the FINISH vertex, with label $n + 1$) have been labelled.

For the current vertex j, calculate, for each arc ij incident to vertex j, the sum

$$e_i + c_{i,j}.$$

Choose the maximum value of these sums for all such arcs ij, and set e_j equal to this value. Set p_j equal to the value of i for which this sum is largest; in the case of a tie, choose any of the appropriate values.

Part B: Backward Scan

STEP 3 Start with the FINISH vertex $n + 1$, and mark the arc joining this vertex to the preceding vertex p_{n+1}.

STEP 4 Consider the vertex j from which the last marked arc is incident. Mark the arc joining this vertex to the preceding vertex p_j. Repeat until the START vertex is reached.
(The marked arcs form a critical path.)

We illustrate the above procedure by an example.

Worked problem using the algorithm for finding a critical path Find a critical path in the following activity network for the building-construction example.

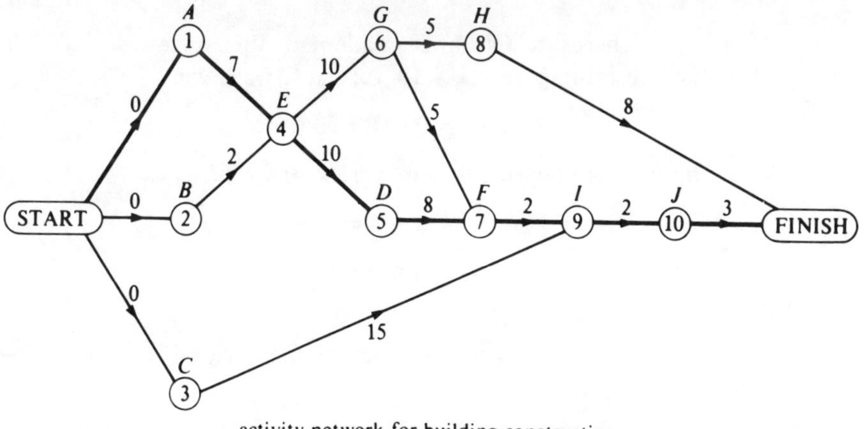

activity network for building construction

Solution The steps involved in applying the algorithm are given below.

Part A: Forward Scan

STEP 1 *Vertex 0* We label the START vertex with $p_0 = 0$ and $e_0 = 0$.

STEP 2 We consider each vertex in turn.

Vertex 1 There is only one arc incident to this vertex, with weight 0. Thus

$$e_1 = e_0 + c_{0,1} = 0 + 0 = 0$$

and

$$p_1 = 0.$$

Vertices 2 and 3 The situation is the same for these vertices as for vertex 1. The current weightings are shown in the following diagram:

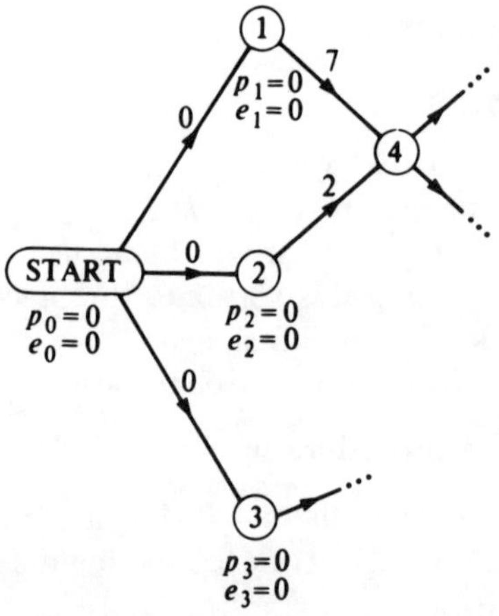

Vertex 4 There are two arcs incident to this vertex.
For the arc joining vertex 1 to vertex 4, we have

$$e_1 + c_{1,4} = 0 + 7 = 7.$$

For the arc joining vertex 2 to vertex 4, we have

$$e_2 + c_{2,4} = 0 + 2 = 2.$$

The largest of these sums is the first, so

$$e_4 = 7,$$

and we set p_4 equal to the number of the vertex from which the corresponding arc is incident, so

$$p_4 = 1.$$

Vertex 5 There is only one arc incident to vertex 5, so

$$e_5 = e_4 + c_{4,5} = 7 + 10 = 17$$

and

$$p_5 = 4.$$

Vertex 6 There is only one arc incident to vertex 6, so

$$e_6 = e_4 + c_{4,6} = 7 + 10 = 17$$

and

$$p_6 = 4.$$

The labelling of vertices 4, 5 and 6 is shown on the following diagram:

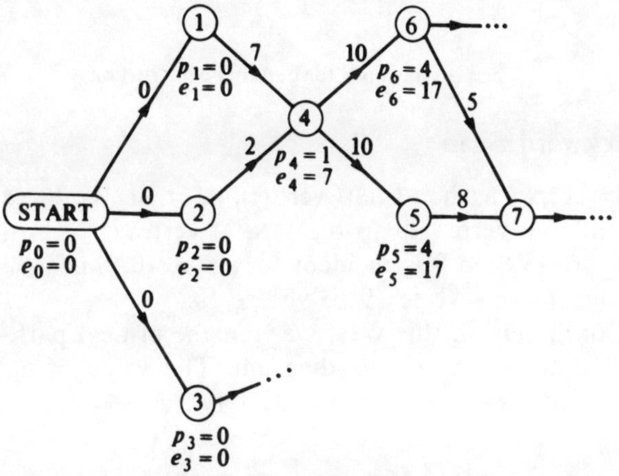

Vertex 7 There are two arcs incident to this vertex.
For the arc joining vertex 5 to vertex 7,

$$e_5 + c_{5,7} = 17 + 8 = 25.$$

For the arc joining vertex 6 to vertex 7,

$$e_6 + c_{6,7} = 17 + 5 = 22.$$

The larger of these sums is the first, so

$$e_7 = 25$$

and

$$p_7 = 5.$$

Continuing in this way, we label all the vertices; the resulting labels are shown in the following diagram:

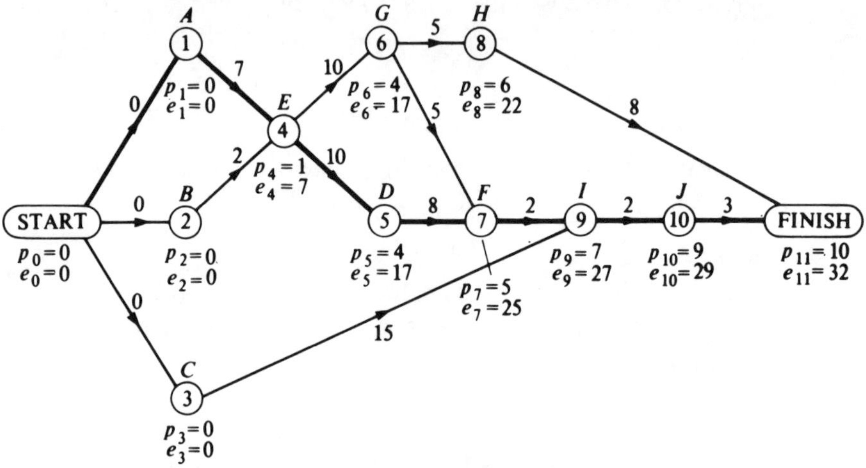

activity network for building construction

Part B: Backward Scan

STEP 3 For vertex 11 (the FINISH vertex), $p_{11} = 10$. We therefore mark the arc joining vertex 10 to the FINISH vertex.

STEP 4 The arc in Step 3 is incident to vertex 10. Since $p_{10} = 9$, we mark the arc joining vertex 9 to vertex 10.

Continuing in this way, we find the critical path shown by the thick lines in the above diagram. The value of e_{11} is 32, so the length of the critical path is 32 days.

15.5 EARLIEST AND LATEST STARTING TIMES

The critical path algorithm assigns a weight e_i to each vertex i. The value of e_i is the length of the longest path from the START vertex to the vertex i. In other words, e_i is the **earliest starting time** of activity i, since activity i cannot be started until all preceding activities have been completed. We can also calculate for each activity i the **latest starting time** l_i, which is the latest time activity i can be started without delaying the whole project. Clearly, the float of activity i (the maximum amount of time by which the activity can be delayed without delaying the project) is just the difference between these two times:

float of activity $i = l_i - e_i$.

Problem 15.3 In the following activity network, the critical path is indicated by thick lines, and the duration times are in days.

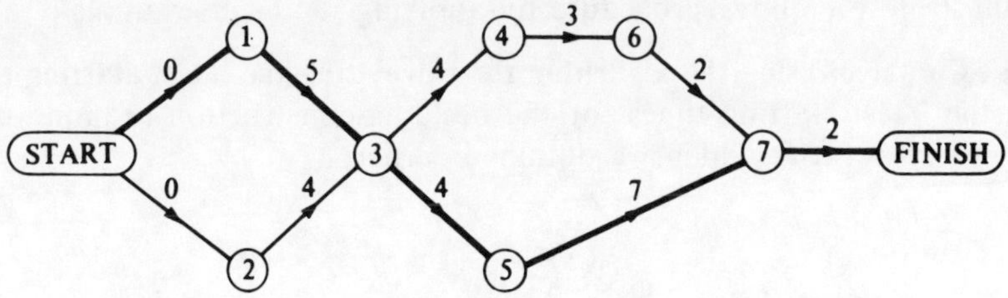

(a) What is the minimum completion time of the project?
(b) Calculate the earliest starting times for activities 4 and 6.
(c) Calculate the latest starting times for activities 4 and 6.
(d) Calculate the floats for activities 4 and 6.
(e) Can both activity 4 and activity 6 be delayed by their floats without delaying the completion of the project?

In the above problem, we calculated the latest starting times and the floats for two activities in a simple activity network. The following algorithm is a systematic method of calculating latest starting times for all the vertices. As in the solution to the above problem, we begin at the FINISH vertex and work back, vertex by vertex, until all the vertices have been considered.

Algorithm for Calculating Latest Starting Times

This algorithm is applied to an activity network (for n activities) for which we have calculated the minimum completion time. It is used to calculate the value of l_i for each vertex i, for $i = n + 1, n, \ldots, 1, 0$.

DESCRIPTION OF THE ALGORITHM FOR CALCULATING LATEST STARTING TIMES

STEP 1 For vertex $n + 1$ (the FINISH vertex), set l_{n+1} equal to the minimum completion time.

STEP 2 Carry out the following procedure for each vertex i, starting with $i = n$ and continuing with $i = n - 1$, $i = n - 2$, and so on, until all the vertices (including the START vertex, with label 0) have been considered.

For the current vertex i, calculate, for each arc ij incident from vertex i, the difference

$$l_j - c_{i,j},$$

where $c_{i,j}$ is the duration of activity i.

Choose the minimum value of these differences for all such arcs ij, and set l_i equal to this value.

(Note that l_o must be zero—if we do not obtain this value, then we know that we have made a mistake somewhere.)

We illustrate the above procedure by applying it to our example.

Worked problem using the algorithm for calculating the latest starting times
Find the latest starting times for the building-construction example using the activity network which we obtained earlier:

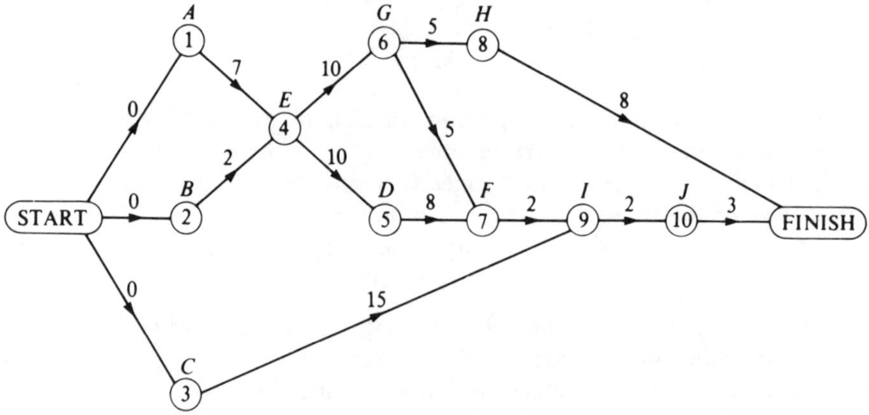

activity network for building construction

Solution The steps involved in applying the algorithm are given below.

STEP 1 For vertex 11 (the FINISH vertex), we set l_{11} equal to the minimum completion time—that is,

$$l_{11} = 32 \text{ days.}$$

STEP 2 We consider each vertex in turn.

Vertex 10 There is only one arc incident from this vertex, so

$$l_{10} = l_{11} - c_{10,11} = 32 - 3 = 29.$$

Vertices 9, 8 and 7 Only one arc is incident from each of these vertices, so

$$l_9 = l_{10} - c_{9,10} = 29 - 2 = 27,$$
$$l_8 = l_{11} - c_{8,11} = 32 - 8 = 24,$$
$$l_7 = l_9 - c_{7,9} = 27 - 2 = 25\,.$$

Vertex 6 There are two arcs incident from vertex 6. For the arc joining this vertex to vertex 8, the difference is

$$l_8 - c_{6,8} = 24 - 5 = 19,$$

and for the arc joining vertex 6 to vertex 7, the difference is

$$l_7 - c_{6,7} = 25 - 5 = 20.$$

The smaller of these differences is the first, so

$$l_6 = 19.$$

We continue in this way until we have calculated the latest starting time for each vertex. These times are shown in the following table:

Activity	Vertex number i	Earliest starting time, e_i	Latest starting time l_i	Float $(l_i - e_i)$
START	0	0	0	0
A	1	0	0	0
B	2	0	5	5
C	3	0	12	12
E	4	7	7	0
D	5	17	17	0
G	6	17	19	2
F	7	25	25	0
H	8	22	24	2
I	9	27	27	0
J	10	29	29	0
FINISH	11	32	32	0

This table also shows the float for each activity. Notice that any activity which is on a critical path has a float of zero—this must be so, since any delay in the completion of an activity on a critical path must delay the project.

Problem 15.4 The activity network for the process of assembling a bicycle is reproduced below.

(a) Use the critical path algorithm to find a critical path in the activity network, and calculate the minimum completion time of the project.
(b) Apply the algorithm for finding latest starting times, and calculate the float for each activity.

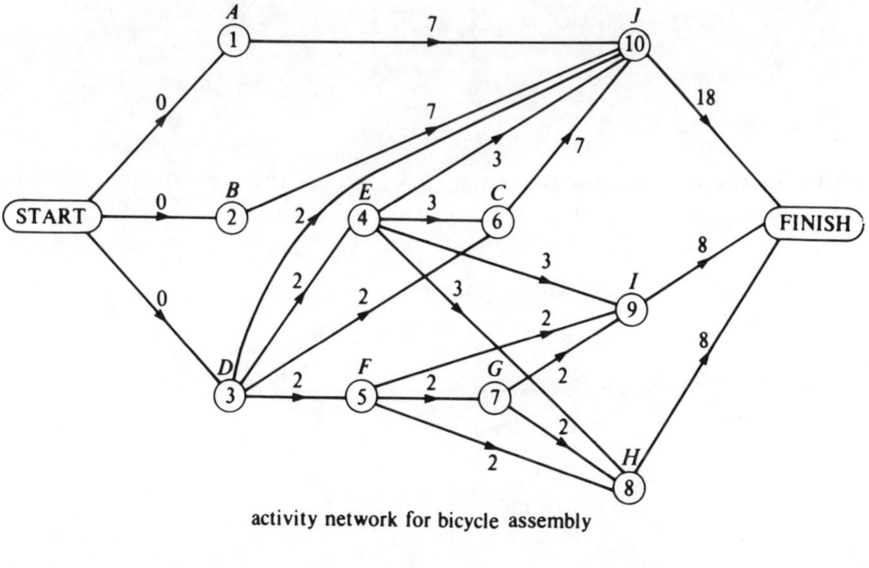

activity network for bicycle assembly

Before leaving the subject of activity networks, we mention an important difficulty which often arises in practice—namely that the estimated durations of the activities may not be accurate, or may vary during the planning and production stages. If estimates change, then the activity network must be updated and, if necessary, the calculations of earliest and latest starting times repeated. In particular, this may result in a different set of activities forming the critical path. The advantage of PERT is that at the planning stage three estimates of the time required may be specified for each activity—the optimistic time, the most likely time, and the pessimistic time. Moreover, most computers have a software package which will handle PERT calculations, thus facilitating the updating of the network, and the monitoring and control of the project once it is underway.

EXERCISES 15

Unless stated otherwise, all exercises refer to activity networks in which the vertices represent activities:

Activity Networks

15.1. Explain why the following network cannot be an activity network:

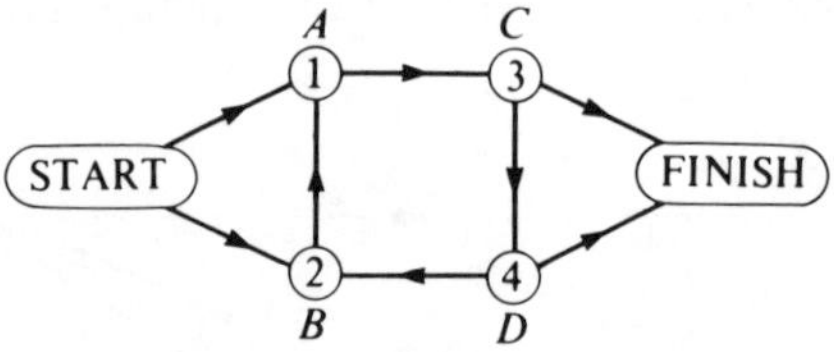

15.2. In the following activity network, two of the arcs are unnecessary. Which are they, and why can they be omitted?

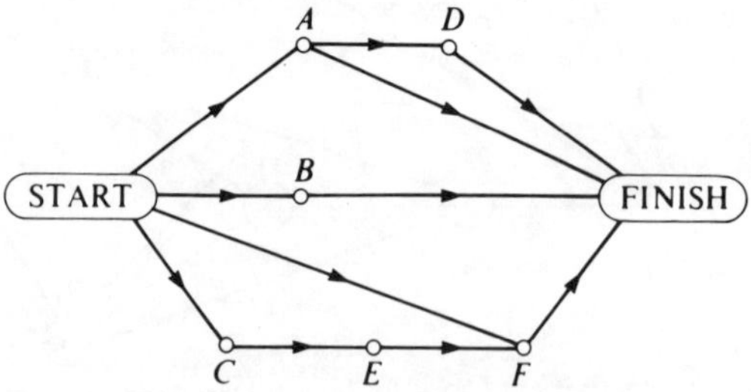

15.3. The following questions refer to the activity network shown below, in which the vertices represent activities and the numbers next to the arcs represent times in days.

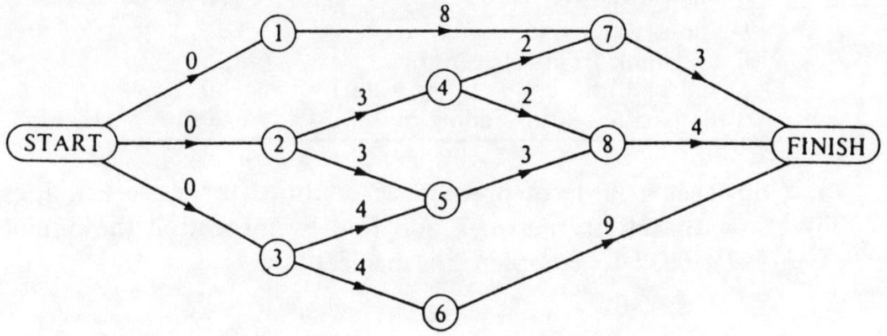

(a) What is the minimum completion time of the project represented by the activity network if only one worker is available?
(b) What is the minimum completion time of the project if an unlimited number of workers are available?
(c) What is the earliest starting time of activity 8?
(d) What is the latest starting time of activity 8?
(e) Which activity has the largest float, and what is the value of this float?

15.4. Suppose that you wish to redecorate a room and put in new self-assembly furniture. The activities to be carried out are as follows:

Activity	Time needed (hours)
C lay carpet	6
W lay wallpaper	10
P paint woodwork	7
F assemble furniture	3
H hang curtains	2

(a) Construct a table of precedence relations for these activities.
(b) Draw an activity network and find by inspection the minimum time needed to complete the project.

15.5. The process of making a cup of instant coffee can be broken up into the following activities:

	Activity	Time needed (seconds)
A	get cup from cupboard	30
B	put coffee and sugar into cup	25
C	boil water	60
D	pour water onto coffee	15
E	get milk from refrigerator	15
F	add milk to coffee (after water)	10
G	stir coffee (after adding milk)	20

(a) Construct a table of precedence relations for these activities.
(b) Draw an activity network and find by inspection the minimum time needed to complete the project.

Use of the Algorithms

15.6. Consider again the example given in the previous question. Using the algorithms given in this chapter:

(a) construct an activity network;
(b) find a critical path and the minimum completion time;
(c) calculate the latest starting time and the float for each activity.

15.7. Suppose that the process of cooking breakfast is split up into the following activities:

	Activity	Time needed (minutes)
A	grind coffee beans	1
B	slice bread	2
C	heat frying pan	3
D	lay table	5
E	make toast	3
F	make coffee	2
G	cook bacon	5
H	cook tomato	2
I	cook eggs	3

The precedence relations of these activities are as follows:

F must follow *A*
E must follow *B* and *C*
G must follow *C*
H must follow *E*, *F* and *I*
I must follow *B*, *C*, *E* and *G*

Using the algorithms given in this chapter:

(a) construct an activity network for this process;
(b) find a critical path and the minimum completion time;
(c) calculate the latest starting time and the float for each activity.

15.8. A project consists of ten activities *A–J*. The duration (in days) of each activity and the activities preceding each of them are as follows:

Activity	Duration	Preceding activities
A	10	—
B	4	—
C	8	*B*
D	6	*C*
E	8	*I*
F	5	—
G	10	*A*, *D*
H	2	*G*
I	4	—
J	10	*D*, *F*, *I*

Using the algorithms given in this chapter:

(a) construct an activity network for this project;
(b) find a critical path in this activity network;
(c) find the latest starting time and the float for each activity.

15.9. Construct an activity network (of the other type), in which vertices represent events, for the example given in Exercise 15.5.

15.10. Construct an activity network in which vertices represent events for the example given in Exercise 15.8.

CHAPTER 16

SCHEDULING

After reading this chapter, you should be able to:

- explain the terms *factory rules*, *heuristic method* and *protection scheme*;
- use the critical path scheduling algorithm to schedule activities for a number of workers.

Introduction

In the previous chapter we saw that the minimum completion time for a project is given by the length of a critical path in the corresponding activity network. This is the shortest time in which the project can be completed if there is no restriction on the number of workers available. However, if there is a limit on the number of workers available, it may not be possible to achieve this minimum completion time. If the product of this minimum completion time and the number of workers is less than the sum of the durations of all the activities, then it is obviously impossible to finish in the minimum completion time. Even if the product is not less than the sum of the durations, the precedence relations may be such that some workers must have idle periods, so that it is again impossible to finish in the minimum completion time.

In this chapter we investigate the problem of scheduling the activities of a project for a given number of workers in the best possible way.

16.1 SCHEDULING A PROJECT FOR A GIVEN NUMBER OF WORKERS

We suppose that an ideal schedule satisfies the following conditions, which we call the *factory rules.*

Factory rules

1. *No worker may be idle if there is some activity which can be done.*
2. *Once a worker starts an activity, that activity must continue until it is completed.*
3. *The project must be completed as soon as possible with the manpower available.*

These rules imply that each worker should have a tightly-packed schedule, and that activities on a critical path should be started as soon as possible.

Suppose, for example, that we wish to schedule the bicycle-assembly project of the previous chapter for two workers according to the factory rules. (This activity network was constructed in Problem 15.1 and the critical path was determined in Problem 15.4.)

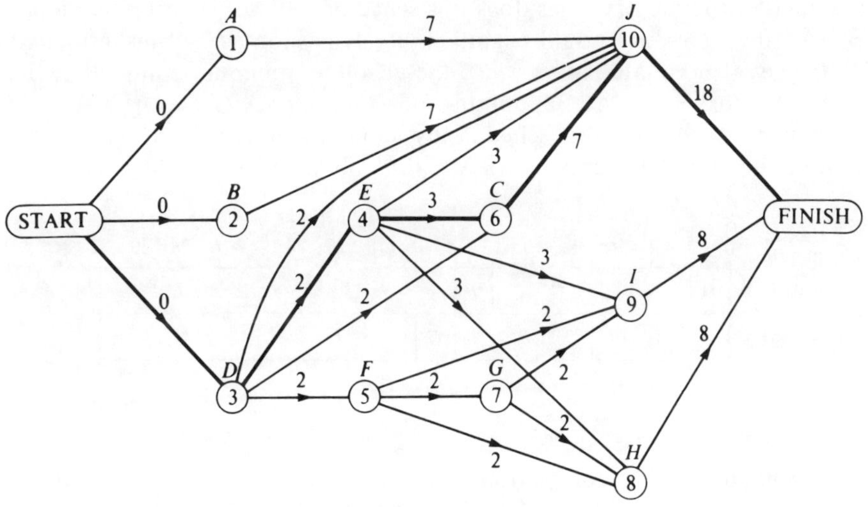

activity network for bicycle assembly

As a first attempt at a schedule, let us assign all the activities on the critical path to worker 1, and all the remaining activities to worker 2. In Problem 15.4, we showed that the activities on the critical path take a total of 30 minutes. Since the sum of the durations for all activities is 64 minutes, the activities assigned to worker 2 take a total of 34 minutes to complete. So, if the activities are scheduled in this way, the shortest possible time for the completion of the project is 34 minutes. The actual time may be greater because of the precedence relations, which mean that some activities cannot be started until others have been completed.

Let us assign the activities in this way—we assign all the activities on the critical path to worker 1, and the remaining activities, in the order of the numbering scheme, to worker 2. In doing this, we must also take into account the precedence relations involving activities done by both workers. The resulting schedule is the following:

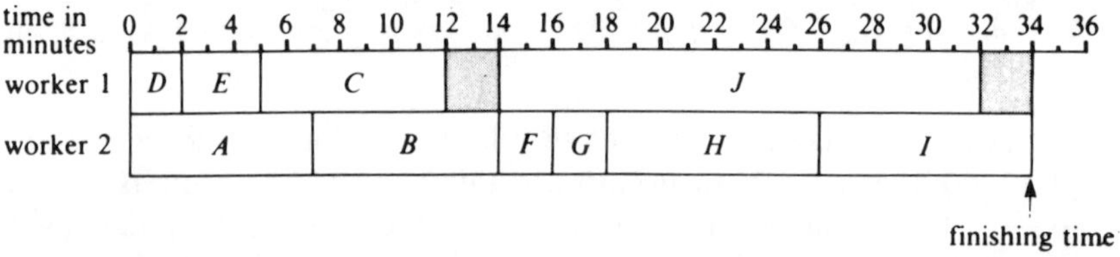

As it happens, this schedule takes 34 minutes—the minimum possible time which we calculated for this way of assigning activities. We notice that worker 1 is idle for two time-intervals.

Which of the factory rules does this schedule violate? It certainly violates rule 1, since worker 1 could begin activity *F* as soon as he finishes activity *C*. It may also violate rule 3, since the absolute minimum completion time with two workers is $\frac{64}{2} = 32$ minutes.

We can improve on this schedule by assigning activity *F* to worker 1 as soon as he completes activity *C*. We thus obtain the following schedule:

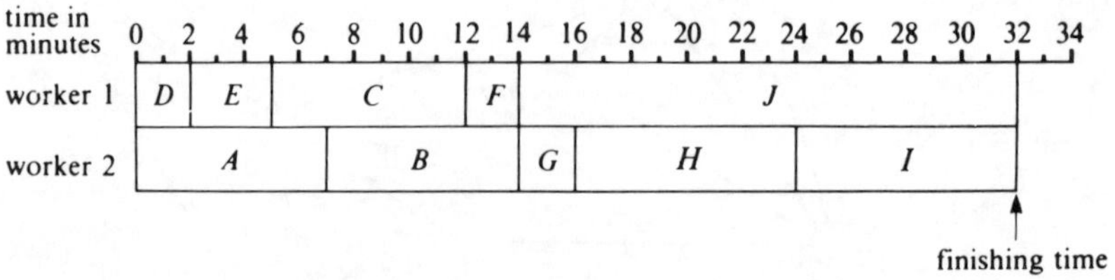

This schedule has a completion time of 32 minutes—it is an optimum schedule in the sense that it has the shortest possible completion time for any schedule for two workers. This schedule also satisfies all the factory rules.

We arrived at this optimum solution by a trial-and-error method. Is it possible to devise a systematic method which will produce an optimum schedule for any activity network with a given number of workers? The answer to this question appears to be NO—there is no known practical algorithm for doing this. You may argue that, in principle, we can list all possible schedules and choose one with a minimum completion time. However, the number of possible schedules grows so rapidly with the number of activities involved that there is no possibility of examining even a small fraction of them, even using a computer, when the number of activities is large. This is another example of the combinatorial explosion.

Since there is no practical algorithm for this type of scheduling problem, we use instead a **heuristic algorithm**—that is, one depending on assumptions based on past experience—which does not necessarily give an optimum solution, but usually gives a reasonable one. The following heuristic algorithm, called the *critical path scheduling algorithm*, has often been used in industry. It produces a schedule which satisfies the first two factory rules, but not necessarily the third rule.

16.2 THE CRITICAL PATH SCHEDULING ALGORITHM

The algorithm assigns activities to a number of processors, which may be people or machines. It is assumed that the activity network has been constructed and the latest starting times calculated.

The algorithm makes use of a hypothetical clock to keep track of time. We call this the *project clock*. At the end of each iteration, we advance the project clock so that it records the amount of time for which the project has been running.

DESCRIPTION OF THE CRITICAL PATH SCHEDULING ALGORITHM

START Set the project clock to 0.

STEP 1 If at least one processor is free, assign to any free processor the most critical unassigned activity which can be started.
(The most critical activity is the one with the smallest latest starting time.)
Repeat until no processor is free or until no unassigned activity can be started.

STEP 2 Advance the project clock until a time is reached when at least one activity is completed, so that at least one processor is free.

STEP 3 If all the activities have been assigned, advance the project clock until all the activities have been completed, and then STOP.
Otherwise, go back to Step 1.

Worked problem using the algorithm to schedule the bicycle-assembly projec Apply the critical path scheduling algorithm to the bicycle-assembly project (introduced in Chapter 15) for two workers. The activity network for this project, and a list of the latest starting times, are given below.

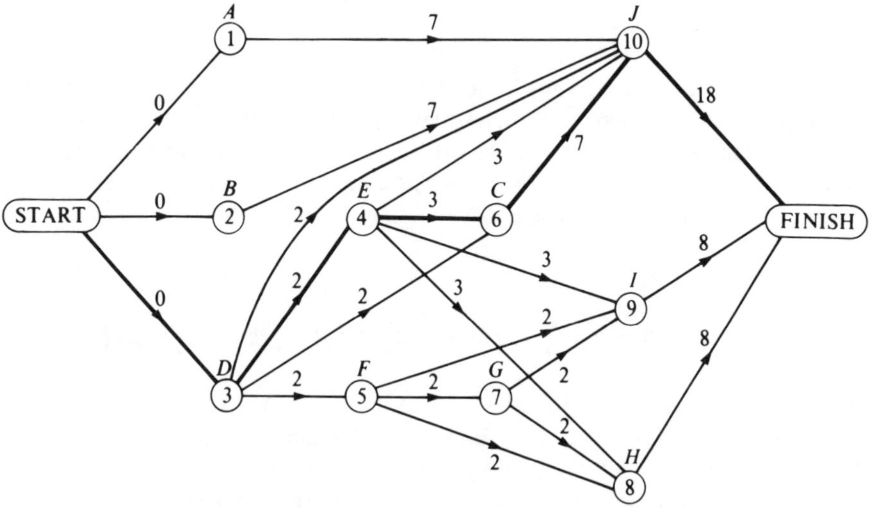

activity network for bicycle assembly

vertex number	1	2	3	4	5	6	7	8	9	10
activity	*A*	*B*	*D*	*E*	*F*	*C*	*G*	*H*	*I*	*J*
latest starting time in minutes	5	5	0	2	18	5	20	22	22	12

Solution The steps involved in applying the algorithm are given below.

START Set the project clock to 0.

STEP 1 The activities which can be started are *A*, *B* and *D*. The most critical of these is activity *D*, since it has the smallest latest starting time. We therefore assign activity *D* to worker 1.

STEP 1 Activities *A* and *B* can be started, and both have the same latest starting time, so either can be chosen. We assign activity *A* to worker 2.

STEP 2 Advance the project clock to 2 minutes.
Activity *D* is now completed and worker 1 is free.

Activities *E* and *F* are now free to be started.

STEP 1 There are now three activities which may be started—activities *B*, *E* and *F*. Activity *E* has the smallest latest starting time, so we assign activity *E* to worker 1.

STEP 2 Advance the project clock to 5 minutes.
Activity *E* is now completed, and so worker 1 is free.
Activity *C* is now free to be started.

The current state of the scheduling of the activities is as follows:

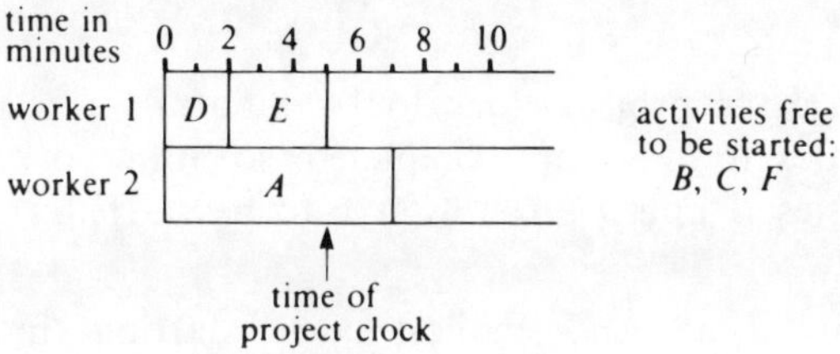

STEP 1 Of the three activities which are free to be started, *B* and *C* are the most critical. Both have a latest starting time of 5 minutes. Since the project clock is now at 5 minutes, and since we cannot assign both of these activities to a worker at this point, there will be a delay in the completion of the project—in other words, the time taken with this schedule will be greater than the length of the critical path.

We assign activity *C* to worker 1.

STEP 2 Advance the project clock to 7 minutes.
Activity *A* is completed, so worker 2 is free.
No further activity is made free to be started by the completion of activity *A*.

STEP 1 Activities *B* and *F* are free to be started. Since activity *B* has the smallest latest starting time, we assign activity *B* to worker 2.

STEP 2 Advance the project clock to 12 minutes.
Activity *C* is completed, so worker 1 is free.
No further activity is made free to be started by the completion of activity *C*.

STEP 1 Activity F is the only activity which is free to be started, so we assign activity F to worker 1.

The current state of the scheduling of the activities is as follows:

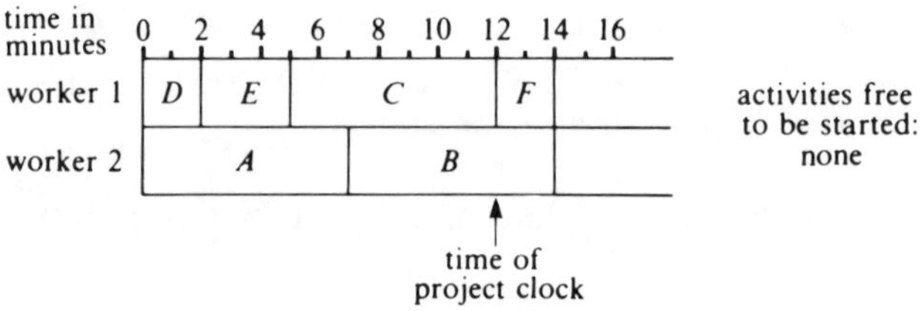

STEP 2 Advance the project clock to 14 minutes.
Activities B and F are completed, so both workers are free.
Activities G and J are now free to be started.

STEP 1 Activity J has the smallest latest starting time, so we assign activity J to worker 1.

The only remaining activity which can be started is G, so we assign activity G to worker 2.

STEP 2 Advance the project clock to 16 minutes.
Activity G is completed, so worker 2 is free.
Activities H and I are now free to be started.

STEP 1 Activities H and I have the same latest starting time. We assign activity H to worker 2.

STEP 2 Advance the project clock to 24 minutes.
Activity H is completed and worker 2 is free.

STEP 1 Activity I is the only remaining activity which is free to be started, so we assign activity I to worker 2.

STEP 2 Advance the project clock to 32 minutes.
All activities have now been completed and both workers are free.

STEP 3 STOP

The resulting schedule is the following:

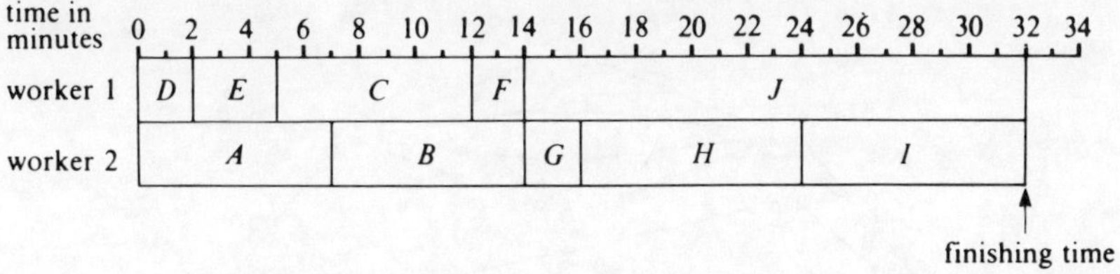

The schedule is the same as the one we obtained by trial and error. In other words, the algorithm produces an optimum schedule in this case.

Problem 16.1 Use the critical path scheduling algorithm to schedule the building-construction project discussed in Chapter 15 for two workers.

The activity network and the latest starting times are as follows:

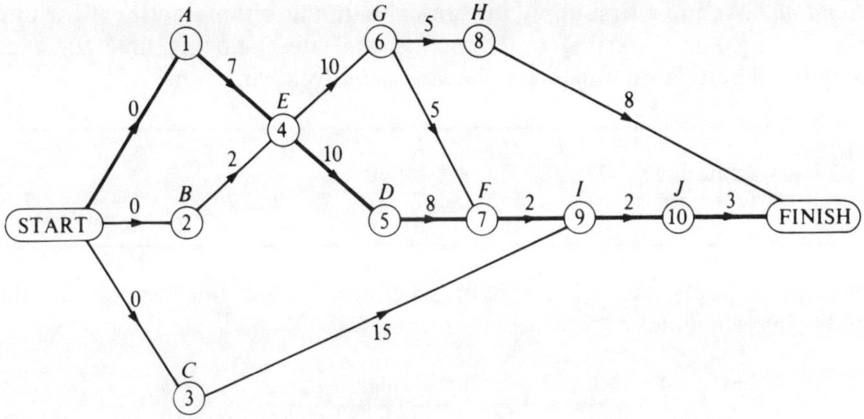

vertex number	1	2	3	4	5	6	7	8	9	10
activity	*A*	*B*	*C*	*E*	*D*	*G*	*F*	*H*	*I*	*J*
latest starting time in days	0	5	12	7	17	19	25	24	27	29

In the previous problem, and in the example discussed earlier, the critical path scheduling algorithm produces optimum schedules. In the next problem, we consider a case in which the algorithm does not produce an optimum schedule. The solution to this problem is given in the following text. We suggest that you attempt the problem for yourself before reading our solution.

Problem 16.2 Apply the critical path scheduling algorithm to the project represented by the following activity diagram, given that four workers are available. (In this example, the vertices are numbered in alphabetical order, so we have omitted the numbers.)

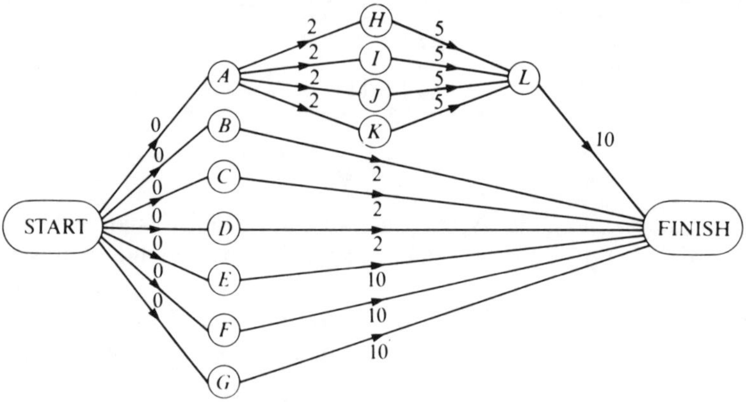

Solution We must first apply the critical path algorithm and the algorithm for finding latest starting times to find the latest starting time for each activity. These latest times are shown in the following table:

activity	*A*	*B*	*C*	*D*	*E*	*F*	*G*	*H*	*I*	*J*	*K*	*L*	FINISH
earliest starting time	0	0	0	0	0	0	0	2	2	2	2	7	17
latest starting time	0	15	15	15	7	7	7	2	2	2	2	7	17
float	0	15	15	15	7	7	7	0	0	0	0	0	0

When we apply the critical path scheduling algorithm, we obtain the following schedule:

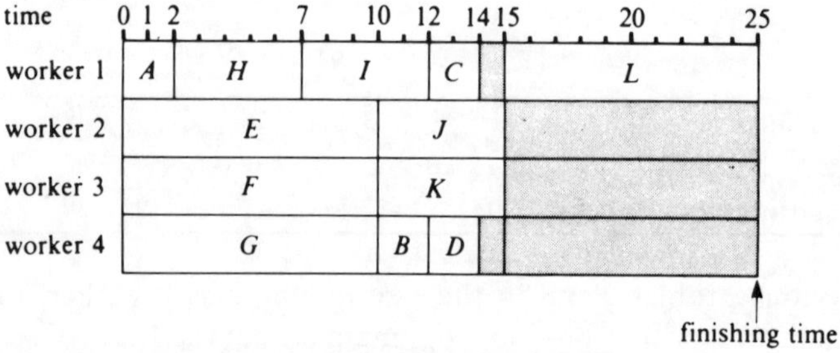

16.3 ALGORITHMS WITH PROTECTION SCHEMES

The schedule obtained in the above problem is far from an optimum one, as can be seen by comparing it with an optimum schedule such as the following.

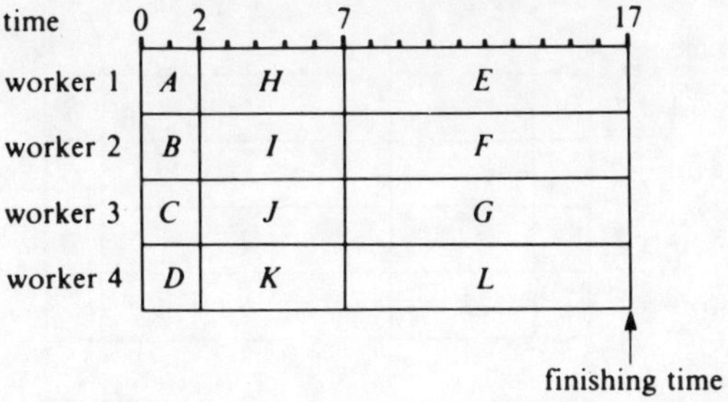

In a situation like this, it is obviously desirable to produce a schedule which is much better than the one obtained by applying the algorithm. To do this in industry, some form of *protection scheme* is usually incorporated in a practical algorithm. There is no general protection scheme, so a network analyst working in a particular situation must develop a protection scheme which is appropriate for the type of scheduling problems involved. For example, for the project discussed in the above problem, a protection scheme may be incorporated in the algorithm as follows.

MODIFIED ALGORITHM INCORPORATING PROTECTION SCHEME FOR THE ABOVE EXAMPLE

STEP 1 Compute for each activity the sum of the earliest starting time and the latest starting time, and rank the activities in ascending order according to the values of these sums.

STEP 2 Assign activities successively to free workers according to this ranking, taking into account the precedence relations.

The following table gives the value of the sum of the earliest and latest starting times for each activity, and ranks the activities according to the values of these sums:

sum	0	4	4	4	4	7	7	7	14	15	15	15
activity	*A*	*H*	*I*	*J*	*K*	*E*	*F*	*G*	*L*	*B*	*C*	*D*

We now go on to Step 2 and assign each activity in turn to a free worker. If an activity cannot be assigned immediately because of the precedence relations, we wait until the appropriate previously assigned activities have been completed. (If a preceding activity has not already been assigned, we must break the ranking order when assigning the activities. However, this difficulty does not arise in this example.)

This procedure produces the following schedule.

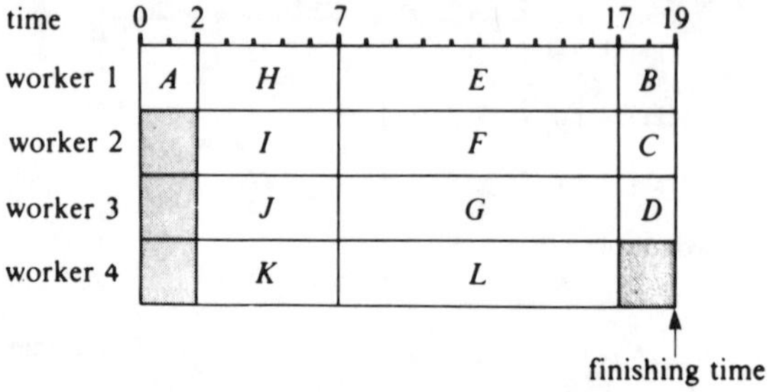

This is very close to an optimum schedule. This schedule does not satisfy the first factory rule, but it comes close to satisfying rule 3.

Once we have produced a schedule using such a protection scheme, we may be able to improve it by moving activities to fill idle time-intervals. For example, in the above schedule, it is a simple matter to fill the initial idle time-intervals of workers 2, 3 and 4 with activities *B*, *C* and *D*. In this way we arrive at the optimum schedule we gave earlier.

This example illustrates the fact that the algorithm incorporating the protection scheme is likely to produce a better solution than the original algorithm when the sum of the completion times for the activities of the project is appreciably greater than the length of a critical path.

EXERCISES 16

The Critical Path Scheduling Algorithm

16.1. Use the critical path scheduling algorithm to schedule the 11 jobs in the following activity network for two workers. There is only one critical path, which is indicated on the diagram by the thick arcs. The latest starting times (in days) for each activity are shown on the next page.

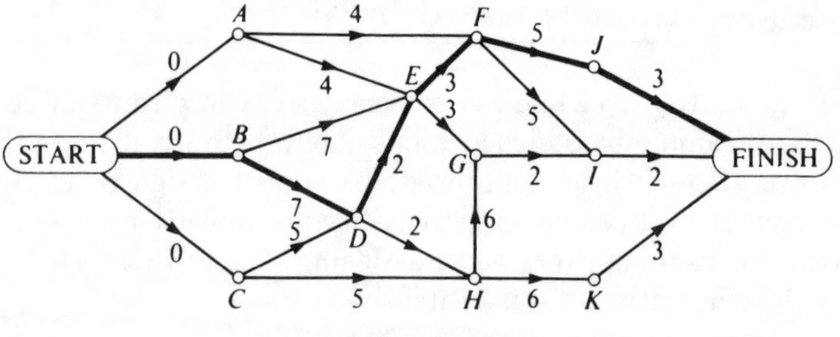

vertex	*A*	*B*	*C*	*D*	*E*	*F*	*G*	*H*	*I*	*J*	*K*
latest starting time	5	0	2	7	9	12	16	10	18	17	17

16.2. For the project described in Exercise 16.1, how much time would be saved if three workers were available? Discuss the advantages and disadvantages of having three workers rather than two.

16.3. Apply the critical path scheduling algorithm to the activity network of Exercise 15.3 if two workers are available.
Find (by trial and error) an optimum solution.

16.4. Use the critical path scheduling algorithm to schedule the project described in Exercise 15.8. Is your schedule an optimum schedule? If so, explain why; if not, find a better one.

16.5. Consider the following activity network, in which the vertices represent activities and the numbers next to the arcs represent time (in days):

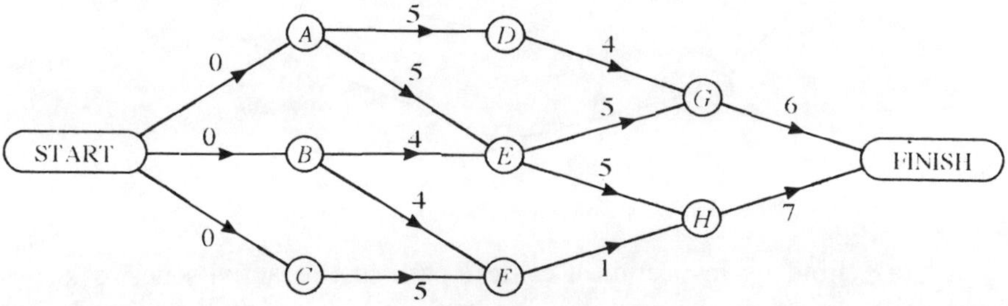

(a) What is the minimum completion time if an unlimited number of workers is available?
(b) What is the minimum completion time if only one worker is available?
(c) Use the critical path scheduling algorithm to schedule the project for two workers. Is your schedule an optimum schedule?

16.6. Apply the critical path scheduling algorithm to the activity network for the process of cooking breakfast obtained in Exercise 15.7 for each of the following cases:

(a) three cooks are available;
(b) two cooks are available.

In each case determine (by trial and error) whether your solution is an optimum schedule, and if not, find an optimum schedule.

Use of Protection Schemes

16.7. In Exercise 16.6, case (b), would it be advantageous to incorporate the protection scheme described in this chapter?

16.8. In the following activity network, the numbers are the completion times in days:

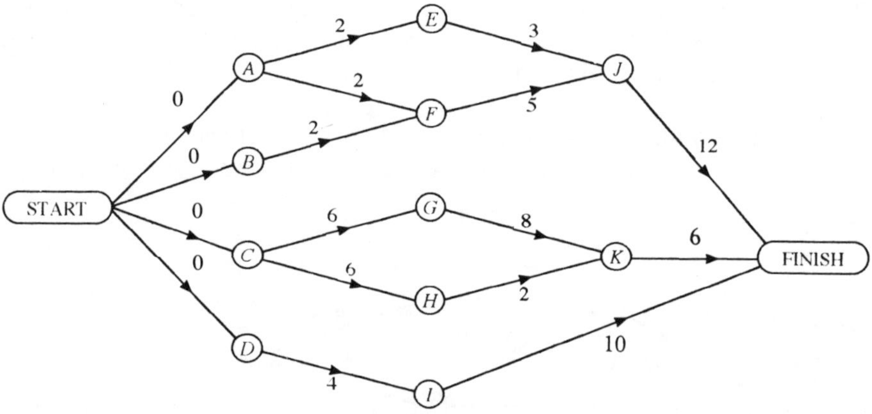

(a) Find by inspection a critical path in this activity network.
(b) Schedule the activities for two workers, using the critical path scheduling algorithm.
(c) Schedule the activities for two workers, using the critical path scheduling algorithm with the protection scheme given in this chapter.

16.9. Explain why the algorithm with the protection scheme given in the text is likely to produce a better solution than the original algorithm for examples in which the sum of the completion times is appreciably greater than the length of a critical path.

CHAPTER 17

PACKING PROBLEMS

After reading this chapter, you should be able to:

- explain what is meant by the *bin-packing problem*, and use the first-fit procedure and the first-fit decreasing procedure to obtain solutions to instances of this problem;
- explain what is meant by the *knapsack problem*, and use the branch-and-bound method to solve instances of this problem.

Introduction

In the previous chapter we discussed scheduling problems in which we wished to schedule the activities of a project for a given number of workers so that the project would be completed in the shortest possible time. We now turn the problem around, and ask for the minimum number of workers required to complete the project within a given period (which must, of course, be greater than or equal to the length of a critical path in the activity network for the project).

17.1 THE BIN-PACKING PROBLEM

Suppose, for example, that a project consists of eleven activities A–K with the following durations (in days) There are no precedence relations.

activity	A	B	C	D	E	F	G	H	I	J	K
duration	8	7	4	9	6	9	5	5	6	7	8

What is the minimum number of workers required to finish the project in 15 days?

This problem is called a **bin-packing problem**, since it amounts to trying to pack the activities in any order into the minimum number of 'bins', each with a capacity of 15 days.

There is no known algorithm which will always give an optimum solution to the bin-packing problem. However, there are a number of simple heuristic algorithms for finding an approximate solution.

One obvious method is **first-fit packing**, in which the items (in this case, activities) are packed in the order in which they occur into a series of bins b_1, b_2, . . . , each of the required capacity (in this case, 15 days). The items are packed according to the following rule:

always place the next item to be packed in the lowest-numbered bin which can take that item.

If we pack the activities in our example according to the first-fit packing procedure, we obtain the following packing:

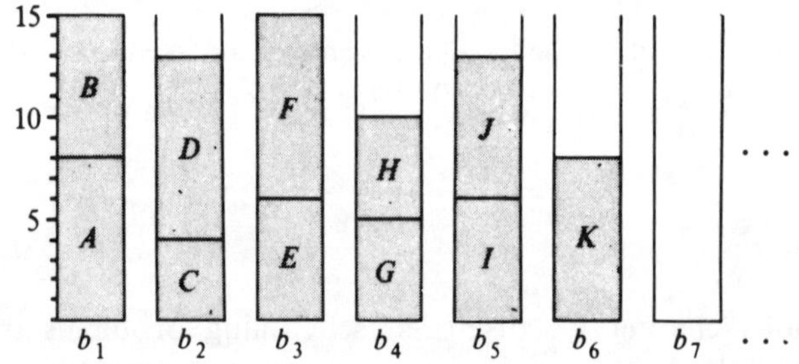

This packing scheme uses 6 bins so, in terms of our original problem, this scheme uses 6 workers to complete the project in 15 days. Notice that the sum of the unfilled capacities of the first 5 bins is 9, which is greater than the duration of activity K. We might suspect that there is a better packing scheme which uses only 5 bins, but we have no means of finding out whether such a scheme exists, apart from trying other packing methods.

How can we improve on the first-fit method? Intuitively, it seems better to pack the activities with the longest durations first, so that we can use activities with smaller durations to fill the spaces in the bins. This idea is the basis of the **first-fit decreasing procedure**, in which we fill the bins according to the following rules.

1. *Reorder the items in decreasing order of size.*
2. *Apply the first-fit procedure to this reordered list.*

Let us try this procedure on our example. First, we reorder the activities in decreasing order of duration: (in days):

activity	*D*	*F*	*A*	*K*	*B*	*J*	*E*	*I*	*G*	*H*	*C*
duration	9	9	8	8	7	7	6	6	5	5	4

When we apply the first-fit procedure, we obtain the following packing scheme:

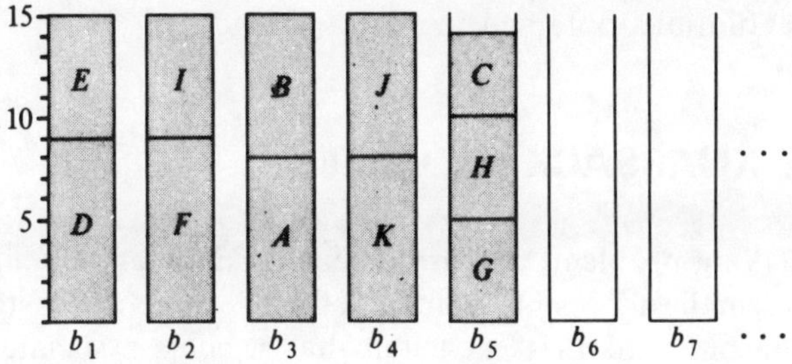

This packing scheme uses only 5 bins, so for our original problem this scheme requires the use of only 5 workers to complete the project within the given time period.

In this case, the first-fit decreasing method has produced an optimum packing (that is, one using the fewest possible bins), but it does not produce an optimum packing in general. However, the first-fit decreasing method is better than the first-fit method, in the sense that it usually produces a packing with fewer bins.

The bin-packing problem occurs in many different practical situations, such as the following.

Example 1 A plumber wishes to minimize the number of pipes of a standard length which he needs in order to cut a given number of lengths of pipe. In this case, the 'bins' are the standard-length pipes, and the problem is to 'pack' the required lengths of pipe into the smallest number of the standard-length pipes.

Example 2 A company wishes to allocate advertisements on commercial television. In this case, the 'bins' are standard-length breaks in programmes, and the problem is to 'pack' a list of advertisements into the smallest possible number of such breaks.

Problem 17.1 A carpenter has some planks of wood, each 12 feet long. He wishes to cut sections from these planks of the following lengths (in feet):

section	*A*	*B*	*C*	*D*	*E*	*F*	*G*	*H*	*I*	*J*	*K*	*L*
length	2	2	3	3	3	3	4	4	4	6	7	7

The problem is to find a way of cutting these sections so that the minimum number of planks is used.

(a) Find a solution using the first-fit method.
(b) Find a solution using the first-fit decreasing method.
(c) By trial and error, find an optimum solution which uses fewer planks than the solutions obtained in parts (a) and (b).

17.2 THE KNAPSACK PROBLEM

In the bin-packing problem, we wish to pack a number of items of various sizes into the smallest possible number of bins. Suppose, instead, that we have only one bin, and that each item has a *value* associated with it as well as a size. Which items should we choose to pack in the bin so that the total value is as large as possible? This problem is called the *knapsack problem*, since it may be formulated in the following terms.

THE KNAPSACK PROBLEM
A hiker going on a journey can carry only a certain total weight in her knapsack. She has a number of items she would like to take with her, each of which has a particular value to her for the journey. Which items should she pack so that the total value is a maximum, subject to the weight restriction?

A more practical interpretation of this problem is the following.

A company has a certain limited resource which can be used for a number of applications. Each application has a certain value, and uses a certain amount of the resource. Which applications should be chosen so that the greatest total value is obtained from the use of the resource?

The technique we give for solving the knapsack problem, called the **branch-and-bound method**, involves a search through a tree of possible solutions. We explain how this method works in the context of the following example.

Suppose that five items are available whose weights and values are given in the following table:

item	*A*	*B*	*C*	*D*	*E*
weight	3	8	6	4	2
value	2	12	9	3	5

The problem is to find a packing of the largest possible total value with a total weight not exceeding 9. In principle, we could list all possible packings and choose one with a maximum total value, subject to the total weight restriction. However, this would obviously not be practical for an example with a large number of items. The branch-and-bound method, which we now describe, is a more efficient method.

The first step is to list the items in decreasing order of value per unit weight, as follows:

order number, i	1	2	3	4	5
item	*E*	*B*	*C*	*D*	*A*
weight, w_i	2	8	6	4	3
value, v_i	5	12	9	3	2
value per unit weight	2.5	1.5	1.5	0.75	0.67

(Note that this ordering of items is not essential for this particular type of problem. However, it is an important feature of the branch-and-bound method when used to obtain approximate solutions. In such cases we do not examine all possible feasible solutions.)

We denote each possible packing by a **solution vector** of the form $(x_1, x_2, x_3, x_4, x_5)$, where each x_i takes the value 1 if item i is packed, and takes the value 0 if that item is not packed. For example, the solution vector (0, 0, 1, 1, 0) denotes the packing which includes only items 3 and 4. The total weight w for this solution vector is $w_3 + w_4 = 10$, and the total value v is $v_3 + v_4 = 12$. A solution which satisfies the weight constraint ($w \leq 9$) is called a **feasible solution**. The *null solution* (0, 0, 0, 0, 0) is obviously a feasible solution, but (0, 0, 1, 1, 0) is infeasible.

The branch-and-bound method uses a branching procedure to search for an optimum solution. Thus, for example, if we take the solution vector (0, 1, 0, 0, 0), we branch out to other solution vectors as follows:

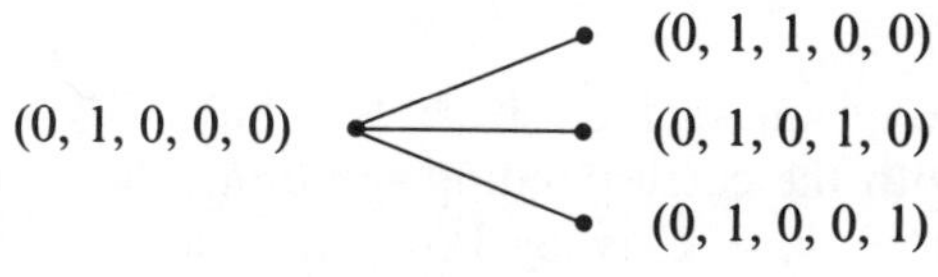

The solutions produced by the branching procedure are those which have one more item than the starting solution, but the number of such new solutions is restricted—we may add a new item only if it has a higher order number than any item already packed—that is, we may change only those positions to the right of the last 1 in the starting solution. Thus, for example, if we start with the solution vector (1, 0, 0, 1, 0), we can add only one item (the fifth), so only one branch is permitted:

The branch-and-bound method uses this branching procedure, starting with the null solution, to examine possible solutions in a systematic way. At each stage, we calculate the total weight for each solution, and the total value for each feasible solution, and branching is then continued from a feasible solution which has not been previously used as a branching point, and which has the highest value of any such solution. Solutions which cannot be branched out further, either because the weight limit has already been reached, or because there is an item at the extreme right position in the solution vector, are marked. The procedure is continued until all end-vertices are marked. A record is kept of the best solution obtained so far, and this is updated if necessary at each stage.

We illustrate this procedure by applying it to our example.

Worked problem using the branch-and-bound method Find an optimum packing for the above example.

Solution First, we branch out from the null solution, as shown below.

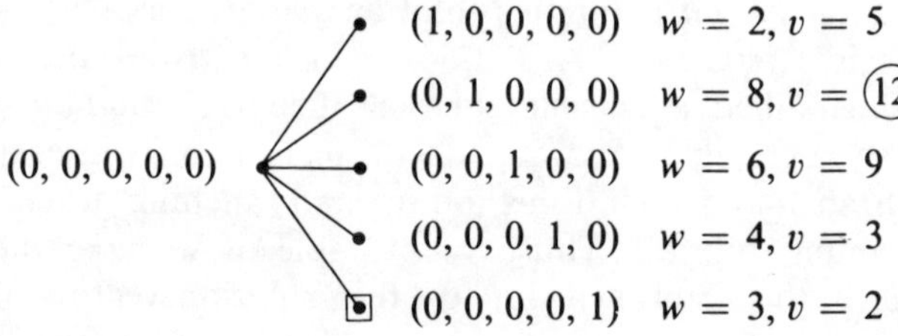

STORE $v = 12$, solution = (0, 1, 0, 0, 0).

The highest value of any of these feasible solutions is 12, so we store this value, together with the corresponding solution. The vector corresponding to the solution (0, 0, 0, 0, 1) is marked with a square, to indicate that we cannot continue the branching process from this vertex. We delete the marked vertex and continue the branching process from the solution with the highest value—that is, (0, 1, 0, 0, 0).

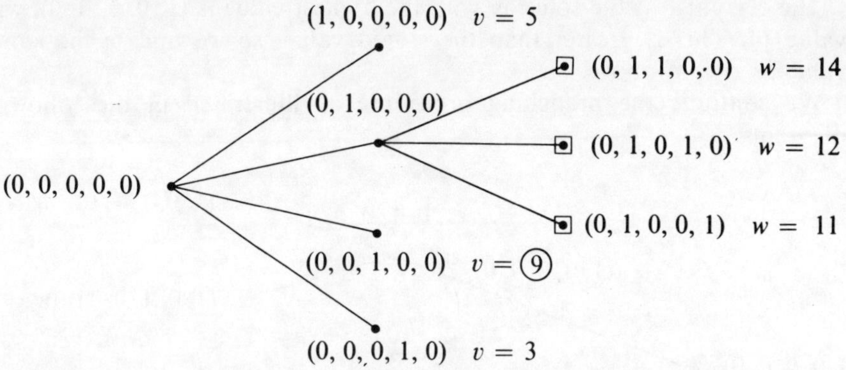

STORE $v = 12$, solution = (0, 1, 0, 0, 0)

All three new solutions are infeasible, so we mark the corresponding vertices with squares. The current feasible solution with the highest value is (0, 0, 1, 0, 0), so we branch out from this vertex.

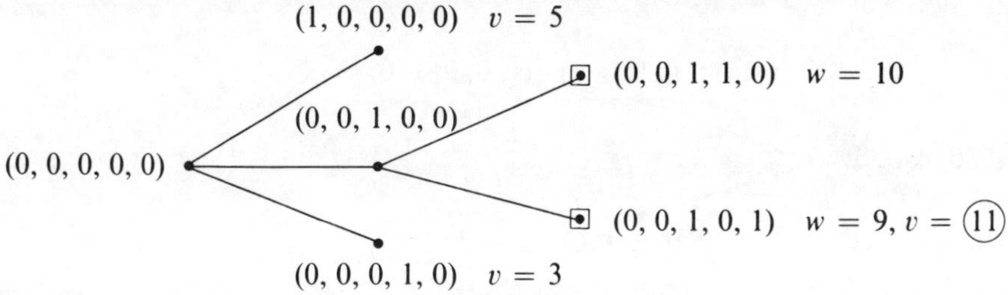

STORE $v = 12$, solution = (0, 1, 0, 0, 0)

Although the new solution (0, 0, 1, 0, 1) is feasible, we cannot branch further from it, because it has an item at the furthermost place to the right, and also because it has already reached the maximum allowable weight. We therefore mark this solution and also the infeasible solution (0, 0, 1, 1, 0).

We continue the branching process from the solution (1, 0, 0, 0, 0).

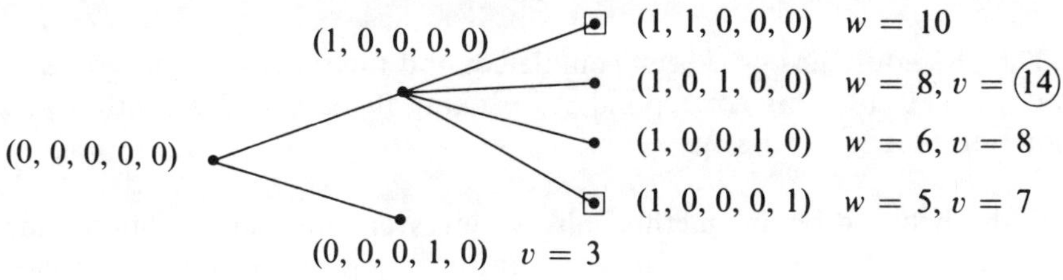

STORE $v = 14$, solution = (1, 0, 1, 0, 0)

The current feasible solution with the highest value is (1, 0, 1, 0, 0), with value 14. This is greater than the stored value, so we update the stored value as shown above.

We continue the branching procedure as illustrated in the following diagrams.

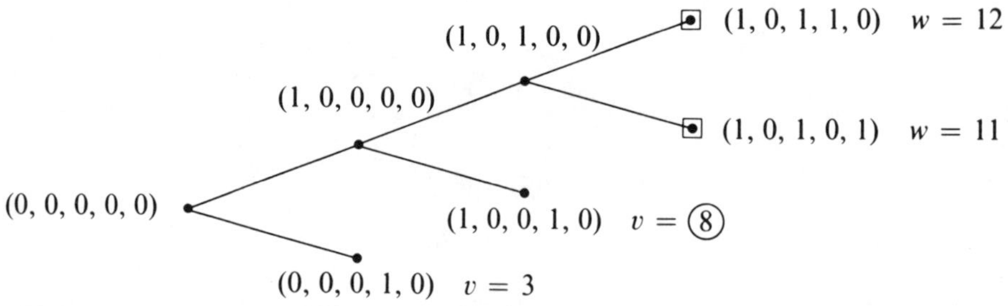

STORE $v = 14$, solution = (1, 0, 1, 0, 0)

At the next step we get:

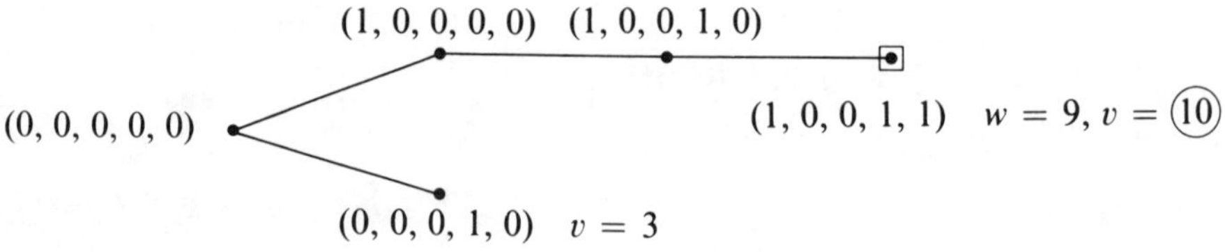

STORE $v = 14$, solution = (1, 0, 1, 0, 0)

Finally, we get:

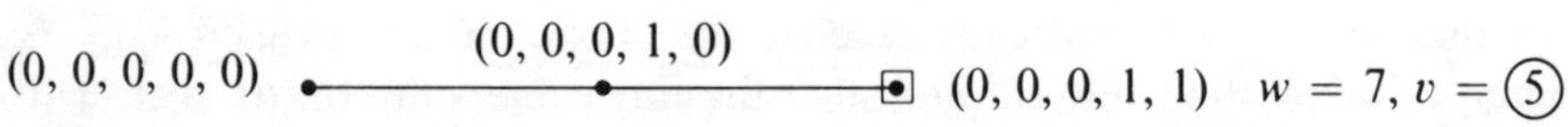

STORE $v = 14$, solution = (1, 0, 1, 0, 0)

The procedure has now been completed, and the required solution vector is (1, 0, 1, 0, 0). This corresponds to packing items *C* and *E*, with a total value of 14.

This branch-and-bound method always gives an optimum solution, since it examines all solutions which occur in the branching process. It is more efficient than simply listing and examining all possible solutions, since the branching process cuts out many solutions which are infeasible. For example, in the solution to the above problem, there are $2^5 = 32$ possible solutions, but only 18 of these are examined in the branch-and-bound procedure.

Problem 17.2 A machine in a factory can be used to make any of five items, *A*, *B*, *C*, *D* and *E*. The time (in days) taken to produce each item, and the value of each item, are shown in the following table:

item	*A*	*B*	*C*	*D*	*E*
production time	3	7	2	4	4
value	3	14	3	7	8

If the machine is available for only 10 days, which of the items should be produced so that the total value is as large as possible?

EXERCISES 17

The Bin-packing Problem

17.1. Explain the difference between the first-fit packing method and the first-fit decreasing method for the bin-packing problem. Does either method always yield an optimum solution?

17.2. Ten examinations (which must not overlap) are to be scheduled in the shortest possible number of days. The maximum allowable number of hours for examinations per day is 6, and the various examinations have the following durations (in hours):

examination	*A*	*B*	*C*	*D*	*E*	*F*	*G*	*H*	*I*	*J*
duration	1	1	1.5	1.5	2	3	3	3	4	4

(a) Find a solution using the first-fit method.
(b) Find a solution using the first-fit decreasing method.
(c) Find an optimum solution by trial and error.

17.3. A project consists of ten activities *A*–*J* with the following durations (in hours). There are no precedence relations.

activity	*A*	*B*	*C*	*D*	*E*	*F*	*G*	*H*	*I*	*J*
duration	7	12	3	9	6	2	11	10	4	8

It is required to find the minimum number of workers needed to complete the project in 16 hours.

(a) What answer is given by the first-fit packing method? (Show your working.)
(b) What answer is given by the first-fit decreasing method? (Show your working.)
(c) What is the optimum solution? (Give reasons for your answer.)

17.4. Repeat Exercise 17.3 for a similar project which has the following durations (in hours) and the same required completion time.

activity	*A*	*B*	*C*	*D*	*E*	*F*	*G*	*H*	*I*	*J*
duration	2	3	4	5	6	7	8	9	10	11

The Knapsack Problem

17.5. A hiker wishes to take some of the following items on a journey.

item	*A*	*B*	*C*	*D*
weight	5	4	3	2
value	5	3	6	1

Use the branch-and-bound method to determine which items he should take if the total value is to be as large as possible and the total weight is not to exceed 9 units.

17.6. A traveller wishes to buy some books to take on a journey. He estimates the time (in hours) it will take him to read each of five books to be as shown in the following table:

book	*A*	*B*	*C*	*D*	*E*
cost (£)	4	6	3	2	5
reading time	5	9	4	4	4

Which of these books should he buy to give him the maximum amount of reading time without spending more than £8? Use the branch-and-bound method to obtain a solution and show all your working.

17.7. A machine can be used to make any of six different items. The time (in hours) needed to produce each item, and the value of each item, are shown in the following table:

item	*A*	*B*	*C*	*D*	*E*	*F*
weight	5	8	3	7	4	6
value	9	7	15	8	12	10

If the machine is available for only 15 hours, which items should be produced so that the total value is as large as possible?

17.8. A hiker wishes to take some of the following items on a journey:

item	*A*	*B*	*C*	*D*
weight	5	3	6	1
value	5	4	3	2

By using a modified branch-and-bound method, determine which items he should take so that the total value is at least 9, but the total weight is a minimum.

CHAPTER 18

LOCATION PROBLEMS

After reading this chapter, you should be able to:

- explain what is meant by the *facility layout problem* and the *set-covering problem*;
- solve simple examples of the following:

  the vertex single-centre location problem;
  the vertex multi-centre location problem;
  the absolute single-centre location problem;
  the absolute p-centre location problem.

Introduction

In Chapters 15 and 16 we discussed the scheduling of the activities involved in a project. If these activities are carried out in a factory, it may be important to consider, in addition to the scheduling of the activities, the layout of machines and other facilities used. This type of problem is called the *facility layout problem.* On a larger scale, examples include the siting of distribution depots, hospitals, postal sorting offices, and so on.

There may be many factors which need to be considered in planning the layout of facilities—for example, we may wish to minimize transit time along a conveyor belt, or to minimize the danger or the noise which may affect people working near machines. The problem may be further complicated by the fact that the same machines may be used for more than one project. Facility layout problems occur, for example, in industry, in hospitals, supermarkets and hotels, in aeroplane cockpits, and in many similar situations.

For example, suppose that an aeroplane cockpit is to be fitted with a set of controls. What factors should a designer take into account in planning their layout?

Obviously, the designer needs to know the regions which a pilot can easily reach with his hands and his feet—these regions form the *reachability envelope*. Emergency controls and frequently-used controls are placed within this envelope. It is also desirable to place in an appropriate order those controls which need to be operated in sequence. Another important consideration is to place controls so that the possibility of error is minimized—for example, vital controls with a similar appearance should not be placed too close to each other. Prototype simulation, possibly preceded by computer simulation, is usually needed to ensure that these and other criteria are satisfied.

Facility layout problems often cannot be formulated precisely, and often there is no clear-cut criterion for an optimum solution. For example, when considering where to locate a power station, criteria based on costs, distances and times are not adequate, since many other factors, such as safety, ecological effects, and political implications, are involved. From now on, we restrict our discussion to a particular class of facility layout problems called *location problems*, which can often be formulated mathematically.

A **location problem** is concerned with a situation which can be represented by a network, and in which we wish to locate some facility or facilities so that some function (called an *objective function*) is maximized or minimized. The resulting location of a facility is called a **centre**. If a centre must be located at a vertex, we shall refer to it as a **vertex centre**. If it is also permitted for a centre to be located along an edge, it is called an **absolute centre**.

Because each centre is to be located either at a vertex or along an edge of a network, the location problem is a much less general type of problem than the facility layout problem, and can often be formulated in mathematical terms and tackled accordingly.

18.1 VERTEX SINGLE-CENTRE LOCATION PROBLEMS

We begin by discussing the simplest type of location problem, that of locating a single centre at a vertex.

Example 1 Suppose that we wish to locate a hospital, police station, fire station or some other service point at one of the towns in a region. In this case, the vertices of the network correspond to the towns, and the edges correspond to roads linking these towns. An example of such a network is the following:

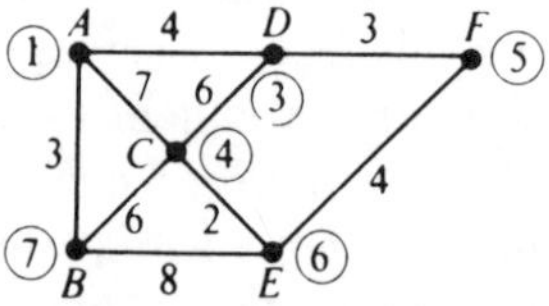

In this example, the weights associated with the edges represent the distances between the towns, and the weights associated with the vertices (shown as circled numbers) represent the population of the towns.

In the above example, we wish to find one vertex centre for a facility. To assess the suitability of each vertex as a centre, we use a function called a **measure**. The numerical value of the measure for each vertex indicates the suitability of that vertex; we calculate this value for each vertex, and then locate the centre at the vertex for which the measure has the optimum value.

For example, suppose that we wish to determine the best location for a health centre, with the criterion that we wish to minimize the inconvenience to the population of any town served by the health centre. We need to find some measure which expresses the rather vague idea of inconvenience to a population. We arrive at a possible measure by the following reasoning. The inconvenience to an individual wanting to visit the health centre is assumed to be proportional to the distance from the town where he lives to the centre. So the inconvenience to the whole population of a town is assumed to be proportional to the distance of that town from the health centre, multiplied by the population of the town. Thus the suitability of vertex i is given by

$$\text{measure for vertex } i = \max_j (v_j d_{ij}),$$

where v_j is the weight associated with vertex j (the population of town j) and d_{ij} is the weight associated with edge ij (the distance between town i and town j).

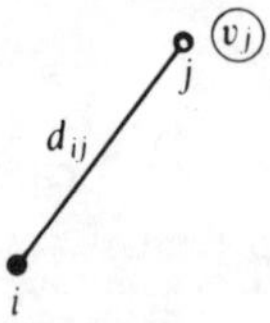

To evaluate $\max_j (v_j d_{ij})$, we calculate the value of the product $v_j d_{ij}$ for each vertex j, and then choose the maximum of these products.

The measure calculated for a vertex i gives the maximum inconvenience suffered by the population of any town. To find the optimum centre, we calculate the value of the measure for each vertex, and choose as a centre the vertex for which this value is the lowest.

Worked problem Find the optimum vertex centre for the network of Example 1, using the measure $\max_j (v_j d_{ij})$.

Solution

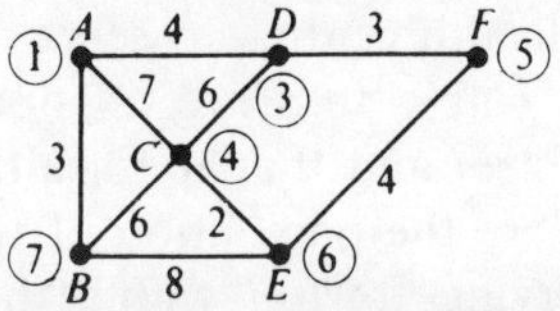

Calculating the value of the measure for each vertex, we obtain the following table:

Vertex	Value of the measure
A	max(21, 28, 12, 54, 35) = 54
B	max(3, 24, 21, 48, 50) = 50
C	max(7, 42, 18, 12, 30) = 42
D	max(4, 49, 24, 42, 15) = 49
E	max(9, 56, 8, 21, 20) = 56
F	max(7, 70, 24, 9, 24) = 70

The vertex with the smallest value of the measure is vertex C, so this is the optimum location for the health centre.

Problem 18.1 A firm wishes to locate a warehouse at one of the towns of the above network. The warehouse is to supply groceries to supermarkets in each of the towns. The delivery costs from the warehouse to a town are assumed to be proportional to the product of the distance to the town and the population of that town.

(a) What measure should be used to work out at which town the warehouse should be located so that the total delivery costs are as low as possible?
(b) Use this measure to find the optimum location for the warehouse.

18.2 VERTEX MULTI-CENTRE LOCATION PROBLEMS

The vertex single-centre location problem is concerned with finding an optimum location for a single facility at one of the vertices of a network. The *vertex multi-centre location problem* is more complicated, since it involves finding a number (usually the minimum possible) of centres for a facility subject to some criteria. In the vertex multi-centre location problem, we assume that there is a capacity constraint associated with each centre for the facility—for example, this may be a limit on the amount of some product which can be supplied by each centre, or the number of people who can be served by each centre.

An example of the vertex multi-centre location problem is the problem of locating a number of health centres to serve n small towns. We assume that each health centre is able to serve a population of M people (the capacity of the health centre) and that the health centres are to be placed at some of the towns, where the population of the jth town is v_j. The cost to the community of serving town j with a health centre at town i is assumed to be proportional to $v_j d_{ij}$—the product of the distance d_{ij} from the health centre to the town, and the population of the town v_j. We further assume that each town is to be served by only one health centre. So the cost to the community for all the towns served by a health centre at town i is

$$\sum_{\substack{\text{all towns}\\ \text{served by}\\ \text{centre } i}} v_j d_{ij},$$

where the summation includes only those towns served by the centre at town i. This can be written as

$$\sum_{j=1}^{n} v_j d_{ij} x_{ij},$$

where the summation now includes all towns, and x_{ij} takes the value 1 if town j is served by a centre at vertex i, and 0 otherwise. This expression is the measure for the appropriateness of locating a single centre at the vertex i. To serve all the towns, we need to locate several health centres at different towns. The measure for the appropriateness of the locations for all these health centres is

$$\sum_{i=1}^{n} \sum_{j=1}^{n} v_j d_{ij} x_{ij},$$

which is just the sum of the values of the previous expression for all the centres. (The actual number of health centres does not appear in this expression, because it is taken care of by the x_{ij}.)

To solve this vertex multi-centre location problem, values of the x_{ij} must be chosen subject to appropriate constraints so that the value of the measure is a minimum. The constraints may, for example, specify that each town j is to be served by only one health centre, and that the total number of health centres should be as small as possible.

For problems which are not too large, the branch-and-bound method can be used. The first step is to decide on the number p of centres which are to be used; this number must be at least $(\Sigma_j v_j)/M$. There are then

$$\binom{n}{p} = \frac{n!}{p!\,(n-p)!}$$

possible ways of locating the centres. However, since those vertices with large values of v_j are favoured as centres, we can order the vertices in decreasing order of v_j and use the branch-and-bound method in such a way that many of these possibilities do not have to be considered explicitly. We shall not discuss the details of the method of solution for this problem of locating health centres; instead, we look at a somewhat simpler example.

Suppose that a number of fire stations are to be located to serve a number of towns. The most important consideration in deciding at which towns they should be located is that it should be possible to send a fire engine to reach a fire in any town before the fire gets out of control. We can take account of this requirement by adding the following constraint to the vertex multi-centre location problem discussed previously:

> the distance from any town to the fire station which serves it should not be more than a specified value λ.

The vertex multi-centre location problem with this constraint is more manageable than the more general problem, since the number of possible locations of the centres (in this case fire stations) is greatly reduced.

Example 2 Consider the following network, which represents four towns A, B, C and D, and the roads connecting them. The length of each road is shown next to the corresponding edge.

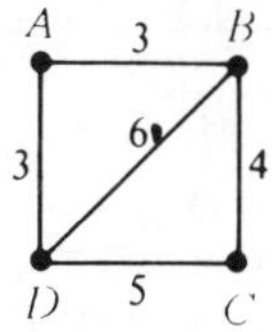

How many centres (fire stations) are needed to serve all the towns if the distance from any town to the nearest centre must not be greater than 4 units? Which towns can be chosen as locations for this minimum number of centres?

For each vertex of the network, we can identify the set of vertices which can be served by a centre located there. We can express this information in the form of a bipartite graph, as follows:

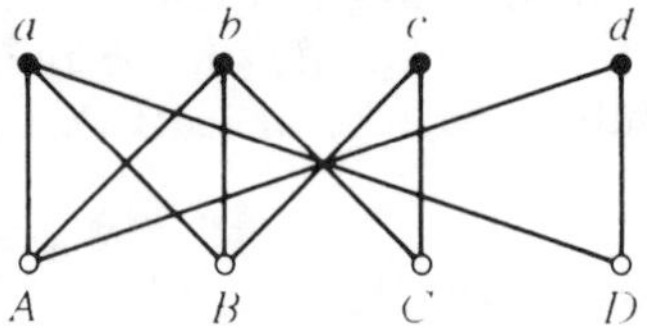

In this graph, vertex *a*, for example, represents town *A* considered as a centre, and this vertex is joined to those vertices (*A*, *B* and *D*) which can be served by it.

From the bipartite graph we see that it is not possible for a single centre located at any of the vertices to serve all four towns. However, two centres located at vertices *A* and *B* can between them serve all the towns, so this is the minimum number of centres. There are in fact four possible pairs of vertices which can serve all four towns, namely

$$\{A, B\}, \quad \{A, C\}, \quad \{B, D\}, \quad \{C, D\}.$$

Formulation as a Set-covering Problem

In the above example we wish to find a collection of vertices from the set $\{a, b, c, d\}$ which can serve every vertex from the set $\{A, B, C, D\}$. This type of problem is called a *set-covering problem.*

Definition *Let* $\mathscr{C} = \{a, b, c, \ldots\}$ *be a collection of subsets of a finite set* $S = \{A, B, C, \ldots\}$. *A* ***cover*** for S is a subset of $\mathscr{C}$ *such that every element of S belongs to at least one member of* $\mathscr{C}$. *A* ***minimum cover*** is a cover which contains the smallest possible number of subsets.

In Example 2, the set S is the set of vertices $\{A, B, C, D\}$ and the collection of subsets $\mathscr{C}$ corresponds to the set of vertices $\{a, b, c, d\}$. Thus

the subset of vertices which can be served by a is $\{A, B, D\}$;
the subset of vertices which can be served by b is $\{A, B, C\}$;
the subset of vertices which can be served by c is $\{B, C\}$;
the subset of vertices which can be served by d is $\{A, D\}$.

Thus the following are some of the covers for S:

$$\{a, b\}, \quad \{a, b, c\}, \quad \{b, c, d\}, \quad \{c, d\}.$$

We wish to find the minimum number of fire stations needed to serve all the towns, so we want a minimum cover for the set S. The minimum number of vertices from $\{a, b, c, d\}$ which provide a cover is two; there are four such minimum covers—namely,

$$\{a, b\}, \quad \{a, c\}, \quad \{b, d\}, \quad \{c, d\}.$$

In general, to solve this type of location problem, we proceed in two stages. First, we find all possible minimum set covers. Secondly, we evaluate the measure for the problem for each minimum set cover, and choose as the centres for the facilities the elements of the cover for which the value of the measure is a minimum.

Problem 18.2 Let

$$S = \{A, B, C, D, E\} \quad \text{and} \quad C = \{a, b, c, d\},$$

where

$$a = \{A, B\}, \quad b = \{B, C, E\}, \quad c = \{A, C, D\}, \quad d = \{A, D\}.$$

Write down all the covers for S and all the minimum covers for S.

Problem 18.3 A number of fire stations are to be located at some of the towns in the following network, in which the numbers next to the edges are distances d_{ij} in miles, and the circled numbers next to the vertices are populations v_j in thousands.

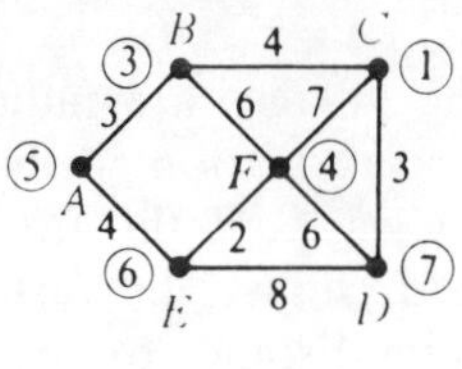

The distance from any town to the fire station which serves it must not be more than $3\frac{1}{2}$ miles. Each fire station has a capacity of 10 (that is, it can serve a population of 10 000). The measure of the appropriateness of vertex i as a centre is $\Sigma_j v_j d_{ij}$.

(a) For each town, find all the towns which can be served by a fire station located there, and check that the capacity of a fire station is sufficient to serve all these towns.

(b) By drawing a bipartite graph, find all possible minimum covers for the set of fire stations which can serve all the towns.
(c) For each cover found in part (b), calculate the sum of the values of the measure for all the vertices in that set, and hence find the optimum locations for the fire stations.

The example considered in the previous problem is particularly simple, not only because of its small size, but also because of the simplicity of the set-covering problem, and because the capacity constraints are satisfied for each minimum cover. In practice, for instances of large size, a heuristic algorithm would be used to obtain a reasonable, but not necessarily optimum, solution.

In general, this type of location problem belongs to a class of problems known as NP-complete problems; these will be discussed in Chapter 20.

18.3 ABSOLUTE CENTRE LOCATION PROBLEMS

Suppose that a fire station is to be built to serve three villages A, B and C represented by the vertices of the following network.

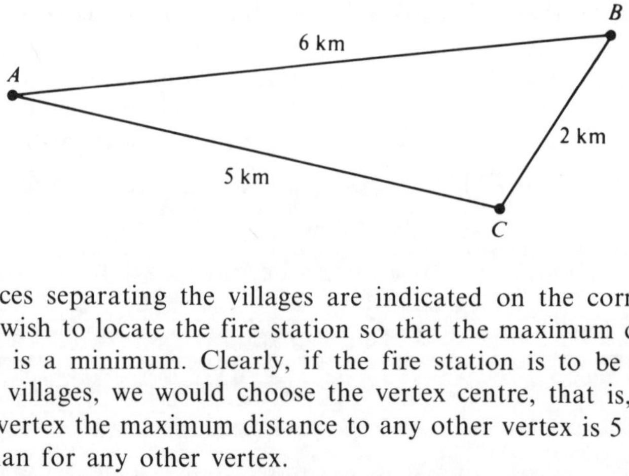

The distances separating the villages are indicated on the corresponding edges. We wish to locate the fire station so that the maximum distance to any village is a minimum. Clearly, if the fire station is to be located in one of the villages, we would choose the vertex centre, that is, vertex C. From this vertex the maximum distance to any other vertex is 5 km, which is better than for any other vertex.

However, we can do better than this if the restriction that the centre be located at a vertex is lifted. We now assume that the fire station can be located anywhere along the connecting roads represented by the edges (as well as at a vertex). The best location is at a point $1\frac{1}{2}$ km from vertex C along the road to vertex A. The maximum distance to any village is then $3\frac{1}{2}$ km, which is the least possible value for any location on the network. This best location is called the **absolute centre** of the network.

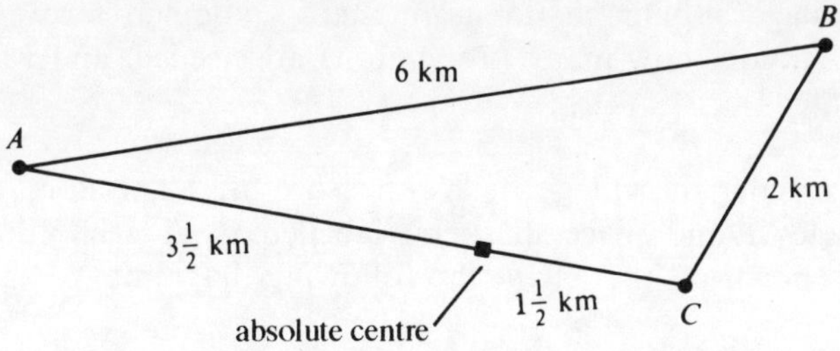

Suppose that, instead of locating just a single fire station, we require to locate a number p of fire stations to minimize the maximum distance from a vertex to the nearest fire station. The problem then becomes that of finding an **absolute p-centre**. In this terminology, the above example is that of finding an absolute 1-centre. For values of p greater than 1 this problem was thought to be intractable for networks of any appreciable size until 1971 when an algorithm was developed by N. Christofides and P. Viola.

In some situations we may wish to find, say, three centres in a network so that the distance to the vertex most remote from any centre is a minimum. This is the problem of finding an absolute 3-centre.

Alternatively, we may wish to know the minimum number of centres and their locations such that the maximum distance from any vertex to a centre is less than a specified value. These two types of problem are closely related and can be solved by the same method. We illustrate this method, which is due to Christofides and Viola, with the following example.

Worked problem In the following network, the vertices *A–F* represent villages and the edges represent roads connecting them. The distances between the villages are given in kilometres.

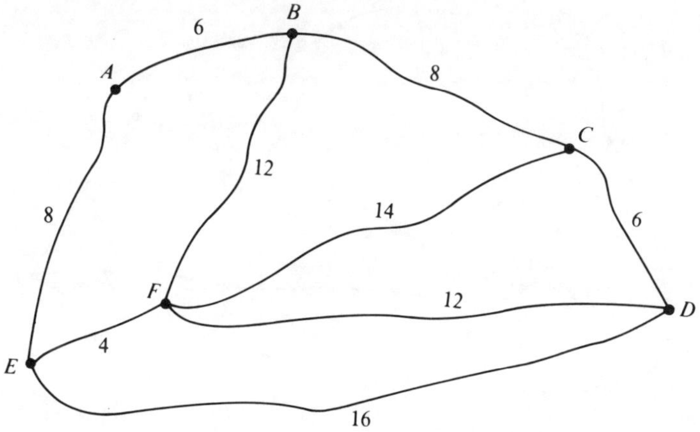

If each village should be no more than 7 kilometres away from the nearest fire station, how many fire stations are needed, and where should they be located?

Solution Consider the village A. We measure off 7 km along every route out of village A and place markers, labelled a, at each furthest point reached, as shown on the left in the following diagram:

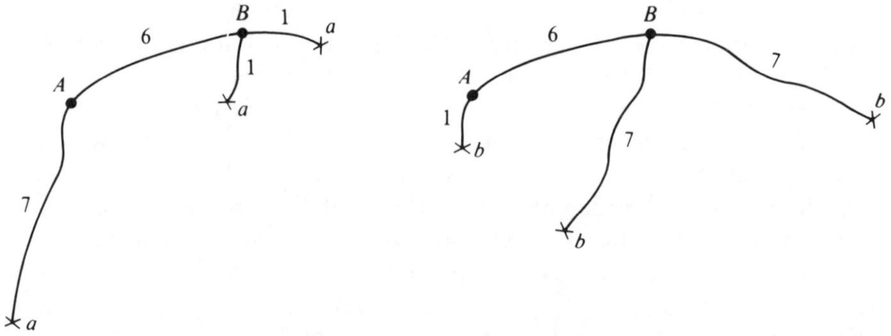

The markers a define the part of the network which is within 7 km of village A. It follows that at least one fire station must lie in this part. Similarly, we place markers b at a distance of 7 km from B on every route out of B, as shown above in the diagram on the right.

If we put markers for all the villages onto the network, we get the following diagram:

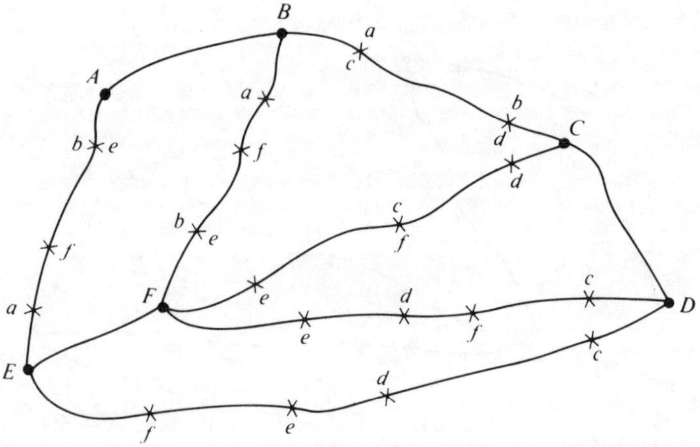

The markers divide the edges of the network into segments. For example, the edge *ED* is divided into five segments: *Ef*, *fe*, *ed*, *dc*, and *cD*.

Where two markers coincide, such as *b* and *e* on the edge *AE*, we get a segment of zero length. In all, including such zero-length segments, there are 33 segments on the network.

We now have to decide which segments we should choose as locations for fire stations. (It clearly does not matter exactly which point within a segment we use.) To do this, for each segment we write down the subset of the villages which are within 7 km of that segment. So, for example, for the segment *Ea* we write down the subset {*E*, *F*} to show that villages *E* and *F* are within 7 km of any point in that segment. Similarly, with the segment of zero length corresponding to the double marker *ac* on *BC* we associate the subset {*A*, *B*, *C*}, and with the segment *ed* on *ED* we associate the empty set, since no village lies within 7 km of this segment.

Some segments have the same subset associated with them. For example, segments *Ab*, *AB*, *Ba* (on edge *BF*) and *Bc* are all within 7 km of villages *A* and *B*, but are further than 7 km from all other villages. Consequently we associate the same subset {*A*, *B*} with each of them. Such a group of segments corresponding to the same subset is called a *region*.

Although there are 33 segments, there are only 18 different regions, that is, only 18 different subsets associated with the segments. These are:

1	{*A*, *B*, *C*}	7	{*D*, *F*}	13	{*B*}
2	{*A*, *B*, *E*}	8	{*A*, *B*}	14	{*C*}
3	{*A*, *E*, *F*}	9	{*A*, *E*}	15	{*D*}
4	{*B*, *C*, *D*}	10	{*B*, *C*}	16	{*E*}
5	{*B*, *F*}	11	{*C*, *D*}	17	{*F*}
6	{*C*, *F*}	12	{*E*, *F*}	18	empty set

Finding the minimum number of fire stations corresponds to finding the minimum number of regions which provide a cover for the set of villages. The first step is to notice that we can discount regions like region 12. A fire station anywhere in region 12 can serve villages *E* and *F*. However, region 3 can serve village *A* as well as villages *E* and *F*, and so region 3 is always a better location than region 12. For similar reasons we can eliminate all of the last 11 regions, and we need consider only regions 1 to 7. Naturally, we look first at regions 1 to 4, each of which can serve three villages. By inspection, we find that the two regions 3 and 4 cover all the villages, and this is our required minimum covering set. The following diagram shows these two regions; region 3 is the segment *af* on edge *EA* and region 4 is the single point at the markers *bd* on edge *BC*.

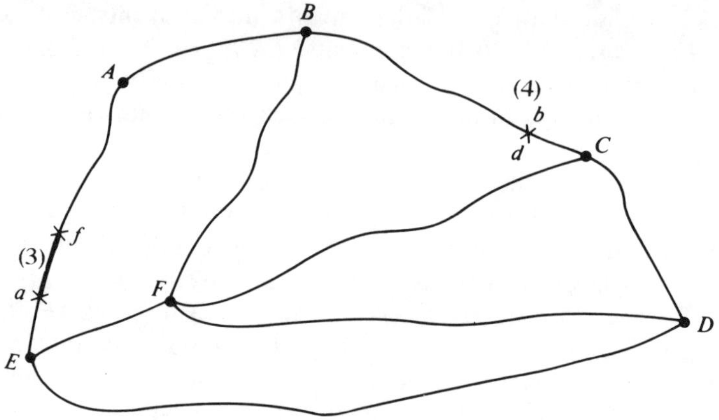

The minimum cover is the absolute 2-centre of the network. Note that if some distance slightly smaller than 7 km had been stipulated, the markers b and d on edge BC would not have coincided and so we would have had to look for an absolute 3-centre.

This method was used successfully in 1972 to locate fire stations in the city of Rotterdam in The Netherlands. It was decided that there should be a fire station within $3\frac{1}{2}$ km of any building in the centre of the city. Because the bridges and tunnels crossing the river Nieuwe Maas which divides the city are frequently blocked at rush hours, the parts of the city north and south of the river were treated as separate problems. It was necessary to use a computer since the calculations were too difficult to do by hand. It was found that a total of six fire stations were needed north of the river, and five south of the river. As a result, two new fire stations were built north of the river, and plans were made to replace another at a different location. South of the river, two fire stations were closed down and three new ones built at different sites.

Problem 18.4 For the following simple network introduced earlier, we found that the absolute 1-centre is not more than $3\frac{1}{2}$ km from any vertex. Use the method of Christofides and Viola to find the location of the absolute 2-centre, and the maximum distance of any vertex from a centre for this optimum location.

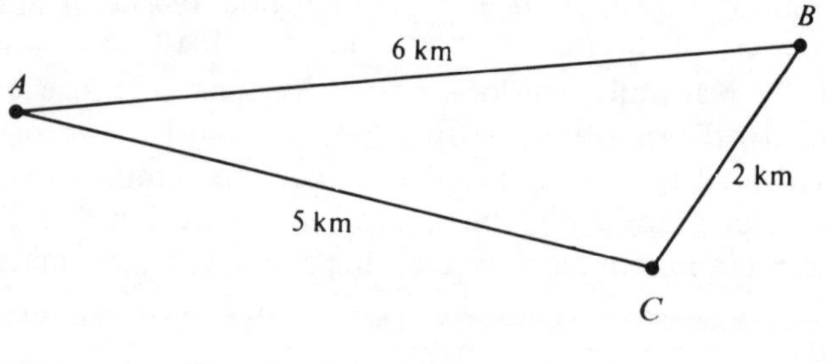

Location problems faced by industry may be considerably more complicated than the examples we have considered in this chapter. For example, if it is decided to set up manufacturing plants to serve a region, then the fixed cost of construction of a plant and the cost of producing each item may be different at different locations, and in such cases these costs have to be taken into account when deciding on the optimum locations.

Finally, we wish to emphasize that the location problem is only part of the more general facility layout problem. Even if we can find an optimum, or near optimum, solution to the location problem, we may still be far from finding an optimum solution to the practical problem. For example, the availability of skilled labour, or other social, political or ecological factors, may affect the choice of location.

EXERCISES 18

Vertex Single-centre Location Problems

18.1. In the following network, the vertices *A*, *B*, *C*, *D* and *E* represent towns and the edges represent roads connecting them. The circled numbers next to the vertices represent town populations (in thousands) and the uncircled numbers next to the edges represent distances.

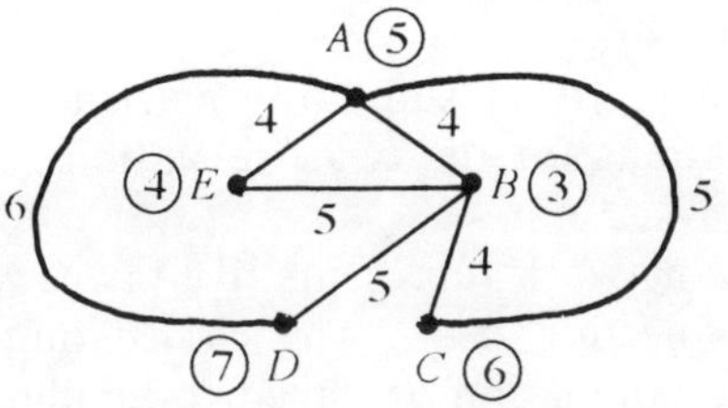

Oil is to be supplied to all five towns from a depot located at either *A* or *B*. The cost of delivery to each town is proportional to the product of the distance from the depot and the population of the town.

(a) Give a suitable measure for rating a town as a possible depot.
(b) Use this measure to determine which of *A* and *B* is the best location for the depot.

18.2. In the following network, the vertices *A*, *B*, *C*, *D* and *E* represent towns and the edges represent roads connecting them. The circled numbers next to the vertices represent town populations (in thousands) and the uncircled numbers next to the edges represent distances (in kilometres).

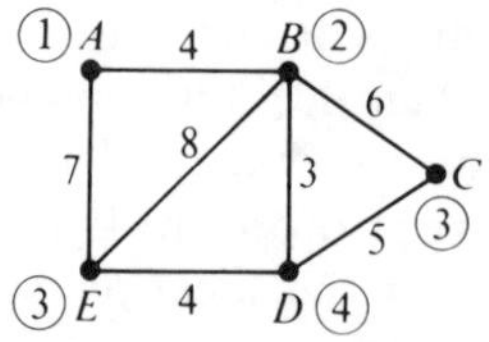

The measure of suitability of vertex i to be a location for a supermarket is

$$\sum_j v_j\, d_{ij},$$

where d_{ij} is the distance (in kilometres) between town i and town j, and v_j is the population of town j.
At which vertex should the supermarket be located?

Vertex Multi-centre Location Problems

18.3. Fire stations are to be located at two towns in the network of Exercise 18.2 in such a way that no town is more than 7 km from a fire station.

Using the measure of suitability given in Exercise 18.2, find the optimum locations for the two fire stations.

18.4. The following network represents five towns A, B, C, D and E, and the roads connecting them. The circled number next to a vertex represents the population of the corresponding town (in thousands) and the number next to an edge represents the distance (in kilometres) between the two towns joined by the corresponding road.

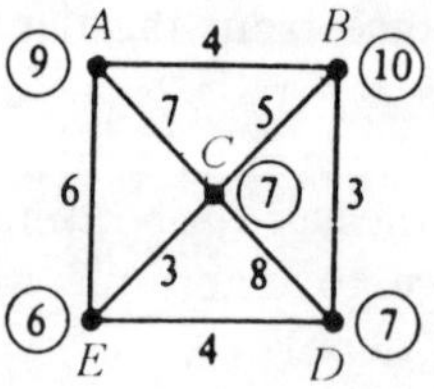

Police stations are to be located at some of the towns so that no town is more than 4 km from a police station. Each town is to be served by only one police station, and it is required that the minimum possible number of police stations be used.

(a) The problem of finding the minimum number of police stations required can be represented as a set-covering problem.

How many different minimum covers appropriate to the conditions are there?

(b) The measure for the appropriateness of a number of locations for the police stations is

$$\sum_{i=1}^{n}\sum_{j=1}^{n} v_j\, d_{ij}\, x_{ij},$$

where d_{ij} is the distance (in kilometres) between town i and town j, the population (in thousands) of town j is v_j and x_{ij} is 1 if town j is served by a centre at vertex i, and 0 otherwise.

Use this measure to find which vertices correspond to the optimum locations for the police stations.

18.5. The following network represents five towns A, B, C, D and E, and the roads connecting them. The number next to an edge represents the distance (in kilometres) between the two towns joined by the corresponding road, and the circled number next to a vertex represents the population of the corresponding town (in thousands).

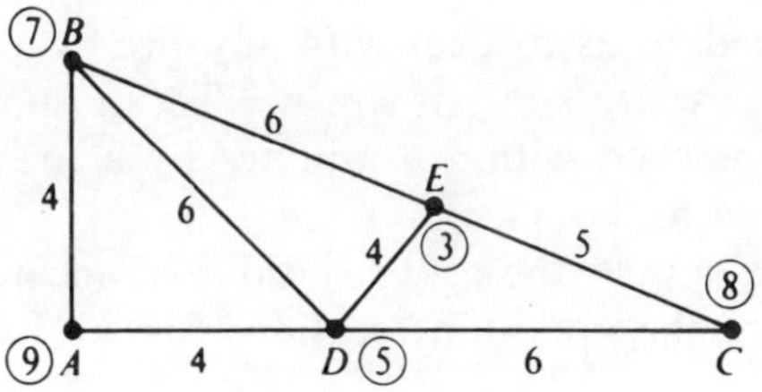

Ambulance depots are to be located in some of the towns in such a way that each town is not more than 5 km from the nearest depot. The minimum possible number of ambulance depots is to be determined.

(a) The problem of finding the minimum number of ambulance depots required can be represented as a set-covering problem.

How many different minimum covers appropriate to the conditions are there?

(b) The measure of appropriateness of the proposed locations of ambulance depots is

$$\sum_{i=1}^{n}\sum_{j=1}^{n} v_j\, d_{ij}\, x_{ij},$$

where d_{ij} is the distance (in kilometres) between town i and town j, the population of town j is v_j, and x_{ij} is 1 if town j is served by a depot at vertex i, and 0 otherwise.

Determine which vertices correspond to the optimum locations of the depots.

Absolute Centre Location Problems

18.6. Four housing estates are linked by major roads as shown in the following diagram, in which distances are marked in kilometres:

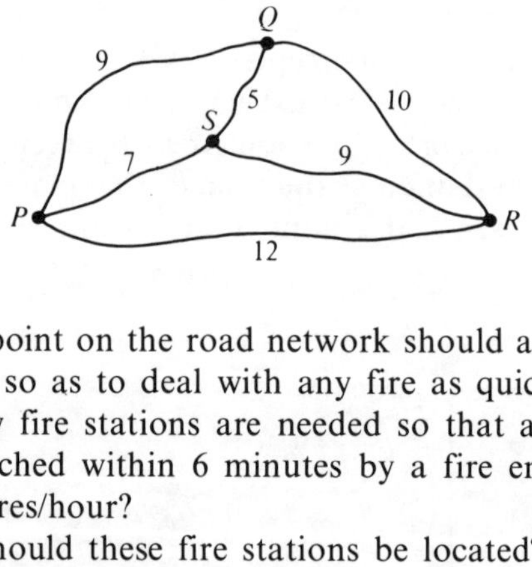

(a) At which point on the road network should a single fire station be located so as to deal with any fire as quickly as possible?

(b) How many fire stations are needed so that any housing estate can be reached within 6 minutes by a fire engine travelling at 30 kilometres/hour?

Where should these fire stations be located?

(c) What are the optimum locations for:

(i) 2 fire stations?
(ii) 3 fire stations?
(iii) 4 fire stations?

CHAPTER 19

THEORY OF NETWORK ANALYSIS

After reading this chapter, you should be able to:

- appreciate the general properties of network models in terms of the constraints on edge (or arc) variables;
- appreciate the importance of vertex variables in the construction of network algorithms;
- express some of the network problems considered in this book as linear programming problems;
- understand what is meant by the *dual* of a network problem, and express the dual of some network problems in linear programming form;
- appreciate the significance of the duality theorem.

Introduction

In this chapter we look at some of the general properties of network models and show how they may be classified according to the types of constraint acting on the variables associated with the edges (or arcs). We show that variables associated with the vertices are of great importance in the construction of algorithms to solve network problems. A number of the central network problems discussed in this book are expressed in the language of linear programming. We introduce the concept of the *dual* of a network problem, a concept which gives great insight into the procedures used in network algorithms.

19.1 PROPERTIES OF NETWORK MODELS

In network analysis we use a graph (or digraph) to represent the underlying structure of interconnections of the elements of the model. There are two ways of doing this:

(i) the elements may be represented by vertices and the interconnections or relations by edges (or arcs);
(ii) the elements may be represented by edges (or arcs) and the interconnections or relations by vertices.

We use the first method for the assignment problem, in which the applicants and the jobs are represented by vertices and the relations between them (possible allocations of people to jobs) are represented by edges.

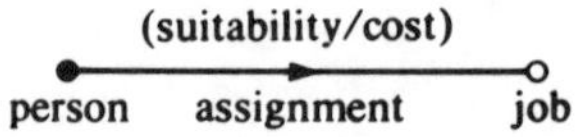

Another example of the use of this method is in the representation of the transportation problem.

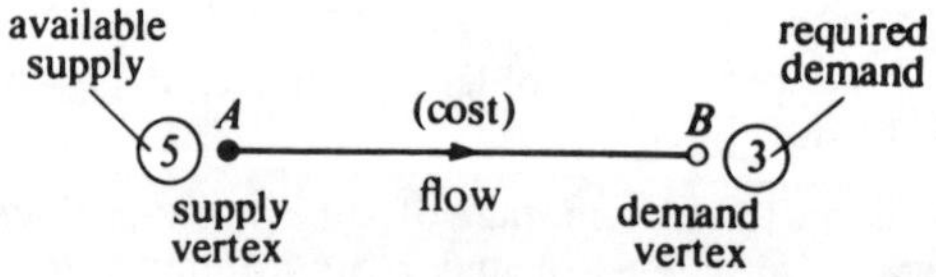

The second method is used widely in network models. An example of its use is in the maximum-flow problem discussed in Chapter 3, where arcs are used to represent the channels (roads or pipes) carrying some flow and vertices represent the connections between them.

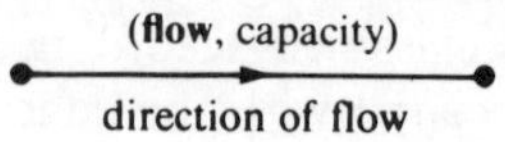

This method is also used in electrical networks where edges represent components and vertices represent the connections between them.

It is sometimes possible to represent a single problem in either way. An example is provided by the two types of activity network used to represent scheduling problems, as described in Chapter 15.

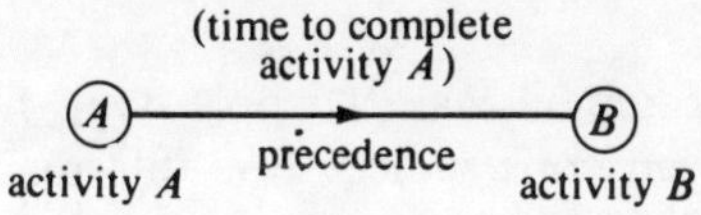

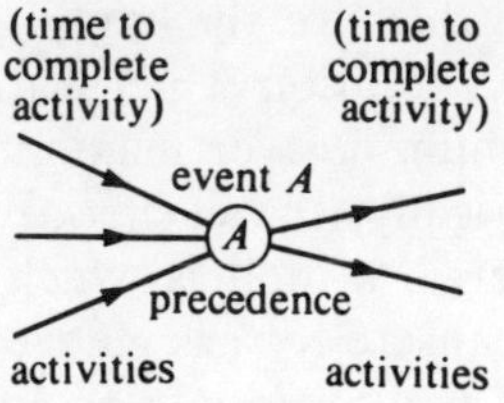

In network problems we usually associate variables with the edges (or arcs in the case of digraphs). These edge variables are of two distinct types. One type corresponds to the size of flow through an edge and is called a *through variable*. The other type corresponds to the difference between quantities at the vertices incident to the edge, and is called an *across variable*. For instance, in electrical networks, currents are through variables and voltages are across variables.

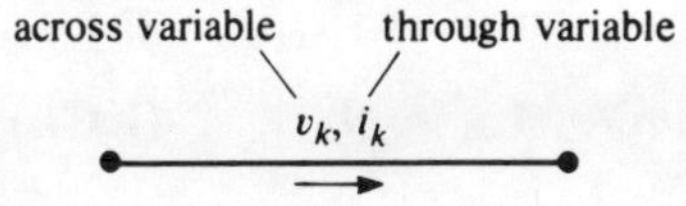

The edge (or arc) variables may be subject to various types of constraint. These constraints may be of four types:

1. conservation of through variables at each vertex—*the vertex law*;
2. conservation of across variables on each cycle—*the cycle law*;
3. relations between through and across variables on a given edge—*the component equations*;
4. upper and lower bounds on the flows—*capacity constraints*.

Electrical networks have severe constraints on all the variables. The through variables (currents) satisfy constraints of type 1 in the form of Kirchhoff's current law. The across variables satisfy constraints of type 2 in the form of Kirchhoff's voltage law. In addition, the through and across variables (currents and voltages) are related by the component equations, a constraint of type 3. These three sets of constraints (expressed as equations relating the edge variables) enable a unique solution to be found to an electrical network problem for given values of the source voltages and currents. However, more general network models may satisfy only one, or neither, of constraints of types 1 and 2, and may also not satisfy a constraint of type 3 involving component equations. For example, the critical path scheduling problem of Chapter 15 satisfies none of the above types of constraint. In general, therefore, the number of equations expressing the constraints in a network problem is not sufficient to give unique values for the edge variables. Usually in such cases we are required to find a solution which satisfies some kind of optimality criterion, such as minimum cost or maximum flow.

Historically, the electrical network model was the prototype for many other network models which have been constructed. For this reason, and also because electrical networks have such severe constraints on the variables, it is appropriate to look first at some of the concepts involved in electrical networks and then to consider the application of some of these ideas to other types of network in which the constraints are less severe.

Electrical Networks

As in Chapter 10, we consider a network for which the oriented graph has n vertices and m edges. We can express the constraints on the variables as linear equations, as follows.

Kirchhoff's current law can be written in the form

$$\mathbf{A}_{n\times m}\mathbf{J}_{m\times 1}=\mathbf{0}_{n\times 1}, \qquad \text{(KCL)}$$

where
$\mathbf{A}$ is the $n\times m$ incidence matrix of the oriented graph of the network and
$\mathbf{J}$ is the $m\times 1$ column vector of currents $i_1, i_2, \ldots, i_m$ associated with the edges of the oriented graph.

In order to write Kirchhoff's voltage law in a similar way, we associate variables with the vertices: we define *vertex potentials* $P_1, P_2, \ldots, P_n$ at each vertex such that the voltage on an edge is the difference between the potentials at the end-vertices:

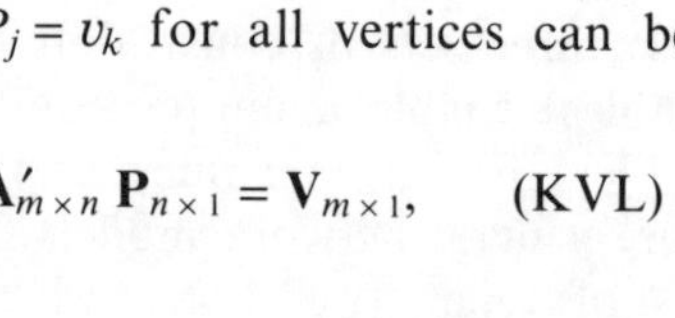

The equations $P_i - P_j = v_k$ for all vertices can be expressed in matrix form:

$$\mathbf{A}'_{m\times n}\,\mathbf{P}_{n\times 1}=\mathbf{V}_{m\times 1}, \qquad \text{(KVL)}$$

where
$\mathbf{A}'$ is the $m\times n$ transpose of the incidence matrix $\mathbf{A}$,
$\mathbf{P}$ is the $n\times 1$ column vector of vertex potentials,
$\mathbf{V}$ is the $m\times 1$ column vector of voltages associated with the m edges of the graph.

In other words, if voltages satisfy Kirchhoff's voltage law, then it is possible to find vertex potentials such that the potential difference across any edge is equal to the difference between the vertex potentials at the end-vertices of that edge. The above matrix equation is therefore equivalent to Kirchhoff's voltage law.

Note that for a given incidence matrix **A** and vector **V**, an infinite number of vectors **P** will, in principle, satisfy the KVL-equations. For instance, if **P** satisfies the equation, then **P** + **C**, where **C** is a column vector with all entries the same, also satisfies the equation. Usually the value of the potential P_j at one vertex j is fixed, so that a unique vector **P** satisfies the equation. This vertex is called the *reference vertex*.

In Chapter 10, we showed that the forms of Kirchhoff's current and voltage laws given above provide a method of deriving Tellegen's theorem, which may be stated as

$$\mathbf{V}'_{1\times m}\,\mathbf{J}_{m\times 1} = \mathbf{0}_{1\times 1} \qquad \text{(TT)}$$

We call **V′J** the *power function* of the network, and the TT-equations express the *law of conservation of energy* in the network.

Tellegen's theorem and the two Kirchhoff laws are closely related, in the sense that *any two imply the third*. In effect, *we have three equivalent statements of the electrical network problem*—namely:

$$\mathbf{A}_{n\times m}\mathbf{J}_{m\times 1} = \mathbf{0}_{n\times 1} \quad \mathbf{A}'_{m\times n}\,\mathbf{P}_{n\times 1} = \mathbf{V}_{m\times 1} \qquad \text{(KCL and KVL)}$$
$$\mathbf{V}'_{1\times m}\,\mathbf{J}_{m\times 1} = \mathbf{0}_{1\times 1} \quad \mathbf{A}_{n\times m}\mathbf{J}_{m\times 1} = \mathbf{0}_{n\times 1} \qquad \text{(TT and KCL)}$$
$$\mathbf{V}'_{1\times m}\,\mathbf{J}_{m\times 1} = \mathbf{0}_{1\times 1} \quad \mathbf{A}'_{m\times n}\,\mathbf{P}_{n\times 1} = \mathbf{V}_{m\times 1} \qquad \text{(TT and KVL).}$$

Any one of these three sets of equations, together with the component equations, provides enough information to solve a given electrical network problem uniquely.

It is important to note that Tellegen's theorem for a particular oriented graph applies to any set of voltages and currents in that graph which satisfy Kirchhoff's laws. So, for example, by considering a part of the graph for which the currents satisfy Kirchhoff's current law, we can apply Tellegen's theorem and obtain an equation relating the corresponding voltages. In this way we can use Kirchhoff's current law and Tellegen's Theorem to generate the Kirchhoff's voltage law equations.

Vertex Variables for other Types of Network

We saw in the above brief presentation of the electrical network model that Kirchhoff's voltage laws can be written in terms of edge variables v_k or equivalently in terms of *vertex potentials* P_i. Vertex potentials are a particular type of *vertex variable* which are applicable for networks which satisfy Kirchhoff's voltage law or an equivalent law. For networks which do not satisfy Kirchhoff's voltage law, we cannot assign a unique potential to each vertex. Nevertheless, for such problems, vertex variables are still useful, and play an important role in algorithms constructed for their solution.

Many of the network problems discussed in this book involve across variables which do not satisfy an equivalent form of Kirchhoff's voltage law. Examples are costs in the max-flow min-cost problem of Chapter 7, costs in the assignment and transportation problems of Chapters 13 and 14, and time in the scheduling problems of Chapter 15.

Many of the network algorithms given in this book introduce vertex variables which guide us to the correct solution. This is no accident, since each such network problem has an equivalent *dual form* (discussed in Sections 19.3 and 19.4) which involves vertex variables. Many algorithms are implicitly using the equivalence of the problem and its dual, and their discovery was in many cases motivated by consideration of the dual form.

Before discussing further the important notion of duality, we show how network problems can be formulated as linear programming problems.

19.2 LINEAR PROGRAMMING FORMULATION OF NETWORK PROBLEMS

Many of the problems involving graphs or networks which we have discussed in this book are optimization problems involving *linear* expressions. Many of these problems can be formulated as *linear programming* problems. Linear programming is concerned with the maximization or minimization of a linear expression—called the *objective function*—subject to *linear constraints*, which may be in the form of linear equations or linear inequalities. This way of formulating network problems is particularly suitable for use with computers. In many cases, the variables involved, such as flow, are restricted to *integer* values. The branch of linear programming which deals with this type of problem is called *integer linear programming*.

We now formulate two of the network problems discussed earlier as integer linear programming problems. We shall formulate other network problems as linear programming problems when we come to discuss duality in the next section. A discussion of methods of solution appropriate for this formulation is, however, beyond the scope of this book.

The Knapsack Problem (Chapter 17)

In the knapsack problem we wish to choose some of a list of n items with values $v_1, v_2, \ldots, v_n$ and corresponding weights $w_1, w_2, \ldots, w_n$ so that we have the maximum possible total value subject to the total weight being not greater than a given value W. We can express this problem as follows:

maximize the linear expression (the objective function)

$$v_1x_1 + v_2x_2 + \ldots + v_nx_n$$

subject to the linear constraint inequality

$$w_1x_1 + w_2x_2 + \ldots + w_nx_n \leqslant W,$$

where each of the variables $x_1, x_2, \ldots, x_n$ takes the value

1 if that item is packed
0 otherwise.

The linear expression which we wish to maximize is just the total value of the packed items, and the linear constraint inequality expresses the fact that the total weight must be less than or equal to W.

We can express this problem concisely by using sigma-notation, as follows:

maximize the objective function

$$\sum_{i=1}^{n} v_i x_i$$

subject to the constraint

$$\sum_{i=1}^{n} x_i v_i \leqslant W,$$

where each x_i takes the value 0 or 1.

The Assignment Problem (Chapter 13)

In the assignment problem, we wish to assign n applicants to n jobs so that the sum of the costs associated with the assignment is a minimum. The objective function is therefore equal to the sum of the costs for an assignment. This can be written as

$$\sum_{i=1}^{n} \sum_{j=1}^{n} \alpha_{ij} x_{ij}$$

where α_{ij} is the cost of assigning applicant i to job j. For a particular assignment, x_{ij} takes the value

1 if applicant i is assigned to job j,
0 otherwise.

We can express the assignment problem concisely as follows:

minimize the objective function

$$\sum_{i=1}^{n} \sum_{j=1}^{n} \alpha_{ij} x_{ij}$$

subject to the constraints:

$$\sum_{j=1}^{n} x_{ij} = 1, \qquad \text{for } i = 1, 2, \ldots, n,$$

$$\sum_{i=1}^{n} x_{ij} = 1, \qquad \text{for } j = 1, 2, \ldots, n.$$

The first set of constraints ensures that each applicant is assigned to exactly one job, and the second set of constraints ensures that each job has exactly one applicant assigned to it.

A solution to the assignment problem is a set of values of the x_{ij} for which the objective function has a minimum value.

19.3 THE DUAL FORM OF A NETWORK PROBLEM

In this section and the next we give the dual forms of the following network problems which have been discussed earlier:

- the assignment problem
- the electrical network problem
- the maximum flow problem
- the max-flow min-cost flow problem
- the transportation problem
- the critical path problem.

To demonstrate what is meant by the *dual* of a network problem, we look in detail at the first example, the dual of the assignment problem.

The Dual of the Assignment Problem (Chapter 13)

Suppose that a particular product is produced by n workers working on n machines, and that the workers are not all equally good at operating the various machines. We can express this in terms of the cost per unit of product, α_{ij}, of employing worker i on machine j. The optimum assignment of workers to machines is obtained by solving the following linear programming problem:

minimize

$$\sum_{i=1}^{n} \sum_{j=1}^{n} \alpha_{ij} x_{ij}$$

subject to the constraints

$$\sum_{j=1}^{n} x_{ij} = 1, \qquad \text{for } i = 1, \ldots, n,$$

$$\sum_{i=1}^{n} x_{ij} = 1, \qquad \text{for } j = 1, \ldots, n.$$

This is, of course, the same as the integer linear programming form of the assignment problem given in the previous section.

Now suppose that the manufacturer has the possibility of hiring machines instead of using his own, and that he can also hire outside workers. He wishes to know at what hire prices it is more economic for him to hire outside machines and workers rather than to use his own. In fact, he wishes to know the maximum economic hire rates.

Let the rate of hiring a worker to replace his worker i be X_i, and the rate of hiring a machine to replace machine j be Y_j. If this is to be as economic or cheaper than using his own resources, then $X_i + Y_j \leqslant \alpha_{ij}$. So the maximum economic hire rates are obtained by solving the linear programming problem:

maximize

$$\sum_{i=1}^{n} X_i + \sum_{j=1}^{n} Y_j$$

subject to the constraints

$$X_i + Y_j \leqslant \alpha_{ij}, \qquad \text{for } i = 1, \ldots, n \quad \text{and} \quad j = 1, \ldots, n.$$

This second formulation is the *dual* of the original problem. Notice that, whereas the original assignment problem was a minimization problem, the dual problem is a maximization problem.

Linear Programming Formulations of Problems and their Duals

It is a general characteristic of a problem and its dual that one of them is a maximization and the other a minimization problem. Also, if the original problem has n variables and m constraints, then the dual problem has m variables and n constraints. (Some care is needed in interpreting this for complex problems—for example, some of the constraints of the problem may not be part of the linear programming formulation.) Furthermore, there is an important theorem, called the *duality theorem of linear programming*, which states that the solution to the dual problem gives the solution to the original problem, and *vice versa*. This means that if it is easier we can choose to solve the dual problem rather than the original one.

The original assignment problem is formulated in terms of edge variables (x_i), whereas the dual is formulated in terms of vertex variables (X_i and Y_j). This again is a general characteristic of the relation between a network problem and its dual.

The Hungarian algorithm for the assignment problem given in Chapter 13 makes use of the dual to find a solution. The dual formulation shows that we want the X_i and Y_j to be as large as possible, subject to the constraint $X_i + Y_j \leqslant \alpha_{ij}$; this suggests that we should try to assign vertex weights $X_1, X_2, \ldots, X_n$ and $Y_1, Y_2, \ldots, Y_n$ such that the sum of the weights on adjacent vertices is as close as possible to the cost α_{ij} on the corresponding edge. This is precisely the aim of the procedure used in the algorithm to obtain the succession of partial graphs (parts A and D of the algorithm). At each stage in the algorithm, a feasible solution to the dual problem identifies a partial graph which is used to obtain an improved matching in the assignment problem. This in turn enables the algorithm to improve the solution to the dual problem. So the algorithm considers the original problem and its dual alternately until a final solution is reached.

Canonical Form of Linear Programming Problem and its Dual

For reference we now give the canonical forms of a linear programming problem and its dual.

The canonical form of a linear programming problem is:

$$\text{minimize} \quad c_1x_1 + c_2x_2 + \ldots + c_nx_n$$

$$\text{subject to:} \quad a_{11}x_1 + a_{12}x_2 + \ldots + a_{1n}x_n \geqslant b_1$$

$$\cdots\cdots\cdots$$

$$a_{m1}x_1 + a_{m2}x_2 + \ldots + a_{mn}x_n \geqslant b_m$$

and

$$x_i \geqslant 0.$$

The dual problem is:

maximize $b_1 Y_1 + b_2 Y_2 + \ldots + b_m Y_m$

subject to: $a_{11} Y_1 + a_{21} Y_2 + \ldots + a_{m1} Y_m \leqslant c_1$

$\ldots\ldots\ldots\ldots$

$a_{1n} Y_1 + a_{2n} Y_2 + \ldots + a_{mn} Y_m \leqslant c_n$

and

$Y_i \geqslant 0.$

These may be conveniently expressed in matrix form as follows. The linear programming problem is:

minimize: $\mathbf{c}_{1\times n}\mathbf{x}_{n\times 1}$

subject to: $\mathbf{A}_{m\times n}\mathbf{x}_{n\times 1} \geqslant \mathbf{b}_{m\times 1}$

and

$\mathbf{x}_{n\times 1} \geqslant \mathbf{0}_{n\times 1}$

where

$$\mathbf{c}_{1\times n} = [c_1 \ldots c_n], \quad \mathbf{x}_{n\times 1} = \begin{bmatrix} x_1 \\ \vdots \\ x_n \end{bmatrix}, \quad \mathbf{b}_{m\times 1} = \begin{bmatrix} b_1 \\ \vdots \\ b_m \end{bmatrix}, \quad \mathbf{A}_{m\times n} = \begin{bmatrix} a_{11} & a_{12} & \ldots & a_{1n} \\ \ldots & \ldots & \ldots & \ldots \\ a_{m1} & a_{m2} & \ldots & a_{mn} \end{bmatrix}.$$

The dual formulation is:

maximize: $\mathbf{b}_{1\times m}\mathbf{Y}_{m\times 1}$

subject to: $\mathbf{A}'_{n\times m}\,\mathbf{Y}_{m\times 1} \leqslant \mathbf{c}'_{n\times 1}$

and

$\mathbf{Y}_{m\times 1} \geqslant \mathbf{0}_{m\times 1}$

where $\mathbf{Y}_{m\times 1} = \begin{bmatrix} Y_1 \\ \vdots \\ Y_m \end{bmatrix}$

Notice that in the canonical form the original formulation has n variables and m constraint equations, and the dual formulation has m variables and n constraint equations. Also, it can be shown that the dual of the dual formulation is the original formulation.

An alternative equivalent form is called the *standard form*. In this, the constraint equations for the original problem are true equalities (rather than inequalities as in the canonical form). We can always write a constraint inequality such as

$$a_{11}x_1 + a_{12}x_2 + \ldots + a_{n1}x_n \geqslant b_1$$

as the equality

$$a_{11}x_1 + a_{12}x_2 + \ldots + a_{n1}x_n - x_s = b_1,$$

where x_s is an additional variable known as a *slack variable*.

The constraints are inequalities in the dual form of the standard form (as for the canonical form) but the dual variables are unrestricted.

The Dual of the Electrical Network Problem (Chapters 9 to 11)

As a very simple example of a network problem and its dual formulation, we return to electrical networks. We saw earlier in this chapter that an electrical network can be considered either as the solution to

$$\mathbf{V}'_{1\times m}\,\mathbf{J}_{m\times 1} = \mathbf{0}_{1\times 1} \qquad \text{(Tellegen's theorem)}$$

subject to

$$\mathbf{A}_{n\times m}\mathbf{J}_{m\times 1} = \mathbf{0}_{n\times 1} \qquad \text{(Kirchhoff's current law)}$$

or as the solution to

$$\mathbf{V}'_{1\times m}\,\mathbf{J}_{m\times 1} = \mathbf{0}_{1\times 1} \qquad \text{(Tellegen's theorem)}$$

subject to

$$\mathbf{A}'_{m\times n}\,\mathbf{P}_{n\times 1} = \mathbf{V}_{m\times 1} \qquad \text{(Kirchhoff's voltage law)}$$

These two formulations are analogous to the formulations of a problem and its dual in linear programming. The product $\mathbf{V}'\mathbf{J}$ does not have to be maximized or minimized as in linear programming, since the electrical network problem has a unique solution.

We illustrate the analogy with linear programming with the following simple example (introduced on page 259):

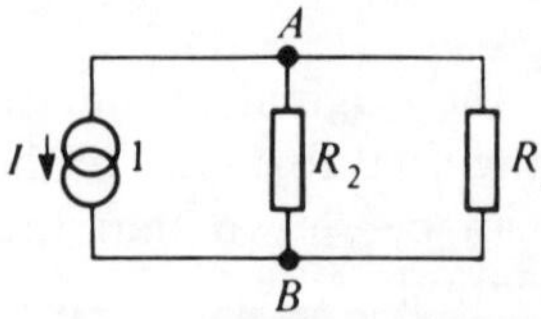

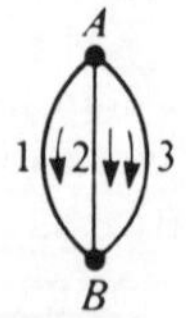

Applying Kirchhoff's current law to vertices A and B, we get

$$\begin{array}{ll} \text{At } A: & i_1 + i_2 + i_3 = 0 \quad \text{or} \\ \text{At } B: & -i_1 - i_2 - i_3 = 0 \quad \text{or} \end{array} \quad \underset{\uparrow \atop \mathbf{A}}{\begin{bmatrix} 1 & 1 & 1 \\ -1 & -1 & -1 \end{bmatrix}} \underset{\uparrow \atop \mathbf{x}}{\begin{bmatrix} i_1 \\ i_2 \\ i_3 \end{bmatrix}} \underset{ \atop =}{=} \underset{\uparrow \atop \mathbf{b}}{\begin{bmatrix} 0 \\ 0 \end{bmatrix}}.$$

Applying Kirchhoff's voltage law to the three possible cycles, we get

$$\left.\begin{array}{l} \text{Cycle 12: } v_1 - v_2 = 0 \\ \text{Cycle 13: } v_1 - v_3 = 0 \\ \text{Cycle 23: } v_2 - v_3 = 0 \end{array}\right\} \text{ or } \begin{array}{l} P_1 - P_2 = v_1 \\ P_1 - P_2 = v_2 \\ P_1 - P_2 = v_3 \end{array} \text{ or } \underset{\uparrow \atop \mathbf{A}'}{\begin{bmatrix} 1 & -1 \\ 1 & -1 \\ 1 & -1 \end{bmatrix}} \underset{\uparrow \atop \mathbf{y}}{\begin{bmatrix} P_1 \\ P_2 \end{bmatrix}} = \underset{\uparrow \atop \mathbf{c}'}{\begin{bmatrix} v_1 \\ v_2 \\ v_3 \end{bmatrix}}$$

The above Kirchhoff's voltage law equations are equivalent to assigning potentials P_1 and P_2 to vertices A and B such that

$$v_1 = v_2 = v_3 = P_1 - P_2.$$

Tellegen's Theorem gives the following equation:

$$v_1 i_1 + v_2 i_2 + v_3 i_3 = 0. \tag{i}$$

If we express v_1, v_2 and v_3 in terms of P_1 and P_2, we get

$$(i_1 + i_2 + i_3)P_1 - (i_1 + i_2 + i_3)P_2 = 0. \tag{ii}$$

We can now formulate these equations in two forms analogous to a linear programming problem and its dual. The canonical form and its dual, as given on page 441 are (omitting the subscripts):

(1) minimize $\mathbf{cx}$ subject to $\mathbf{Ax} \geqslant \mathbf{b}$
(2) (dual) maximize $\mathbf{b'y}$ subject to $\mathbf{A'y} \leqslant \mathbf{c'}$.

The analogue of (1) for our problem is to solve the first form of Tellegen's Theorem equation subject to the Kirchhoff's current law euqations, or:

solve: $v_1 i_1 + v_2 i_2 + v_3 i_3 = 0$ or $\mathbf{cx} = \mathbf{0}$

subject to: $i_1 + i_2 + i_3 = 0$, $-i_1 - i_2 - i_3 = 0$ or $\mathbf{Ax} = \mathbf{b}$

where

$$\mathbf{c} = [v_1, v_2, v_3], \quad \mathbf{x} = \begin{bmatrix} i_1 \\ i_2 \\ i_3 \end{bmatrix}, \quad \mathbf{A} = \begin{bmatrix} 1 & 1 & 1 \\ -1 & -1 & -1 \end{bmatrix}, \quad \mathbf{b} = \begin{bmatrix} 0 \\ 0 \end{bmatrix}.$$

The analogue of (2)(the dual) for our problem is to solve the second form of Tellegen's Theorem equation given above, subject to the Kirchhoff's voltage law equations, or:

solve: $(i_1 + i_2 + i_3)P_1 - (i_1 + i_2 + i_3)P_2 = 0$ or $\mathbf{b}'\mathbf{y} = \mathbf{0}$

subject to: $P_1 - P_2 = v_1$
$P_1 - P_2 = v_2$ or $\mathbf{A}'\mathbf{y} = \mathbf{c}'$.
$P_1 - P_2 = v_3$

where

$$\mathbf{y} = \begin{bmatrix} P_1 \\ P_2 \end{bmatrix}.$$

19.4 DUAL FORMS OF OTHER NETWORK PROBLEMS

We now formulate some of the other central network problems of this book as linear programming problems and state the dual formulations. We indicate how the dual formulation motivates the construction of the algorithm used for the solution of the original problem. We do not discuss the algorithms in detail, but indicate an emerging pattern which identifies the construction of vertex variables as the crucial factor in the algorithmic procedures for solution. We shall see that the algorithms generally contain procedures for considering both the original formulation of the problem and its dual. The algorithms alternate between successive approaches to a solution for each formulation, and finally obtain solutions to both the original problem and its dual. Note that these algorithms do not solve the general linear programming problem. The general linear programming problem is very difficult in the sense that the time taken by any known algorithm to solve a given instance of a problem increases exponentially with the number of edges and vertices. The network problems we consider here have a particular structure which enables us to construct more efficient algorithms. We shall have more to say about the efficiency of algorithms in the next chapter.

The Dual of the Maximum Flow Problem (Chapter 3)

The maximum flow problem is to find the maximum total flow $f_{\max}$ in a network in which each arc has a given capacity. We suppose that there are n vertices, including the SOURCE (vertex 1) and the SINK (vertex n). We can express this problem as follows:

maximize

$$f_{\max} = \sum_{j=1}^{n} f_{1j}$$

subject to the constraints:

$$\sum_{j=1}^{n} f_{ij} - \sum_{j=1}^{n} f_{ji} = \begin{cases} f_{\max} & \text{if } i = 1 \\ 0 & \text{if } i \neq 1 \text{ or } n \\ -f_{\max} & \text{if } i = n \end{cases} \quad \text{for } i = 1, \ldots, n$$

$$0 \leqslant f_{ij} \leqslant c_{ij}, \qquad \text{for } i = 1, \ldots, n \quad \text{and} \quad j = 1, \ldots, n$$

where

f_{ij} is the flow along arc ij ($f_{ij} = 0$ if there is no arc ij)

and

c_{ij} is the capacity of arc ij.

The first constraint ensures that, except for the SOURCE (vertex 1) and the SINK (vertex n), the flow out of a vertex is equal to the flow into that vertex—this is the flow conservation law. The second constraint ensures that the flow in an arc cannot exceed the capacity of that arc.

The dual problem is:

minimize

$$\sum_{i=1}^{n} \sum_{j=1}^{n} c_{ij} F_{ij}$$

subject to the constraints

$$X_1 - X_n \geqslant 1,$$

$$X_j - X_i + F_{ij} \geqslant 0,$$

$$F_{ij} \geqslant 0,$$

where the X_i are dual vertex variables, and the F_{ij} are dual arc variables.

The dual formulation suggests that we try to assign quantities $X_1, X_2, \ldots, X_n$ to the vertices such that the differences between quantities on adjacent vertices are as small as possible. This is precisely the aim of the *vertex labelling procedure* explained in detail in Chapter 3 which identifies a flow-augmenting path. At each stage of the algorithm for finding a maximum flow, a feasible solution to the dual problem identifies a flow-augmenting path. (A feasible solution to a linear programming problem satisfies the inequality constraints, but is not necessarily optimal.) This allows us to improve the feasible solution to the flow problem, which in turn allows us to improve the solution to the dual problem. Thus the algorithm considers the problem and its dual alternately.

The Dual of the Max-flow Min-cost Problem (Chapter 7)

The max-flow min-cost problem, which is to find a maximum flow $f_{\max}$ at minimum cost, can be expressed as follows:

minimize

$$\sum_{i=1}^{n} \sum_{j=1}^{n} \alpha_{ij} f_{ij}$$

subject to the constraints

$$\sum_{j=1}^{n} f_{ij} - \sum_{j=1}^{n} f_{ji} = \begin{cases} f_{\max} & \text{if } i = 1 \\ 0 & \text{if } i \neq 1 \text{ or } n, \\ -f_{\max} & \text{if } i = n \end{cases} \qquad \text{for } i = 1, \ldots, n$$

$$0 \leqslant f_{ij} \leqslant c_{ij}, \qquad \text{for } i = 1, \ldots, n \text{ and } j = 1, \ldots, n,$$

where
α_{ij} is the cost of sending unit flow along arc ij;
f_{ij} is the flow along arc ij;
c_{ij} is the capacity of arc ij.

Equivalently, by inserting an extra arc n, 1 with capacity $c_{n,1} = f_{\max}$, flow $f_{n,1} = f_{\max}$ and $\alpha_{n,1} = 0$, we have the following problem:

maximize

$$-\sum_{i=1}^{n} \sum_{j=1}^{n} \alpha_{ij} f_{ij}$$

subject to the constraints

$$\sum_{j=1}^{n} f_{ij} - \sum_{j=1}^{n} f_{ji} = 0,$$

$$f_{n,1} = c_{n,1} = f_{\max},$$

$$0 \leqslant f_{ij} \leqslant c_{ij}, \text{ for each arc } ij \text{ except arc } n, 1.$$

The dual problem is

minimize

$$\sum_{i=1}^{n} \sum_{j=1}^{n} c_{ij} F_{ij}$$

subject to the constraints

$$X_j - X_i - F_{ij} \leqslant \alpha_{ij}, \qquad \text{for } i = 1, \ldots, n \text{ and } j = 1, \ldots, n,$$

$$F_{ij} \geqslant 0, \text{ for each arc } ij \text{ except arc } n, 1,$$

where the X_i are dual vertex variables and the F_{ij} are dual arc variables.

The dual formulation suggests that we try to assign vertex quantities $X_1, X_2, \ldots, X_n$ such that the differences between quantities on adjacent vertices are as near as possible to the costs of the corresponding arcs. This is precisely the aim of the procedures used in the algorithm of Chapter 7 for assigning potentials to the vertices which identify minimum-cost paths. At each stage in the algorithm for finding a minimum-cost flow, a feasible solution of the dual problem identifies a minimum-cost path. This allows us to improve the solution to the flow problem. As in the maximum flow problem, the algorithm considers the problem and its dual alternately. Note that the vertex quantities $X_1, X_2, \ldots, X_n$ are not true potentials, since the costs do not satisfy the cycle law.

The Dual of the Transportation Problem (Chapter 14)

The transportation problem is concerned with sending goods from a number of supply points to a number of demand points at minimum total cost.

Suppose that there are n supply points which can supply amounts of goods $s_1, s_2, \ldots, s_n$, and m demand points with demands $r_1, r_2, \ldots, r_m$, and the cost of supplying one unit from supply point i to demand point j is α_{ij}. Let x_{ij} be the number of units of the goods sent from the supply point i to the demand point j. In our formulation of this problem we assume the use of a dummy vertex, so that we may take the total demand to be equal to the total supply.

We can express the transportation problem as follows:

minimize

$$\sum_{i=1}^{n} \sum_{j=1}^{m} \alpha_{ij} x_{ij}$$

subject to the constraints:

$$\sum_{j=1}^{m} x_{ij} = s_i, \qquad \text{for } i = 1, 2, \ldots, n,$$

$$\sum_{i=1}^{n} x_{ij} = r_j, \qquad \text{for } j = 1, 2, \ldots, m.$$

The first set of constraint equations ensures that, for each value of i, the supply point i supplies a total of s_i units; the second set of constraint equations ensures that, for each value of j, the demand point j receives r_j units. The solution is a set of values for the flows x_{ij} between each supply point and each demand point for which the objective function has a minimum value subject to the constraints.

Note that we must also have

$$\sum_{i=1}^{n} s_i = \sum_{j=1}^{m} r_j,$$

which represents a feasibility condition on the supplies and demands. This equation is not an inherent part of the linear programming formulation.

The dual problem is:

maximize

$$\sum_{i=1}^{n} s_i X_i + \sum_{j=1}^{m} r_j Y_j$$

subject to the constraints

$$X_i + Y_j \leqslant \alpha_{ij}, \qquad \text{for } i = 1, \ldots, n \text{ and } j = 1, \ldots, m,$$

where the X_i and Y_j are dual vertex variables.

The dual formulation suggests that we try to assign vertex quantities $X_1, X_2, \ldots, X_n$ to the supply vertices, and vertex quantities $Y_1, Y_2, \ldots, Y_m$ to the demand vertices, such that the sum of quantities on adjacent vertices is as close as possible to the cost on the corresponding arcs. This is precisely the aim of the procedure to obtain a succession of modified partial graphs in the Hungarian algorithm for the transportation problem given in Chapter 14. At each stage in the algorithm for finding a minimum-cost transportation, a feasible solution to the dual problem identifies a partial graph which is used to improve the feasible solution for the flow problem. Again, the algorithm considers the problem and its dual alternately.

The Dual of the Critical Path Problem (Chapter 15)

The scheduling problem as discussed in Chapter 15 consists in essence of finding a critical path in an activity network. Finding a critical path in an activity network with n vertices together with a START vertex (vertex 0) and a FINISH vertex (vertex $n + 1$) can be expressed as:

maximize

$$\sum_{i=0}^{n+1} \sum_{j=0}^{n+1} \alpha_{ij} x_{ij}$$

subject to the constraints

$$\sum_{i=0}^{n+1} x_{ij} = \sum_{i=0}^{n+1} x_{ji}, \qquad \text{for } j = 1, 2, \ldots, n,$$

$$\sum_{j=0}^{n+1} x_{0j} = 1,$$

$$\sum_{i=0}^{n+1} x_{i,n+1} = 1,$$

where
x_{ij} is equal to 0 or 1
and
α_{ij} is the length of the arc incident from vertex i to vertex j.

The first set of constraint equations ensures that if we arrive at any vertex, except the START or FINISH vertices, then we must leave it; the second constraint equation ensures that we must leave the START vertex along one arc; the third constraint ensures that we must arrive at the FINISH vertex along one arc.

The solution is a set of values of x_{ij} which satisfy the constraints. This set of values defines a path through the network from the START vertex to the FINISH vertex for which the objective function (the length of the path) is a maximum.

The dual problem is:

minimize

$$X_0 + X_{n+1} + \sum_{i=0}^{n+1} \sum_{j=0}^{n+1} Y_{ij}$$

subject to the constraints

$$X_i - X_j + Y_{ij} \geqslant \alpha_{ij}, \quad \text{for } i = 0, \ldots, n+1 \text{ and } j = 0, \ldots, n+1,$$

$$Y_{ij} \geqslant 0, \quad \text{for } i = 0, \ldots, n+1 \text{ and } j = 0, \ldots, n+1,$$

where the X_i are dual vertex variables and Y_{ij} are dual arc variables.

The formulation of the dual problem suggests that we try to assign quantities $X_0, X_1, \ldots, X_{n+1}$ to the vertices such that the difference between the quantities on adjacent vertices is as close as possible to the cost on the corresponding arcs. This is precisely the aim of the labelling procedure used in Chapter 15 to obtain the critical path, where the notation e_i is used for the vertex variables which correspond to earliest starting times. Thus this algorithm also considers the problem and its dual alternately.

CHAPTER 20

ALGORITHMS AND NP-COMPLETENESS

After reading this chapter, you should be able to:

- explain what is meant by *a problem instance*, *decision problem*, *time complexity function*, *polynomial-time algorithm*, *exponential-time algorithm* and *polynomial-time reducibility*;
- explain why polynomial-time algorithms are usually considered to be efficient, but exponential-time algorithms are usually considered to be inefficient;
- describe the classification of problems as problems in the classes P, NP and NP-complete.

20.1 EFFICIENCY OF ALGORITHMS

For many of the problems considered in this book, we have given the method of solution in the form of algorithms. Such algorithms can be converted into computer programs so that examples of the problems can be solved using a computer. Most of the examples discussed have been simple ones which we were able to solve by implementing the steps of the appropriate algorithm by hand. For more complex examples, it would be very tedious and time-consuming to apply the procedures by hand, whereas we can often obtain a solution very quickly using a computer. However, because of the phenomenon known as the *combinatorial explosion*, even a computer would take an unacceptably long time to solve even moderately large examples of some problems.

Once an algorithm has been constructed for a particular problem, a number of questions about the performance of the algorithm present themselves. Can we predict how long it will take a computer using the algorithm to solve a particular example? Can moderately large examples of the problem be solved in a reasonable time? Is it possible to construct other algorithms which are more efficient—that is, appreciably faster? The answers to such questions are often not clear-cut, and, for some problems, nobody has been able to give definite answers to them. However, we can make a useful distinction between algorithms which are likely to be useful in practice and those which are likely not to be. To do this, we must first clarify what we mean by a *problem* and what we mean when we say that an algorithm is *efficient*.

For our purposes, we consider a *problem* to be a general question involving a number of parameters whose values are left unspecified, together with a statement of what properties the solution should have. An *instance* of the problem is obtained by specifying particular values for all the parameters.

Example 1 An example of a problem in the above sense is the travelling salesman problem. The parameters consist of a set of cities and the set of distances between all pairs of these cities. A solution is an ordered list of cities which corresponds to a Hamiltonian cycle of minimum length in the corresponding graph.

(The travelling salesman problem was introduced in Chapter 1, and discussed further in Chapter 8.)

An instance of the travelling salesman problem is a particular set of cities for which the distance between each pair of cities is specified. For example, one instance is represented by the following graph:

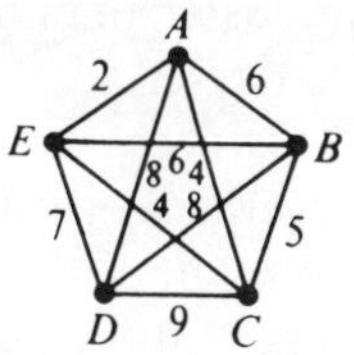

The solution to this instance of the problem is *ACBDEA* (or *AEDBCA*), corresponding to a cycle of length 26.

Example 2 Another example of a problem is the *graph isomorphism problem*: given two labelled graphs G and H, each with n vertices and m edges, are G and H isomorphic?—that is, does there exist a one-to-one correspondence between the vertices of G and those of H, such that the number of edges joining any pair of vertices in G is equal to the number of edges joining the corresponding pair of vertices in H?

For this problem, the parameters may consist of the vertex-set and edge-list of each of the graphs G and H, or the adjacency matrices of G and H, or the incidence matrices of G and H; alternatively, the relevant information may be represented diagrammatically. The solution is either YES or NO.

An instance of the graph isomorphism problem is provided by the following graphs:

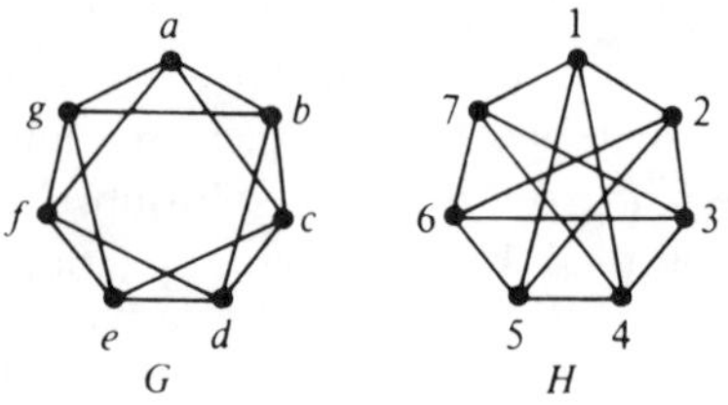

It is easy to check that a one-to-one correspondence with the required property is

$$a \leftrightarrow 1, \quad b \leftrightarrow 4, \quad c \leftrightarrow 7, \quad d \leftrightarrow 3, \quad e \leftrightarrow 6, \quad f \leftrightarrow 2, \quad g \leftrightarrow 5,$$

so in this instance the answer is YES.

A problem for which the solution is either YES or NO is called a **decision problem**.

The idea of *efficiency* when applied to an algorithm normally refers to the time taken by a computer to solve problem instances—a more efficient algorithm is, roughly speaking, faster than a less efficient one. (In general, the idea of efficiency may also refer to other computing resources—for example, the computer store. However, the time taken is usually the dominant concern.) It is clear, however, that the time taken to solve a particular problem instance depends (among other things) on how large that instance is. We therefore specify the time requirement of an algorithm as a function of a single parameter, called the *size* of the problem instance. For example, for the travelling salesman problem, it is usual to take the size of a problem instance to be the number of cities; similarly, for the graph isomorphism problem we may choose to take the size to be the number of vertices.

The **time complexity function $f(n)$** for an algorithm is the longest time taken to solve any instance of size n. It is defined in terms of the longest time, because different instances of a problem which are of the same size may require very different times for a solution to be obtained using the algorithm. In other words, the function $f(n)$ gives an *upper bound* for the time taken to solve an instance of size n.

The time taken by an algorithm to solve a problem instance depends on many factors—for example, on which computer is used and on how the computer program is written—so it is difficult to specify a time complexity function precisely. Fortunately, this does not matter greatly, since we are interested primarily in the general way in which the time taken by an algorithm increases with the size of a problem instance, so that we can gauge whether the time taken for instances of large size is likely to be reasonable, and so that we can compare the likely time performances of different algorithms. In particular, we are interested in the asymptotic form of $f(n)$—that is, its form for large values of n.

We illustrate these ideas by working out the form of the time complexity function for an algorithm which uses the branch-and-bound method to solve the knapsack problem (discussed in Chapter 17).

The branch-and-bound algorithm examines some of the possible selections of items which can be packed in the knapsack. We take the size of an instance of the knapsack problem to be equal to the number of items available. Clearly, the time taken by the algorithm is roughly proportional to the number of selections which are examined. For different problem instances of the same size, the number of selections examined may be different, since different numbers of selections may be infeasible. In the worst case for an instance of size n, all possible selections must be examined. The number of such selections is 2^n, so we say that

$$f(n) \text{ is of the order of } 2^n,$$

which we write as

$$f(n) \text{ is } O(2^n).$$

(A function $f(n)$ is said to be $O(g(n))$ if there exists a constant c and a function $g(n)$ such that $|f(n)| \leq c\,|g(n)|$ for all values of $n > 0$.)

There are many different time complexity functions, but the crucial distinction as far as the solvability of a problem is concerned is the distinction between a function which is *polynomial* in n (such as n^2, n^3, etc.) and one which is *exponential* in n (such as 2^n, e^n, etc.). To see why this distinction is important, consider the following table, which compares the times taken for various values of n for different time complexity functions.

Table 1 Comparison of computation times for different sizes of problem instance

		size n			
		10	30	50	70
polynomial time complexity function	n	0.00001 second	0.00003 second	0.00005 second	0.00007 second
	n^3	0.001 second	0.027 second	0.125 second	0.343 second
	n^5	0.1 second	24.3 seconds	5.2 minutes	28 minutes
exponential time complexity function	2^n	0.001 second	17.9 minutes	35.7 years	37 million years
	3^n	0.059 second	6.5 years	2.3×10^{10} years	8×10^{13} million years

In Table 2, we list the time complexity functions for some of the algorithms described in this book. We do not include most of the network algorithms because of their complicated nature, and because their time complexity functions depend crucially on the details of the algorithms.

Table 2 Time complexity functions for algorithms based on graphs or digraphs with n vertices and m edges

Shortest path algorithm	$O(n^2)$
Longest path algorithm	$O(n^2)$
DFS algorithm	$O(m)$
BFS algorithm	$O(m)$
Algorithm for constructing a spanning tree	$O(n^2)$
Greedy algorithm for constructing a minimum spanning tree	$O(n^2)$
Prim's algorithm for constructing a minimum spanning tree	$O(n^2)$
Hungarian algorithm for the assignment problem	$O(n^3)$
† Fleury's algorithm for constructing an Eulerian trail	$O(m)$
† Algorithm for solving the Chinese postman problem	$O(n^3)$

† These algorithms are introduced in R. J. Wilson and J. J. Watkins, *Graphs: an Introductory Approach*, John Wiley & Sons, New York, 1990.

The times taken by exponential-time algorithms for problem instances of moderate size are so long that problems for which no polynomial-time algorithm exists are generally considered to be intractable. So we can make the very useful distinction between *polynomial-time algorithms*, which are normally efficient, and *exponential-time algorithms*, which are usually so inefficient as to be of little practical use except for problem instances of small size.

Some caution must be exercised, however, in applying this general conclusion to particular problems. For example, it is possible for a polynomial-time algorithm to have a time complexity function with a very large coefficient (such as $10^{100}n$) or a very large exponent (such as n^{100}), so that the algorithm is not efficient in practice. However, such cases are rare, and most polynomial-time algorithms met with in practice have a reasonably small coefficient and an exponent not greater than 2 or 3. Further, some exponential-time algorithms are quite efficient in practice. An example of this is the branch-and-bound algorithm for the knapsack problem, for which the time taken in practice for most problem instances is much better than the worst-case time predicted by the time-complexity function.

20.2 NP-COMPLETE PROBLEMS

Suppose that, for some problem, we have constructed an algorithm which has an exponential time complexity function. From the previous discussion, we know that the algorithm is not likely to be efficient—that is, it will almost certainly take an inordinately large time to solve a problem instance of moderate size. Obviously, we should prefer to have a polynomial-time algorithm, but suppose that after much effort we are unable to construct one. To save further wasted effort, we should like to be able to prove that no such algorithm exists; we could then look for a heuristic algorithm which might produce an acceptable, although probably not the best, solution.

Nobody has yet been able to prove whether or not polynomial-time algorithms can be constructed for a large class of problems (including the travelling salesman problem) for which only exponential-time algorithms are known. However, the theory of NP-completeness, which we now discuss, provides strong evidence to suggest that polynomial-time algorithms for such problems do not exist.

NP-problems

We remarked earlier that nobody has been able to find a polynomial-time algorithm for the travelling salesman problem. Suppose that, instead of the usual travelling salesman problem:

> find a sequence of cities which form a Hamiltonian cycle of minimum length,

we consider the related problem:

> is there a Hamiltonian cycle of total length less than L?

where L is some number. This is called the *corresponding decision problem.*

Suppose that we have a particular instance of this problem, and that we are *given* a cycle whose total length is less than L. If we wish to check that the given cycle really does have total length less than L, then it is quite straightforward: this checking can certainly be done in polynomial time. Any problem like this whose solution *when given* can be *checked* in polynomial time (even if it took exponential time to find that solution originally) is called a **non-deterministic polynomial-time problem**, abbreviated to **NP-problem**. The class of all such problems is denoted by NP.

For example, any problem which can be solved in polynomial time is also an NP-problem, because we can certainly *check* the given solution in polynomial time if it took only polynomial time to find it in the first place! Problems whose solution can be found in polynomial time using an algorithm form the class of **polynomial-time problems**, denoted by P.

In order to compare the performance of algorithms for problems in the class NP with algorithms for problems in P in a more formal manner, logicians have introduced the concept of a *non-deterministic computer*. This is a type of model of computation which is deliberately unrealistic and does not correspond to any existing physical computing device. An ordinary computer is a *deterministic* machine—that is, one in which the state at a particular time is determined in a predictable manner by the state and input to the machine at a previous time. A non-deterministic computer is a hypothetical device which has the remarkable ability to 'guess' the answer to a problem.

For the decision-problem form of the travelling salesman problem, the 'guess' takes the form of a sequence of cities. The computer can then verify that the 'guess' is a solution—that is, that the sequence forms a cycle which has a total length less than L. This *checking stage* can be carried out in polynomial time by a deterministic process. Thus the non-deterministic computer operates in two stages:

(i) a 'guessing' stage;
(ii) a checking stage, in which the computer verifies that the 'guess' is in fact a solution.

The non-deterministic computer always 'guesses' correctly. The reason for having a checking stage in the non-deterministic computer is so that the performance of an algorithm for a non-deterministic computer can be compared with the performance of an algorithm for an ordinary deterministic machine.

It is assumed that the 'guessing' stage takes no appreciable time, so in this case, since the checking stage can be carried out in polynomial time, the non-deterministic computer can answer the decision form of the travelling salesman problem in polynomial time. Thus an NP-problem (non-deterministic polynomial-time problem) is one which can be solved in polynomial time on a non-deterministic computer.

We can use the non-deterministic computer to solve a decision problem—if the answer is YES, the computer produces a 'guess' and verifies it in polynomial time; if the answer is NO, the computer either produces this answer or does not stop running.

Let us take another example of a problem which cannot be solved in polynomial time—the graph isomorphism decision problem: given two graphs G and H, are they isomorphic? To date, the best known algorithm for solving this problem is an exponential-time algorithm. But suppose that, for a particular instance of this problem, an isomorphism between G and H *is given*. Then it takes only polynomial time to verify that this one-to-one correspondence really is an isomorphism. A non-deterministic computer can therefore solve the graph isomorphism decision problem by producing a one-to-one correspondence in the 'guessing' stage, and by verifying that this is an isomorphism in the checking stage. Hence this problem is an NP-problem.

The class P of polynomial-time problems is a subclass of the class of problems which can be solved in polynomial time using an ordinary deterministic computer.

Some Important Results

We now state and discuss three important results concerning the classes of problems.

1. *The class P is contained in NP* Every decision problem in P is also in NP. This is obvious, since if a problem is solvable in polynomial time on a deterministic computer, we can solve it using a non-deterministic computer by simply ignoring the guessing stage, and using the deterministic algorithm instead of the checking stage. The question arises as to whether there are any problems in NP which are not in P. Algorithms for a non-deterministic computer would seem to be very much more powerful than algorithms for a deterministic computer, so we should intuitively expect NP to contain more problems than P. However, nobody has yet been able to prove this, so it remains a conjecture, although one that is generally accepted to be true.

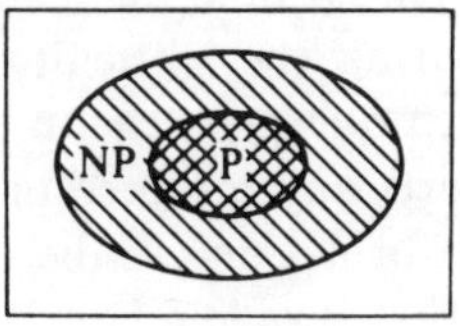

$P \subseteq NP$

Is P a proper subclass of NP?

2. *A polynomial-time algorithm for a non-deterministic computer can be converted to an exponential-time algorithm for a deterministic computer* To convert an algorithm for a non-deterministic computer to one for a deterministic machine, we must replace the guessing stage by a deterministic process. The only obvious way of doing this is to try all possible guesses. Unfortunately, the number of possible guesses is usually an exponential function of the problem size. It can be shown that a problem of size n in NP can be solved by an algorithm for a deterministic computer which has time-complexity $2^{p(n)}$, where $p(n)$ is a polynomial function. It seems likely, therefore, that there are problems in NP for which the only possible algorithms for a deterministic machine have exponential time complexity.

3. *A decision problem is no harder than the corresponding optimization problem* We can associate a decision problem with any optimization problem. For example, if an optimization problem requires a solution which has minimum cost or minimum length, we can associate with it a decision problem which asks whether there is a solution whose cost or length is not more than a given bound L. Decision problems can be associated in a similar way with maximization problems by replacing 'not more than' by 'not less than'.

Provided that the cost or length of a solution is easy to evaluate, a decision problem can be no harder than the corresponding optimization problem. For example, if we have solved an instance of the travelling salesman problem, all we have to do to answer the corresponding decision problem is to compare the length of a minimum cycle with the bound L in the decision problem. So, although the theory of NP-complete problems applies to decision problems, we can extend many of the results about the difficulty of a problem to the corresponding optimization problems.

Polynomial Time Reducibility

We have seen that polynomial-time algorithms are generally considered to be efficient, and that exponential-time algorithms are generally considered to be inefficient. In comparing the difficulty of two problems, a useful technique is to try to reduce one problem to the other—that is, to find a transformation which converts any instance of one problem to an instance of the other. If such a transformation can be carried out by a polynomial-time algorithm, and if the first problem can be solved in polynomial time, then so can the second. Polynomial-time reducibility plays an important part in the theory of NP-complete problems, as we shall see shortly.

NP-complete Problems

The basis of the theory of NP-completeness was provided in 1971 by Stephen Cook of the University of Toronto in a paper entitled *The complexity of theorem-proving procedures*. In this paper, he proved that one particular problem in NP, called the *satisfiability problem*, has the property that every other problem in NP can be polynomially reduced to it. The implication of this is that if the satisfiability problem can be solved in polynomial time (on a deterministic computer), then so can every other problem in NP (and so NP = P). Also, if any problem in NP is intractable, in the sense that it can be solved only in exponential time, then the satisfiability problem must also be intractable (and so NP is strictly larger than P).

Subsequently, a large number of problems (including the decision form of the travelling salesman problem) have been shown to share this property of the satisfiability problem. This class of problems is called the class of **NP-complete problems**.

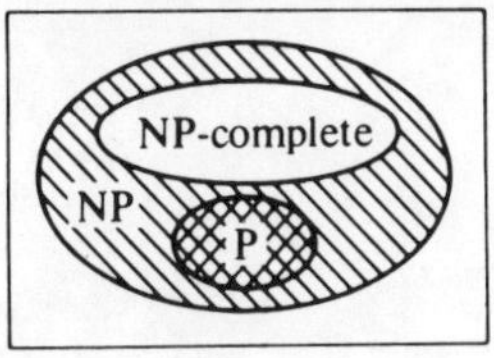

It follows that if a polynomial-time algorithm can be found for any one of these NP-complete problems, then every one of them must be solvable in polynomial time, and conversely, if one can be proved to be intractable, then they must all be intractable. The question of whether NP-complete problems are intractable is generally considered to be one of the principal unsolved problems of the theory of algorithms. However, as the number of problems shown to be NP-complete grows, and no polynomial-time algorithm is found for any of them, it becomes more and more likely that NP-complete problems are intractable.

The following list gives some combinatorial problems for which the optimization or the decision form (or both) are known to be NP-complete:

- the travelling salesman problem;
- the graph isomorphism problem;
- the Hamiltonian cycle problem (does a given graph G have a Hamiltonian cycle?);
- the bin-packing problem;
- the knapsack problem.

Historical Note

In October 1979, British mathematicians were amazed to read, on the front page of a leading daily newspaper, a report that the Soviet mathematician L. G. Khachiyan had found a polynomial-time algorithm for solving the travelling salesman problem. As the travelling salesman problem is an NP-complete problem, this would of course imply that *all* the NP-problems could be solved in polynomial time and that NP is just the class P. However, when the full details of Khachiyan's work reached the West, it transpired that the report was inaccurate, and that Khachiyan had in fact found a polynomial-time algorithm for a class of linear programming problems which are *not* NP-complete problems. Nevertheless, Khachiyan's work has caused much excitement, because previously the best algorithm for doing these linear programming problems was the so-called *simplex algorithm*, which requires exponential time. If this class of linear programming problems can be solved in polynomial time, then is it possible that a polynomial-time algorithm exists for the NP-complete problems? The issue remains undecided.

SUGGESTIONS FOR FURTHER READING

References have been grouped under the main subject areas of this book although many of the references cited will also be found useful for subject areas other than the ones under which they are listed.

Graph Theory (Chapter 2)

R. J. Wilson and J. J. Watkins, *Graphs: An Introductory Approach*, John Wiley & Sons, New York, 1990.

This is a companion volume to the present book and covers all the material of Chapter 2 together with further graph theory and applications. It also contains a useful bibliography on graphs.

Network Flows (Chapters 3–5)

This material is covered in a number of books—for example:

N. Deo, *Graph Theory with Applications to Engineering and Computer Science*, Prentice-Hall, Englewood Cliffs, New Jersey, 1974,
N. Christofides, *Graph Theory: An Algorithmic Approach*, Academic Press, London 1975.

The classic treatment of network flow problems is in the book by Ford and Fulkerson:

L. R. Ford, Jr. and D. R. Fulkerson, *Flows in Networks*, Princeton University Press, Princeton, New Jersey, 1962.

Paths and Trees (Chapters 6–8)

Further material on trees, connectivity and on Eulerian and Hamiltonian graphs can be found in standard books on graph theory—for example:

J. A. Bondy and U. S. R. Murty, *Graph Theory with Applications*, Macmillan, London, 1976,
R. J. Wilson, *Introduction to Graph Theory*, 3rd edition, Longmans, Harlow, Essex, 1985,
G. Chartrand and L. Lesniak, *Graphs and Digraphs*, 2nd edition, Wadsworth and Brooks/Cole, Monterey, California, 1986.

A discussion of the shortest-path algorithm can be found in the above book by Bondy and Murty or in the book by Christofides (see under Network Flows).

Tree-searching algorithms and other applications of trees to computing are covered in:

E. S. Page and L. B. Wilson, *Information Representation and Manipulation in a Computer*, 2nd edition, Cambridge Computer Science Texts 2, Cambridge University Press, Cambridge 1978.

The material on gas pipeline networks is discussed in:

H. Frank and I. T. Frisch, 'Network analysis', in *Scientific American*, July 1970, pp. 94–103,
B. Rothfarb, H. Frank, D.M. Rosenbaun, K. Steiglitz and D. J. Kleitman, 'Optimum design of offshore natural gas pipeline systems,' in *Operations Research*, Vol. 18, No. 6 (Nov.–Dec. 1970), pp. 992–1020.

Physical Networks (Chapters 9–11)

The theory of electrical networks is discussed in:

C. A. Desoer and E. S. Kuh, *Basic Circuit Theory*, McGraw-Hill, New York, 1969,
S. W. Director, *Circuit Theory: A Computational Approach*, John Wiley & Sons, New York, 1975,
W. A. Blackwell, *Mathematical Modelling of Physical Networks*, Macmillan, New York, 1968.

S. Seshu and M. B. Reed, *Linear Graphs and Electrical Networks*, Addison-Wesley, Reading, Mass. 1961.

Further information about physical networks in general may be found in:

P. H. Roe, *Networks and Systems*, Addison-Wesley, London, 1966.

Assignment and Transportation Problems (Chapters 12–14)

Discussion of the material of Chapter 12 can be found in many books on combinatorics, for example:

R. J. Wilson, *Introduction to Graph Theory*, 3rd edition, Longmans Harlow, Essex, 1985,
C. L. Liu, *Introduction to Combinatorial Mathematics*, McGraw-Hill, New York, 1968.

Most books on network theory include a section on the assignment and transportation problems—for example:

L. R. Ford, Jr. and D. R. Fulkerson, *Flows in Networks*, Princeton University Press, Princeton, New Jersey, 1962,
E. Lawler, *Combinatorial Optimization: Networks and Matroids*, Holt, Rinehart & Winston, New York, 1976.

Scheduling and Location Problems (Chapters 15–18)

Activity networks for project planning are discussed in the following:

H. R. Hoare, *Project Management using Network Analysis*, McGraw-Hill, London, 1973,
A. Battersby, *Network Analysis for Planning and Scheduling*, 3rd edition, Macmillan, London, 1970,
K. G. Lockyer, *An Introduction to Critical Path Analysis*, Pitman, London, 1977.

The following book deals with the subject from the point of view of the manager who wishes to understand operations research, but does not wish to specialize in it. It also covers the transportation problem of Chapter 14.

E. S. Buffa and J. S. Dyer, *Management Science/Operations Research*, John Wiley & Sons, New York, 1981.

Location problems are discussed in the book by Christofides (see under Network Flows).

Network Theory and Algorithms (Chapters 19–20)

Many books deal with linear programming and duality—for example:

S. Vajda, *Linear Programming: Algorithms and Applications*, Chapman & Hall, London and New York, 1981,
B. Van der Veen, *Introduction to the Theory of Operational Research*, N. V. Philips, Eindhoven, The Netherlands, 1967,
T. C. Hu, *Integer Programming and Network Flows*, Addison-Wesley, Reading, Mass., 1969,
M. S. Bazaraa and J. J. Jarvis, *Linear Programming and Network Flows*, John Wiley & Sons, New York, 1977.

The following book gives a more theoretical treatment of the subject:

A. Schriver, *Theory of Linear and Integer Programming*, John Wiley & Sons, Chichester, 1986.

A full discussion of graph algorithms, including a chapter on NP-completeness is given in:

S. Even, *Graph Algorithms*, Pitman, London, 1979.

The following book gives a full account of the theory of NP-completeness and contains an extensive list of NP-complete problems.

M. R. Garey and D. S. Johnson, *Computers and Intractability: A Guide to the Theory of NP-Completeness*, W. H. Freeman, San Francisco, 1979.

A comprehensive mathematical treatment of network algorithms, including a discussion of NP-completeness is given in:

C. H. Papadimitriou and K. Steiglitz, *Combinatorial Optimization: Algorithms and Complexity*, Prentice-Hall, New Jersey, 1982.

General

A useful survey of graph and network problems with the emphasis on the practical aspects is:

V. Chachra, P. M. Ghare and J. M. Moore, *Applications of Graph Theory Algorithms*, Elsevier North-Holland, New York, 1979.

POSTSCRIPT

The final test of a theory is its capacity to solve the problems which originated it.

George B. Dantzig, *Linear Programming and Extensions*

One must understand clearly how real-life situations are different from computer situations, exactly because they do not have the precise solutions that chess or engineering calculations do.

J. Bronowski, *The Ascent of Man*

In all human traditions there has been a very great antagonism against all this counting business. . . . The first chap who arranged the census was King David, and when he arranged the census the Lord was utterly furious. . . . There was something wrong in having a census which treats people as if they were units, whereas they are not. Each is a universe.

E. E. Schumacher, *Good Work*

So down he came; for loss of time,
Although it griev'd him sore,
Yet loss of pence, full well he knew,
Would trouble him much more.

William Cowper

SOLUTIONS TO PROBLEMS IN THE TEXT

CHAPTER 1

1.1. (a) Since the Bexleyheath and Sidcup lines join at Lewisham, the number of trains on these two lines together is limited by the capacity of the single line from Lewisham to London Bridge, which is only 10 trains per hour. A further 10 trains per hour can run via Woolwich, giving a total of 20 trains per hour.

(b) If the capacity of the section from Lewisham to London Bridge were to be doubled, both the Bexleyheath and Sidcup lines could run at full capacity and the total number of trains would be increased to 30 per hour.

1.2. The route must go through A, B or C. There is no way from A to T in less than 120 units (along $A \to D \to T$) and hence the shortest path from S to T via A is 150 units. It is not hard to see that the shortest path from B to T is via the edge BD; the length of this path is 100 units, making the path $S \to B \to D \to T$ 140 units long. The shortest path from C to T is $C \to F \to T$; the length of this path is 110 units, and so the shortest path is $S \to B \to D \to T$; its length is 140 units.

1.3. (a) existence; (b) optimization; (c) enumeration; (d) optimization; (e) construction.

CHAPTER 2

2.1. (a) vertex-set: {0, 1, 2, 3, 4, 5, 6, 7, 8, 9}
edge-list: 01, 12, 23, 34, 40, 57, 68, 79, 85, 96, 05, 16, 27, 38, 49
(b) vertex-set: {*a*, *b*, *c*, *d*, *e*, *f*, *g*}
edge-list: *ab*, *bb*, *bb*, *cd*, *dd*, *de*, *ef*, *ef*, *fc*, *fc*, *fc*.

2.2. Neither (a) nor (b) is a subgraph of the given graph;
(c) is a subgraph—for example, the 'pentagon' 012340;
(d) is a subgraph—for example, the 'hexagon' 0127940.

2.3. (i) Graphs (a) and (c) are the same;
graph (d) is isomorphic to (a) and (c), as can be seen by interchanging the labels 3 and 6;
graph (b) is not isomorphic to any of the other three, since it contains no 'triangles'.
(ii) No. To see this, look at the four vertices of degree 2. In the first graph they are joined in pairs, whereas in the second graph none of them is joined to any other.

2.4. (a), (d), (g) and (h) are TRUE;
(b), (c), (e) and (f) are FALSE.
Note that (c) and (f) are both meaningless since the term *adjacent* is defined only for a pair of vertices, and the term *incident* is defined only for a vertex and an edge.

2.5. (i) (a) (1, 1, 1, 1, 1, 1, 2, 4, 4); (b) (4, 4, 4, 4, 4);
(c) (0, 1, 3, 4, 4, 5, 5).

(ii)

		graph (a)	graph (b)	graph (c)
(1)	number of edges	8	10	11
(2)	sum of degrees	16	20	22

In each case the sum of the degrees is exactly twice the number of edges. The reason for this is given on page 25.

2.6. (a) trail, 5, *x*, *y*; (b) walk, 4, *v*, *v*; (c) path, 1, *v*, *w*;
(d) trail, 6, *u*, *u*.
(Note that alternative answers are possible in parts (a), (c) and (d); for example, we could have given the answer *walk* in each case, since every trail is a walk. In each case we have chosen the most restrictive term.)

2.7. length 3: *svwz*
length 4: *stvwz* and *svwyz*
length 5: *stuvwz*, *stvwyz* and *svwxyz*
length 6: *stuvwyz* and *stvwxyz*
length 7: *stuvwxyz*.

2.8. (a) There are various possibilities—for example, *vuzvwzv*.
(b) Again, there are various possibilities—for example, *vuuzv*.
(c) length 1: the loop *uu*,
length 2: the multiple edges *uzu*,
length 3: the triangle *vwzv*, and both of the triangles *uvzu*,
length 4: both of the 'quadrilaterals' *uvwzu*.

2.9. (a)

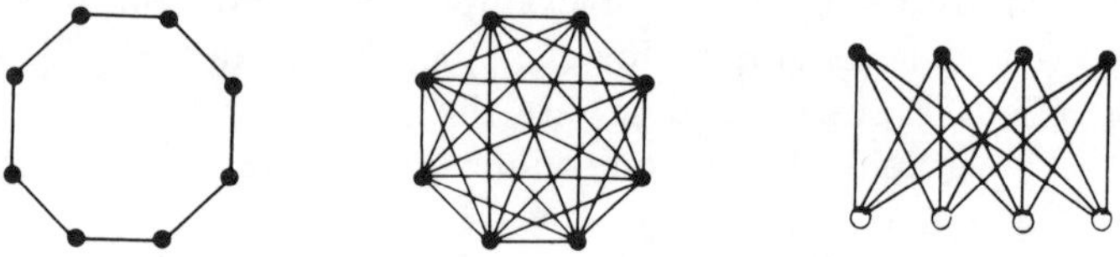

(b) If we colour the vertices of the bipartite graph black and white, then the vertices in each cycle must alternate between these two colours. This implies that the number of edges in each cycle must be even.

2.10. (a)

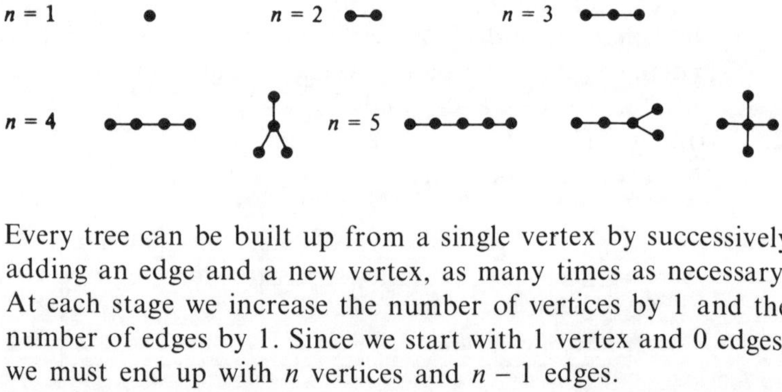

(b) Every tree can be built up from a single vertex by successively adding an edge and a new vertex, as many times as necessary. At each stage we increase the number of vertices by 1 and the number of edges by 1. Since we start with 1 vertex and 0 edges, we must end up with n vertices and $n-1$ edges.
(c) Let T be a tree with at most one vertex of degree 1. Then T has at least $n-1$ vertices of degree 2 or more, where n is the number of vertices. It follows that the sum of the vertex-degrees is at least $2(n-1)+1$, and so, by the handshaking lemma, the number of edges is at least $(n-1)+\frac{1}{2}$. This contradicts the fact that T has exactly $n-1$ edges, and the result follows immediately.
(d) Choose any vertex v and colour it black. Colour all vertices adjacent to v white. Next, colour all uncoloured vertices adjacent to these black. Continue this process until every vertex has been coloured. Since T is connected and contains no cycles, there is one and only one path between any two vertices. Thus no two adjacent vertices have the same colour, so T is bipartite.

2.11.

$$\begin{bmatrix} 1 & 0 & 0 & 0 & 1 & 0 & 0 & 0 \\ 1 & 1 & 0 & 0 & 0 & 1 & 1 & 1 \\ 0 & 1 & 1 & 0 & 0 & 0 & 0 & 0 \\ 0 & 0 & 1 & 1 & 0 & 1 & 1 & 1 \\ 0 & 0 & 0 & 1 & 1 & 0 & 0 & 0 \end{bmatrix}$$

2.12.

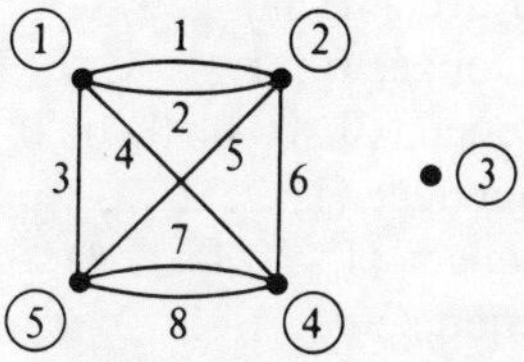

2.13.

$$\begin{bmatrix} 0 & 1 & 0 & 0 & 1 \\ 1 & 0 & 1 & 3 & 0 \\ 0 & 1 & 0 & 1 & 0 \\ 0 & 3 & 1 & 0 & 1 \\ 1 & 0 & 0 & 1 & 0 \end{bmatrix}$$

2.14.

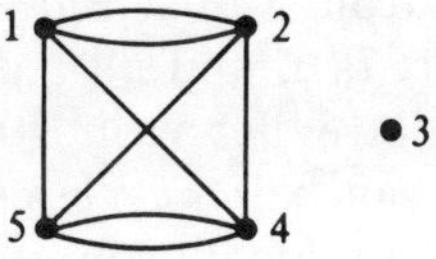

2.15. (a) vertex-set: {1, 2, 3, 4, 5}
arc-list: 21, 25, 42, 43, 51, 52, 54

(b) vertex-set: {*a*, *b*, *c*, *d*, *e*, *f*, *g*}
arc-list: *ab*, *bb*, *bb*, *cd*, *cf*, *dd*, *ed*, *ef*, *ef*, *fc*, *fc*.

2.16. (a) is a subdigraph—for example, the subdigraph whose arcs are 42, 43, 52 and 54;

(b) is a subdigraph—for example, the subdigraph whose arcs are 21, 51, 52 and 54;

(c) is not a subdigraph.

2.17. (i) Digraphs (b) and (d) are the same;
digraph (c) is isomorphic to (b) and (d), as can be seen by interchanging the labels 1 and 4;
digraph (a) is not isomorphic to any of the other three, since it contains a vertex (2) with three arcs out of it.

(ii) No. The first digraph contains a 'directed triangle' (shown overleaf). The second digraph contains no such triangle.

2.18. (a), (c), (d) and (f) are TRUE;
(b), (e), (g) and (h) are FALSE.

2.19. (a) out-degree sequence: (0, 1, 1, 1, 1, 1, 1, 1, 1),
in-degree sequence: (0, 0, 0, 0, 0, 0, 1, 3, 4);
(b) out-degree sequence: (1, 2, 2, 2, 3),
in-degree sequence: (1, 2, 2, 2, 3);
(c) out-degree sequence: (1, 1, 2, 2, 2, 3),
in-degree sequence: (0, 0, 2, 3, 3, 3).

2.20. (a) length 5: *stvwyz*
length 6: *stuvwyz* and *stvwxyz*
length 7: *stuvwxyz*;
(b) length 3: *zuvs* and *zwxs*
length 5: *zuvwxs*;
(c) *stvwyzuvs* and *stvwyzwxs*.
Any cycle containing both s and z must consist of a path from s to z followed by a path from z to s. But all paths from s to z contain both v and w (by part (a)), and all paths from z to s contain v or w (by part (b)), so that either v or w must occur twice. Since this is not allowed, there can be *no* cycle containing both s and z.

2.21. (a) connected, but not strongly-connected (since there are no paths from the centre vertex to any other);
(b) strongly-connected;
(c) disconnected;
(d) connected, but not strongly-connected (since there are no paths from the top right-hand vertex to any other).

2.22.
$$\begin{bmatrix} 1 & 0 & 0 & 0 & -1 & 0 & 0 & 0 \\ -1 & 1 & 0 & 0 & 0 & 1 & -1 & -1 \\ 0 & -1 & -1 & 0 & 0 & 0 & 0 & 0 \\ 0 & 0 & 1 & -1 & 0 & -1 & 1 & 1 \\ 0 & 0 & 0 & 1 & 1 & 0 & 0 & 0 \end{bmatrix}$$

2.23.

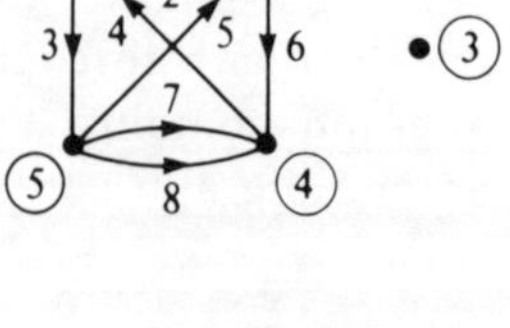

2.24.

$$\begin{bmatrix} 0 & 1 & 0 & 0 & 0 \\ 0 & 0 & 1 & 1 & 0 \\ 0 & 0 & 0 & 0 & 0 \\ 0 & 2 & 1 & 0 & 0 \\ 1 & 0 & 0 & 1 & 0 \end{bmatrix}$$

2.25.

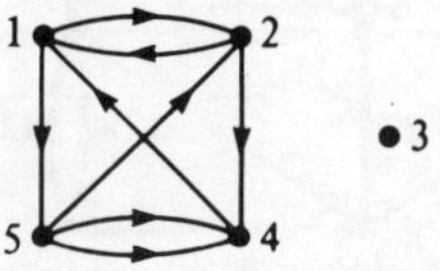

CHAPTER 3

3.1. (a) The flow can be increased by 2, thus:

(b) The flow can be increased by 1, thus:

(c) The flow can be increased by 3, thus:

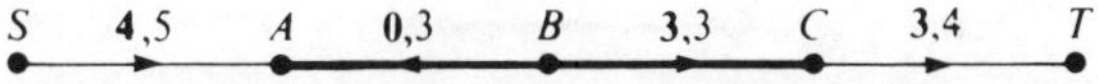

3.2. (a) First, for each of the *forward* arcs, calculate the largest amount by which the flow can be *increased* without exceeding the capacity, and let the smallest of these numbers be r. Next, for each of the *backward* arcs, calculate the largest amount by which the flow can be *decreased* without 'going below zero', and let the smallest of these numbers be s. The required amount is the smaller of these numbers r and s.

(b) For the given flow-augmenting path,
r is the smallest of the numbers 3, 3 and 7, which is 3;
s is the smallest of the numbers 3, 2 and 5, which is 2.
The required amount is therefore 2.

3.3. A flow-augmenting path is *SBEFT*:

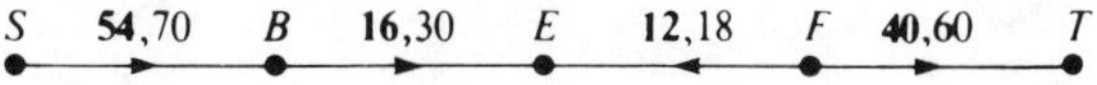

The flow along this path can be increased by 12 to give the flow of value 136 shown below:

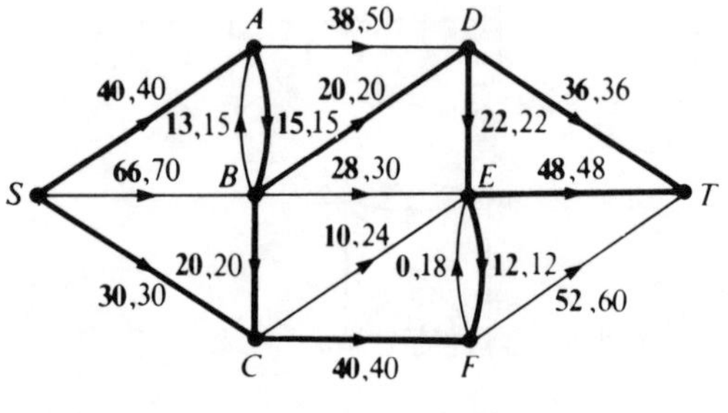

There are no further flow-augmenting paths in this network.

3.4. (a) The value of the maximum flow is 2:

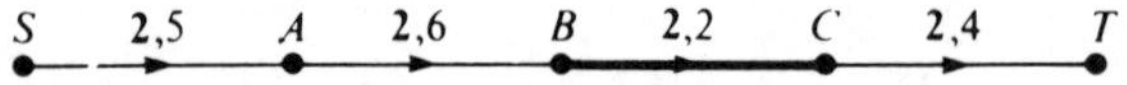

The worst bottle-neck is the arc BC, whose presence restricts the value of the flow to at most 2.

(b) The value of a maximum flow is 3; for example:

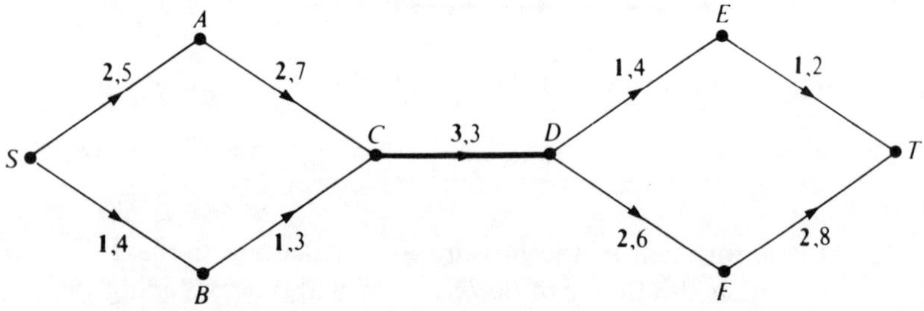

In this case the worst bottle-neck is the arc CD, whose presence restricts the value of every flow to at most 3.

(c) The value of a maximum flow is 3; for example:

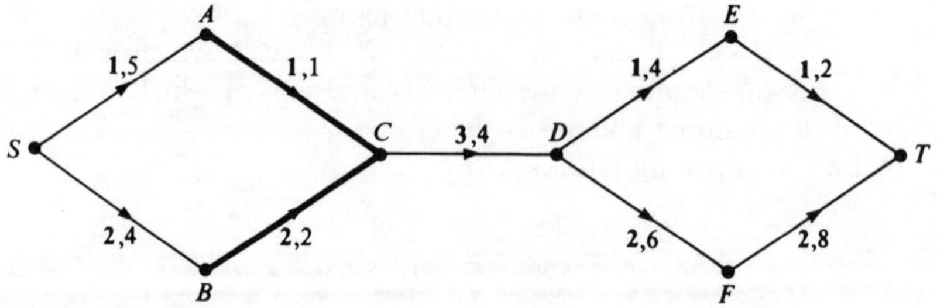

The worst bottle-neck comprises the arcs AC and BC, which restrict the value of every flow to at most $1 + 2 = 3$.

(d) The value of a maximum flow is again 3; for example:

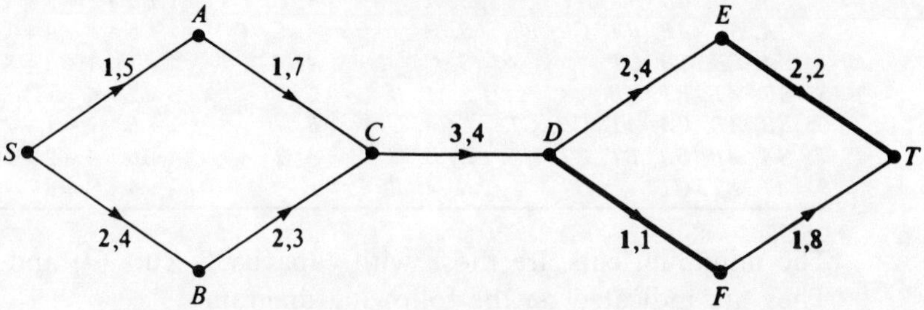

The worst bottle-neck comprises the two arcs DF and ET, which restrict the value of every flow to at most $1 + 2 = 3$.

3.5. (a) The minimum cut consists of the arc BC, with capacity 2;

$$X = \{S,\ A,\ B\}, \quad Y = \{C,\ T\}.$$

(b) The minimum cut consists of the arc CD, with capacity 3;

$$X = \{S,\ A,\ B,\ C\}, \quad Y = \{D,\ E,\ F,\ T\}.$$

(c) The minimum cut comprises the arcs AC and BC, with capacity 3;

$$X = \{S,\ A,\ B,\},\quad Y = \{C,\ D,\ E,\ F,\ T\}.$$

(d) The minimum cut comprises the arcs DF and ET, with capacity 3;

$$X = \{S,\ A,\ B,\ C,\ D,\ E\}, \quad Y = \{F,\ T\}.$$

3.6. (a) There are many possible answers, the simplest of which is:

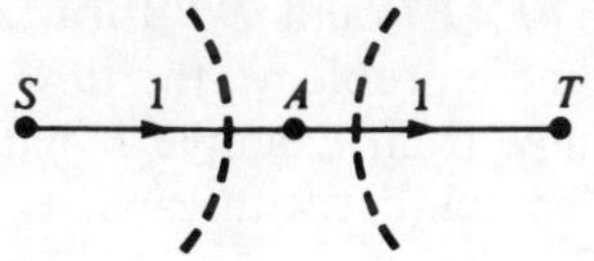

Other examples are:

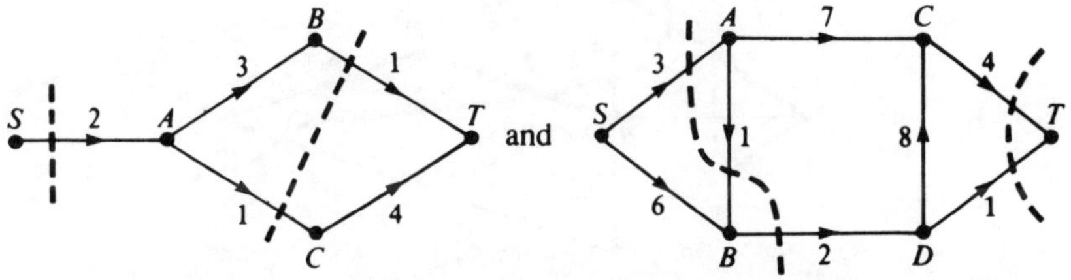

(b)

	arcs in cut	vertices in X	vertices in Y	capacity of cut
1	SA, SB, SC	S	A, B, C, T	$4+1+5=10$
2	SB, SC, AB, BA, AT	S, A	B, C, T	$1+5+5+2=13$
3	SA, SC, AB, BA, CB, BT	S, B	A, C, T	$4+6+5+4=19$
4	SA, SB, CB, CT	S, C	A, B, T	$4+1+1+2=8$
5	SC, AT, BT, CB	S, A, B	C, T	$5+2+4=11$
6	SB, AB, BA, AT, CB, CT	S, A, C	B, T	$1+5+2+1+2=11$
7	SA, AB, BA, BT, CT	S, B, C	A, T	$4+6+4+2=16$
8	AT, BT, CT	S, A, B, C	T	$2+4+2=8$

The minimum cuts are those with capacity 8: cuts (4) and (8). They are indicated on the following diagram:

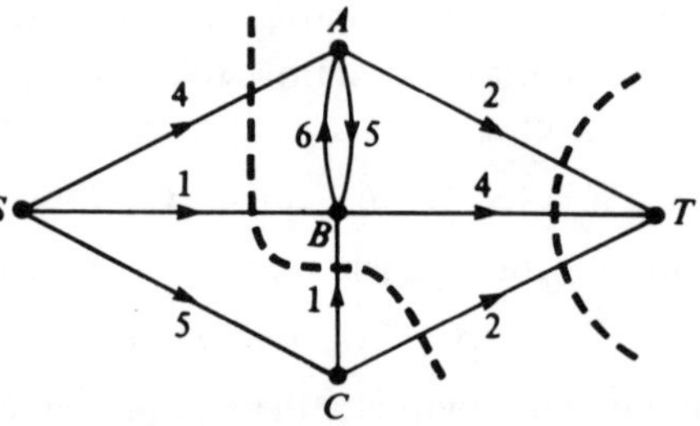

3.7. (a) Its value must be 7 or less.
(b) Its capacity must be 11 or more.
(c) The value of a maximum flow and the capacity of a minimum cut must lie between 4 and 6 (inclusive).
(d) You have made a mistake in your calculations (because the value of a flow cannot exceed the capacity of a cut).

3.8. The arcs CF, EF, FE, ET and DT form a cut of capacity 136. By the rule preceding these problems, the flow must be a maximum flow.

3.9. (a) We already know from Example 1 that the arcs SA, AB, BD and ET form a minimum cut with capacity 12, and it follows that the value of any flow cannot exceed 12. The following diagram shows a flow of value 12, which must therefore be a maximum flow:

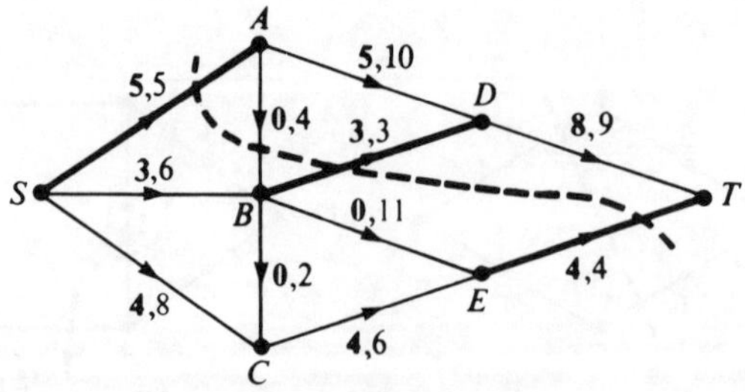

(b) We already know from Example 2 that the arcs *BD*, *CD* and *CT* form a minimum cut with capacity 6, and it follows that the value of any flow cannot exceed 6. The following diagram shows a flow of value 6, which must therefore be a maximum flow:

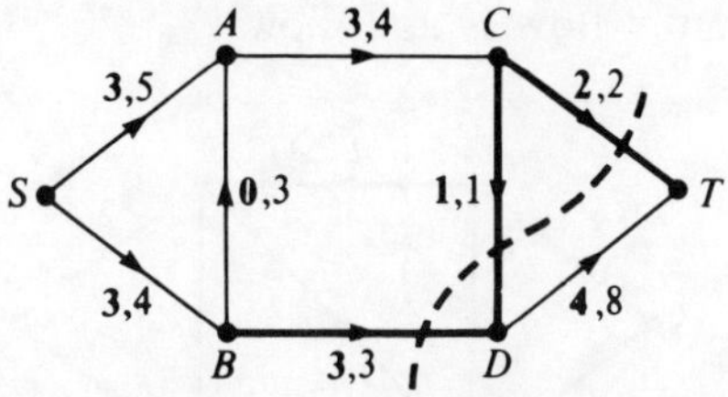

(c) We already know from Problem 3.6 (b) that {*SA*, *SB*, *CB*, *CT*} and {*AT*, *BT*, *CT*} are minimum cuts with capacity 8, and it follows that the value of any flow cannot exceed 8. The following diagram shows a flow of value 8, which must therefore be a maximum flow:

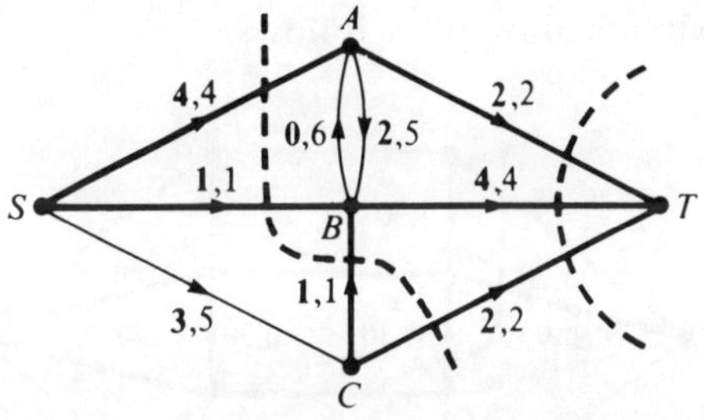

3.10. (a) The statement is FALSE. To see this, consider the example of Problem 3.8 in which the minimum cut *CF*, *EF*, *FE*, *ET*, *DT* contains the unsaturated arc *FE*. A correct statement is:

every minimum cut dividing a network carrying a maximum flow into two parts *X* and *Y* consists entirely of saturated arcs directed from *X* to *Y* and arcs directed from *Y* to *X* carrying a zero flow.

(b) The statement is TRUE. Since the capacity of every arc is an integer, the capacity of every cut is an integer (since it is a sum of integers). In particular, the capacity of any minimum cut is an integer, and so, by the max-flow min-cut theorem, the value of any maximum flow is also an integer.

3.11. For each part, we examine all the cuts and thus find a minimum cut; we then find a maximum flow and observe that its value is the same as the capacity of the minimum cut.

(a) The cut {*SB*, *BA*, *AC*}, is a minimum cut with capacity $4 + 2 = 6$.

A maximum flow is as follows:

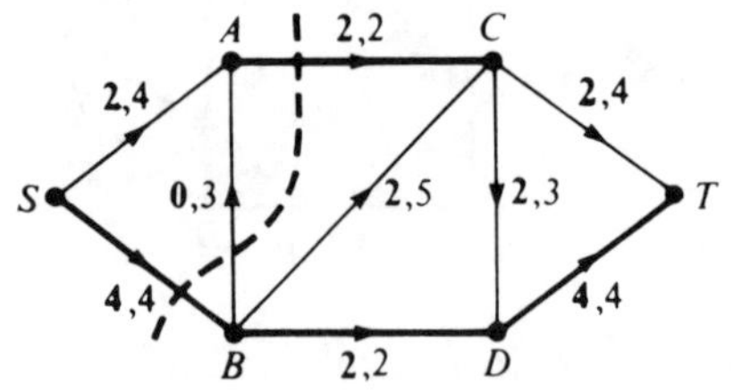

The value of the maximum flow is 6, the same as the capacity of the minimum cut.

(b) The cut {*SA*, *SB*, *CB*, *CG*, *DH*} is a minimum cut with capacity 140.

A maximum flow is as follows:

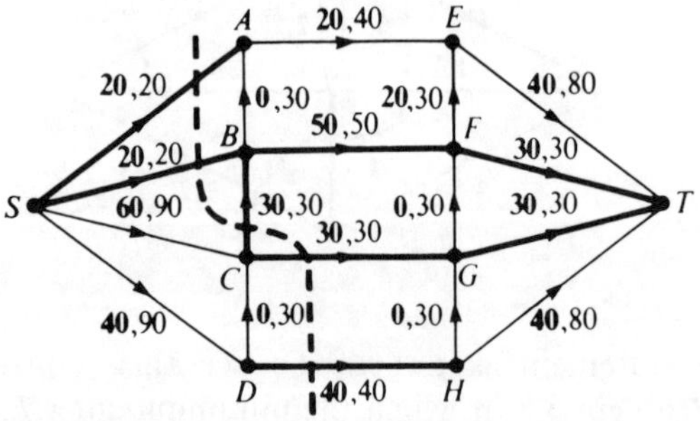

The value of the maximum flow is 140, the same as the capacity of the minimum cut.

Note that, in each case, if we can find any flow which has a value equal to the capacity of any cut, then it follows from the theorem that the flow must be a maximum flow and the cut must be a minimum cut (see the discussion on page 63).

3.12.

Start with zero flow in the network.

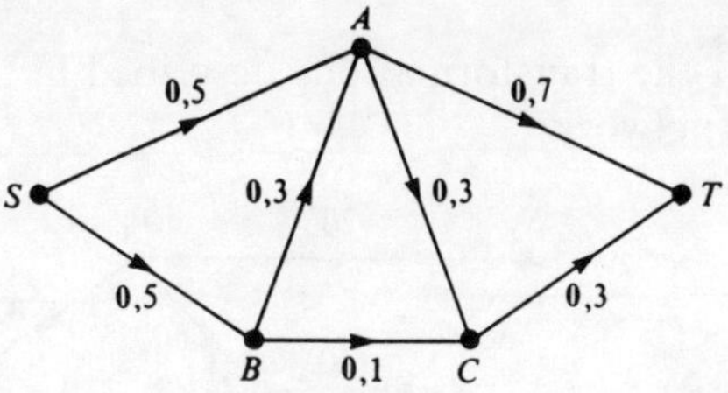

Increase the flow from S to T via A and C.

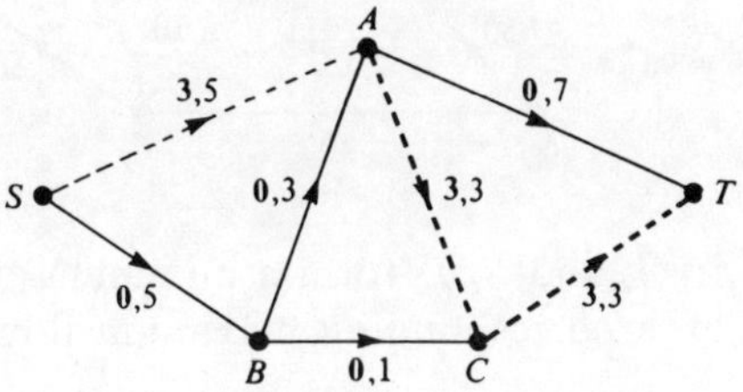

The flow along $SACT$ is increased by 3.

Increase the flow from S to T via A.

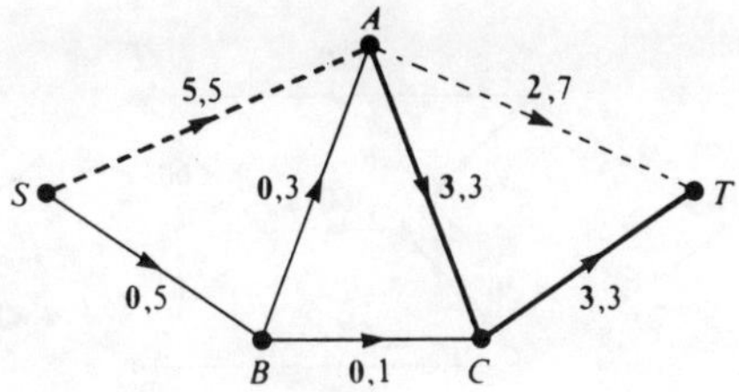

The flow along SAT is increased by 2.

Increase the flow from S to T via B and A.

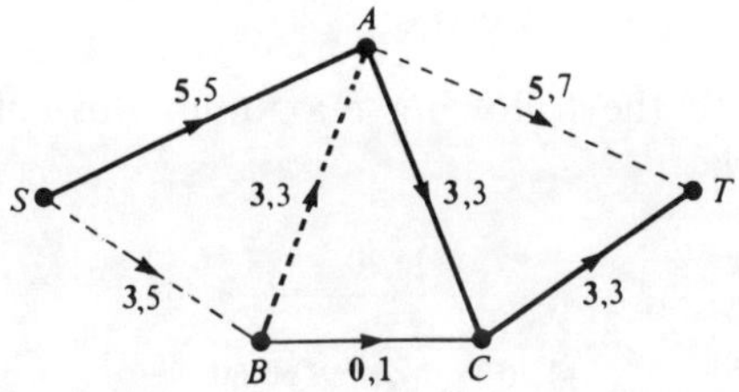

The flow along $SBAT$ is increased by 3.

Increase the flow from S to T via B, C and A.

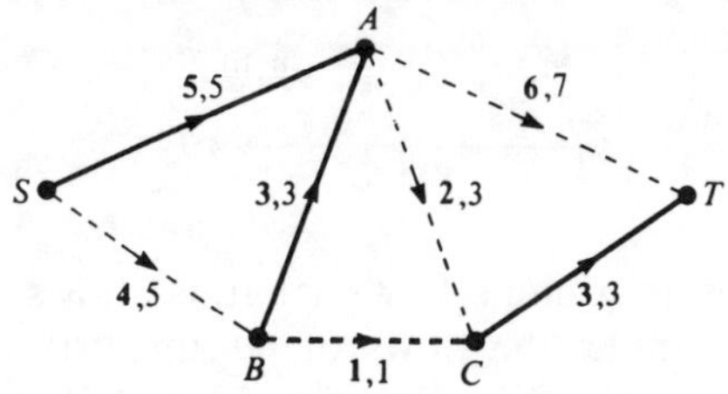

The flow along $SBCAT$ is increased by 1.

STOP: no further increase is possible.

Maximum flow is $3 + 2 + 3 + 1 = 9$.

CHAPTER 4

4.1. Applying two of the transformations described in the text, we get the following basic network:

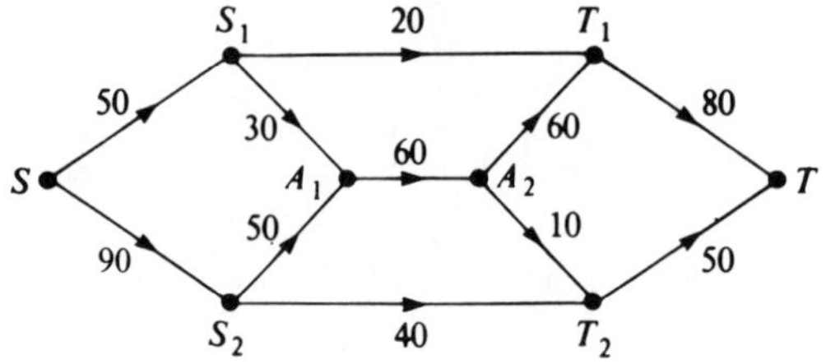

The arcs S_1T_1, A_1A_2 and S_2T_2 form a minimum cut in this network, with capacity 120; a corresponding maximum flow is as follows:

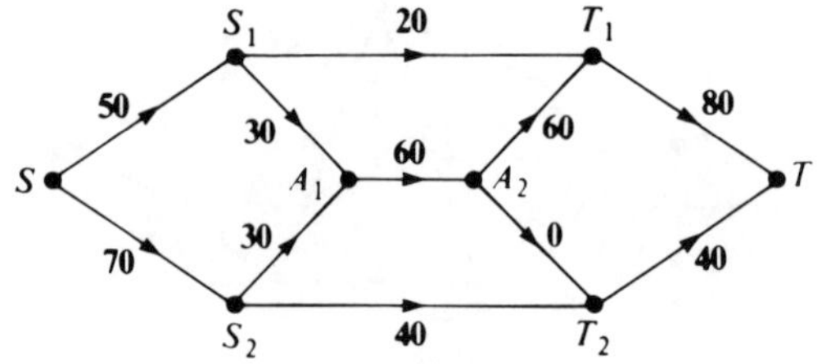

This gives rise to the following maximum flow of value 120 in the original network:

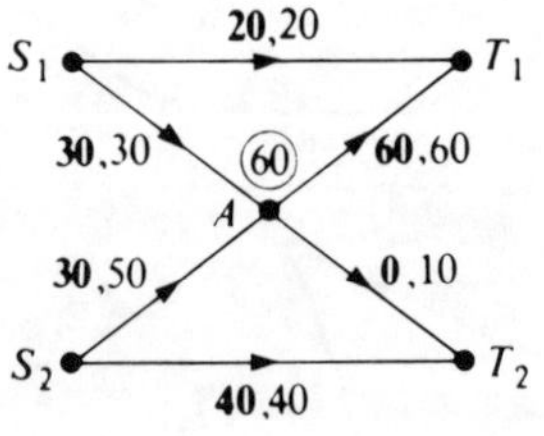

Note that other maximum flows (that is, flows of value 120) are possible—for example, the flows of 60 and 0 through the arcs AT_1 and AT_2 can be replaced by flows of 50 and 10 respectively.

4.2. We carry out the steps of the Feasibility Algorithm as shown in the following diagram, in which each figure corresponds to the box in the same position on the page as in the flow chart for the algorithm, given on page 90.

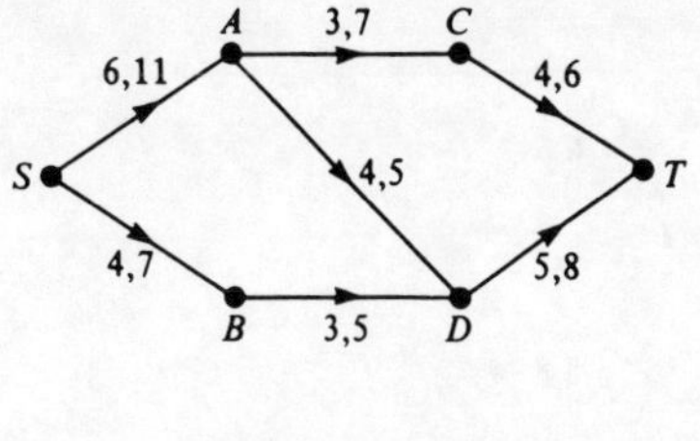

the network N

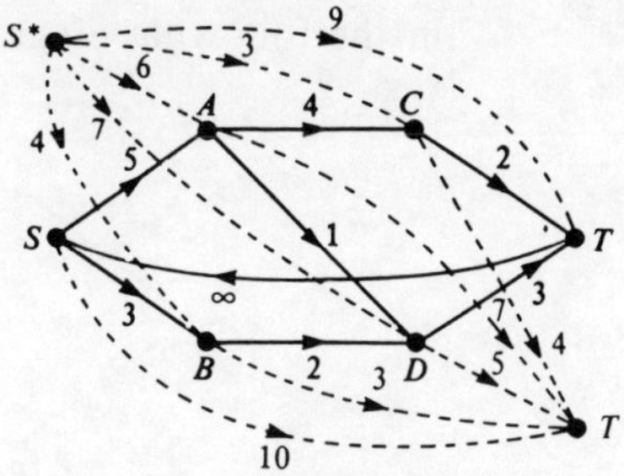

the basic network N*

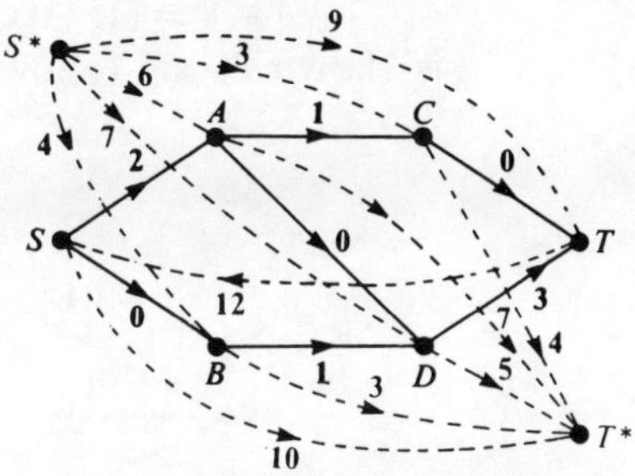

a suitable flow in N*

remove S*, T*, and dashed arcs

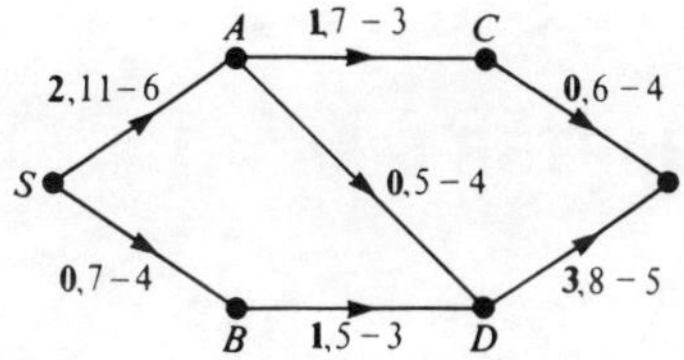

add back lower capacities

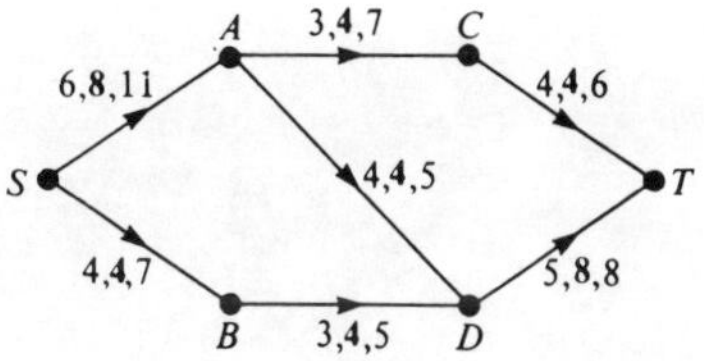

a flow in N

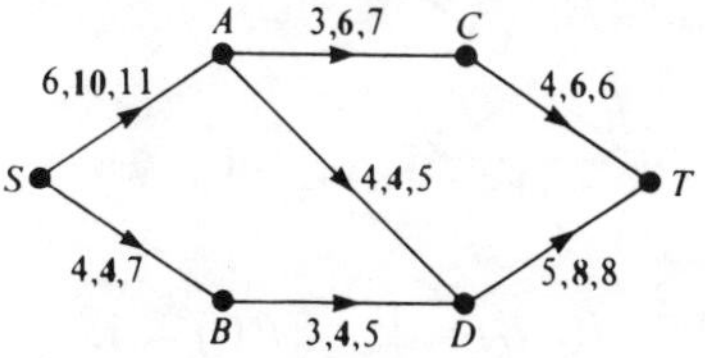

a maximum flow in N

Thus the value of a maximum flow is 14.

4.3. (a) The arcs CT and DT form a minimum cut of capacity $6 + 8 = 14$. A corresponding maximum flow of value 14 is shown in the following diagram:

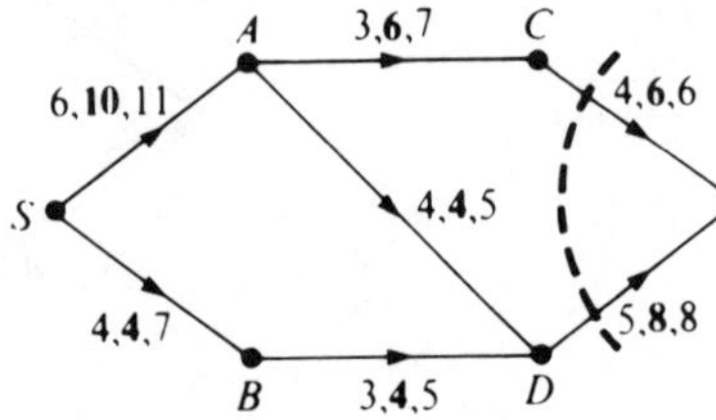

(b) The arcs SA, AB, BA and CT form a minimum cut of capacity $4 + 3 - 4 + 8 = 11$. A corresponding maximum flow of value 11 is shown in the following diagram:

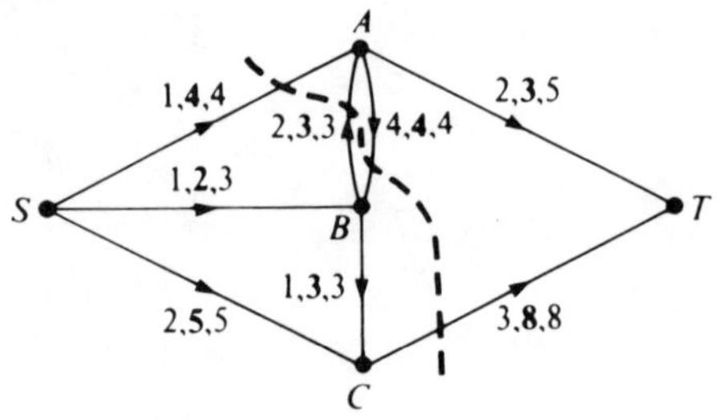

CHAPTER 5

5.1.

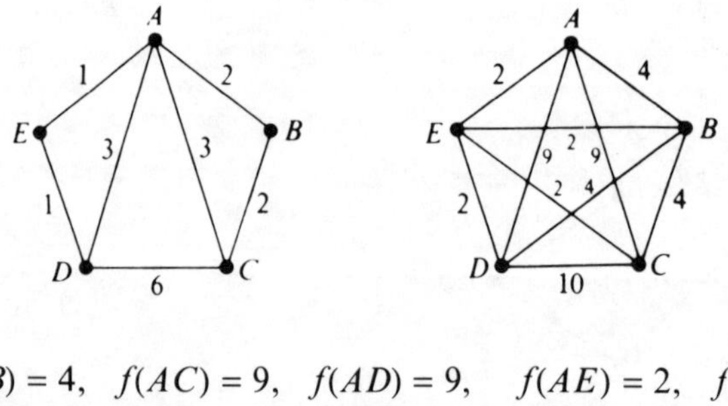

(a) $f(AB) = 4$, $f(AC) = 9$, $f(AD) = 9$, $f(AE) = 2$, $f(BC) = 4$, $f(BD) = 4$, $f(BE) = 2$, $f(CD) = 10$, $f(CE) = 2$, $f(DE) = 2$.

(b) 1. in each cycle the smallest capacity occurs more than once; 2. $C(N)$ has only four different capacities (2, 4, 9 and 10).

5.2.

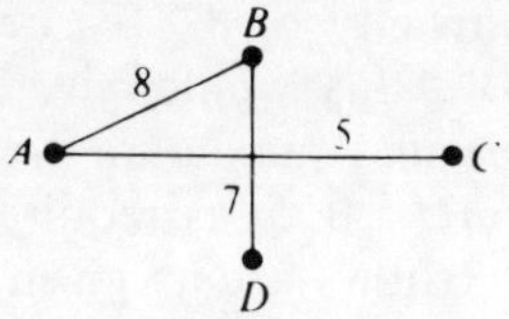

By the capacity rule, we have:

$$f(AD) = \min(f(AB),\ f(BD)) = \min(8,\ 7) = 7,$$

$$f(BC) = \min(f(BA),\ f(AC)) = \min(8,\ 5) = 5,$$

$$f(CD) = \min(f(CA),\ f(AB),\ f(BD)) = \min(5,\ 8,\ 7) = 5,$$

as required.

5.3.

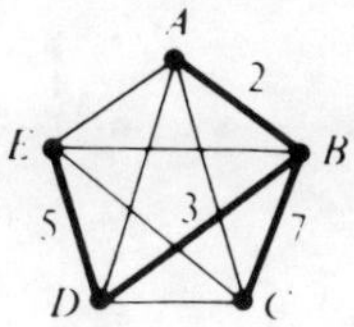

By the capacity rule, we have:

$$f(AC) = \min(f(AB),\ f(BC)) = \min(2, 7) = 2,$$
$$f(AD) = \min(f(AB),\ f(BD)) = \min(2, 3) = 2,$$
$$f(AE) = \min(f(AB),\ f(BD),\ f(DE)) = \min(2, 3, 5) = 2,$$
$$f(BE) = \min(f(BD),\ f(DE)) = \min(3, 5) = 3,$$
$$f(CD) = \min(f(CB),\ f(BD)) = \min(7, 3) = 3,$$
$$f(CE) = \min(f(CB),\ f(BD),\ f(DE)) = \min(7, 3, 5) = 3.$$

5.4.

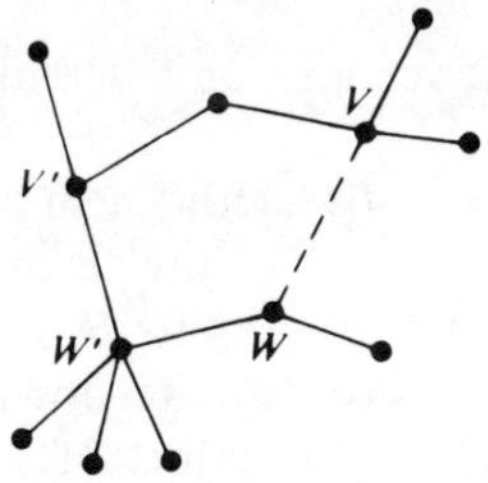

Let $V'\ W'$ be an edge of smallest capacity in the path joining V and W in the maximum tree. If we now add the edge VW to the maximum tree, we get a cycle containing both VW and $V'\ W'$. By applying Theorem 5.1 to this cycle, we see that the capacity of VW cannot be smaller than the capacity of $V' W'$. But the capacity of VW cannot be *larger* than the capacity of $V' W'$ because if it were we could replace $V' W'$ by VW; this would give us a tree of greater total capacity than our original maximum tree, which is impossible. It follows that $f(VW) = f(V' W')$, as required.

5.5. (a) In each case, max flows are as given in the following table:

	A	*B*	*C*	*D*
A	—	8	5	7
B	8	—	5	7
C	5	5	—	5
D	7	7	5	—

Thus all the trees are flow-equivalent to N.

(b) Min cuts for max flow from A to B are:

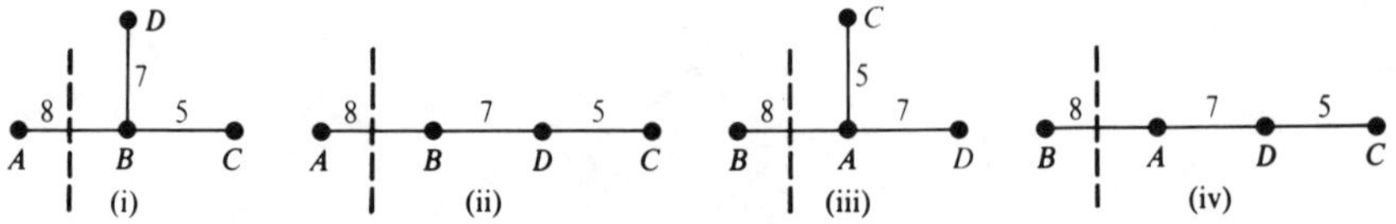

However, in N, a max flow of 8 from A to B corresponds to the min cut $\{A, D\}$, $\{B, C\}$, so none of the trees is a cut tree for N.

5.6.

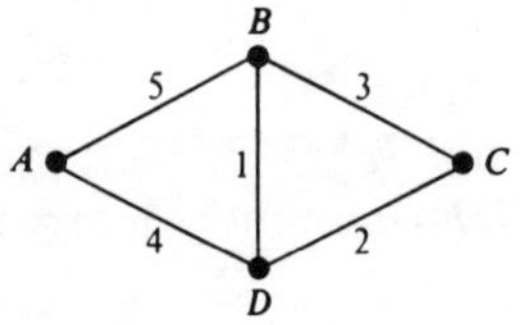

STEP 1 Choose any 2 vertices, say A and B.

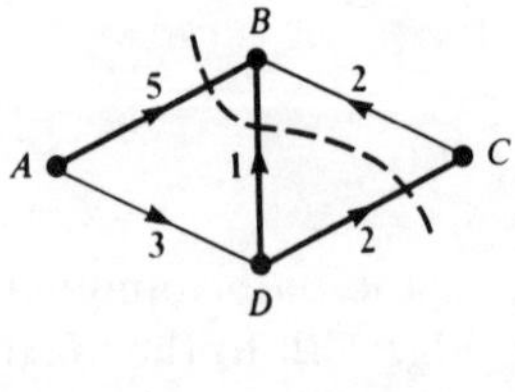

(i) find a max flow from A to B — max flow is 8

(ii) write down effect of min cut — min cut separates $\{A, D\}$, $\{B, C\}$

(iii) draw branch

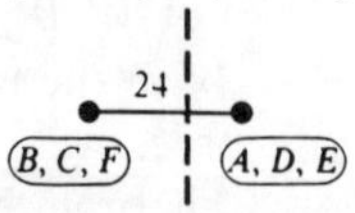

STEP 2 Consider vertices *B* and *C*.

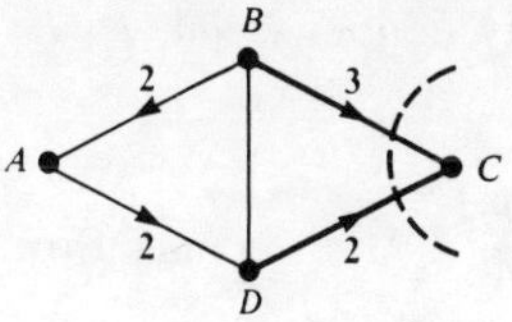

(i) find a max flow from *B* to *C* — max flow is 5

(ii) write down effect of min cut — min cut separates $\{C\}$, $\{A, B, D\}$

(iii) draw branch

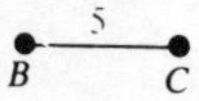

(iv) join branch to tree

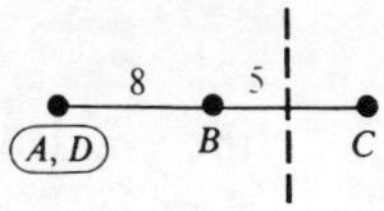

STEP 3 Consider vertices *A* and *D*.

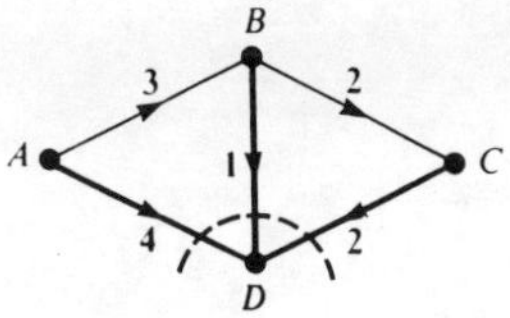

(i) find a max flow from *A* to *D* — max flow is 7

(ii) write down effect of min cut — min cut separates $\{D\}$, $\{A, B, C\}$

(iii) draw branch

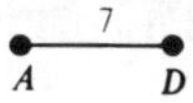

(iv) join branch to tree

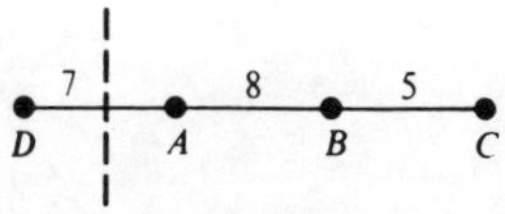

After 3 steps, we obtained a cut tree.

5.7. The solution is given in the text as part of the solution to the worked problem at the end of the chapter (page 114).

5.8. We begin by finding a maximum flow from *A* to *B*:

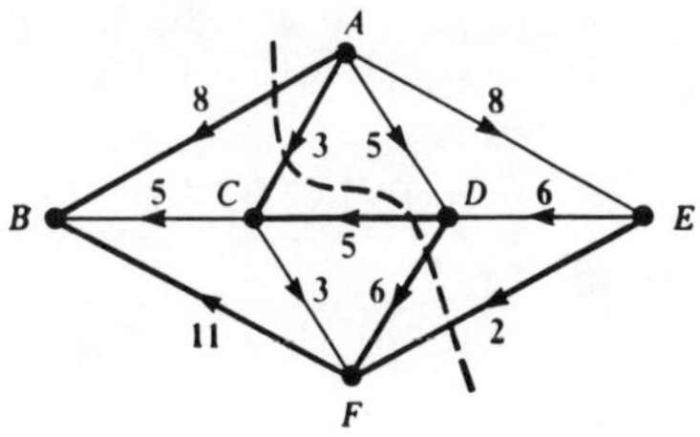

max flow is 24

min cut separates $\{B, C, F\}$, $\{A, D, E\}$

The first branch of the tree is

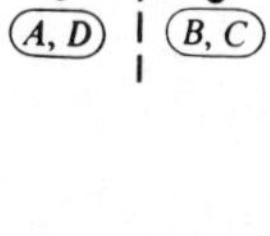

First, we split up the generalized vertex (*B*, *C*, *F*).

To find a maximum flow from *B* to *C*, we may condense (*A*, *D*, *E*).

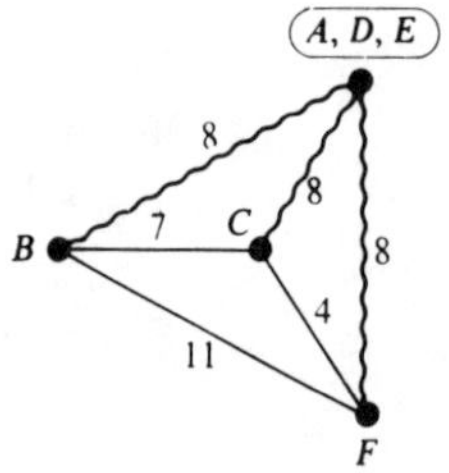

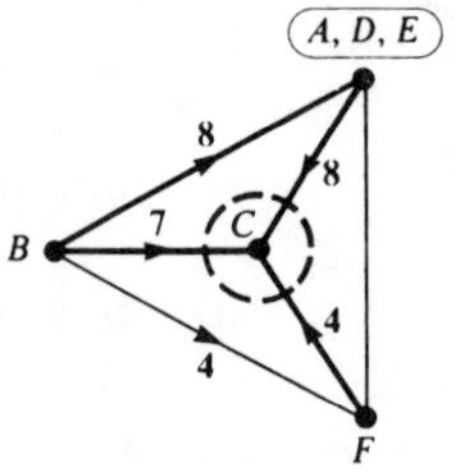

max flow is 19

min cut separates $\{C\}$, $\{B, F, A, D, E\}$

branch is

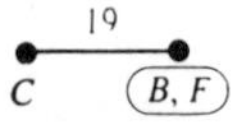

the condensed network   a max flow from *B* to *C*

The min cut separates *C* from the other vertices, so the tree becomes:

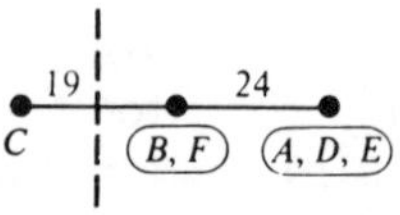

Next, we find a maximum flow from *B* to *F*, again using the network with *A*, *D*, *E* condensed:

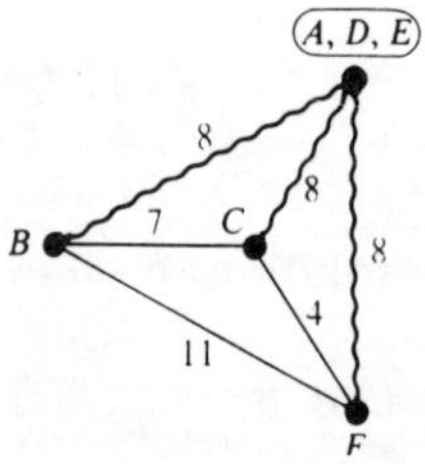

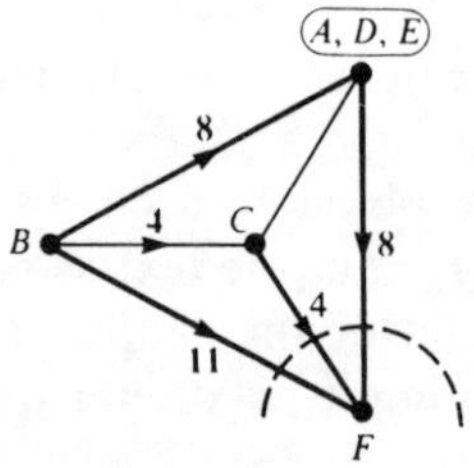

max flow is 23

min cut separates $\{F\}$, $\{B, C, A, D, E\}$

branch is

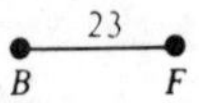

the condensed network   a max flow from *B* to *F*

The min cut separates *F* from the other vertices, so the tree becomes:

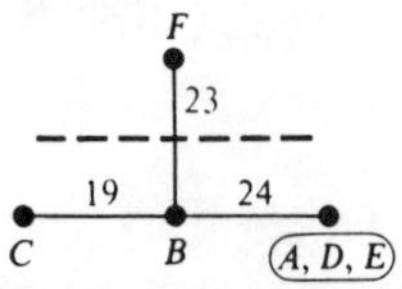

We now split up the generalized vertex $\boxed{A, D, E}$.

To find a maximum flow from A to D, we may condense $\boxed{B, C, F}$

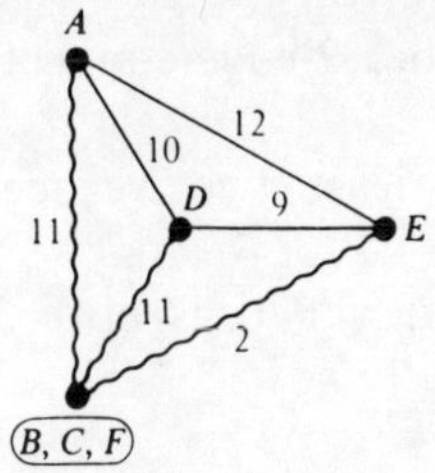

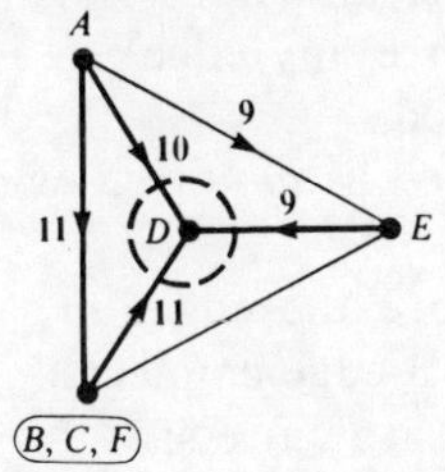

max flow is 30

min cut separates $\{D\}$, $\{A, E, B, C, F\}$

branch is

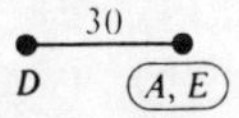

the condensed network   a max flow from A to D

The min cut separates D from the other vertices, so the tree becomes:

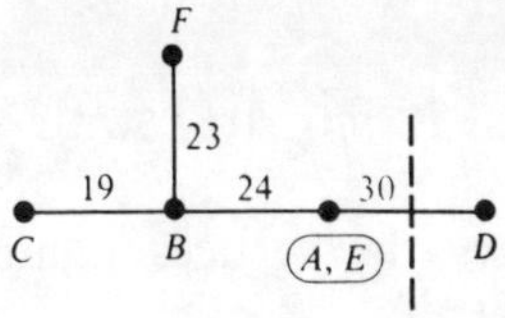

Finally, we find a maximum flow from A to E, again using the network with B, C, F condensed.

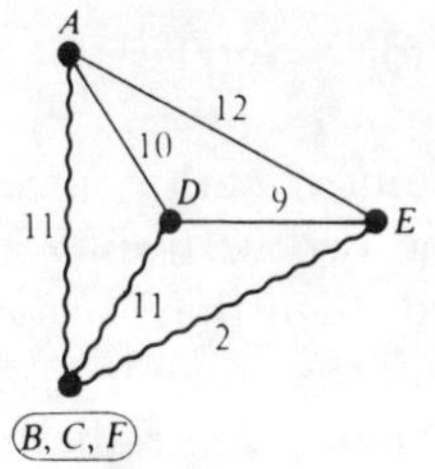

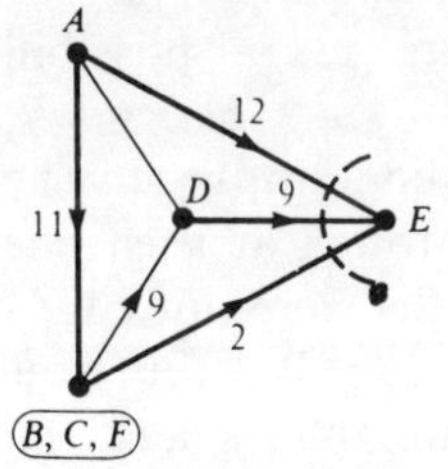

max flow is 30

min cut separates $\{E\}$, $\{A, D, B, C, F\}$

branch is

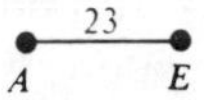

the condensed network   a max flow from A to E

The min cut separates E from the other vertices, so the cut tree is

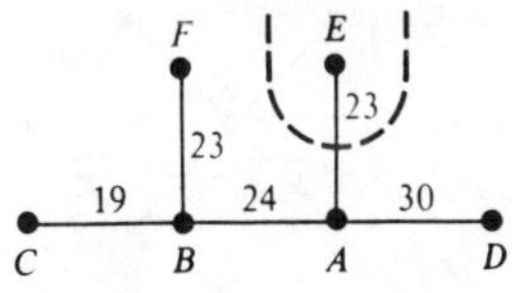

CHAPTER 6

6.1. (a) $\lambda(G) = 2$ (for example, remove the edges *vw* and *xy*); *G* is 2-edge-connected but not 3-edge-connected.

(b) $\lambda(G) = 1$ (remove any edge); *G* is neither 2-edge-connected nor 3-edge-connected.
Note that if *G* is *any* tree with at least two vertices then $\lambda(G) = 1$.

(c) $\lambda(G) = 3$ (remove the edges *uv*, *uy* and *xy*); *G* is both 2-edge-connected and 3-edge-connected.

6.2. (a), (c), (d) and (f) are cutsets;

(b) is not a cutset, since its removal does not disconnect the graph;

(e) is not a cutset, since we can disconnect the graph by removing just *xz* and *yz*.

6.3. (a) $\kappa(G) = 2$ (for example, remove the vertices *v* and *x*); *G* is 2-connected, but not 3-connected.

(b) $\kappa(G) = 1$ (remove the vertex *u*); *G* is neither 2-connected nor 3-connected.
Note that if *G* is *any* tree with at least two vertices then $\kappa(G) = 1$.

(c) $\kappa(G) = 2$ (remove the vertices *v* and *y*); *G* is 2-connected, but not 3-connected.

6.4. (a) and (d) are vertex-cutsets;

(b) is not a vertex-cutset, since its removal does not disconnect the graph.

(c) is not a vertex-cutset, since we can disconnect the graph by removing just *u* and *x*, or just *y*.

6.5. In each case there are several possibilities—for example:
(a) *saet*, *sbdt*, *sceft;* (b) *sbet*, *sabdt;* (c) *saet*, *sbft*
This graph does not contain three vertex-disjoint *st*-paths, since every *st*-path must pass through at least one of the two vertices *b* and *e*.

6.6. (a) If the two *st*-paths were not edge-disjoint, then they would have an edge in common. But this would mean that they had at least one vertex (other than *s* and *t*) in common, contradicting the fact that they are vertex-disjoint.

(b) There are many possibilities—for example,

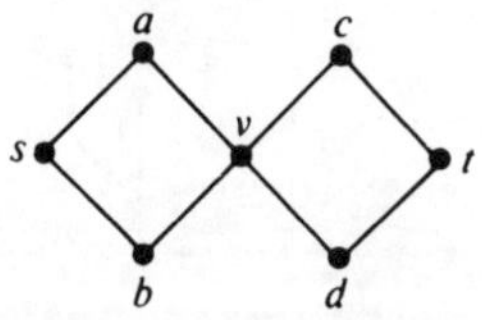

In the graph shown, the only pairs of edge-disjoint st-paths are $savct$ and $sbvdt$, and $savdt$ and $sbvct$. In neither case are the paths vertex-disjoint, since they all pass through the vertex v.

6.7. (a) In this case, $k = 2$; two edge-disjoint st-paths are $sact$ and $sbdt$, and two edges separating s from t are sa and sb. Thus the maximum number of edge-disjoint st-paths and the minimum number of edges separating s from t are both equal to 2.

(b) Again, $k = 2$; two edge-disjoint st-paths are $svxt$ and $swyt$, and two edges separating s from t are vx and wy. Thus the maximum number of edge-disjoint st-paths and the minimum number of edges separating s from t are both equal to 2.

(c) In this case, $k = 3$; three edge-disjoint st-paths are $suwzt$, syt, and $svxt$, and three edges separating s from t are su, sv and sy. Thus the maximum number of edge-disjoint st-paths and the minimum number of edges separating s from t are both equal to 3.

6.8. (a) In this case, $k = 2$; two arc-disjoint st-paths are $svxt$ and $swyt$, and two arcs separating s from t are vx and wy. Thus the maximum number of arc-disjoint st-paths and the minimum number of arcs separating s from t are both equal to 2.

(b) In this case, $k = 3$; three arc-disjoint st-paths are $sadt$, $sbft$ and $scet$, and three arcs separating s from t are sa, sb and sc. Thus the maximum number of arc-disjoint st-paths and the minimum number of arcs separating s from t are both equal to 3.

6.9. In this case, $k = 2$; two vertex-disjoint st-paths are $svxt$ and $swyt$, and two vertices separating s from t are v and w. Thus the maximum number of vertex-disjoint st-paths and the minimum number of vertices in a vertex-cutset separating s from t are both equal to 2.

6.10. In this case, $k = 3$; three vertex-disjoint st-paths are $suwzt$, syt and $svxt$, and three vertices separating s from t are u, v and y. Thus the maximum number of vertex-disjoint st-paths and the minimum number of vertices in a vertex-cutset separating s from t are both equal to 3.

6.11. (i) (a) For C_n, we have

$$\kappa(C_n) = 2, \quad \lambda(C_n) = 2, \quad \delta(C_n) = 2, \quad (2m/n) = 2n/n = 2,$$

so C_n has optimal connectivity.

(b) For K_n, the number of edges is $\frac{1}{2}n(n-1)$, by consequence 3 of the handshaking lemma. We have

$$\kappa(K_n) = n - 1, \quad \lambda(K_n) = n - 1, \quad \delta(K_n) = n - 1,$$

$$(2m/n) = 2 \times \tfrac{1}{2}n(n-1)/n = n - 1$$

so K_n has optimal connectivity.

(c) For $K_{r,r}$, we have

$$\kappa(K_{r,r}) = r, \quad \lambda(K_{r,r}) = r, \quad \delta(K_{r,r}) = r, \quad (2m/n) = 2r^2/(r+r) = r,$$

so $K_{r,r}$ has optimal connectivity.

6.12. (a) There are only two regular graphs with 6 vertices and 9 edges:

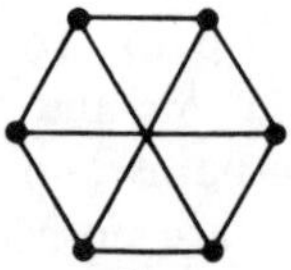

$K_{3,3}$

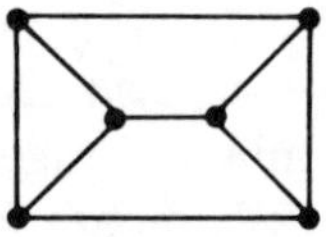

For these two graphs we have

$$\kappa(G) = \lambda(G) = \delta(G) = (2m/n) = 3,$$

so the graphs have optimal connectivity.

(b) There is only one possibility:

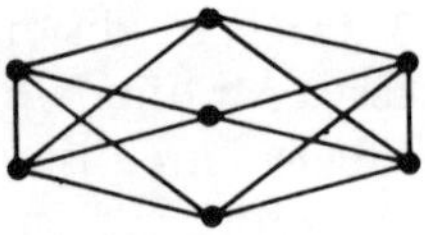

The removal of the middle three vertices disconnects the graph, so $\kappa(G) = 3$; since $\kappa(G) < \delta(G)$, the graph does not have optimal connectivity.

CHAPTER 7

7.1.

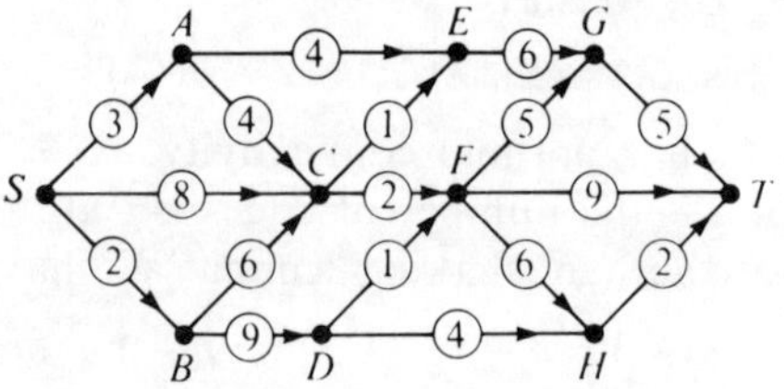

For the above network, we obtain the following table:

step	vertex started from	A	B	C	D	E	F	G	H	T
STEP 1	S	$\boxed{3}$	$\boxed{2}$	8						
STEP 2	B			8	$\boxed{11}$					
STEP 3	A			$\boxed{7}$		$\boxed{7}$				
STEP 4	C						$\boxed{9}$			
	E							$\boxed{13}$		
STEP 5	F							$\boxed{13}$	$\boxed{15}$	18
STEP 6	D								$\boxed{15}$	
STEP 7	G									18
STEP 8	H									$\boxed{17}$

We see from the table that there are two shortest paths of length 17:

SACFHT

and

SBDHT.

7.2. Using the algorithm, we carry out the following steps:

STEP 1 Vertices reached directly from *S*

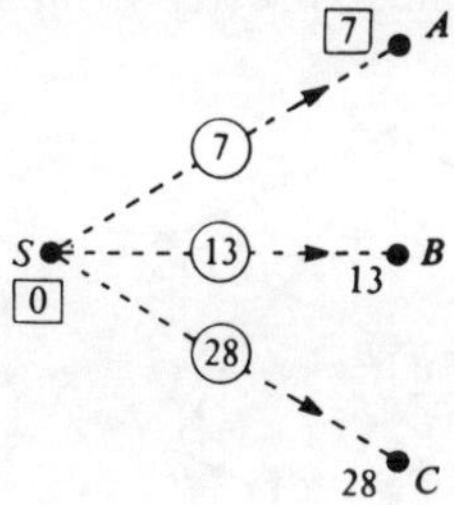

A has potential 7

STEP 2 Vertices reached directly from *A*

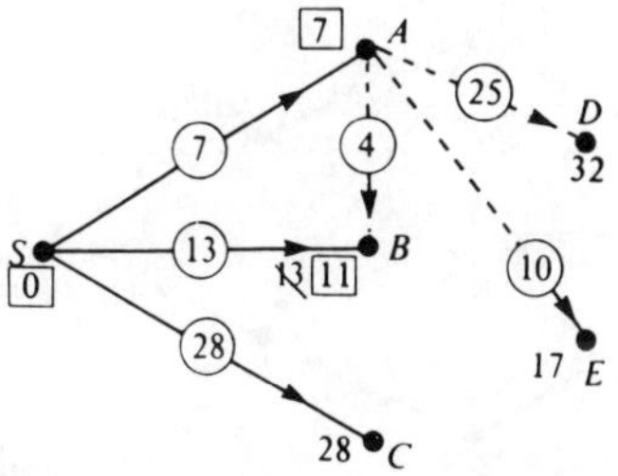

B has potential 11

STEP 3 Vertices reached directly from B

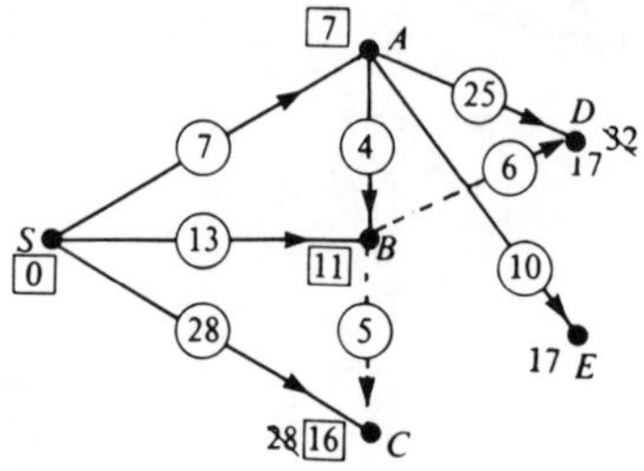

C has potential 16

STEP 4 Vertices reached directly from C

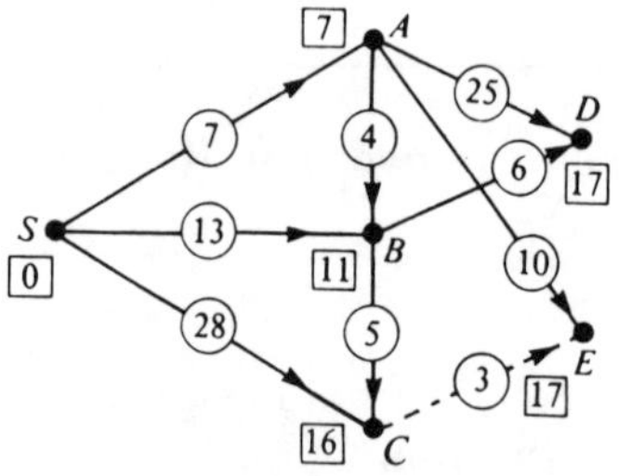

D and E have potential 17

STEP 5 Vertices reached directly from D and E

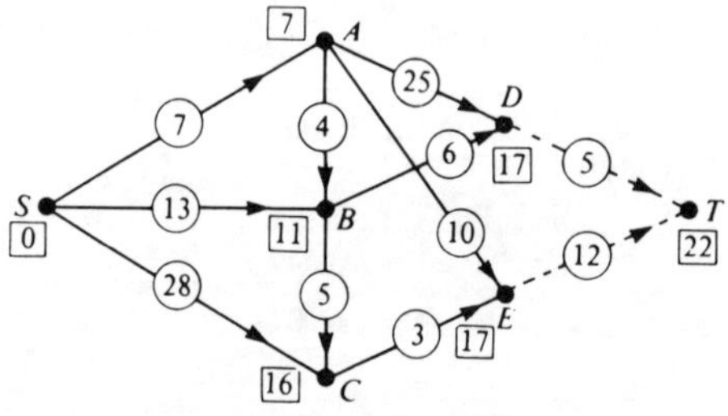

T has potential 22

We can now read off the shortest path:

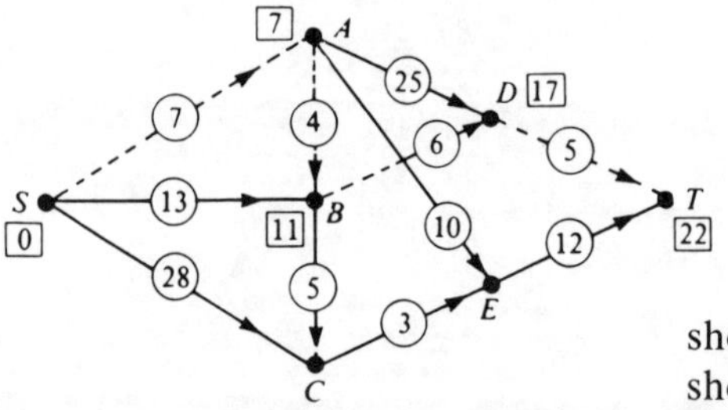

shortest distance is 22
shortest path is $SABDT$

Alternatively, using the tabular method, we obtain the following table:

step	vertex started from	A	B	C	D	E	T
STEP 1	S	[7]	13	28			
STEP 2	A		[11]		32	[17]	
STEP 3	B			[16]	[17]		
STEP 4	C						
STEP 5	D						[22]
	E						29

We can now read off the shortest path:

T has potential 22 assigned from D,
D has potential 17 assigned from B,
B has potential 11 assigned from A,
A has potential 7 assigned from S,

so

the shortest path is $SABDT$ with length 22.

7.3. Using the algorithm, we carry out the following steps:

STEP 1 Vertices reached only from S:

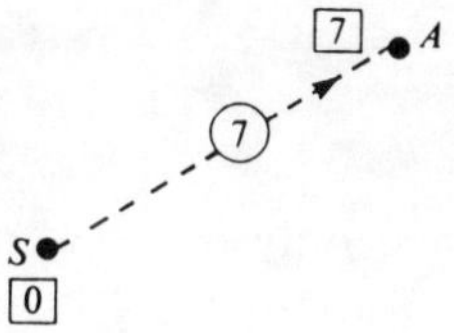

A has potential 7

STEP 2 Vertices reached only from S and A:

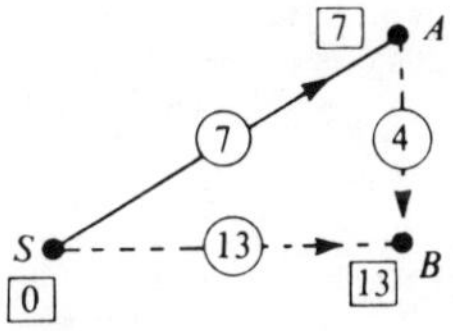

B has potential 13

STEP 3 Vertices reached only from *S*, *A* and *B*:

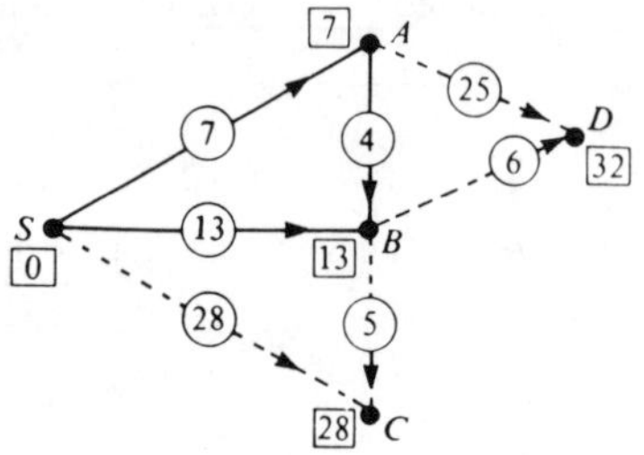

C has potential 28
D has potential 32

STEP 4 Vertices reached only from *S*, *A*, *B*, *C* and *D*:

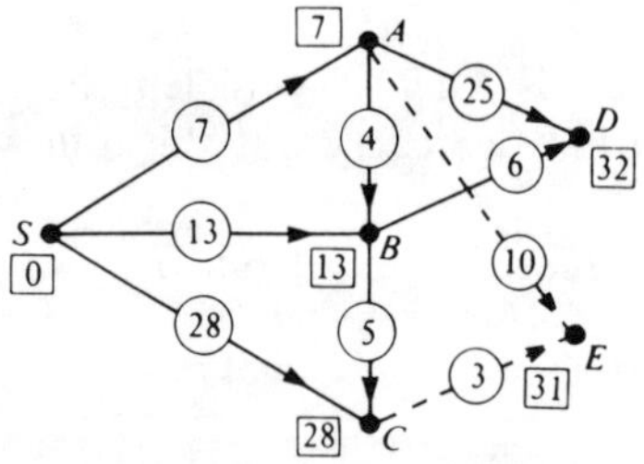

E has potential 31

STEP 5 Vertices reached only from *S*, *A*, *B*, *C*, *D* and *E*:

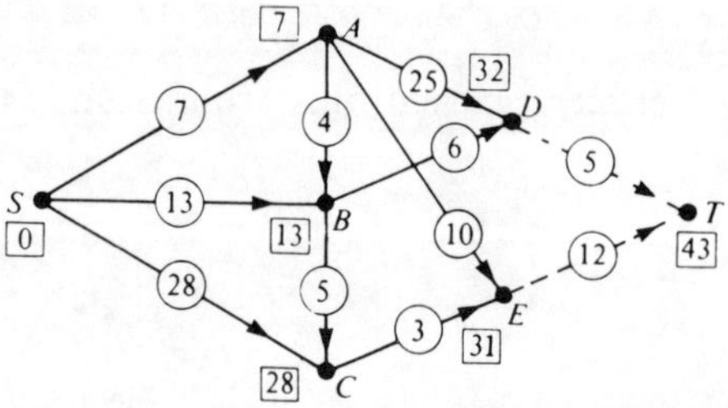

T has potential 43

We can now read off the longest path:

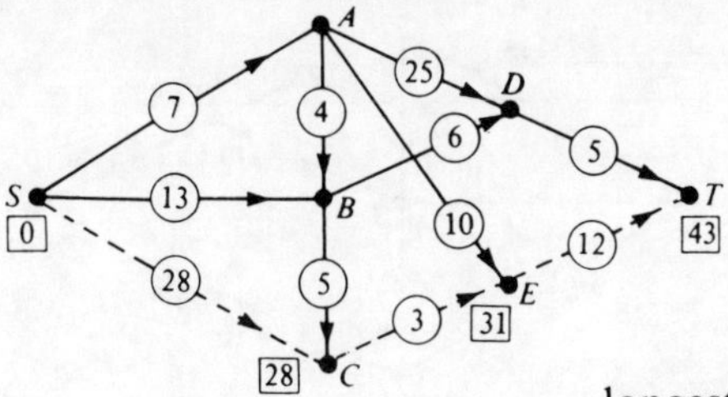

longest distance is 43
longest path is *SCET*

Alternatively, using the tabular method, we obtain the following table:

step	vertex assigned labels	origin vertices *S*	*A*	B	*C*	*D*	*E*
STEP 1 (*S*)	*A*	$\boxed{7}$					
STEP 2 (*S*, *A*)	*B*	$\boxed{13}$	11				
STEP 3 (*S*, *A*, *B*)	*C*	$\boxed{28}$		18			
	D		$\boxed{32}$	19			
STEP 4 (*S*, *A*, *B*, *C*, *D*)	*E*		17		$\boxed{31}$		
STEP 5 (*S*, *A*, *B*, *C*, *D*, *E*)	*T*					37	$\boxed{43}$

We can now read off the longest path:

T has potential 43 assigned from *E*,
E has potential 31 assigned from *C*,
C has potential 28 assigned from *S*,

so

the longest path is *SCET* with length 43.

7.4. Using the algorithm, we carry out the following steps:
STEP 1 First flow-augmenting path: *SABDT*

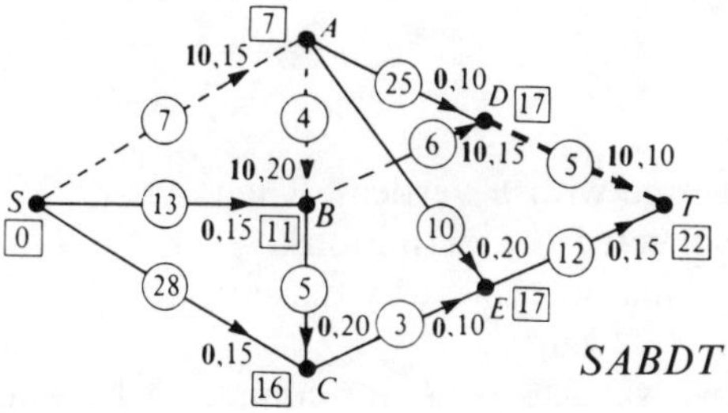

SABDT has flow 10

STEP 2 Second flow-augmenting path *SAET*

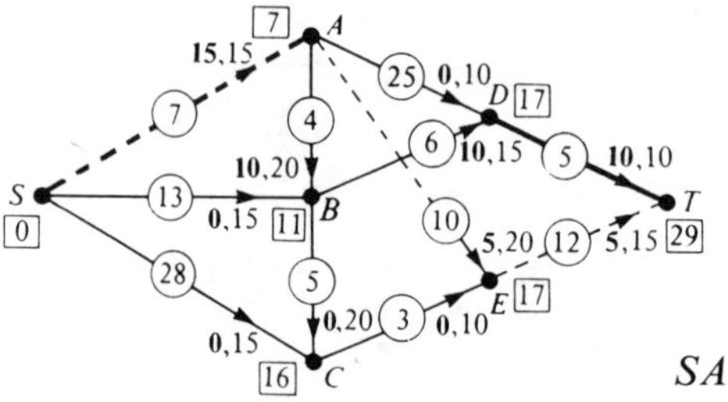

SAET has flow 5

STEP 3 Third flow-augmenting path: *SBAET*
(Note that *BA* is a backward arc.)

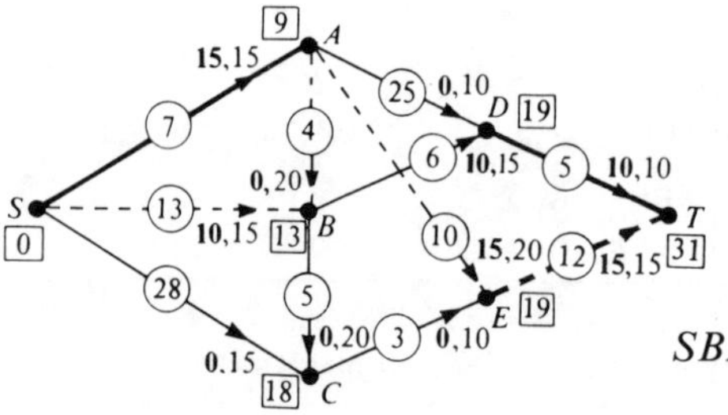

SBAET has flow 10

Value of flow is 25 Min cut is $\{DT, ET\}$

total cost is $(15 \times 7) + (10 \times 13) + (15 \times 10) + (10 \times 6) + (10 \times 5) + (15 \times 12) = 675$

SA *SB* *AE* *BD* *DT* *ET*

CHAPTER 8

8.1. Let T be a tree with n vertices. Then
(a) T is conected (by definition);
(b) T contains no cycles (by definition);
(c) T has $n - 1$ edges;
(d) any two vertices of T are connected by exactly one path;

(e) Each edge of T is a bridge (that is, its removal disconnects the tree);
(f) Adding an edge between any two vertices of T creates exactly one cycle;
(g) T is bipartite. (See Chapter 2, Problem 2.10 (d).)
(Properties (c)–(f) are discussed in the text.)

8.2. (a) $n = 5$ 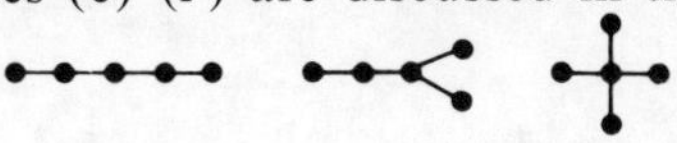

(b) $n = 6$ 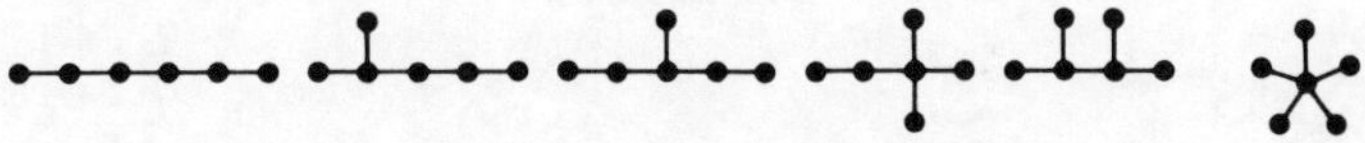

8.3. There are 18 spanning trees other than the three depicted in the text:

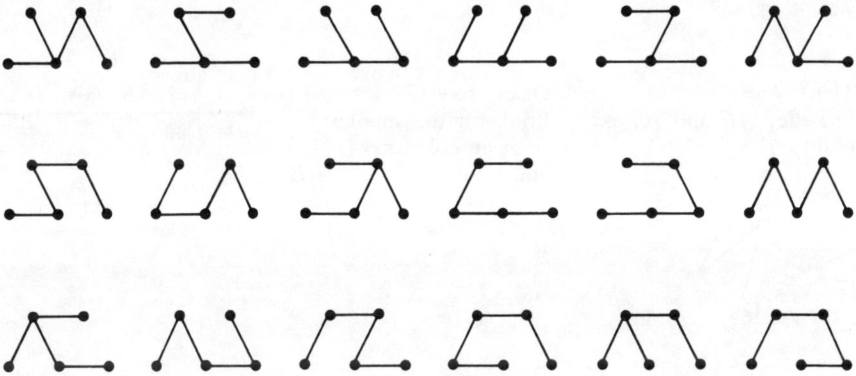

8.4. We start by choosing the edges AB and AC. We can then choose either of the edges AD and CD, and then either of the edges DE and BE. These choices give us four minimum spanning trees, all of weight 49:

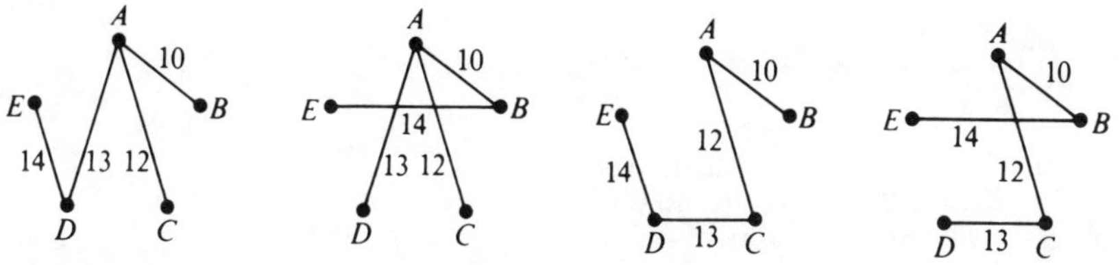

8.5.

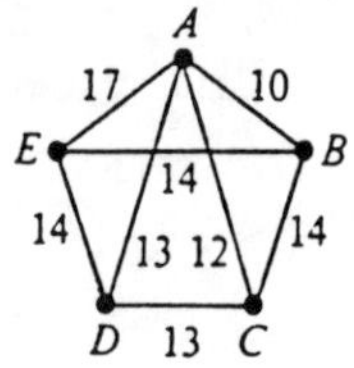

	A	B	C	D	E
A	—	10	12	13	17
B	10	—	14	—	14
C	12	14	—	13	—
D	13	—	13	—	14
E	17	14	—	14	—

First choice
Put vertex *A* into *T*.

Delete row *A*.
Look for the smallest entry in column *A*; it is 10 in row *B*.

	A	B	C	D	E
B	⑩	—	14	—	14
C	12	14	—	13	—
D	13	—	13	—	14
E	17	14	—	14	—

Second choice
Put edge *AB* and vertex *B* into *T*.

Delete row *B*.
Look for the smallest entry in columns *A* and *B*; it is 12 in row *C*.

	A	B	C	D	E
C	⑫	14	—	13	—
D	13	—	13	—	14
E	17	14	—	14	—

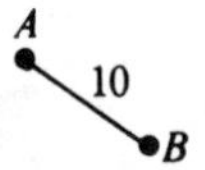

Third choice
Put edge *AC* and vertex *C* into *T*.

Delete row *C*.
Look for the smallest entry in columns *A*, *B* and *C*; it is 13 in row *D*.

	A	B	C	D	E
D	⑬	—	⑬	—	14
E	17	14	—	14	—

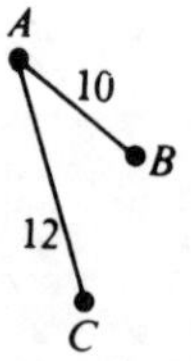

Fourth choice
Put edge *AD* (or *CD*) and vertex *D* into *T*.

Delete row *D*.
Look for the smallest entry in columns *A*, *B*, *C* and *D*; it is 14 in row *E*.

	A	B	C	D	E
E	17	⑭	—	⑭	—

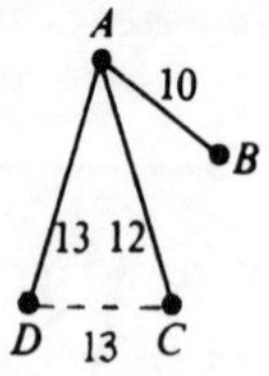

Fifth choice
Put edge *BE* (or *DE*) and vertex *E* into *T*.

All the vertices are now in the tree.

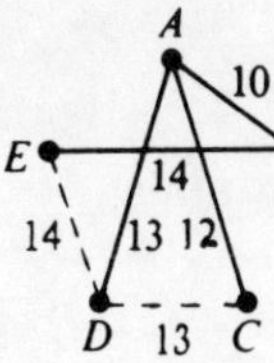

This completes the spanning tree, which is a minimum spanning tree of weight

$$10 + 12 + 13 + 14 = 49.$$

Note that there are four possible minimum spanning trees, as there are two possibilities for the fourth choice (*AD* or *CD*) and also for the fifth choice (*BE* or *DE*).

8.6. (a)

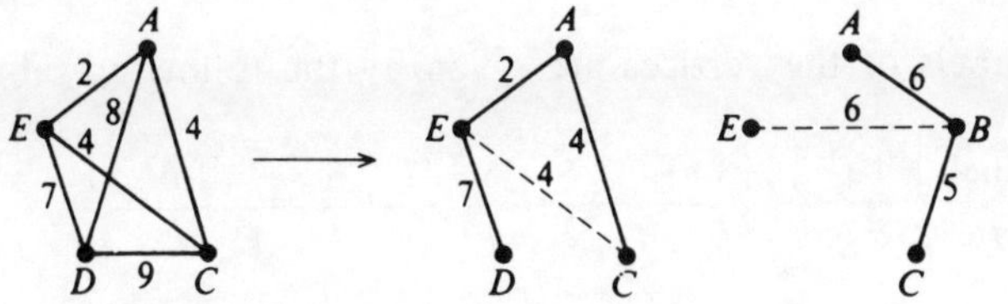

The minimum-weight spanning tree joining the vertices *A*, *C*, *D* and *E* is the tree with edges *AE*, *DE*, and *AC* or *CE*, with total weight 13. The two edges of smallest weight incident to *B* are *BC* and *BA* or *BE*, with total weight 11. The lower bound is therefore 13 + 11 = 24.

(b)

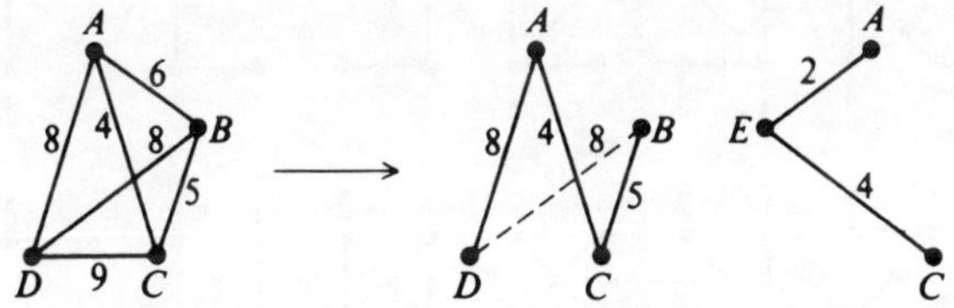

The minimum-weight spanning tree joining the vertices *A*, *B*, *C* and *D* is the tree with edges *AC*, *BC*, and *AD* or *BD*, with total weight 17. The two edges of smallest weight incident to *E* are *EA* and *EC*, with total weight 6. The lower bound is therefore 17 + 6 = 23.

8.7. There are several DFS orderings, depending on how we choose the vertices at each stage. If we choose the left-hand vertex whenever there is a choice, we get the following list:

a, ab, abe, ab, abf, ab, a, ac, acg, ac, a, ad,
adh, ad, adi, adij, adi, adik, adi, ad, a.

The labels of the vertices are given by the following table:

Label	1	2	3	4	5	6	7	8	9	10	11
Vertex	*a*	*b*	*e*	*f*	*c*	*g*	*d*	*h*	*i*	*j*	*k*

The abbreviated list is *abefcgdhijk*.

8.8. There are several BFS orderings, depending on how we choose the vertices at each stage. If we move from left to right on each level, we get the following list:

a, ab, abc, abcd, bcd, bcde, bcdef, cdef, cdefg, defg,
defgh, defghi, efghi, fghi, ghi, hi, i, ij, ijk, jk, k.

The labels of the vertices are given by the following table:

Label	1	2	3	4	5	6	7	8	9	10	11
Vertex	*a*	*b*	*c*	*d*	*e*	*f*	*g*	*h*	*i*	*j*	*k*

The abbreviated list is *abcdefghijk*.

8.9. (a) By Cayley's theorem, the number of labelled trees with 4 vertices is $4^{4-2} = 16$.
The sixteen labelled trees are as follows.

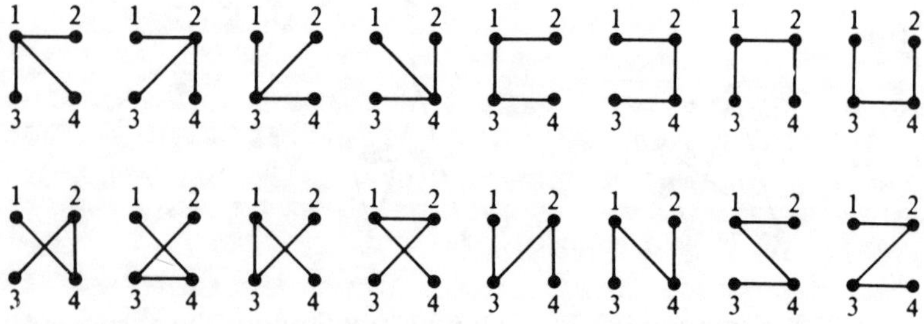

(b) There are 2 non-isomorphic unlabelled trees with 4 vertices:

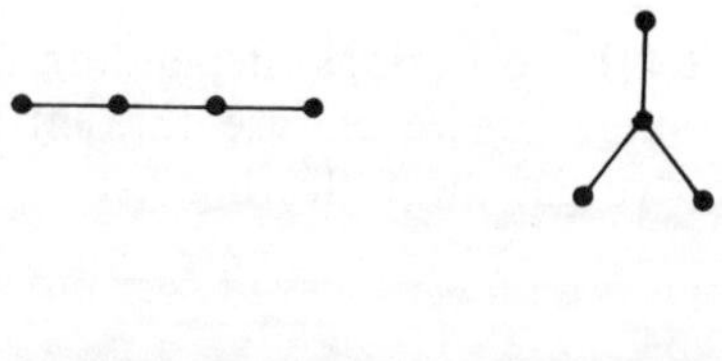

CHAPTER 9

9.1. There are several possible oriented graphs, depending on how we label the components and which direction we assign to each edge. One possibility is:

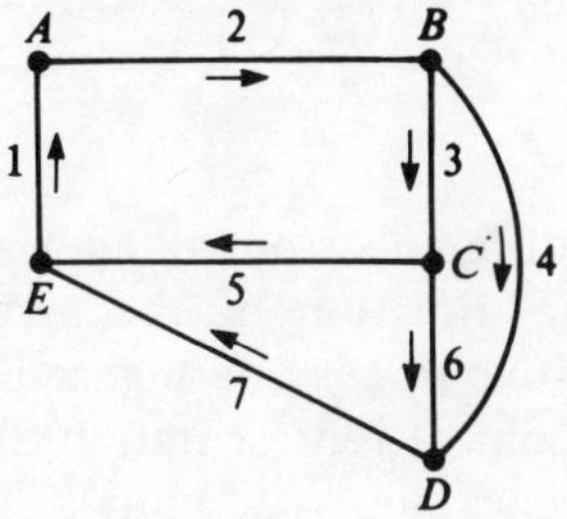

9.2. Differentiating the given equation, we obtain

$$\frac{df}{dt} = k\frac{dx}{dt} = ku$$

or

$$u = \frac{1}{k}\frac{df}{dt}.$$

In this case, f is the through variable and u is the across variable. The above equation is of the same form as the component equation for an inductor,

$$v = L\frac{di}{dt},$$

so the spring is analogous to an *inductor*, and the constant $1/k$ (the reciprocal of the spring stiffness) corresponds to the inductance L. The analogous quantities for the two components are shown in the following table:

	Mechanical	*Electrical*
component	spring	inductor
through variable	f	i
across variable	u	v
component equation	$u = \frac{1}{k}\frac{df}{dt}$	$v = L\frac{di}{dt}$

9.3. (a) Applying Kirchhoff's current law to the vertex A, we obtain
current flowing into $A = i_a$.
Applying Kirchhoff's current law to the vertex A for the original graphical representation, we obtain
current flowing into $A = i_1 + i_2$.

Hence

$i_a = i_1 + i_2$.

Applying Kirchhoff's current law to the vertex C, we obtain

current flowing into $C = -i_b$.

Applying Kirchhoff's current law to the vertex C for the graphical representation in the text, we obtain

current flowing into $C = -i_2$.

Hence

$i_b = i_2$.

The potential difference (or voltage) across terminals A and B is just v_a. (The direction of the arrow indicates that this is considered positive if A is at a higher potential than B.) Looking back at the graphical representation in the text, we see that the potential difference across terminals A and B is v_1. Hence

$v_a = v_1$.

The potential difference across terminals B and C is v_b. Looking at the graphical representation in the text, we see that the potential difference across terminals B and C is $v_2 - v_1$ (that is, the potential at B minus the potential at C). Hence

$v_b = v_2 - v_1$.

(b) Using the same method as in part (a), we obtain the results:

$$i_a = i_1 + i_2$$
$$i_b = -i_1$$
$$v_a = v_2$$
$$v_b = v_2 - v_1.$$

9.4. (a) To find a graphical representation of this system, we replace each 2-terminal component by an edge, each 3-terminal component by a tree with two edges, and the 4-terminal component by a tree with three edges. Since each tree can be chosen in several different ways, there are many possible ways of representing this system. One possible way is the following. The tree edges are shown by thick lines; the edges corresponding to 2-terminal components are shown by ordinary lines.

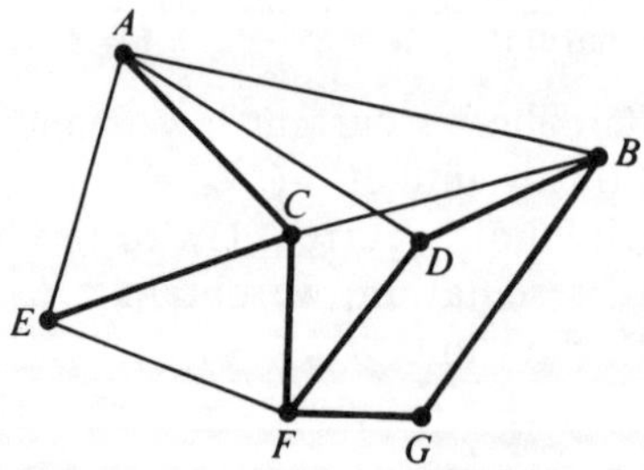

Component		Edges
ACE	(3-terminal)	*AC*, *CE*
BDG	(3-terminal)	*DB*, *GB*
CDFG	(4-terminal)	*CF*, *DF*, *FG*

(b) For any component, the number of component equations is equal to the number of edges in its graphical representation. So, to find the number of component equations associated with the system, we simply count the number of edges in the graph—there are 12 edges, so there are 12 component equations.

CHAPTER 10

10.1. (a)

cycle	voltage equations	
ABCEA	$v_2 + v_3 + v_5 + v_1 = 0$	(1)
BDCB	$v_4 - v_6 - v_3 = 0$	(2)
CDEC	$v_6 + v_7 - v_5 = 0$	(3)
ABDCEA	$v_2 + v_4 - v_6 + v_5 + v_1 = 0$	(4)
ABCDEA	$v_2 + v_3 + v_6 + v_7 + v_1 = 0$	(5)
BDECB	$v_4 + v_7 - v_5 - v_3 = 0$	(6)
ABDEA	$v_2 + v_4 + v_7 + v_1 = 0$	(7)

(b) The maximum number of linearly independent equations is 3. For example, equations (1), (2) and (3) are linearly independent, since each contains at least one term not included in either of the other two, and so none can depend on the other two. However, the remaining equations (4)–(7) all depend on the first three, as follows:

equation (4) = equation (1) + equation (2)
equation (5) = equation (1) + equation (3)
equation (6) = equation (2) + equation (3)
equation (7) = equation (1) + equation (2) + equation (3).

It follows that, if we wish to study the network, we need choose only equations (1), (2) and (3) (or any other 3 linearly independent equations), since the remaining equations give us no further information.

10.2. (a) There are 49 other spanning trees—for example:

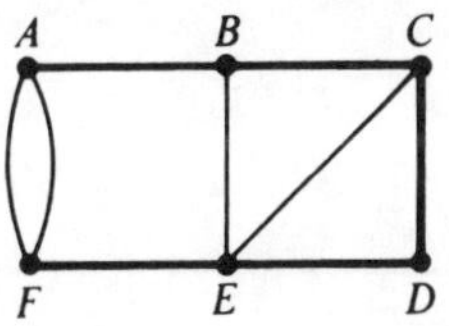

(i)

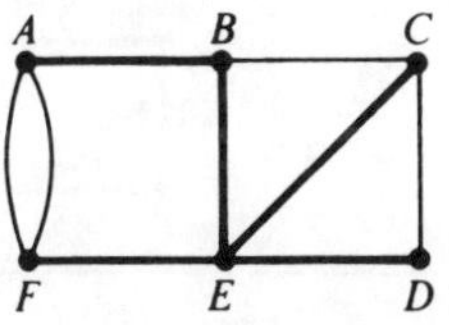

(ii)

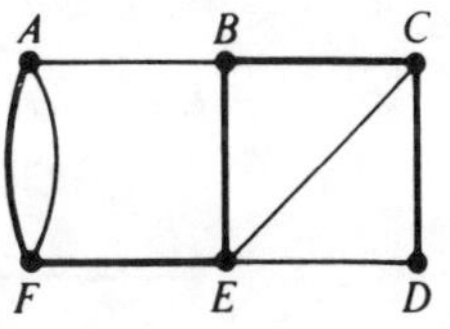

(iii)

tree (i)		tree (ii)		tree (iii)	
branches	chords	branches	chords	branches	chords
AB	*AF*	*AB*	*AF*	*BC*	*AB*
BC	*BE*	*BE*	*BC*	*BE*	*AF*
CD	*CE*	*CE*	*CD*	*CD*	*CE*
DE	*FA*	*DE*	*FA*	*EF*	*DE*
EF		*EF*		*FA*	

(b) Every tree with n vertices has just $n - 1$ edges (see Chapter 8). Thus there are $n - 1$ branches, and so there are $m - (n - 1) = m - n + 1$ chords associated with each spanning tree.

(c) If we add a chord to a spanning tree, we obtain a single cycle. For example, if we add each chord separately to the spanning tree in our example, we get the following cycles:

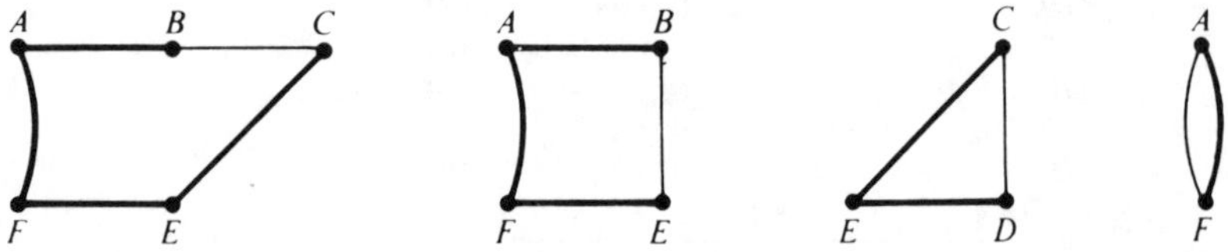

chord *BC* chord *BE* chord *CD* chord *FA*

10.3. (a)

<table>
<tr><th>chord</th><th>fundamental cycle</th><th>diagram</th><th>fundamental cycle equation</th></tr>
<tr><td>BD</td><td>BDCB</td><td>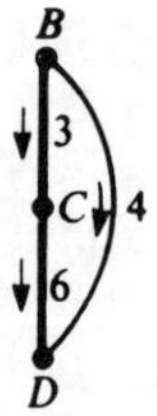</td><td>$v_4 - v_6 - v_3 = 0$</td></tr>
<tr><td>DE</td><td>CDEC</td><td>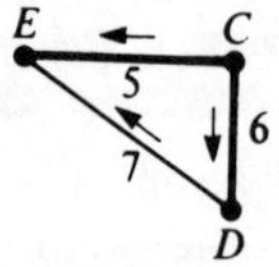</td><td>$v_6 + v_7 - v_5 = 0$</td></tr>
<tr><td>EA</td><td>ABCEA</td><td>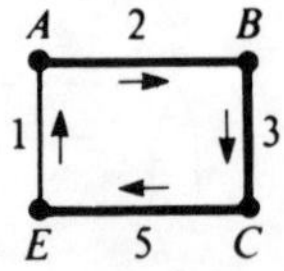</td><td>$v_2 + v_3 + v_5 + v_1 = 0$</td></tr>
</table>

The cycle *ABCDEA* contains the chords *DE* and *EA*. We obtain the voltage law equation for this cycle by combining the cycle equations for these two chords, giving

$$v_2 + v_3 + v_6 + v_7 + v_1 = 0$$

Note that in combining these two equations we must eliminate v_5, since the edge *CE* does not appear in the cycle *ABCDEA*. We therefore *add* the two equations.

(b)

chord	fundamental cycle	diagram	fundamental cycle equation
AB	*ABDEA*	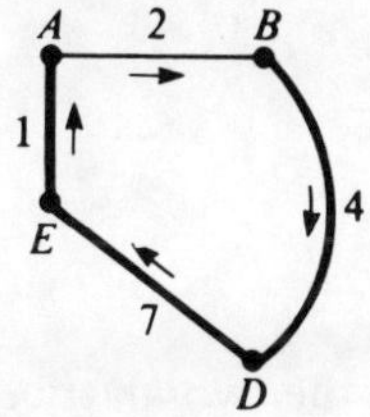	$v_2 + v_4 + v_7 + v_1 = 0$
CD	*BDCB*	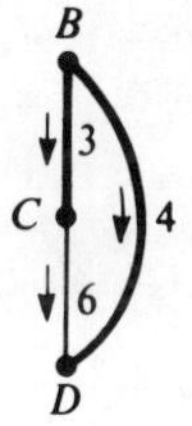	$v_4 - v_6 - v_3 = 0$
CE	*BDECB*	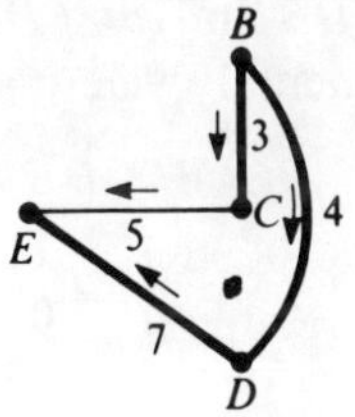	$v_4 + v_7 - v_5 - v_3 = 0$

The cycle *ABCDEA* contains the chords *AB* and *CD*. We obtain the voltage law equation for this cycle by combining the cycle equations for these two chords, giving

$$v_2 + v_6 + v_3 + v_7 + v_1 = 0$$

In this case we wish to eliminate v_4, so we combine the two equations by *subtraction*.

10.4. (a) The chords are *BD*, *DE* and *EA*, and the fundamental cycles are *BDCB*, *CDEC* and *ABCEA*. The fundamental cycle matrix is therefore

$$\begin{array}{c} \\ BD \\ DE \\ EA \end{array}\begin{array}{c} \begin{array}{ccccccc} AB & BC & CD & CE & BD & DE & EA \end{array} \\ \begin{bmatrix} 0 & -1 & -1 & 0 & 1 & 0 & 0 \\ 0 & 0 & 1 & -1 & 0 & 1 & 0 \\ 1 & 1 & 0 & 1 & 0 & 0 & 1 \end{bmatrix} \end{array}$$

and the fundamental cycle equations are given by

$$\begin{bmatrix} 0 & -1 & -1 & 0 & 1 & 0 & 0 \\ 0 & 0 & 1 & -1 & 0 & 1 & 0 \\ 1 & 1 & 0 & 1 & 0 & 0 & 1 \end{bmatrix}\begin{bmatrix} v_2 \\ v_3 \\ v_6 \\ v_5 \\ v_4 \\ v_7 \\ v_1 \end{bmatrix} = \begin{bmatrix} 0 \\ 0 \\ 0 \end{bmatrix}.$$

Multiplying the two matrices on the left, we obtain the equation

$$\begin{bmatrix} -v_3 - v_6 + v_4 \\ v_6 - v_5 + v_7 \\ v_2 + v_3 + v_5 + v_1 \end{bmatrix} = \begin{bmatrix} 0 \\ 0 \\ 0 \end{bmatrix},$$

which is equivalent to the three equations on page 502.

(b) The chords are *AB*, *CD* and *CE*, and the fundamental cycles are *ABDEA*, *BCDB* and *BCEDB*. (Note that the cycles are taken in the same direction as the chords they contain. Thus we have written *BCDB* and *BCEDB*, rather than *BDCB* and *BDECB*.) The fundamental cycle matrix is therefore

$$\begin{array}{c} \\ AB \\ CD \\ CE \end{array}\begin{array}{c} \begin{array}{ccccccc} BC & BD & DE & EA & AB & CD & CE \end{array} \\ \begin{bmatrix} 0 & 1 & 1 & 1 & 1 & 0 & 0 \\ 1 & -1 & 0 & 0 & 0 & 1 & 0 \\ 1 & -1 & -1 & 0 & 0 & 0 & 1 \end{bmatrix} \end{array}$$

and the fundamental cycle equations are given by

$$\begin{bmatrix} 0 & 1 & 1 & 1 & 1 & 0 & 0 \\ 1 & -1 & 0 & 0 & 0 & 1 & 0 \\ 1 & -1 & -1 & 0 & 0 & 0 & 1 \end{bmatrix}\begin{bmatrix} v_3 \\ v_4 \\ v_7 \\ v_1 \\ v_2 \\ v_6 \\ v_5 \end{bmatrix} = \begin{bmatrix} 0 \\ 0 \\ 0 \end{bmatrix}.$$

Multiplying the two matrices on the left, we obtain the equation

$$\begin{bmatrix} v_4 + v_7 + v_1 + v_2 \\ v_3 - v_4 + v_6 \\ v_3 - v_4 - v_7 + v_5 \end{bmatrix} = \begin{bmatrix} 0 \\ 0 \\ 0 \end{bmatrix},$$

which is equivalent to the three equations on page 503.

10.5. (a)

branch	vertices in X	vertices in Y	fundamental cutset	diagram	fundamental cutset equation
AB	A	B, C, D, E	$\{AB, EA\}$	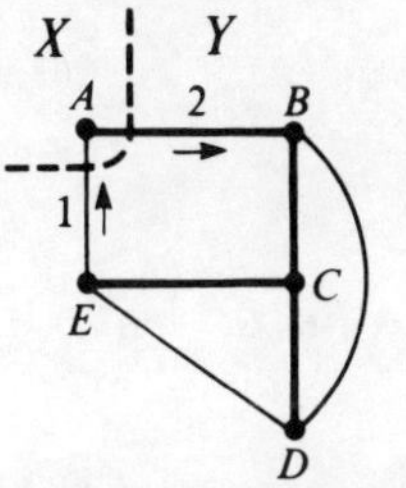	$i_2 - i_1 = 0$
BC	A, B	C, D, E	$\{BC, BD, EA\}$	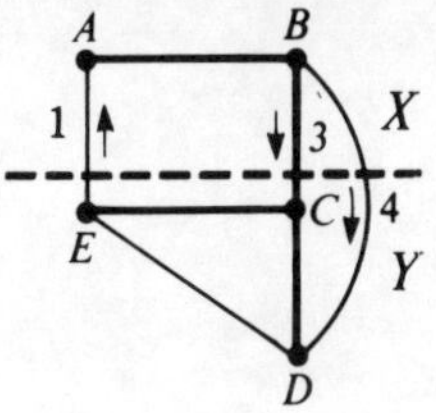	$i_3 + i_4 - i_1 = 0$
CD	A, B, C, E	D	$\{BD, CD, DE\}$	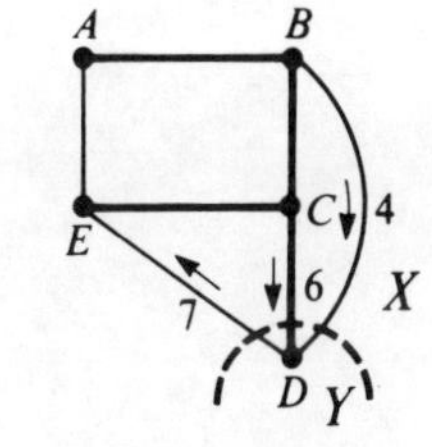	$i_4 + i_6 - i_7 = 0$
CE	A, B, C, D	E	$\{CE, DE, EA\}$	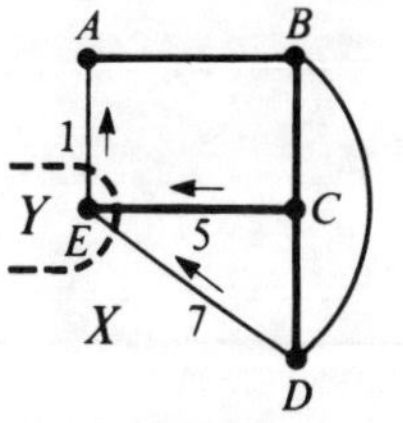	$i_5 + i_7 - i_1 = 0$

The cutset $\{AB, BC, BD\}$ contains the branches AB and BC. We obtain the current law equation for this cutset by combining the cutset equations for these two branches, giving

$$i_3 + i_4 - i_2 = 0.$$

(b)

branch	vertices in X	vertices in Y	fundamental cutset	diagram	fundamental cutset equation
BC	A, B, D, E	C	$\{BC, CD, CE\}$	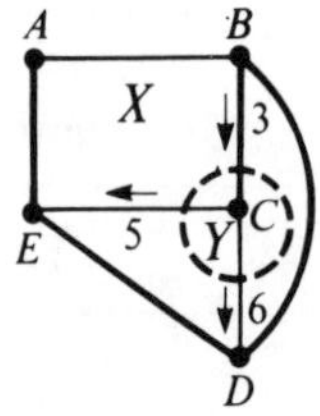	$i_3 - i_5 - i_6 = 0$
BD	B, C	A, D, E	$\{AB, BD, CD, CE\}$	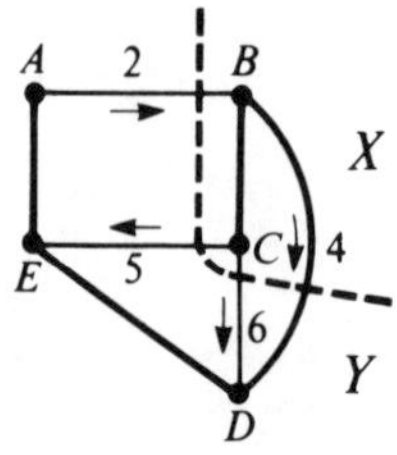	$i_4 + i_5 + i_6 - i_2 = 0$
DE	B, C, D	A, E	$\{AB, CE, DE\}$	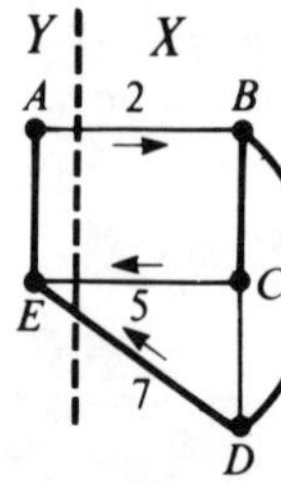	$i_5 + i_7 - i_2 = 0$
EA	B, C, D, E	A	$\{AB, EA\}$	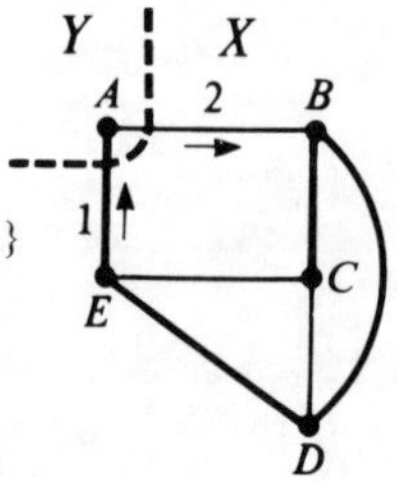	$i_1 - i_2 = 0$

The cutset $\{AB, BC, BD\}$ contains the branches BC and BD. We obtain the current law equation for this cutset by combining the cutset equations for these two branches, giving

$$i_3 + i_4 - i_2 = 0.$$

10.6. (a) The branches are AB, BC, CD and CE, and the fundamental cutsets are $\{AB, EA\}$, $\{BC, BD, EA\}$, $\{BD, CD, DE\}$ and $\{CE, DE, EA\}$. The fundamental cutset matrix is therefore

$$\begin{array}{c} \\ AB \\ BC \\ CD \\ CE \end{array}\begin{array}{c} \begin{array}{ccccccc} AB & BC & CD & CE & BD & DE & EA \end{array} \\ \begin{bmatrix} 1 & 0 & 0 & 0 & 0 & 0 & -1 \\ 0 & 1 & 0 & 0 & 1 & 0 & -1 \\ 0 & 0 & 1 & 0 & 1 & -1 & 0 \\ 0 & 0 & 0 & 1 & 0 & 1 & 1 \end{bmatrix} \end{array}$$

and the fundamental cutset equations are given by

$$\begin{bmatrix} 1 & 0 & 0 & 0 & 0 & 0 & -1 \\ 0 & 1 & 0 & 0 & 1 & 0 & -1 \\ 0 & 0 & 1 & 0 & 1 & -1 & 0 \\ 0 & 0 & 0 & 1 & 0 & 1 & -1 \end{bmatrix} \begin{bmatrix} i_2 \\ i_3 \\ i_6 \\ i_5 \\ i_4 \\ i_7 \\ i_1 \end{bmatrix} = \begin{bmatrix} 0 \\ 0 \\ 0 \\ 0 \end{bmatrix}.$$

Multiplying the two matrices on the left, we obtain the equation

$$\begin{bmatrix} i_2 - i_1 \\ i_3 + i_4 - i_1 \\ i_6 + i_4 - i_7 \\ i_5 + i_7 - i_1 \end{bmatrix} = \begin{bmatrix} 0 \\ 0 \\ 0 \\ 0 \end{bmatrix}$$

which is equivalent to the four equations on page 505.

(b) The branches are BC, BD, DE and EA, and the fundamental cutsets are $\{BC, CD, CE\}$, $\{AB, BD, CD, CE\}$, $\{AB, CE, DE\}$ and $\{AB, EA\}$. The fundamental cutset matrix is therefore

$$\begin{array}{c} \\ BC \\ BD \\ DE \\ EA \end{array}\begin{array}{c} \begin{array}{ccccccc} BC & BD & DE & EA & AB & CD & CE \end{array} \\ \begin{bmatrix} 1 & 0 & 0 & 0 & 0 & -1 & -1 \\ 0 & 1 & 0 & 0 & -1 & 1 & 1 \\ 0 & 0 & 1 & 0 & -1 & 0 & 1 \\ 0 & 0 & 0 & 1 & -1 & 0 & 0 \end{bmatrix} \end{array}$$

and the fundamental cutset equations are given by

$$\begin{bmatrix} 1 & 0 & 0 & 0 & 0 & -1 & -1 \\ 0 & 1 & 0 & 0 & -1 & 1 & 1 \\ 0 & 0 & 1 & 0 & -1 & 0 & 1 \\ 0 & 0 & 0 & 1 & -1 & 0 & 0 \end{bmatrix} \begin{bmatrix} i_3 \\ i_4 \\ i_7 \\ i_1 \\ i_2 \\ i_6 \\ i_5 \end{bmatrix} = \begin{bmatrix} 0 \\ 0 \\ 0 \\ 0 \end{bmatrix}.$$

Multiplying the two matrices on the left, we obtain the equation

$$\begin{bmatrix} i_3 - i_6 - i_5 \\ i_4 - i_2 + i_6 + i_5 \\ i_7 - i_2 + i_5 \\ i_1 - i_2 \end{bmatrix} = \begin{bmatrix} 0 \\ 0 \\ 0 \\ 0 \end{bmatrix}$$

which is equivalent to the four equations on page 506.

10.7. (a)

$$\begin{array}{c} \\ A \\ B \\ C \\ D \\ E \end{array} \begin{array}{c} \begin{array}{ccccccc} AB & BC & BD & CD & CE & DE & EA \end{array} \\ \begin{bmatrix} 1 & 0 & 0 & 0 & 0 & 0 & -1 \\ -1 & 1 & 1 & 0 & 0 & 0 & 0 \\ 0 & -1 & 0 & 1 & 1 & 0 & 0 \\ 0 & 0 & -1 & -1 & 0 & 1 & 0 \\ 0 & 0 & 0 & 0 & -1 & -1 & 1 \end{bmatrix} \end{array}$$

(b)

$$\begin{array}{c} \\ A \\ B \\ C \\ E \end{array} \begin{array}{c} \begin{array}{ccccccc} AB & BC & BD & CD & CE & DE & EA \end{array} \\ \begin{bmatrix} 1 & 0 & 0 & 0 & 0 & 0 & -1 \\ -1 & 1 & 1 & 0 & 0 & 0 & 0 \\ 0 & -1 & 0 & 1 & 1 & 0 & 0 \\ 0 & 0 & 0 & 0 & -1 & -1 & 1 \end{bmatrix} \end{array}$$

(c)

$$\mathbf{A}_t = \begin{array}{c} A \\ B \\ C \\ E \end{array} \overset{\begin{array}{cccc} AB & BC & CD & CE \end{array}}{\begin{bmatrix} 1 & 0 & 0 & 0 \\ -1 & 1 & 0 & 0 \\ 0 & -1 & 1 & 1 \\ 0 & 0 & 0 & -1 \end{bmatrix}} \qquad \mathbf{A}_c = \begin{array}{c} A \\ B \\ C \\ E \end{array} \overset{\begin{array}{ccc} BD & DE & EA \end{array}}{\begin{bmatrix} 0 & 0 & -1 \\ 1 & 0 & 0 \\ 0 & 0 & 0 \\ 0 & -1 & 1 \end{bmatrix}}$$

10.8. We first draw the spanning tree and indicate D as the chosen reference vertex.

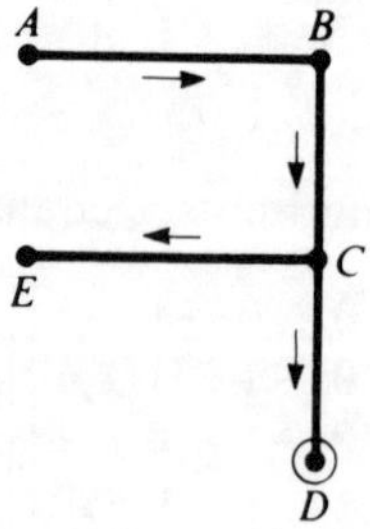

Tracing the path from each vertex to the reference vertex, we obtain the matrix

$$\mathbf{A}_t^{-1} = \begin{array}{c} \\ AB \\ BC \\ CD \\ CE \end{array} \begin{array}{c} \begin{array}{cccc} A & B & C & E \end{array} \\ \begin{bmatrix} 1 & 0 & 0 & 0 \\ 1 & 1 & 0 & 0 \\ 1 & 1 & 1 & 1 \\ 0 & 0 & 0 & -1 \end{bmatrix} \end{array}.$$

We have

$$\mathbf{A}_t^{-1}\mathbf{A}_t = \begin{bmatrix} 1 & 0 & 0 & 0 \\ 1 & 1 & 0 & 0 \\ 1 & 1 & 1 & 1 \\ 0 & 0 & 0 & -1 \end{bmatrix} \begin{bmatrix} 1 & 0 & 0 & 0 \\ -1 & 1 & 0 & 0 \\ 0 & -1 & 1 & 1 \\ 0 & 0 & 0 & -1 \end{bmatrix} = \begin{bmatrix} 1 & 0 & 0 & 0 \\ 0 & 1 & 0 & 0 \\ 0 & 0 & 1 & 0 \\ 0 & 0 & 0 & 1 \end{bmatrix} = \mathbf{I}_4,$$

$$\mathbf{A}_t\mathbf{A}_t^{-1} = \begin{bmatrix} 1 & 0 & 0 & 0 \\ -1 & 1 & 0 & 0 \\ 0 & -1 & 1 & 1 \\ 0 & 0 & 0 & -1 \end{bmatrix} \begin{bmatrix} 1 & 0 & 0 & 0 \\ 1 & 1 & 0 & 0 \\ 1 & 1 & 1 & 1 \\ 0 & 0 & 0 & -1 \end{bmatrix} = \begin{bmatrix} 1 & 0 & 0 & 0 \\ 0 & 1 & 0 & 0 \\ 0 & 0 & 1 & 0 \\ 0 & 0 & 0 & 1 \end{bmatrix} = \mathbf{I}_4,$$

as required.

10.9. (a)

$$\mathbf{A}_c = \begin{bmatrix} 0 & 0 & -1 \\ 1 & 0 & 0 \\ 0 & 0 & 0 \\ 0 & -1 & 1 \end{bmatrix} \qquad \mathbf{A}_t^{-1} = \begin{bmatrix} 1 & 0 & 0 & 0 \\ 1 & 1 & 0 & 0 \\ 1 & 1 & 1 & 1 \\ 0 & 0 & 0 & -1 \end{bmatrix}.$$

Hence

$$\mathbf{A}_t^{-1}\mathbf{A}_c = \begin{bmatrix} 1 & 0 & 0 & 0 \\ 1 & 1 & 0 & 0 \\ 1 & 1 & 1 & 1 \\ 0 & 0 & 0 & -1 \end{bmatrix} \begin{bmatrix} 0 & 0 & -1 \\ 1 & 0 & 0 \\ 0 & 0 & 0 \\ 0 & -1 & 1 \end{bmatrix} = \begin{bmatrix} 0 & 0 & -1 \\ 1 & 0 & -1 \\ 1 & -1 & 0 \\ 0 & 1 & -1 \end{bmatrix}$$

and so

$$\mathbf{B}_f = \left[\begin{array}{cccc:ccc} 1 & 0 & 0 & 0 & 0 & 0 & -1 \\ 0 & 1 & 0 & 0 & 1 & 0 & -1 \\ 0 & 0 & 1 & 0 & 1 & -1 & 0 \\ 0 & 0 & 0 & 1 & 0 & 1 & -1 \end{array}\right],$$

which agrees with the answer given in the solution of Problem 10.6. Also,

$$(\mathbf{A}_t^{-1}\mathbf{A}_c)' = \begin{bmatrix} 0 & 1 & 1 & 0 \\ 0 & 0 & -1 & 1 \\ -1 & -1 & 0 & -1 \end{bmatrix}$$

and so

$$\mathbf{C}_f = \left[\begin{array}{cccc|ccc} 0 & -1 & -1 & 0 & 1 & 0 & 0 \\ 0 & 0 & 1 & -1 & 0 & 1 & 0 \\ 1 & 1 & 0 & 1 & 0 & 0 & 1 \end{array}\right],$$

which agrees with the answer given in the solution of Problem 10.4.

(b)

$$\mathbf{C}_f\mathbf{B}_f' = \left[\begin{array}{cccc|ccc} 0 & -1 & -1 & 0 & 1 & 0 & 0 \\ 0 & 0 & 1 & -1 & 0 & 1 & 0 \\ 1 & 1 & 0 & 1 & 0 & 0 & 1 \end{array}\right] \left[\begin{array}{cccc} 1 & 0 & 0 & 0 \\ 0 & 1 & 0 & 0 \\ 0 & 0 & 1 & 0 \\ 0 & 0 & 0 & 1 \\ \hline 0 & 1 & 1 & 0 \\ 0 & 0 & -1 & 1 \\ -1 & -1 & 0 & -1 \end{array}\right] = \begin{bmatrix} 0 & 0 & 0 & 0 \\ 0 & 0 & 0 & 0 \\ 0 & 0 & 0 & 0 \end{bmatrix}.$$

This is a special case of the following general result:
for an oriented graph with n vertices and m edges the matrix product $\mathbf{C}_f\mathbf{B}_f' = \mathbf{0}_{m-n+1,\,n-1}$ (that is, the zero matrix with $m - n + 1$ rows and $n - 1$ columns).

CHAPTER 11

11.1. This example is similar to Example 2. The matrix equation is

$$\underset{\uparrow \atop \mathbf{T}}{\left[\begin{array}{ccccccc|ccccccc} 1 & 0 & 0 & 0 & 0 & 0 & 0 & -R_2 & 0 & 0 & 0 & 0 & 0 & 0 \\ 0 & 1 & 0 & 0 & 0 & 0 & 0 & 0 & -R_3 & 0 & 0 & 0 & 0 & 0 \\ 0 & 0 & 1 & 0 & 0 & 0 & 0 & 0 & 0 & -R_4 & 0 & 0 & 0 & 0 \\ 0 & 0 & 0 & 1 & 0 & 0 & 0 & 0 & 0 & 0 & -R_5 & 0 & 0 & 0 \\ 0 & 0 & 0 & 0 & 1 & 0 & 0 & 0 & 0 & 0 & 0 & -R_6 & 0 & 0 \\ 0 & 0 & 0 & 0 & 0 & 1 & 0 & 0 & 0 & 0 & 0 & 0 & -R_7 & 0 \\ 0 & 0 & 0 & 0 & 0 & 0 & 1 & 0 & 0 & 0 & 0 & 0 & 0 & 0 \end{array}\right]} \underset{\uparrow \atop \left[\begin{array}{c}\mathbf{v}\\ \hline \mathbf{i}\end{array}\right]}{\left[\begin{array}{c} v_2 \\ v_3 \\ v_4 \\ v_5 \\ v_6 \\ v_7 \\ v_1 \\ \hline i_2 \\ i_3 \\ i_4 \\ i_5 \\ i_6 \\ i_7 \\ i_1 \end{array}\right]} = \underset{\uparrow \atop =\ \mathbf{f}}{\begin{bmatrix} 0 \\ 0 \\ 0 \\ 0 \\ 0 \\ 0 \\ -V \end{bmatrix}}$$

11.2. (a)

fundamental cycles:

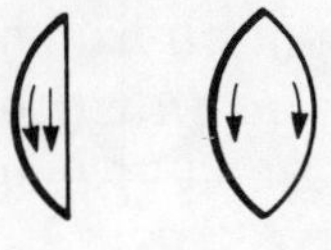

$$\mathbf{C}_f = \begin{array}{c} \begin{matrix} 1 & 2 & 3 \end{matrix} \\ \left[\begin{array}{c|cc} -1 & 1 & 0 \\ -1 & 0 & 1 \end{array}\right] \end{array}$$

fundamental cutset:

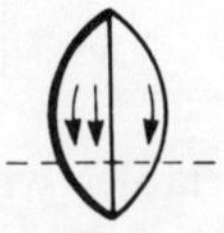

$$\mathbf{B}_f = [1 \mid 1 \;\; 1]$$

Combining these matrices with the matrix on page 260, and re-ordering the variables to agree with the matrices $\mathbf{B}_f$ and $\mathbf{C}_f$, we get

$$\underset{\uparrow \atop \mathbf{H}}{\left[\begin{array}{ccc|ccc} 0 & 0 & 0 & 1 & 0 & 0 \\ 0 & 1 & 0 & 0 & -R_2 & 0 \\ 0 & 0 & 1 & 0 & 0 & -R_3 \\ \hline -1 & 1 & 0 & 0 & 0 & 0 \\ -1 & 0 & 1 & 0 & 0 & 0 \\ \hline 0 & 0 & 0 & 1 & 1 & 1 \end{array}\right]} \underset{\uparrow \atop \mathbf{x}}{\left[\begin{array}{c} v_1 \\ v_2 \\ v_3 \\ \hline i_1 \\ i_2 \\ i_3 \end{array}\right]} \underset{\atop =}{=} \underset{\uparrow \atop \mathbf{y}}{\left[\begin{array}{c} I \\ 0 \\ 0 \\ \hline 0 \\ 0 \\ \hline 0 \end{array}\right]}$$

Notice that we have made the current source the first component in the column vector. It does not matter where the *independent* sources appear as long as the network voltages and currents appear in the same order in the component, cycle and cutset equations.

(b)

$$\left[\begin{array}{cccc|ccc|cccc|ccc} 1 & 0 & 0 & 0 & 0 & 0 & 0 & -R_2 & 0 & 0 & 0 & 0 & 0 & 0 \\ 0 & 1 & 0 & 0 & 0 & 0 & 0 & 0 & -R_3 & 0 & 0 & 0 & 0 & 0 \\ 0 & 0 & 1 & 0 & 0 & 0 & 0 & 0 & 0 & -R_6 & 0 & 0 & 0 & 0 \\ 0 & 0 & 0 & 1 & 0 & 0 & 0 & 0 & 0 & 0 & -R_5 & 0 & 0 & 0 \\ 0 & 0 & 0 & 0 & 1 & 0 & 0 & 0 & 0 & 0 & 0 & -R_4 & 0 & 0 \\ 0 & 0 & 0 & 0 & 0 & 1 & 0 & 0 & 0 & 0 & 0 & 0 & -R_7 & 0 \\ 0 & 0 & 0 & 0 & 0 & 0 & 1 & 0 & 0 & 0 & 0 & 0 & 0 & 0 \\ \hline 0 & -1 & -1 & 0 & 1 & 0 & 0 & 0 & 0 & 0 & 0 & 0 & 0 & 0 \\ 0 & 0 & 1 & -1 & 0 & 1 & 0 & 0 & 0 & 0 & 0 & 0 & 0 & 0 \\ 1 & 1 & 0 & 1 & 0 & 0 & 1 & 0 & 0 & 0 & 0 & 0 & 0 & 0 \\ \hline 0 & 0 & 0 & 0 & 0 & 0 & 0 & 1 & 0 & 0 & 0 & 0 & 0 & -1 \\ 0 & 0 & 0 & 0 & 0 & 0 & 0 & 0 & 1 & 0 & 0 & 1 & 0 & -1 \\ 0 & 0 & 0 & 0 & 0 & 0 & 0 & 0 & 0 & 1 & 0 & 1 & -1 & 0 \\ 0 & 0 & 0 & 0 & 0 & 0 & 0 & 0 & 0 & 0 & 1 & 0 & 1 & -1 \end{array}\right] \left[\begin{array}{c} v_2 \\ v_3 \\ v_6 \\ v_5 \\ v_4 \\ v_7 \\ v_1 \\ \hline i_2 \\ i_3 \\ i_6 \\ \hline i_5 \\ i_4 \\ i_7 \\ i_1 \end{array}\right] = \left[\begin{array}{c} 0 \\ 0 \\ 0 \\ 0 \\ 0 \\ 0 \\ -V \\ \hline 0 \\ 0 \\ 0 \\ \hline 0 \\ 0 \\ 0 \\ 0 \end{array}\right]$$

11.3. (a)

$$\left[\begin{array}{cccccc|cc} 1 & 0 & 0 & 0 & 0 & 0 & 0 & 0 \\ 0 & 1/L & 0 & 0 & 0 & 0 & 0 & 0 \\ 0 & 0 & 1 & 0 & 0 & -R_4 & 0 & 0 \\ 0 & 0 & 0 & 0 & 1/C & 0 & 0 & 0 \end{array}\right] \left[\begin{array}{c} v_1 \\ v_3 \\ v_4 \\ i_1 \\ i_2 \\ i_4 \\ \hline v_2 \\ i_3 \end{array}\right] = \left[\begin{array}{c} k \\ 0 \\ 0 \\ 0 \end{array}\right] + \left[\begin{array}{cccccc|cc} 0 & 0 & 0 & 0 & 0 & 0 & 0 & 0 \\ 0 & 0 & 0 & 0 & 0 & 0 & 0 & 1 \\ 0 & 0 & 0 & 0 & 0 & 0 & 0 & 0 \\ 0 & 0 & 0 & 0 & 0 & 0 & 1 & 0 \end{array}\right] \left[\begin{array}{c} \dot{v}_1 \\ \dot{v}_3 \\ \dot{v}_4 \\ \dot{i}_1 \\ \dot{i}_2 \\ \dot{i}_4 \\ \hline \dot{v}_2 \\ \dot{i}_3 \end{array}\right]$$

Note that the variables whose derivatives occur in the component equations (that is, v_2 and i_3) appear last in the column vectors.

(b) The fundamental cycle and cutset matrices can be rewritten as follows:

$$\left[\begin{array}{cccccc|cc} 1 & 1 & 0 & 0 & 0 & 0 & 1 & 0 \\ 0 & -1 & 1 & 0 & 0 & 0 & 0 & 0 \\ 0 & 0 & 0 & 1 & -1 & 0 & 0 & 0 \\ 0 & 0 & 0 & 0 & -1 & 1 & 0 & 1 \end{array}\right] \left[\begin{array}{c} v_1 \\ v_3 \\ v_4 \\ i_1 \\ i_2 \\ i_4 \\ \hline v_2 \\ i_3 \end{array}\right] = \left[\begin{array}{c} 0 \\ 0 \\ 0 \\ 0 \end{array}\right]$$

(Note that we have rearranged the columns of the matrix so as to put the edge-variables in the same order as in the component-equation matrix.)

Combining this with the component-equation matrix, we obtain the **H**-matrix equation $\mathbf{Hx} = \mathbf{y} + \mathbf{K}\dot{\mathbf{x}}$, as follows:

$$\left[\begin{array}{cccccc|cc} 1 & 0 & 0 & 0 & 0 & 0 & 0 & 0 \\ 0 & 1/L & 0 & 0 & 0 & 0 & 0 & 0 \\ 0 & 0 & 1 & 0 & 0 & -R_4 & 0 & 0 \\ 0 & 0 & 0 & 0 & 1/C & 0 & 0 & 0 \\ 1 & 1 & 0 & 0 & 0 & 0 & 1 & 0 \\ 0 & -1 & 1 & 0 & 0 & 0 & 0 & 0 \\ 0 & 0 & 0 & 1 & -1 & 0 & 0 & 0 \\ 0 & 0 & 0 & 0 & -1 & 1 & 0 & 1 \end{array}\right] \left[\begin{array}{c} v_1 \\ v_3 \\ v_4 \\ i_1 \\ i_2 \\ i_4 \\ \hline v_2 \\ i_3 \end{array}\right] = \left[\begin{array}{c} k \\ 0 \\ 0 \\ 0 \\ 0 \\ 0 \\ \hline 0 \\ 0 \end{array}\right] + \left[\begin{array}{cccccc|cc} 0 & 0 & 0 & 0 & 0 & 0 & 0 & 0 \\ 0 & 0 & 0 & 0 & 0 & 0 & 0 & 1 \\ 0 & 0 & 0 & 0 & 0 & 0 & 0 & 0 \\ 0 & 0 & 0 & 0 & 0 & 0 & 1 & 0 \\ 0 & 0 & 0 & 0 & 0 & 0 & 0 & 0 \\ 0 & 0 & 0 & 0 & 0 & 0 & 0 & 0 \\ 0 & 0 & 0 & 0 & 0 & 0 & 0 & 0 \\ 0 & 0 & 0 & 0 & 0 & 0 & 0 & 0 \end{array}\right] \left[\begin{array}{c} \dot{v}_1 \\ \dot{v}_3 \\ \dot{v}_4 \\ \dot{i}_1 \\ \dot{i}_2 \\ \dot{i}_4 \\ \hline \dot{v}_2 \\ \dot{i}_3 \end{array}\right]$$

(c) To find the state equations we start with rows 5 and 8 and perform row operations in such a way that the first six entries in these rows are all zero.

First state equation: Take row 5, and subtract row 1 (to eliminate the 1 in column 1) and $L \times$ row 2 (to eliminate the 1 in column 2). Performing the same operations on the right-hand side of the equation yields the state equation

$$v_2 = -k - L\dot{i}_3$$

Second state equation: We can eliminate the entries in columns 5 and 6 of row 8 by taking

$$(\text{row } 8) + (C \times \text{row } 4) + ((1/R_4) \times \text{row } 3) - ((1/R_4) \times \text{row } 6) - ((L/R_4) \times \text{row } 2)$$

This yields the state equation

$$i_3 = C\dot{v}_2 - \frac{L}{R_4}\dot{i}_3.$$

CHAPTER 12

12.1. (a) The bipartite graph representing this situation is:

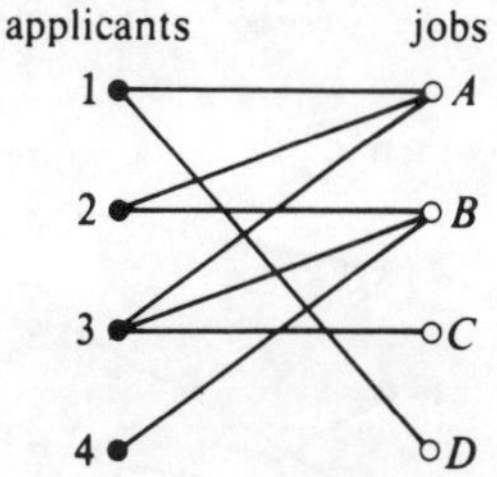

(b) The only possible matching containing four edges is the following:

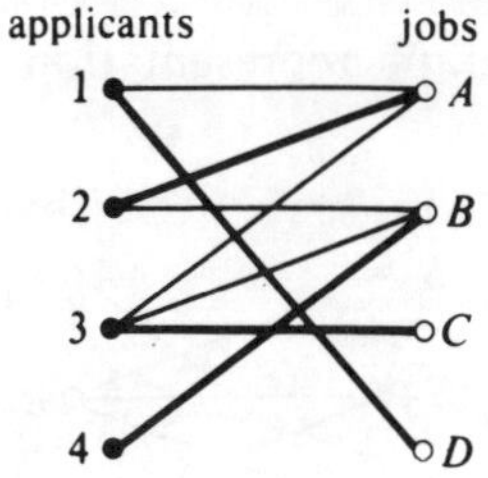

This is a maximum matching, since there are clearly no matchings possible with more than four edges. The matching corresponds to all four workers being assigned to jobs for which they are qualified.

12.2. (a)

subset of men	women known	number of men	number of women known	is marriage condition satisfied?
$\{A_1, A_2, A_3\}$	B_1, B_2, B_3, B_4	3	4	yes
$\{A_1, A_2\}$	B_1, B_3, B_4	2	3	yes
$\{A_1, A_3\}$	B_1, B_2, B_3, B_4	2	4	yes
$\{A_2, A_3\}$	B_2, B_3, B_4	2	3	yes
$\{A_1\}$	B_1, B_3, B_4	1	3	yes
$\{A_2\}$	B_4	1	1	yes
$\{A_3\}$	B_2, B_3	1	2	yes

The marriage condition is satisfied for every subset, so, by the marriage theorem, all three men can marry women they know.

(b) The number of non-empty subsets of a set of size n is $2^n - 1$. For this problem, $n = 3$, so we need to consider $2^3 - 1 = 7$ subsets—this agrees with the number of entries in the table.

(c) The bipartite graph representation of the solution is shown below. A maximum matching of three edges (corresponding to all three men being married) is indicated by heavy lines. Other matchings with three edges are also possible.

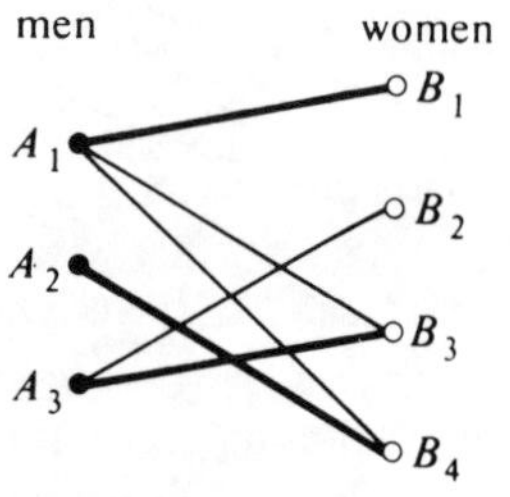

12.3. (a) The bipartite graph representation is:

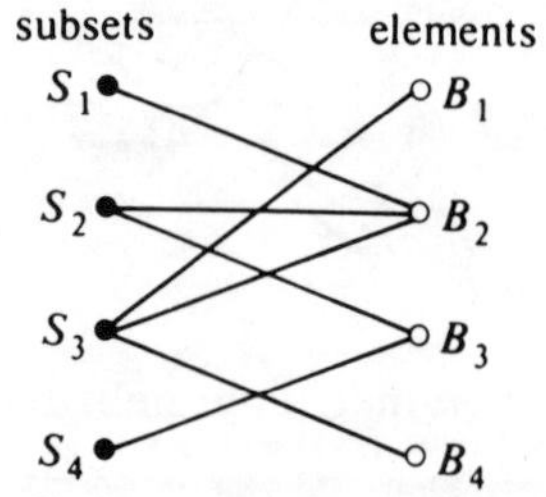

(b) The only subset of men for which the marriage condition is not satisfied is the subset $\{A_1, A_2, A_4\}$, which contains three men who collectively know only two women, B_2 and B_3.

(c) Since we know from part (b) that the marriage condition is not satisfied, the marriage theorem tells us that not all of the four men can marry women they know. It follows that if we can find a matching containing three edges, then this is a maximum matching. There are several such matchings, two of which are:

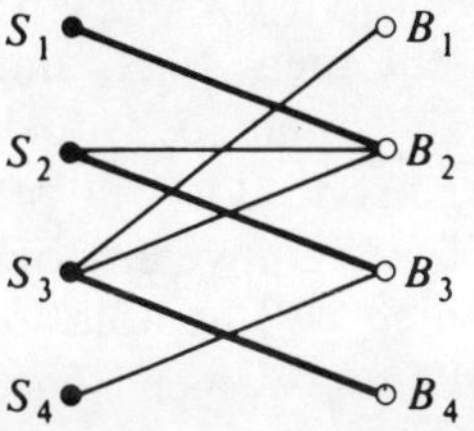

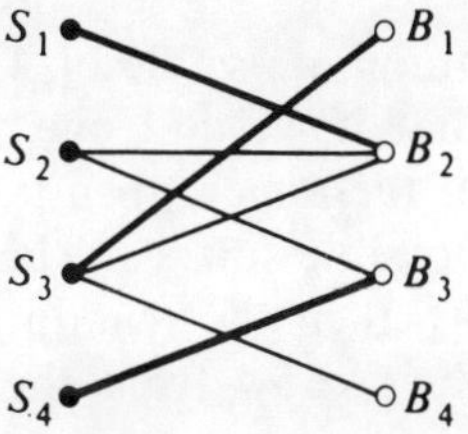

Since a maximum matching contains three edges, the maximum number of men who can marry is three.

(d) It follows from the answer to part (c) that the family $\mathcal{F}$ does not have a transversal.

12.4. (a) The marriage condition is clearly not satisfied, since the five men between them know only four women.

(b) The bipartite graph representation of the problem is as follows:

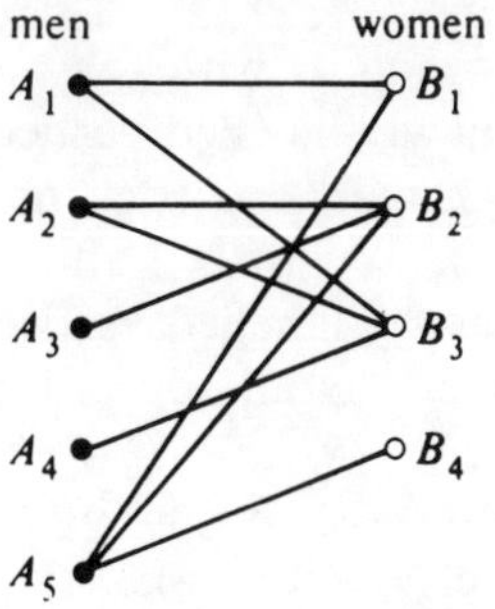

(c) The subsets of men for which the marriage condition is not satisfied are listed overleaf, together with the value of the expression in the modified marriage theorem.

subset of men	women known	number of men (m)	number of women known (p)	$p + (n - m)$
$\{A_2, A_3, A_4\}$	B_2, B_3	3	2	4
$\{A_1, A_2, A_3, A_4\}$	B_1, B_2, B_3	4	3	4
$\{A_1, A_2, A_3, A_4, A_5\}$	B_1, B_2, B_3, B_4	5	4	4

(d) From the above table, we see that the minimum value of the expression $p + (n - m)$ of the modified marriage condition is 4. By the theorem, the largest number of men who can marry women they know is therefore 4.

All the subsets of men for which $p + (n - m)$ is equal to its minimum value of 4 contain the men A_2, A_3 and A_4. It follows that if we add one additional woman B_5 who knows any one of these men, then the marriage condition will be satisfied for all possible subsets. Therefore, if we remove any one of A_2, A_3 and A_4, then the remaining four men will be able to marry women they know from the original four women. The list of subsets of four men who can all marry is therefore as follows:

$$\{A_1, A_3, A_4, A_5\}, \{A_1, A_2, A_4, A_5\}, \{A_1, A_2, A_3, A_5\}$$

12.5. (a) (i) This is not an alternating path, since the initial vertex x_3 and the final vertex y_1 are both incident with edges in M.

(ii) This is an alternating path.

(iii) This is not an alternating path, since the initial vertex x_3 is incident with an edge in M.

(b) There are six possible alternating paths—namely, $x_3y_3x_2y_1x_1y_2$, $x_3y_4x_4y_5$, $x_5y_4x_4y_5$, $x_3y_4x_4y_3x_2y_1x_1y_2$, $x_5y_4x_4y_3x_2y_1x_1y_2$ and x_3y_2.

12.6. Part A: The labelling procedure

STEP 1 We label vertices x_1 and x_4 with (*).

STEP 2 (a) For the two newly-labelled vertices in X, we label vertices in Y joined to them by an edge not in the matching—that is, we label vertex y_1 with (x_1) and vertex y_3 with (x_4).

(b) For the two newly-labelled vertices in Y, we label vertices in X joined to them by an edge which is in the matching—that is, we label vertex x_2 with (y_1) and vertex x_3 with (y_3).

(a) We return to part (a). We choose x_2 as a newly-labelled vertex in X, and label y_2 with (x_2).

No more labelling is possible, so we go on to Step 3.

STEP 3 Vertex y_2 is a labelled vertex in Y which is not incident with an edge in the matching, so breakthrough has occurred.

We go on to Part B, Step 4.

Part B: Matching improvement procedure

STEP 4 The alternating path is $y_2x_2y_1x_1$. We go on to Step 5.

STEP 5 The new matching consists of the three edges x_1y_1, x_2y_2 and x_3y_3 Remove all labels and go back to Step 1.

STEP 1 No further labelling is possible. STOP. (No further improvement is possible.)

Note that if in the second time through Step 2 part (a) you chose vertex x_3 as the newly-labelled vertex in X, you would have obtained a different alternating path ($y_2x_3y_3x_4$) and consequently a different matching (x_2y_1, x_3y_2, x_4y_3) with the same number of edges (three) as above.

CHAPTER 13

13.1. Part A: Construction of the initial partial graph

STEP 1 We assign weights to the vertices as follows. Looking across the first row, we see that the lowest cost is 4, so this is the weight for x_1. Similarly, the weights for x_2, x_3 and x_4 are 8, 6 and 8 respectively. We subtract these weights from the edge costs on the corresponding row of the matrix, and then repeat the procedure with the vertices in Y. There are now zeros in the columns y_2, y_3 and y_4, but the column y_1 reads 4, 2, 3, 4. So vertex y_1 has a weight of 2 and the other vertices in Y have weights of zero. Subtracting these weights fron the edge costs gives the first revised cost matrix, shown below.

STEP 2 We form the partial graph containing the edges with zero current cost—this corresponds to the zero entries in the revised cost matrix.

				(x_3)		
			2	0	0	0
			y_1	y_2	y_3	y_4
(y_2)	4	x_1	2	**0**	2	3
	8	x_2	0	4	**0**	6
(*)	6	x_3	1	0	5	9
	8	x_4	2	0	6	**0**

first revised cost matrix

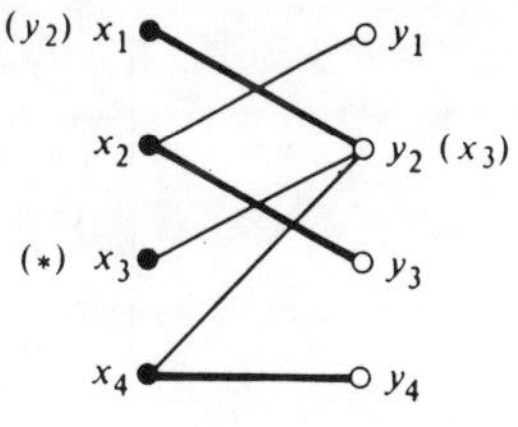

partial graph
no breakthrough

Parts B and C: The labelling procedure and matching improvement

When the labelling procedure is carried out we obtain a matching with three edges. For brevity we do not show the steps involved. We have chosen the edges x_1y_2, x_2y_3 and x_4y_4, indicated in the cost matrix on the previous page by the zeros in bold type. Other matchings with three edges are possible at this stage, but the final solution will be the same.

Part B: The labelling procedure

We now carry out the labelling procedure again to see whether the matching can be improved further.

STEP 3 We label vertex x_3 with (*).
STEP 4 We label vertex y_2 with (x_3).
STEP 5 We label vertex x_1 with (y_2).

(The labels of steps 3 and 4 are shown on the matrix and also on the partial graph on the previous page.)
No breakthrough has been achieved, so we go on to Part D, Step 8.

Part D: Modification of the partial graph

STEP 8 To find the cost δ we look at non-zero costs on all edges (not just those on the partial graph) that join a labelled vertex in X to an unlabelled vertex in Y. So we look at the edges joining x_1 and x_3 to y_1, y_3 and y_4. The lowest of all these non-zero costs is that on edge x_3y_1, namely a cost of one, so this is the value of δ.

STEP 9 (a) We increase by one the weights on x_1 and x_3.
(b) We decrease by one the weight on y_2.
(c) We decrease by one the costs on edges x_1y_1, x_1y_3, x_1y_4, x_3y_1, x_3y_3 and x_3y_4.
(d) We increase by one the costs on edges x_2y_2 and x_4y_2.

Thus we obtain the second revised cost matrix shown below, which has one new edge x_3y_1 in the partial graph.

			(x_3)			
			2	−1	0	0
			y_1	y_2	y_3	y_4
	5	x_1	1	**0**	1	2
	8	x_2	0	5	**0**	6
(*)	7	x_3	**0**	0	4	8
	8	x_4	2	0	6	**0**

second revised cost matrix

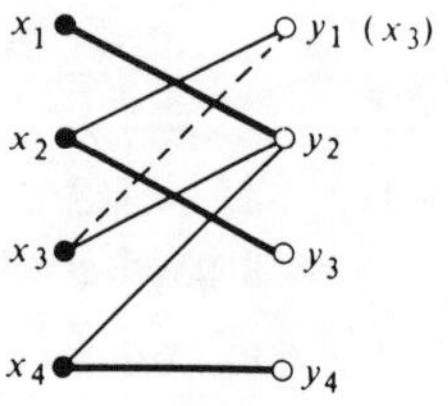

partial graph

Part B: The labelling procedure

STEP 3 We label vertex x_3 with (*).
STEP 4 We label vertex y_1 with (x_3).
Breakthrough has been achieved, so we go on to Part C, Step 6.

Part C: Matching improvement

STEP 6 The alternating path is simply $y_1 x_3$.

STEP 7 The new matching is simply the current matching plus the edge $y_1 x_3$. Remove labels, return to Step 3. STOP.

This is an optimum assignment which consists of the edges $x_1 y_2$, $x_2 y_3$, $x_3 y_1$ and $x_4 y_4$. The total cost (obtained by adding the original edge costs) is

$$4 + 8 + 9 + 8 = 29.$$

(This cost can also be obtained by adding all the vertex weights on the final revised cost matrix.)

CHAPTER 14

14.1. A flow chart for the Hungarian algorithm for the transportation problem is given below.

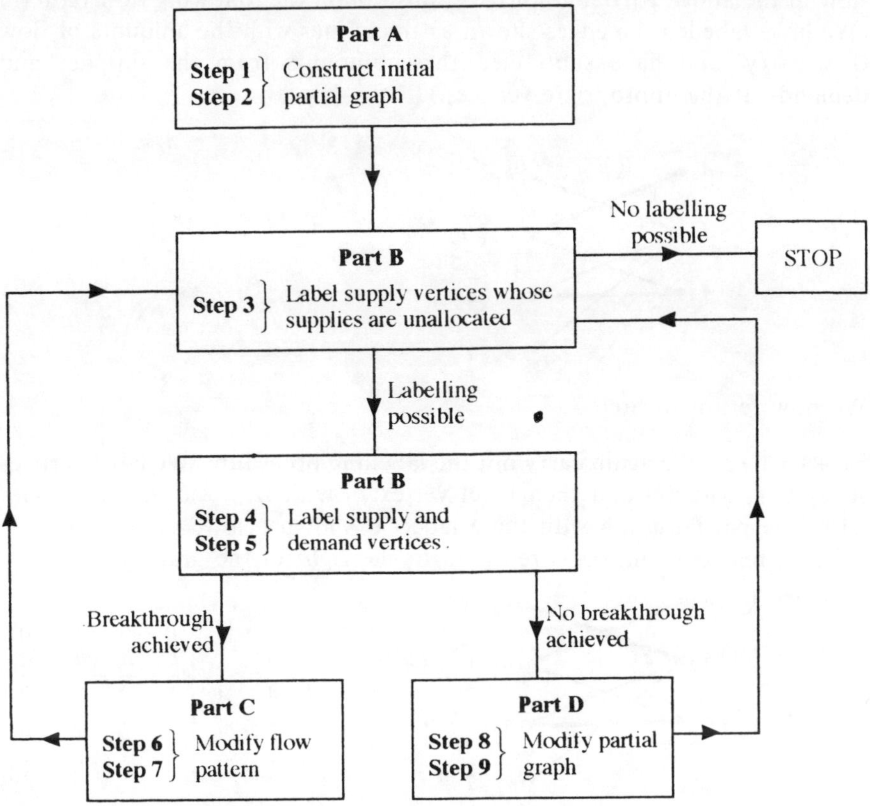

14.2. The step-by-step application of the algorithm is given below. For convenience, we have omitted the circled numbers representing supplies and demands from the cost matrices.

STEPS 1 AND 2 We assign weights to the vertices, and construct the partial graph. The resulting cost matrix and partial graph are given below.

(The vertex weights are shown next to the appropriate rows and columns of the cost matrix.)

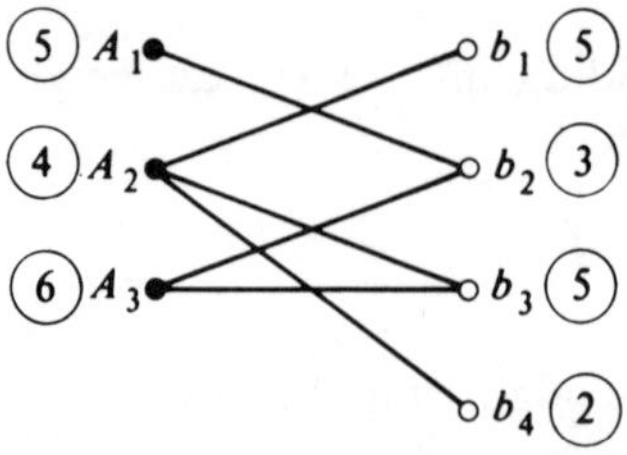

		0	0	1	0
		b_1	b_2	b_3	b_4
4	A_1	1	0	2	2
2	A_2	0	3	0	0
3	A_3	3	0	0	1

STEPS 3 TO 7 Following the procedure of Steps 3 to 6, we find a maximum flow in the above partial graph. We thus obtain the following flow pattern. (We have labelled the edges shown as thick lines with the amounts of flow they carry, and have subtracted these amounts from the supplies and demands at the appropriate vertices.)

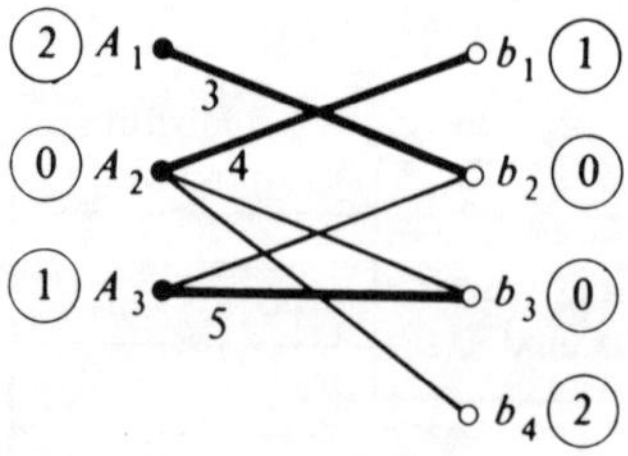

We now return to Step 3.

STEPS 3 TO 5 We again carry out the labelling procedure. We label vertices A_1 and A_3 with (*) and then label vertex b_2 with (A_1) and vertex b_3 with (A_3). The partial graph with these labels is shown below.
(The current cost matrix is repeated to the right of the graph.)

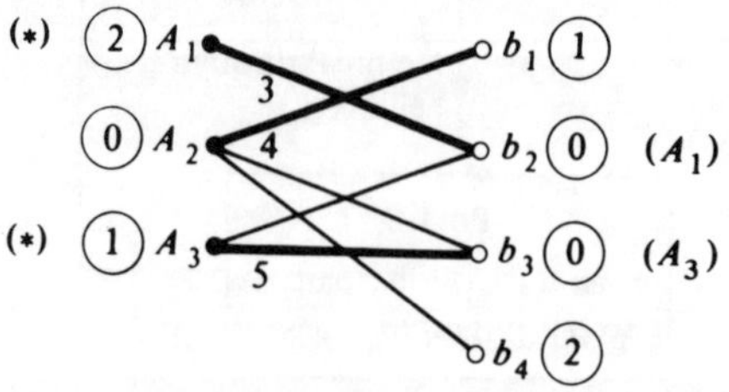

		0	0	1	0
		b_1	b_2	b_3	b_4
4	A_1	1	0	2	2
2	A_2	0	3	0	0
3	A_3	3	0	0	1

No more labelling is possible, and breakthrough has not been achieved, so we proceed to Step 8.

STEPS 8 AND 9 When we apply Steps 8 and 9, we obtain the following revised cost matrix and partial graph.

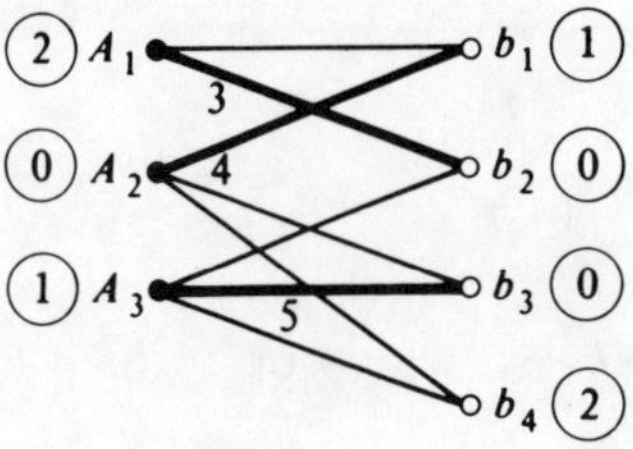

		0	−1	0	0
		b_1	b_2	b_3	b_4
5	A_1	0	0	2	1
2	A_2	0	4	0	0
4	A_3	2	0	0	0

We now return to Step 3.

STEPS 3 TO 5 When we carry out the labelling procedure, we obtain the following labels on the partial graph:

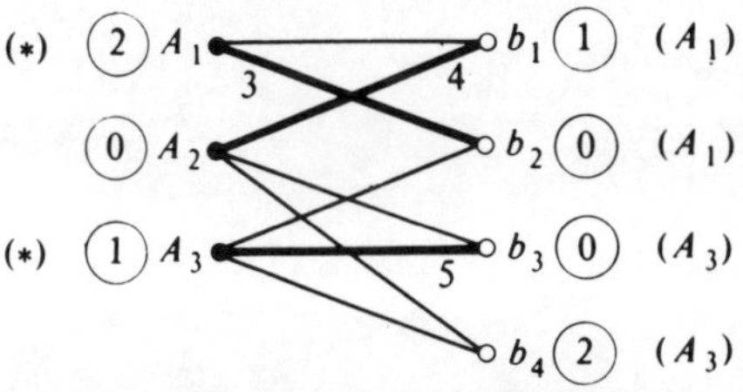

Vertices b_1 and b_4 are both cases of breakthrough, so we proceed to Step 6.

STEPS 6 AND 7 The procedure of Step 6 results in a flow of 1 being sent from A_1 to b_1, and a flow of 1 being sent from A_3 to b_4. This results in the following flow pattern:

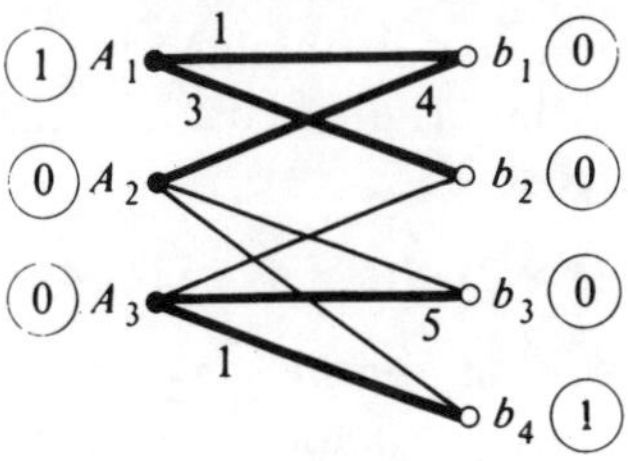

We now return to Step 3.

STEPS 3 TO 5 We carry out the labelling procedure, and thus obtain the following labels on the partial graph:

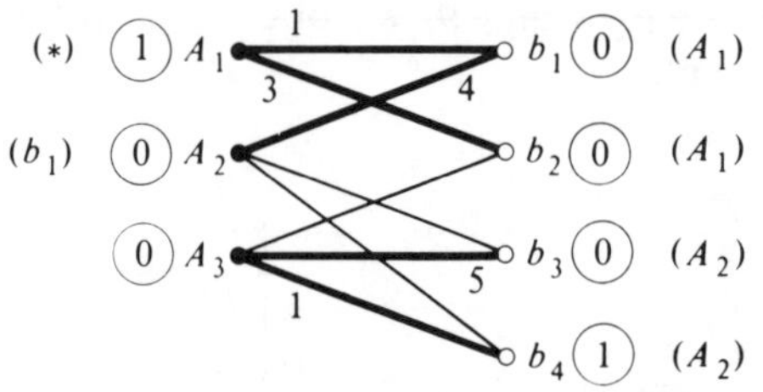

Breakthrough is achieved at vertex b_4, so we go on to Step 6.

STEPS 6 AND 7 Backtracking through the labels, we obtain the flow-augmenting path $A_1b_1A_2b_4$. The minimum of the flows assigned to even edges is 4, along edge A_2b_1. (There is only one such edge.) The available supply at the START vertex A_1 is 1, and the required demand at the end vertex b_4 is also 1. The maximum flow F is therefore equal to min $(4, 1, 1) = 1$. Sending this flow along the flow-augmenting path, we obtain the following flow pattern:

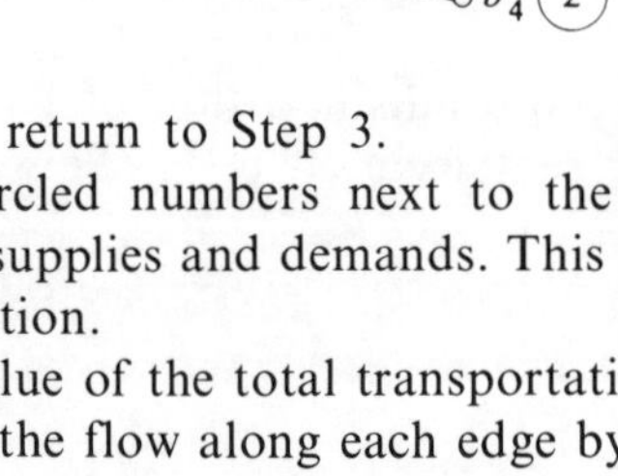

We now return to Step 3.

The circled numbers next to the vertices in the above graph are the original supplies and demands. This flow pattern is the required minimum-cost solution.

The value of the total transportation cost is obtained by multiplying the value of the flow along each edge by the original cost associated with that edge. Thus the total transportation cost is

$$(2 \times 5) + (3 \times 4) + (3 \times 2) + (1 \times 2) + (5 \times 4) + (1 \times 4) = 54.$$

14.3. The steps in applying the algorithm are the same as for the solution to Problem 14.2, except that now we indicate the various flows on the cost matrix as circled numbers adjacent to the costs, and put the labels near the vertices on the matrix.

STEPS 1 AND 2 The revised cost matrix is shown below. The edges of the partial graph correspond to the zero entries in the matrix.

				(A_2)	(A_1)	(A_2)	(A_2)
				⑤	③	⑤	②
				0	0	1	0
				b_1	b_2	b_3	b_4
(*)	⑤	4	A_1	1	0 ③	2	2
(*)	④	2	A_2	0 ④	3	0	0
(*)	⑥	3	A_3	3	0	0	1

In the above matrix, the vertex weights are shown immediately next to the appropriate vertex labels. The demand or supply for each matrix is shown as a circled number.

STEPS 3 TO 7 Following Steps 3 to 7, we label the vertices as shown above, and send a flow of 4 units along A_2b_1, and a flow of 3 units along A_1b_2, as indicated above by the circled numbers.

We return to Step 3 and continue the labelling process as shown below.

					(A_1)	(A_3)	
				①	⓪	⑤	②
				0	0	1	0
				b_1	b_2	b_3	b_4
(*)	②	4	A_1	1	0 ③	2	2
	⓪	2	A_2	0 ④	3	0	0
(*)	⑥	3	A_3	3	0	0 ⑤	1

We send a flow of 5 units along A_3b_3, and return again to Step 3. The labelling is the same as previously (shown above). No more labelling is possible, and breakthrough has not occurred, so we go on to Step 8.

STEPS 8 AND 9 The revised cost matrix is shown below.

				(A_1)	(A_1)	(A_3)	(A_3)
				①	⓪	⓪	②
				0	−1	0	0
				b_1	b_2	b_3	b_4
(*)	②	5	A_1	0	0 ③	2	1
	⓪	2	A_2	0 ④	4	0	0
(*)	①	4	A_3	2	0	0 ⑤	0

We now return to Step 3.

STEPS 3 TO 5 The labelling procedure results in the labelling shown above. Breakthrough is achieved at vertices b_1 and b_4, so we proceed to Step 6.

STEPS 6 AND 7 We send a flow of 1 along A_1b_1 and a flow of 1 along A_3b_4, resulting in the flow pattern shown below.

				(A_1)	(A_1)	(A_2)	(A_2)
				⓪	⓪	⓪	①
				0	−1	0	0
				b_1	b_2	b_3	b_4
(*)	①	5	A_1	0 ①	0 ③	2	1
(b_1)	⓪	2	A_2	0 ④	4	0	0
(b_3)	⓪	4	A_3	2	0	0 ⑤	0 ①

We now return to Step 3.

STEPS 3 TO 5 These produce the labels shown above.

STEPS 6 AND 7 We send 1 unit along $A_1b_1A_2b_4$, resulting in the flow pattern shown below, which is the required minimum cost solution.

		0	−1	0	0
		b_1	b_2	b_3	b_4
5	A_1	0 ②	0 ③	2	1
2	A_2	0 ③	4	0	0 ①
4	A_3	2	0	0 ⑤	0 ①

The total transportation cost, obtained by adding the products of the flow value multiplied by the original cost for each arc is 54.

CHAPTER 15

15.1. To start Part A of the algorithm, we draw a bipartite graph representing the precedence relations:

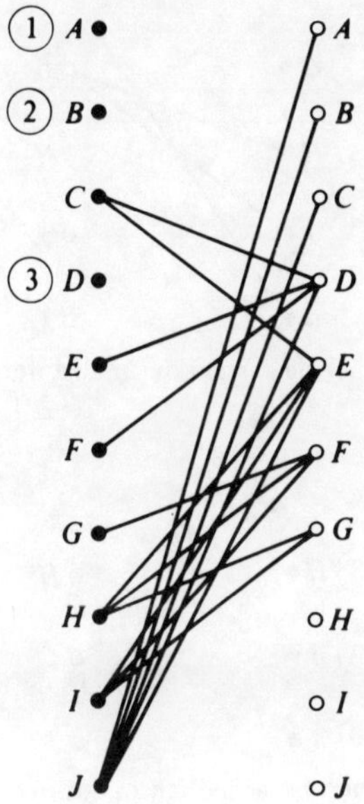

In the first iteration, vertices A, B and D are numbered, as indicated on the above diagram. The numbering of vertices carried out in the second iteration is shown on the following diagram:

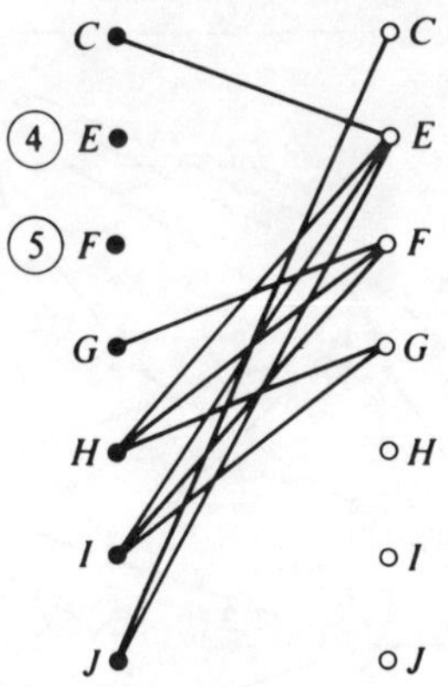

numbering carried out in 2nd iteration

At the next stage we get

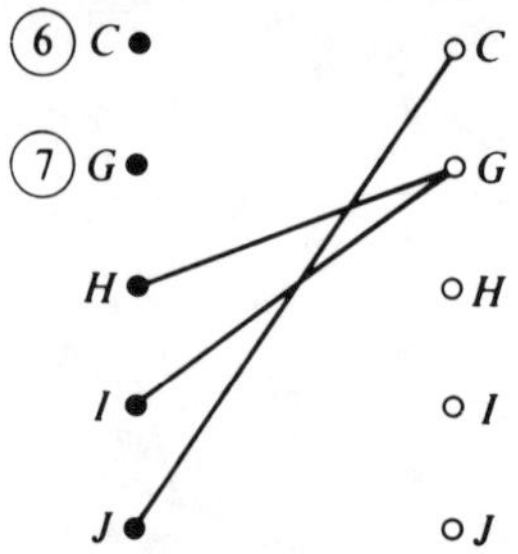

numbering carried out in 3rd iteration

Finally, we get

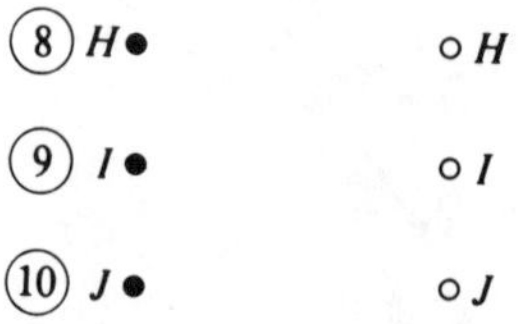

numbering carried out in 4th iteration

We now construct the activity network following the procedure of Part B of the algorithm. The resulting activity network is shown below.

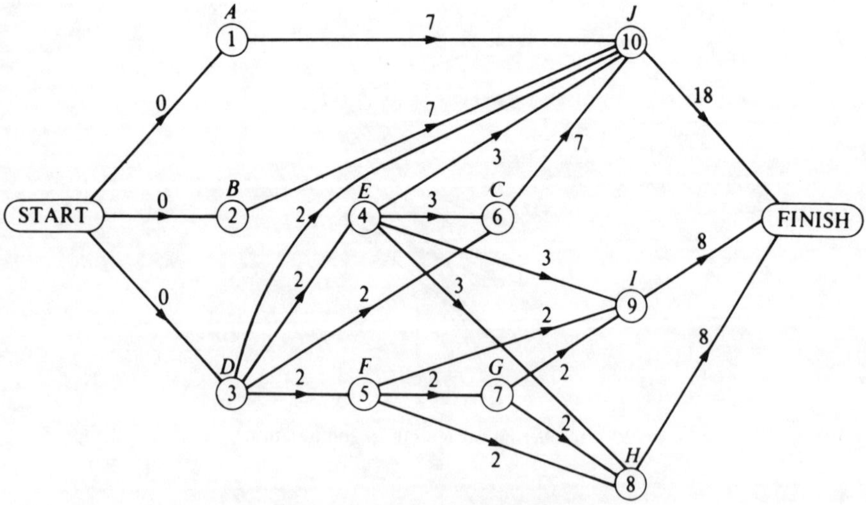

activity network for bicycle assembly

15.2. Let us imagine that we steadily increase the delay in completing an activity which is not on a critical path. To begin with, this delay will not affect the time needed to complete the project. However, as the delay increases, there will come a point when the delay becomes so large that the activity becomes part of a new critical path, so delays larger than this will delay the whole project.

15.3. (a) The minimum completion time is equal to the length of the critical path, which is

$$0 + 5 + 4 + 7 + 2 = 18 \text{ days.}$$

(b) The earliest starting time of activity 4 is equal to the length of the longest path from the START vertex to vertex 4. This longest path passes through vertices 1 and 3, and its length is $0 + 5 + 4 = 9$, so

$$e_4 = 9 \text{ days.}$$

Activity 6 cannot start until activity 4 has been completed, and the duration of activity 4 is 3 days, so the earliest starting time for activity 6 is

$$e_6 = 9 + 3 = 12 \text{ days.}$$

(c) We can find the latest starting time for activity 6 by working backwards from the FINISH vertex. The minimum completion time is 18 days, so activity 7 must begin after $18 - 2 = 16$ days, and hence activity 6 must begin at the latest after $16 - 2 = 14$ days, so

$$l_6 = 14 \text{ days.}$$

Activity 4 must begin not later than 3 days before the latest starting time of activity 6, so

$$l_4 = 14 - 3 = 11 \text{ days.}$$

(d) The float of activity 4 = $l_4 - e_4 = 11 - 9 = 2$ days.
The float of activity 6 = $l_6 - e_6 = 14 - 12 = 2$ days.

(e) If activity 4 is delayed by 2 days, then the path through vertices 3, 4, 6 and 7 becomes part of a new critical path. So if activity 6 is also delayed by 2 days, this causes a delay of 2 days in the completion of the whole project. In other words, the two activities cannot both be delayed by their float times without delaying the project.

15.4. (a) Using the critical path algorithm, we obtain the vertex-weights shown on the following diagram.

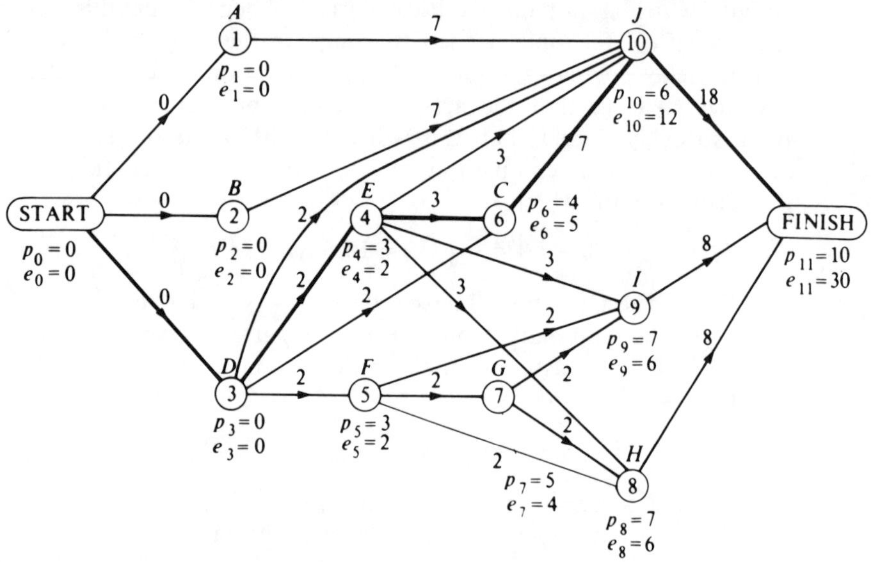

activity network for bicycle assembly

The critical path is indicated by heavy lines.

The minimum completion time is the length of the critical path, which is

$$0 + 2 + 3 + 7 + 18 = 30 \text{ minutes.}$$

(b) The values of the latest starting time and the float for each activity are shown in the following table:

activity	vertex number i	earliest starting time, e_i	latest starting time, l_i	float $(l_i - e_i)$
START	0	0	0	0
A	1	0	5	5
B	2	0	5	5
D	3	0	0	0
E	4	2	2	0
F	5	2	18	16
C	6	5	5	0
G	7	4	20	16
H	8	6	22	16
I	9	6	22	16
J	10	12	12	0
FINISH	11	30	30	0

CHAPTER 16

16.1. The steps involved in applying the algorithm are given briefly below.

START Set project clock to 0 days.
STEP 1 Activity *A* is assigned to worker 1.
Activity *B* is assigned to worker 2.
STEP 2 Advance project clock to 2 days.
Activity *B* is completed; worker 2 is free.
STEP 1 Activity *C* is assigned to worker 2.
STEP 2 Advance project clock to 7 days.
Activity *A* is completed; worker 1 is free.
Activity *E* is free to be started.
STEP 1 Activity *E* is assigned to worker 1.
STEP 2 Advance project clock to 17 days.
Activities *C* and *E* are completed; both workers are free.
Activities *D* and *G* are free to be started.
STEP 1 Activity *D* is assigned to worker 1.
Activity *G* is assigned to worker 2.
STEP 2 Advance project clock to 22 days.
Activity *G* is completed; worker 2 is free.
Activity *H* is free to be started.
STEP 1 Activity *H* is assigned to worker 2.
The current state of the scheduling of activities is as follows:

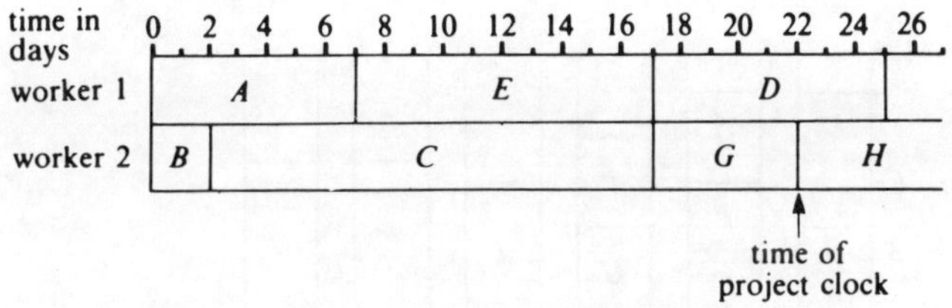

STEP 2 Advance project clock to 25 days.
Activity *D* is completed; worker 1 is free.
Activity *F* is free to be started.
STEP 1 Activity *F* is assigned to worker 1.
STEP 2 Advance project clock to 27 days.
Activity *F* is completed; worker 1 is free.
Activity *I* is free to be started.
STEP 1 Activity *I* is assigned to worker 1.
STEP 2 Advance project clock to 29 days.
Activity *I* is completed; worker 1 is free.
Activity *J* is free to be started.

STEP 1 Activity J is assigned to worker 1.
All the activities have been assigned.
STEP 2 Advance project clock to 30 days.
Activity H is completed; worker 2 is free.
STEP 3 Advance project clock to 32 days.
All the activities have been completed.
STOP

The final schedule is shown below. It is an optimum schedule—no other schedule has a shorter completion time.

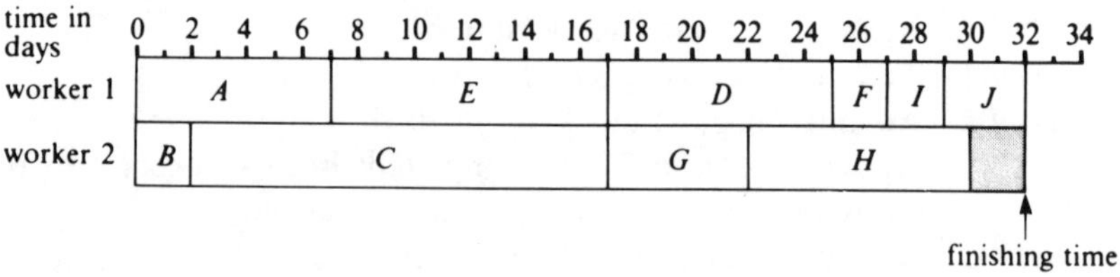

16.2. The solution appears in the text immediately after the problem.

CHAPTER 17

17.1. (a)

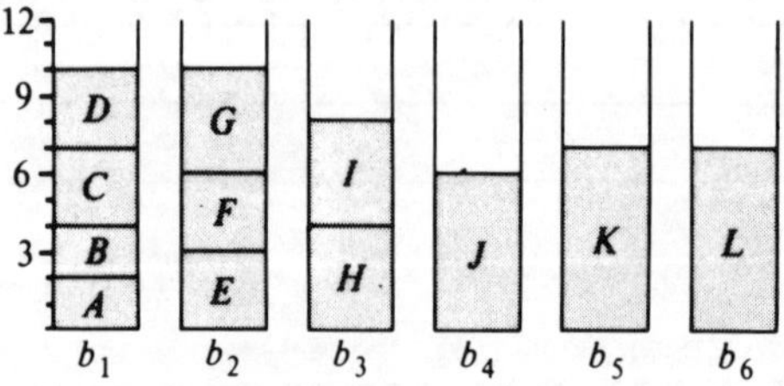

(b)

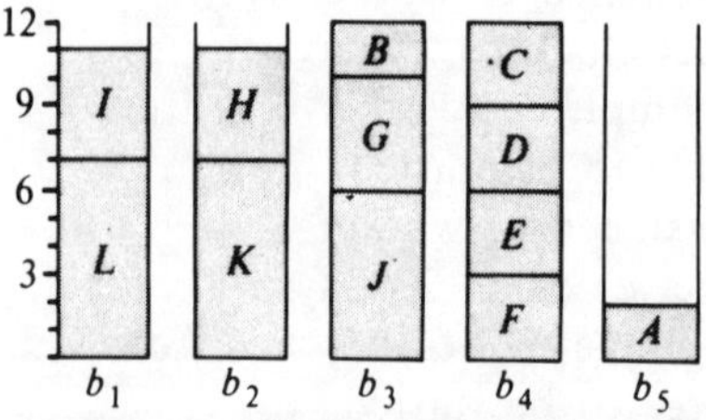

(c)

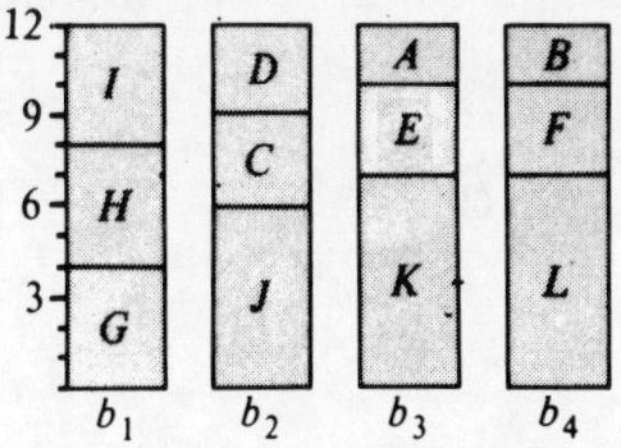

The above solution is the only optimum one, apart from solutions obtained by interchanging sections of equal length (assuming that the order of packing within a bin, and the order of the bins, do not matter).

17.2. First, we order the items in decreasing order of value per unit weight, as shown below. (In this case, the weight of an item is the time needed to produce that item.)

order number, i	1	2	3	4	5
item	B	E	D	C	A
weight, w_i	7	4	4	2	3
value, v_i	14	8	7	3	3
value per unit weight	2	2	1.75	1.5	1

We begin by branching out from the null solution, as shown below:

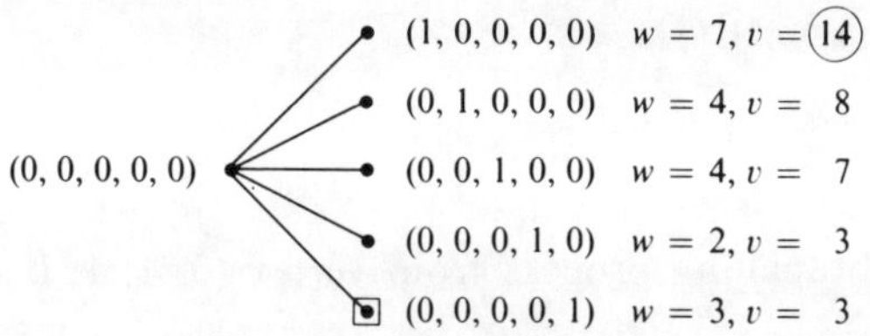

STORE $v = 14$, solution = (1, 0, 0, 0, 0)

Next we branch out from the solution (1, 0, 0, 0, 0) as follows:

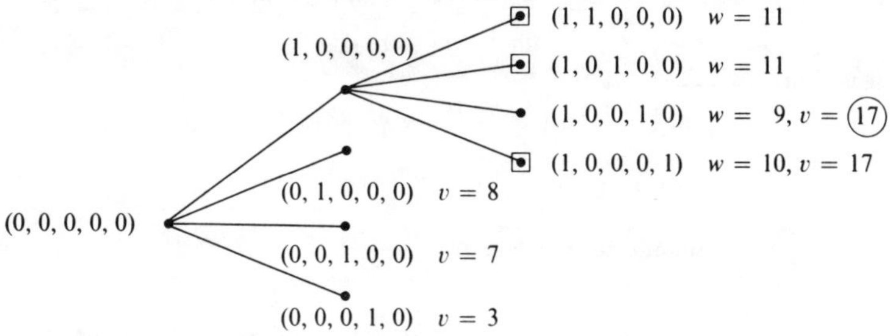

STORE $v = 17$, solution = (1, 0, 0, 1, 0)

We delete the marked vertices and continue the branching process from vertex (1, 0, 0, 1, 0).

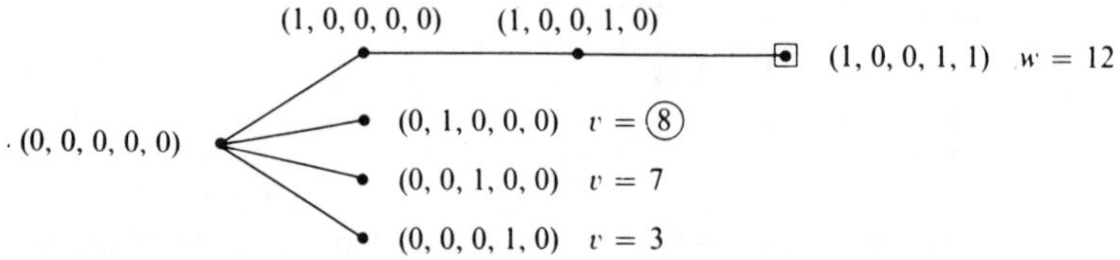

STORE $v = 17$, solution = (1, 0, 0, 1, 0)

The new solution is infeasible, so we continue the branching process from the current feasible solution with the highest value—namely, (0, 1, 0, 0, 0).

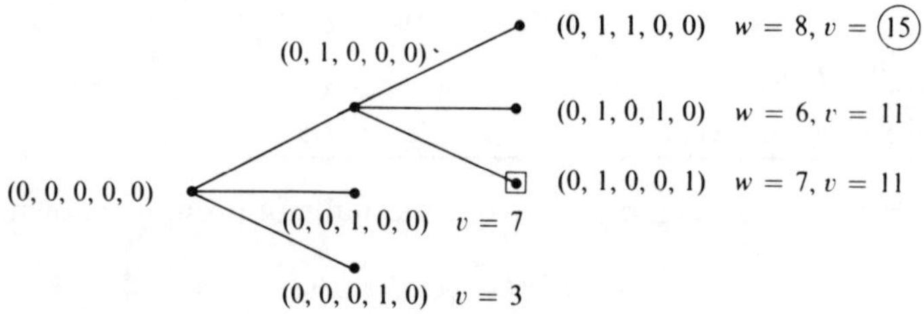

STORE $v = 17$, solution = (1, 0, 0, 1, 0)

We continue the branching process from vertex (0, 1, 1, 0, 0).

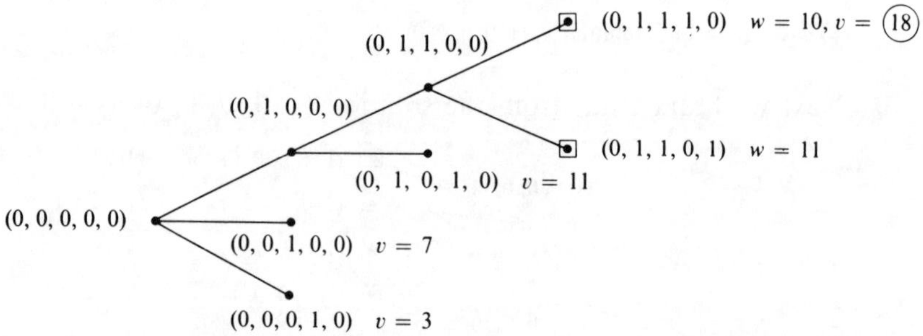

STORE $v = 18$, solution = (0, 1, 1, 1, 0)

The solution (0, 1, 1, 1, 0) is a feasible solution with a value greater than the previous stored value, so we update the stored value, as shown above.

We continue the branching process as shown in the following diagrams.

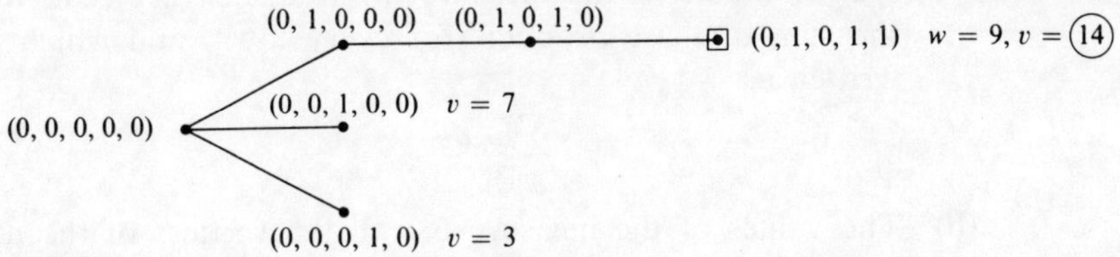

STORE $v = 18$, solution = (0, 1, 1, 1, 0)

At the next step we get:

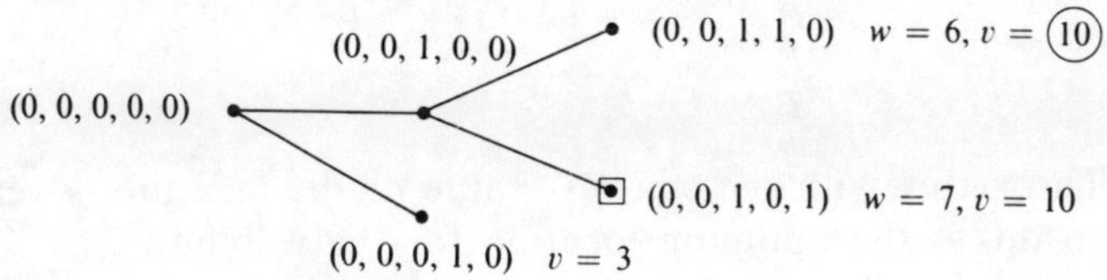

STORE $v = 18$, solution = (0, 1, 1, 1, 0)

At the next step we get:

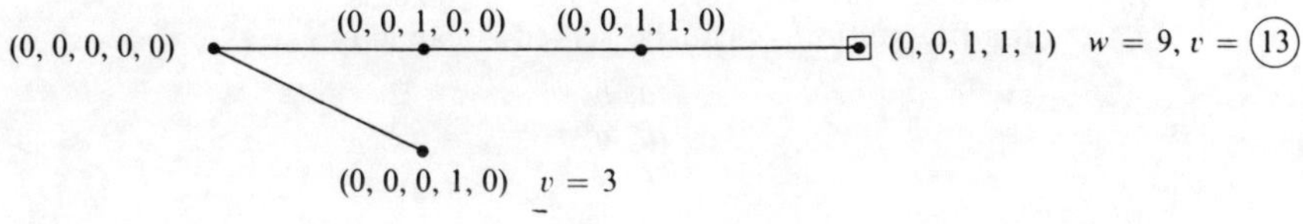

STORE $v = 18$, solution = (0, 1, 1, 1, 0)

Finally, we get

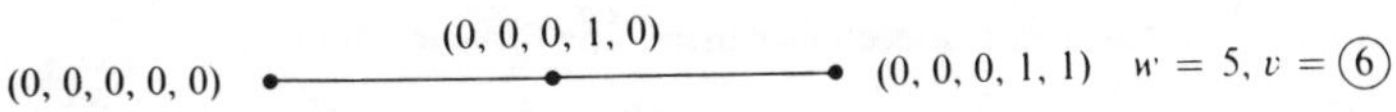

STORE $v = 18$, solution = (0, 1, 1, 1, 0)

The branch-and-bound procedure has now been completed, so the optimum solution vector is (0, 1, 1, 1, 0), corresponding to items E, D and C, with a total value of 18.

CHAPTER 18

18.1. (a) The cost of delivery to a town j from a centre at town i is $v_j d_{ij}$. We take as the measure the total delivery cost, which is the sum of these products for every town, and which can be written as

$$\sum_j v_j d_{ij}.$$

(b) The values of the measure for all the vertices of the network are as follows:

Vertex	Value of the measure
A	150
B	146
C	109
D	134
E	114
F	134

The vertex with the smallest value of the measure is vertex C, so this is the optimum location for the warehouse.

18.2. The covers for S are

$$\{a, b, c, d\}, \{a, b, c\}, \{a, b, d\}, \{b, c, d\}, \{b, c\}, \{b, d\}.$$

The minimum covers for S are

$$\{b, c\}, \{b, d\}.$$

18.3. (a)

vertex centre for fire station	vertices which can be served	excess capacity
a	a, b	2
b	b, a	2
c	c, d	2
d	d, c	2
e	e, f	0
f	f, e	0

From this table, we see that the capacity of a single fire station is sufficient in each case.

(b)

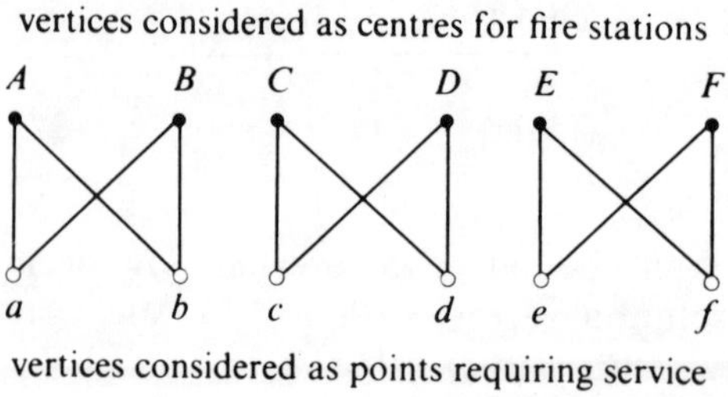

From the bipartite graph representation on the previous page, it is clear that to serve all the towns we must choose one vertex from the set $\{A, B, C, D, E, F\}$ for each component of the graph. The eight possible minimum set covers are

$$\{A, C, E\}, \{A, C, F\}, \{A, D, E\}, \{A, D, F\},$$

$$\{B, C, E\}, \{B, C, F\}, \{B, D, E\}, \{B, D, F\}.$$

(c) For each minimum set of centres, we must calculate the sum of the values of the measure for all the vertices in that set.

For the elements of the set $\{A, C, E\}$, the values of the measure are as follows:
Vertex A: $5 \times 0 + 3 \times 3 = 9$
Vertex C: $1 \times 0 + 7 \times 3 = 21$
Vertex E: $6 \times 0 + 2 \times 4 = 8$
The sum of the values of the measure = 38.

The sum of the values of the measures calculated for all the eight minimum sets are as follows:

minimum set of centres	sum of values of the measure for each centre
$\{A, C, E\}$	38
$\{A, C, F\}$	42
$\{A, D, E\}$	20
$\{A, D, F\}$	24
$\{B, C, E\}$	44
$\{B, C, F\}$	48
$\{B, D, E\}$	26
$\{B, D, F\}$	30

The set of vertices with the lowest sum is $\{A, D, E\}$, so the fire stations should be located at the corresponding towns, a, d and e.

We could have arrived at this answer more directly by noting that a fire station located at any vertex can serve only two vertices—the vertex at which it is located, and one other. So we can consider the vertices in pairs, and a fire station located at either of the vertices in a pair can serve them both. We have therefore only to place each centre at the vertex from each pair which has the higher population v_j.

18.4. For this problem we are asked to find the absolute 2-centre, so we do not yet know the maximum distance of any vertex from a centre. We begin by letting this maximum distance be x km. If $x = 0$, then we must have a centre at each vertex, and these form the absolute 3-centre. We now increase x until we find the smallest value of x for which there is an absolute 2-centre. At each stage, for each value of x tried, we must put down markers and study the resulting regions. For small values of x, we obtain the following:

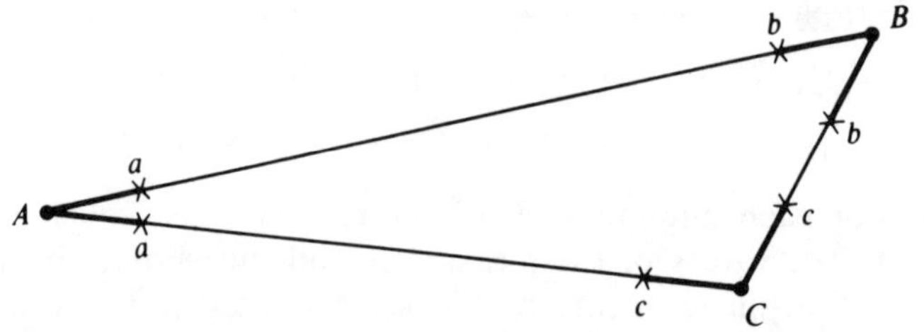

We still need three fire-stations, one in each of the marked regions. There will be no significant change until two markers coincide. This happens when we increase x to a value of 1 km when the markers on the shortest edge, BC, coincide.

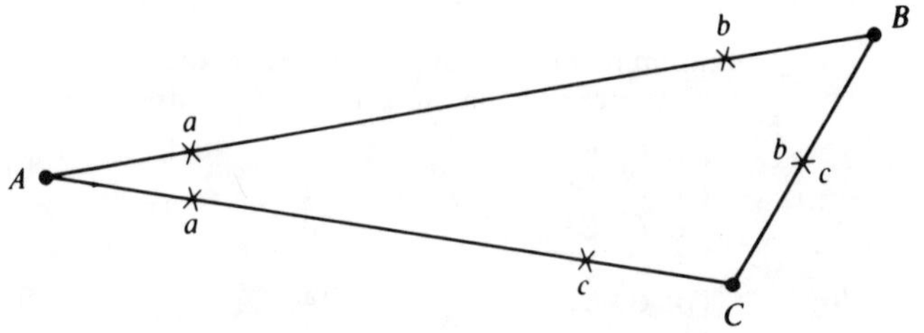

We now have a new region, the single-point region bc. The regions and the corresponding subsets of vertices are as follows:

1. bc $\{B, C\}$
2. a $\{A\}$
3. b $\{B\}$
4. c $\{C\}$

(We have ignored segments like ab which are more than 1 km from any vertex.)
A cover is provided by region 1 and region 2, so we have found the absolute 2-centre. One centre must be located in region 2 (that is, within 1 km of vertex A), and the other centre is in region 1 at the midpoint of BC. The maximum distance of any vertex from a centre is 1 km.

INDEX

Where more than one page reference is given, the page on which an entry is defined or explained is shown in **bold**; other important references are indicated in roman. Generally, where a topic is discussed on a number of consecutive pages, only the first page is indexed.